Interdisciplinary Biotechnological Advances

This series is an authoritative resource that explains recent advances and emerging directions in biotechnology, reflecting the forefront of research clearly and reliably, without excessive hype. Each volume is written by authors with excellent reputations and acknowledged expertise in the topic under discussion. The volumes span the entire field from an interdisciplinary perspective, covering everything from biotechnology principles and methods to applications in areas including genetic engineering, transgenic plants and animals, environmental problems, genomics, proteomics, diagnosis of disease, gene therapy, and biomedicine. The significance of these applications for the achievement of UN Sustainable Development Goals is highlighted. The series will be highly relevant for Master's and PhD students in Biotechnology, Nanochemistry, Biochemical Engineering, and Microbiology, medical students, academic and industrial researchers, agricultural scientists, farmers, clinicians, industry personnel, and entrepreneurs.

Angana Sarkar • Idris Adewale Ahmed
Editors

Microbial products for future industrialization

Springer

Editors
Angana Sarkar
Department of Biotechology and Medical Engineering
National Institute of Technology
Rourkela, India

Idris Adewale Ahmed
Department of Biotechnology, Faculty of Applied Science
Lincoln University College
Petaling Jaya, Selangor, Malaysia

ISSN 2730-7069 ISSN 2730-7077 (electronic)
Interdisciplinary Biotechnological Advances
ISBN 978-981-99-1736-5 ISBN 978-981-99-1737-2 (eBook)
https://doi.org/10.1007/978-981-99-1737-2

This Springer imprint is published by the registered company Springer Nature Singapore Pte Ltd.
The registered company address is: 152 Beach Road, #21-01/04 Gateway East, Singapore 189721, Singapore

Foreword

We are elated to write an introductory message for *Microbial products for future industrialization*, a timely volume on the rapidly evolving field of microbial biotechnology focusing on the exploitation of various advanced microbial and molecular biology technologies and their associated processes (microbial/molecular/chemical) for future industrialization of emerging new microbial products engineering.

This book covers all the technical, economic, ethical, and societal aspects which impact on bioprocessing of microbial products at the industrial level along with biotechnological intervention for better production of microbial products in the near future.

This book encompasses advanced and updated information as well as future directions for young researchers and scientists, who are working in the field of microbial product production in a sustainable manner. The entire journey of new emerging microbial products from laboratory to the industry with all the hurdles are emphasized in this book. This book covers key areas of microbial biotechnology that would allow the readers to easily understand the concepts and implement them.

This comprehensive book contains 20 chapters with several aspects of microbial biotechnology from basic to advanced level and from laboratory scale to industrial scale production of industrial microbes. It covers a number of major topics including microbial product commercialization; assessment of microbes and microbial products for future industrialization; design and operation of new microbial product bioprocessing system; industrial aspect of marine bioprocessing; application of cutting-edge molecular biotechnological tools in microbial bioprocessing; bio-refinery for microbial products; bioprospecting of microbes for value generation from wastes; emerging microbial enzymes for future industrialization; bioethanol from microbial fermentation of prospecting biomass; microbial biodiesel for future commercialization; microbial production of bioactive compounds; future marine microbial products for pharmaceuticals industry; microbial production of polyhydroxyalkanoate; organic acids and solvents production from microbial fermentation; microbial biomaterials and their industrial application; green synthesis of microbial nanoparticles; electrochemically active microorganisms; techno-economic feasibility analysis of microbial product commercialization along ethical issues of microbial products for industrialization all of which have been presented using an easy-to-understand yet elaborative narration.

We are pleased to recognize the valuable efforts of all the authors and contributors including Dr. Angana Sarkar, Dr. Sabina Khanam, Dr. Ravi Ranjan, Dr. Jayanthi Abraham, Dr. Abhishek S. Dhoble, Dr. Ashish Prabhu, Dr. Taniya Sengupta Rathore, Dr. Gayathri. M, Dr. J. Basilio Heredia, Dr. Soma Ghosh, Dr. Anupama Shrivastav, Dr. Shaikh Yasmeen, Dr. Chitra Bhattacharya, Dr. Rabiya Tabbassum Khan, Dr. Shafaq Rasool, Dr. Ahmed A. Mohamed, Dr. Soumya Pandit, Dr. S. Muthu Kumar, and Dr. Idris Adewale Ahmed who have collaboratively brought out an excellent volume through the support from Springer Nature.

We strongly believe that this book will be really useful for students, researchers, scientists, stakeholders, and policymakers, among many others.

Department of Biotechology and Medical Engineering, National Institute of Technology, Rourkela, India
Angana Sarkar

Department of Biotechnology, Faculty of Applied Science, Lincoln University, Petaling Jaya, Selangor, Malaysia
Idris Adewale Ahmed

Preface

Microbes are the key players, who ultimately govern life on Earth. Numerous microorganisms were exploited to serve the requirements of mankind from the ancient past. Starting from fermentation, several biochemical processes are controlled by microbial activities. At the beginning of the twentieth century, the concept of Industrial Microbiology came when the production of different microbiological products started at the industrial level. The intervention of Biotechnology at the later stage made this process more improved and multi-dimensional. Industrialization of microbiological products through biotechnological roots will not only offer the production of conventional commercial products in a greener way but also help in exploring the novel basis of biosynthesis of different value-added products. A series of microbial products including fermented foods, beverages, pharmaceuticals, solvents, enzymes, etc. are already produced in the industry successfully in a sustainable manner. In the past few decades, the major goal of the research has shifted toward microbial production from waste materials. Hence, microbes are equipped with diverse metabolic and genetic mechanisms, and the potential for new microbial product formation is enormous. Therefore, there will be always new scopes and huge opportunities for introducing new microbial products for future industrialization. This book focuses on the area of commercialization of new microbial products including the improvement of traditional industrial production. Working at the interface of the multidisciplinary area including chemistry, microbiology, metabolic engineering, molecular biology, and synthetic biology with the industrialization of new potential microbial products for next generation, this book also enlightens all the microbial techniques, ethical clearances, industrial setup, techno-economic feasibility, etc. required to accomplish the new microbial products commercialization.

We hope that this book will not only provide a better understanding of the evolving field of microbial biotechnology but also trigger unanswered questions. Despite the fact that immense efforts have been invested to make this book user-friendly, we are aware that the first version always comes with bugs. This book will be an excellent basis and proverbial springboard from which scientific knowledge can grow, widen, and accelerate the industrialization of useful microbial products.

Rourkela, India Angana Sarkar
Petaling Jaya, Malaysia Idris Adewale Ahmed

Contents

About the Editors

Angana Sarkar Ph.D. Assistant Professor, Department of Biotechnology & Medical Engineering. She pursued her graduation in Agricultural Engineering from Bidhan Chandra Krishi Viswasvidyalaya, West Bengal, India, followed by post-graduation in Biotechnology and Biochemical Engineering from Indian Institute of Technology, Kharagpur, India, 2008. Later, she completed her Ph.D. in Environmental Biotechnology from Indian Institute of Technology, Kharagpur, India. Subsequently, she joined National Institute of Technology, Rourkela, India, in the year 2015 as an Assistant Professor in the Department of Biotechnology and Medical Engineering. Her research area is mainly focused on (1) Pollutant detection using biosensors, (2) Groundwater bioremediation, (3) Waste water (domestic and industrial) treatment, (4) Solid waste management by bio-refinery approach to produce environmental waste to value like bioethanol, pigment, biofertilizers, etc., (5) Hydrocarbon and other organic pollutants degradation, (6) Waste to value generation, (7) Bioenergy production, etc. She is supervising a number of doctoral dissertations, master thesis, and advised many undergraduate students. She has a combined publication of >90 in various SCI cited journals of high repute, book chapters, and peer-reviewed conferences. She is also serving as a reviewer in various journals of national and international repute. She has worked as handling editor in International Biodeterioration and Biodegradation (ELSEVIER) and Biocatalysts and Agricultural Biotechnology (ELSEVIER), Environmental Science and Pollution Research (Springer), Ecotoxicology (Springer), and Environmental Quality Management (Wiley Online Library) for special issue for International Conference on Bioprocess for Sustainable Environment and Energy (ICBSEE_2018, ICBSEE_2020 and ICBSEE_2022).

Idris Adewale Ahmed Ph. D. is a distinguished Associate Professor at the Department of Biotechnology, Faculty of Applied Science, Lincoln University College (LUC), Malaysia. Prior to this, he was a Postdoctoral Fellow and Visiting Research Fellow at the prestigious Center for Natural Products and Drug Discovery, Universiti Malaya, and also collaborated with International Islamic University Malaysia (IIUM) for some projects. He received his Ph.D. in Health Sciences (2015) from IIUM with a research focus on Food and Nutrition by evaluating the phytochemical composition and antioxidant properties of *Baccaurea angulata*—an underutilized fruit from Sarawak—and testing its biological effects on cardiovascular disease biomarkers in diet-induced atherosclerotic rabbits. He also had his M. Sc. degree in Biotechnology Engineering (2011) with a research focus on Food Processing and Food Microbiology by studying the proximate composition analysis and disinfection kinetics of Nahar (*Mesua ferrea*) seed kernel oil. He has more than 8 years of research and teaching experience in Natural Products and Medicinal Plants, Biotechnology, and Industrial Microbiology. He has published 42 articles, 33 opinion letters, and 16 book chapters. He is currently the chairperson of Pure Heart Relief Organization (RC 132277) championing women's empowerment, child education, potable water infrastructure, and healthy food distribution to disadvantaged families. He is a member of the "Council of Experts" for the Halal Compliance and Food Safety Ltd (RC number: 1507999); a registered member of the Biochemical Society (https://biochemistry.org/), the Malaysian Natural Products Society, and the Malaysian Association of Sports Medicine.

Chapter 1
Microbial Product Commercialization from Lab to Industry

Puja Dokania, Tapaswini Nayak, Sohom Roy Chawdhury, and Angana Sarkar

Abstract Products obtained from microorganisms have always been of foremost industrial importance. But there has been a significant gap, in the number of microbial products produced at the lab scale and the same being scaled up for industrial production. The process of introduction of a novel microbial product produced at the lab scale into the industry constitutes the process of microbial product commercialization. The commercialization process again consists of both upstream and downstream procedures, right from formulation, research and development, and product harvesting to product isolation and purification. But the commercialization step is not hurdle free, as there are several impediments, like the time factor required for successful translation of products to the industrial scale. This also adds to the economic risk and even could lead to the failure of the commercialization of the particular product. It is also absolutely essential to obtain ethical clearance from the concerned authorities in the lead-up to the technology transfer and commercialization of the products. It is imperative that the ethical clearance step adds to the already existing time shortage factor and further delays the commercialization process. The impediments faced during the scale-up procedure of specific products could also be due to the lack of viable energy alternatives. However ample opportunities exist to try and narrow the existing gap between the laboratory table and the industrial production belt. A combination of both biological and chemical procedures is the key in the way forward to achieving appreciable technology transfer from the labs to the industries.

Keywords Microbial products · Industrial transfer · Scale-up · Microbial product commercialization · Product isolation · Ethical clearance

---

P. Dokania · T. Nayak · S. R. Chawdhury · A. Sarkar (✉)
Department of Biotechnology and Medical Engineering, National Institute of Technology, Rourkela, India
e-mail: sarkara@nitrkl.ac.in

A. Sarkar, I. A. Ahmed (eds.), *Microbial products for future industrialization*, Interdisciplinary Biotechnological Advances,
https://doi.org/10.1007/978-981-99-1737-2_1

1.1 Introduction

Microbial products are defined as a product derived from a number of microscopic organisms. It may consist of the organisms themselves or the metabolites they produce. Microbial products are derived from microorganisms such as bacteria, fungi, or protozoa. In large-scale industrial processes, microorganisms are comprehensively used to combine an assortment of products that are valuable to people (Gupta et al. 2019). There are a few industrial items that are gotten from microorganisms like food added substances, alcoholic and non-alcoholic beverages, biofuels, metabolites, biofuels and bio-fertilizers, a few chemicals, enzymes, and different bioactive molecules, various antibiotics, and vaccines to kill or retard the development of illness-causing organisms (Hansen et al. 2019). Microorganisms are utilized in some large-scale industrial processes. They produce ethanol, which is utilized as a fuel, dissolvable, and for the vast majority of different purposes (Park and Lee 2013; Park et al. 2013). Microorganisms are also used for the production of metabolites, bio-fertilizers, and synthetic insulin. Microbial product commercialization is the process of introducing a new microbial product from the lab to the industry (Parnell et al. 2016). Microorganisms that exhibit good characteristics in the lab may lack the key traits for industrial application. This chapter offers an overview of imperative issues in taking microbial products from the lab to the industry. In addition, this chapter gives an idea about some of the most tremendous microbial products, which are already in the market today and some of the primary challenges in bringing these microbial products to industry and commercialization of really helpful microbial products for extensive industrial application.

This chapter mainly focuses to improve the likelihood of commercialization of effective microbial products for mass production in an industrial scale. The industries like food and Dairy industry, pharmaceutical, and chemical industries are using a broad range of microorganisms for producing various microbial products that are beneficial to human beings. The most useful products include antibiotics, anticancer agents, food additives, alcoholic and nonalcoholic beverages, biofuels, metabolites, nutraceuticals, polymers, surfactants, and vaccines, which have been commercialized. The most economically significant mixtures created by microorganisms (Demain 2000) other than chemicals and recombinant proteins are the low molecular weight primary and secondary metabolites. Essential metabolites are those mixtures associated with the development of microorganisms through secondary metabolites. The most significant primary metabolites of industry are amino acids, purine nucleotides, vitamins, and organic acids. With respect to secondary metabolites, the most financially significant are antitumor agents, antibiotics, cholesterol-lowering agents, immune-suppressants, and antiparasitic drugs. Microorganisms isolated from nature ordinarily produce incredibly low degrees of such metabolites. Overproduction should be accomplished in the lab before the pilot plant increase endeavors. This chapter presents the natural as well as recombinant microbial products and their various industry types, processes for microbial product transfer from lab to industry, hurdles or drawbacks, ethical clearance as well as the future prospects of microbial product commercialization from lab to industry. The main focus is on the microbial products, the development of inoculum, different industries, and the parameters required for the commercialization process from lab to industry.

1.2 Natural and Recombinant Microbial Products and Various Industry Types

The below table shows an overview of various industry types and the microbial products manufactured by the industry. For example, microbial products like alcoholic beverages vinegar, soy sauce, wine, and coffee are manufactured in the food industry using microorganisms such as *Saccharomyces cerevisiae*, *Acetobacter aceti*, *Aspergillus oryzae*, *Lactobacillus buchneri*, *Acetobacter fabarum*, respectively. Apart from the food industry, there are other industries as well that produce microbial product and those are the dairy industry, pharmaceutical industry, and chemical industry (Table 1.1).

Today, in excess of 151 exceptional recombinant therapeutics have been supported by the FDA as well as by the European Medicines Agency for various clinical signs. Thirty-three percent of these endorsed protein therapeutics are created in *Escherichia coli*, demonstrating that it is a significant workhorse for recombinant restorative creation. Other recombinant microorganisms, such as *E. coli*, and *Saccharomyces cerevisiae* stay the most appealing because of their well-characterized genetics, expression systems, and versatile cloning tools, and as a result of which they have been introduced as a potential recombinant microbial candidate that can produce the microbial product at the lab scale as well as at the industrial scale (Belwal et al. 2020). It is advantageous to use these microorganisms for industrial-scale production because of their ease of scale-up, low-cost media, rapid growth, and capability to produce therapeutics with high quality and yield. Microbial products can be broadly classified into natural and recombinant on the basis of the source microorganism from which it has been derived. Natural products are derived from naturally occurring microorganisms. On the other hand, through recombinant DNA technology, microorganisms have been created that are successful in synthesizing human insulin, growth hormone, alpha interferon, hepatitis B vaccine, and different medically beneficial substances (Patnaik 2008). These products are known to be recombinant as they are produced using microorganisms, which have been created using recombinant DNA technology (Table 1.2).

Table 1.1 List of various industry types and natural microbial products produced using different microbial strains

Sl. no.	Type of industry	Microbial products	Microbial strain	Name of the company	References
1	Food Industry	Alcoholic beverages	*Saccharomyces cerevisiae*	Som Distilleries and Breweries Limited	Hansen et al. (2019)
		Vinegar	*Acetobacter aceti*	Fleischmann's Vinegar Company	
		Soy sauce	*Aspergillus oryzae*	G D Foods Mfg India Pvt. Ltd.	
		Wine	*Lactobacillus buchneri*	Grover Vineyards Limited	
		Coffcc	*Acetobacter fabarum*	Nescafe	
2	Dairy Industry	Cheese	*Streptococcus thermophilus*	Milky Mist Dairy Foods Pvt. Ltd.	Rezac et al. (2018)
		Yogurt	*Streptococcus thermophilus*	Amul	Riley (1988)
		Curd	*Lactobacillus lactis*	Parag Milk Foods Ltd.	
		Butter	*Leuconostoc* spp.	Mother Dairy	
3	Pharmaceutical Industry	Cyclosporin A Statins	*Tricoderma polysporum*	MITS Healthcare Pvt. Ltd	Delihas (1997)
		Statins	*Monascus purpureus*	Cipla Ltd.	Schwartz (1962)
		Penicillin	*Penicillium chrysogenum*	Hindustan Antibiotics Limited (HAL)	Tareq et al. (2015)
		Gentamycin	*Micromonospora purpurea*	Akorn Inc.	Demain and Vaishnav (2011)
		Amphotercin B	*Streptomyces nodosus*	Sun Pharma Ltd.	Schwartz (1962)
		Ivermectin	*Streptomyces avermitilis*	Merck	Shen (2015)
		Antipain	*Streptomyces yokosukaensis*	Keva Industries	Tareq et al., (2015)
		Leupeptin	*Streptomyces roseochromogenes*	Tocris Bioscience	Demain and Vaishnav (2011)
		Chymostatin	*Streptomyces hygroscopicus*	Antozyme biotech	Schwartz (1962)
4	Chemical Industry	Citric acid	*Aspergillus niger*	Meru Chem pvt. Ltd. Bajaj	Dashen et al. (2014)

(continued)

Table 1.1 (continued)

Sl. no.	Type of industry	Microbial products	Microbial strain	Name of the company	References
		Ethanol	*Saccharomyces cerevisiae*	Hindusthan Limited	Park et al. (2013)
		Xanthan gum	*Xanthomonas campestris*	Sarda Biopoly-mers Pvt. Ltd.	Elander (2003)
5	Cosmetics Industry	Beta carotene	*Erwinia uredovora*	Akhil Healthcare (P) Ltd.	Shafiei et al. (2022)
		Beta carrageenan	*Chondrus crispus*	Marine hydrocolloids	
		Fucoxanthin	*Laminaria japonica*	Manus Aktteva Biopharma LLP	

Table 1.2 List of various industry types and recombinant microbial products, produced using genetically modified different microbial strains

Type of industry	Microbial products	Microbial strain	Name of the company	References
Pharmaceutical Industry	Humulin	*Escherichia coli*	Biocon	Bell et al. (1980)
	Protropin/ Humatrope		Novo Nordisk India Pvt. Ltd.	Bengtsson (1993)
	Intron A	*Saccharomyces cerevisiae*	Schering-Plough	Bahr and Wilson (2012)
	Recombinax HB/Engerix		Merck	Berlec and Štrukelj (2013)
	Erythropoitein		Zydus	
	Glucagon		Novo Nordisk India Pvt. Ltd.	

1.3 In Lab But Not in Industry

Academic researchers face a number of challenges that hinder commercialization of products from lab to industry. The challenges in the commercialization process include, but are not limited to, risk reduction; obstacles in the time of wisdom; lack of financial support; policy/regulatory barriers; lack of infrastructure; lack of common understanding of the importance of commercial research; lack of business thinking among intellectuals; and a lack of collaboration and collaboration between universities and industry. Previous research found that costs, time constraints, inadequate infrastructure, and a lack of industry cooperation were the most common factors that researchers face that hinder the commercialization of products from the lab to the industry (Bermudez-Aguirre 2018). There are many barriers where microbial products have been upgraded to laboratory scales but have not yet been commercialized on an industrial scale. Some of the obstacles are, for example, the time required from the laboratory scale to the industrial scale is usually 3–10 years,

Table 1.3 Microbial products (Natural/Recombinant) that have not yet been commercialized

Sl. no.	Type of product	Product name	Microbial strain	References
1	Natural product	Polyketide pederin	*Pseudomonas symbiont*	Bok et al. (2006)
		Terpene	*Aspergillus nidulans*	
2	Recombinant product	Chloramphenicol acetyltransferase	*Escherichia coli*	Kenny et al. (1991)
		Beta galactosidase		
		Alkaline phosphatase		Gentschev et al. (1990)
		Dihydrofolate reductase		Nakano et al. (1992)

under these circumstances, economic risk is high, so the deterioration of the overall performance over time will increase and disrupt, which will no doubt lead to failure, no matter how well you prepare, there will be problems during the growing season. Typical examples include resource disruption, bacterial contamination, variable raw fabric quality, biofouling of strategic equipment, equipment failures, and sudden process performance. The major challenges in the commercialization of products can be divided into four components, namely, sufficient funding, skilled staff, marketing strategy, and knowledge gap and skills between academics and industry.

The main challenges that a researcher faces in commercializing products from the lab to the industry include:

1. Measuring production to meet commercial needs
2. Ensuring compliance with products
3. To earn enough money to develop and produce a product
4. Protect intellectual property

The below table shows some examples of natural as well as recombinant products that have not been commercialized yet based totally on data drawn from the literature. For example, Polyketide pederin and terpene are natural products that have been derived from *Pseudomonas symbiont* and *Aspergillus nidulans* (Bok et al. 2006), respectively, and Chloramphenicol acetyltransferase, Beta Galactosidase, Alkaline phosphatase, Dihydrofolate reductase are the recombinant products derived from the *E. coli* (Kenny et al. 1991) are there in the lab-scale but are not yet been able to commercialize into industrial scale (Table 1.3).

1.4 Basic Processes for Microbial Product Transfer from Lab to Industry

Scaling up requires the art of designing large-scale apparatus with the use of the data obtained from the laboratory studies. In many industries, such an important role has been played by the bioreactor, for example, in wastewater treatment, fermentation, food, medicine, etc. Recently, translating a small-scale product into a larger-scale design has been a major challenge in terms of the growth of a bioreactor (Crater and Lievense 2018). While scaling-up of cell culture processes, declining productivity was reported many times, which can be attributed to a few factors, including shear pressure, oxygen supply, and gas composition.

1.4.1 Preparation of Inoculum

The inoculum microbe should be adapted to the preservation culture for the purpose that it is usually used for fermentation. The process includes many ways to make sure of the maximum yield. The first culture is ready from the culture of preservation in agar slants and then used to create a functional culture. Now, microorganisms begin to grow. In the processes of small fermentation, the use of a culture, functioning as an inoculum, however at large scale fermentation, additional steps are being involved for inoculum production. Glass beads which contain liquid additive or saline water that is sterilized has been added to the agar slant and stirred for the purpose of microbial suspension (Wang et al. 2019).

This suspension is delivered in a flat bottle containing an agar medium, which is sterile. In the bottle, the microorganism can grow. Then, from the flask-bed bottles, microbial cells are transported to the shake flask, which contains a nutrient medium of liquid that is placed on a rotating shaker bed inside the incubation. In terms of growing, faster air circulation is required. The motive for this is the further development of microbial biomass, which contributes to the end of the fermentation process as yield is determined by the biomass ratio and substrate weight.

1.4.2 Developmental Process of Inoculum for Fermentation

This inoculation development process for fermentation is one of the most important processes involved in the commercialization of microbial products from lab to industry (Ciani et al. 2016). This process involves the following steps,

Improvement of strain: The product yield will be substantially less when normally accessible (natural) microorganisms are utilized. Giving the growth conditions that is optimum expands the productivity barely. Accordingly, in terms of

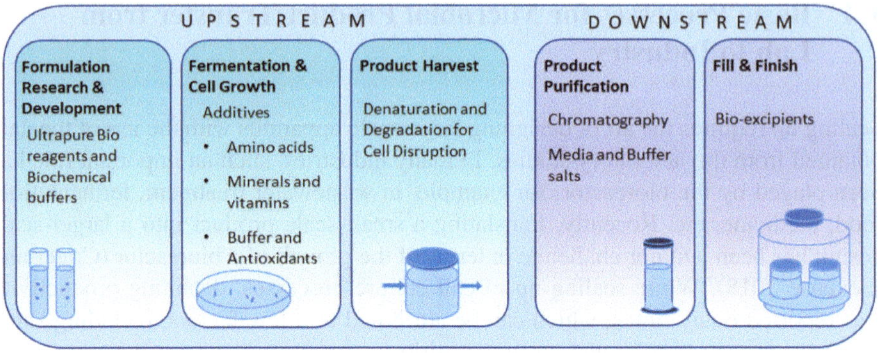

Fig. 1.1 Upstream and downstream processes for microbial product scale-up

increasing the efficiency of microorganisms, it is important to alter their genetic structure.

Change in the structure of genes likewise impact the culture medium and requirements of nutrition. Changes at the genetic level in microorganisms can be instigated by different techniques like improvement of a strain, which is a classical strain, transformation, and determination, or by the utilization of recombinant DNA technology.

Mutation: Each time a cell of a microbe undergoes differentiation, there is a chance of the species getting mutated, which changes it from the genetic level, making it different from the wild type. Mutation potential can be enhanced by exposing the wild type to mutagenic agents, for example, non-ionizing ultraviolet (UV) radiation, ionizing radiation, and various synthetic agents, for example, nitrous acid and others.

Recombinant DNA (RDT) technology: This is a method that supports years of new genetic mixes, which were previously present in two different organisms, in vivo or in vitro.

In vitro, RDT is used in organisms, for example, *Streptomycetes* and filamentous fungi to produce a better yield. In terms of developing the advanced forms of synthetic antibiotics, which are useful in the medical field, in vitro DNA recombination has been added.

Natural variant selection: Microorganisms undergo small genetic mutations in each cell differentiation. After a few differentiations, the cultural medium incorporates microorganisms with a broad scope of genetics. From these variants, the most productive varieties can be selected for fermentation or mutation (Fig. 1.1).

1.4.3 Development of Inoculum Monitoring

The establishment of cultural conditions and a framework for observation are necessary in order to determine the optimal transmission time, compliance with stored physical conditions, and a well-designed production process (Sharma et al. 2019).

- Biomass is known as an important component of the fermentation process, which directly affects the introduction of the fermentation system and the quality and yield of the system.
- Biomass levels can be measured by looking at parameters such as full cell volume, wet weight, dry weight, respiration, a saturation of nutrients, turbidity, and morphology.
- Another new generation of highly specific biosensors was created by the interaction of immobilized enzymes that can travel with electrochemical sensors, i.e., sensitive alcohols as well as glucose (Sharma et al. 2019). For example, based on the glucose oxidase enzyme the concentration is determined by the glucose sensor.

1.4.3.1 Inoculum Transfer to the Fermenter Vessel or the Process of Scaling Up

The inoculum is transferred aseptically to the production reactor. To do this, the inoculum (which must be approximately 5–10% of the total volume of the cultural volume) has been transferred through a sterile syringe of the appropriate size (Wynn et al. 2016).

- The syringe is quickly pushed through the silicon membrane and the vessel is inoculated. The vessel is effectively kept at positive pressure through sparging of the sterile air to reduce the risk of contamination into the vessel. The syringe is removed quickly and the silicon membrane recurs.
- During the process of scaling up when microorganisms are transferred from the smaller to larger reactors in subsequent cycles, there is a problem with homogeneity in the larger frameworks.
- This happens may be due to the development of the ratio of surface-to-volume or the cultural exchange itself due to the extended cultural period. Similarly, in the process of scale-up, many factors contribute to the growth of bacteria and the yield of the process of fermentation, where heat dissipation and oxygen supply are important factors.

1.4.4 Biological Performances Affected Due To Different Parameters of the Process

The primary purpose of bioreactor choice, design, as well as control, is to provide a good or ideal environment for the bioreactor system. The optimal conditions such as pH, temperature, oxygen transfer as well as mixing, and concentration of substrate must be given by the bioreactor, in addition to its basic control volume (Meramo et al. 2021).

Temperature: Microorganisms are much of the time characterized by their temperature for growth as either psychrophiles (temperature for growth: <20 °C), or mesophiles (temperature for growth: from 20 °C to 50 °C), or thermophiles (temperature for growth: >50 °C).

pH: Ranges of pH differ optimally for different biological systems. Most microorganisms develop best in the range between pH 5 and 7. In the process of fermentation, the change of pH can be observed. Metabolites are delivered into the medium. As the cells develop, substrate utilization likewise causes pH change. Various analysts have researched the impact of pH on the kinetics of growth of microorganisms, enzymatic activities, and the synthesis of products.

Oxygen Transfer: Transfer of oxygen is generally a worry in the biological system of an aerobic condition. Many nutrients needed for the proliferation of cells and their metabolism are exceptionally water soluble. An adequate and supply of timely basis of these supplements can be accomplished in a thoroughly mixed bioreactor. Nonetheless, the transfer of oxygen frequently turns into a restricting step for the presentation of the optimal basis of the biological system as well as furthermore for scaling up since oxygen has been just sparingly dissolvable in an aqueous system. At the point when the oxygen supply is restricted, both the growth of cells as well as the formation of products can be seriously impacted.

Blending: To guarantee a satisfactory nutrient supply and to forestall the amassing of harmful metabolites, sufficient mixing is fundamental in bioreactors. For a bioreactor intended for the frame of suspension, blending or mixing time is an important parameter to be considered as well as assessed. The hydrodynamics of fluid, impeller type, rheology of fluid, vessel size, and the input of power input can all impact the conditions for mixing or blending (Fig. 1.2).

1.5 Drawbacks or Hurdles

One can't really amplify lab-scale tools and duplicate lab-scale stipulations at large scale. Without apperception of large-scale instruments and how the change of scale-dependent parameters happens, a person is likely to get into huge trouble. There are so many hurdles for which the microbial products are already scaled up in lab scale but still have not commercialized in industrial scale (Goyal et al. 2020).

Some of the hurdles are as follows:

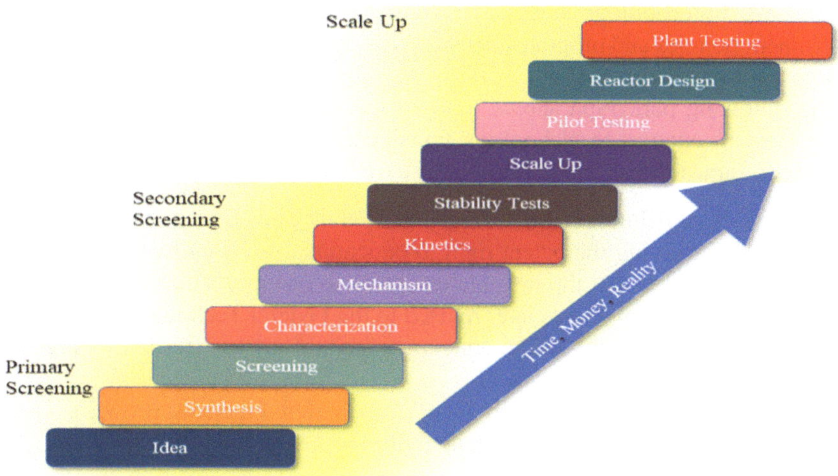

Fig. 1.2 Step by step processes for lab to industry commercialization

- The transition of lab scale to industrial scale requires the time of generally 3–10 years. The economic risk is high under these circumstances, so deterioration in overall performance at some stage in scaling up will be highly priced and disruptive, doubtlessly even leading to failure.
- Regardless of how nicely you prepare, there will be problems that occur throughout the process of scaling up. Common examples include encompassing utility interruptions, variable raw fabric quality, microbial contamination, tools failure, biofouling of technique equipment, and sudden terrible process performance at scale.

The challenges in the commercialization process include risk reduction; obstacles in the time of wisdom; lack of financial support; policy/regulatory barriers; lack of infrastructure; lack of common understanding of the importance of commercial research; lack of business thinking among intellectuals; and a lack of collaboration and collaboration between universities and industry. Previous research found that costs, time constraints, inadequate infrastructure, and a lack of industry cooperation were the most common factors that researchers face that hinder the commercialization of products from the lab to the industry.

1.6 Ethical Clearance

During the process of technology transfer and commercialization of products from lab to industry, some of ethical issues will be raised. The ethical issues are nothing but the potential conflicts of interest that broadly impacts individual and organizational integrity. In order to fill the gap from research to product commercialization some of the biotechnology companies were established and also some of the new government policies and programs helped in the commercialization of products from lab to industry (Zucker 2011).

The following paragraph shows some of the acts that have been passed by the government to encourage technology transfer and commercialization of products from lab to industry.

- Congress passed Public Law 96-517, the Patent and Trademark Law Amendments Act for giving the patent right to educational institutions for government-funded research so that the institute can work to commercialize them.
- Innovation Act of 1980 was later revised to the Federal Technology Transfer Act of 1986; Public Law 99-502 to encourage technology transfer between government labs and private companies.
- In medicine, the FDA's current Good Production Practices (cGMP). Dietary requirements and kosher rules can apply to food and nutritional products.

Drugs and other products produced for human consumption must comply with government or industry regulations. Even if the custom combination is an active pharmaceutical ingredient (API), and therefore does not require a cGMP certificate, the provision of specialized intermediaries and forensic applications for health science applications may require the issuance of an ISO certificate on the commercial scale. Conflicts of interest are a key concern for the commercialization of products in any research. It puts a negative impact on the integrity and objectivity of the research (Romain 2015). When a researcher is an industry person this conflict of interest (COI) does not put that much of an impact on his/her work but for an academic researcher this thing gets even more complicated and badly impacts his/her work.

In order to resolve this and promote technology transfer and research commercialization, there must be proper trust and integrity between the universities and the industries.

1.7 Future Prospects

At present, there is a very prominent difference between the lab-scale production of novel biotechnological products and their eventual transfer to the industry. This can be due to various reasons, like impediments in the scale-up procedure of specific products, lack of energy alternatives, or even due to the fact that biological products are less suitable for downstream processing procedures, unlike chemical products. This is the primary factor behind the significant gap in the number of new innovations coming up in the laboratories and those being actually commercialized after industrial production (Gross et al. 2018). However, there are ample opportunities to optimize the industrial biotechnology production processes to try and narrow this gap between the laboratory table and the industrial production belt. Focusing on the molecular biology and synthetic biology aspects, like selecting the most productive and optimizable strain for the particular product or improving the bioprocess parameters to achieve the desirable scale-up and employing easier downstream processes to ultimately have product isolation comparable to that of chemical products is the key. The key problems identified during the industrial production of biological products include the in-general slow growth rate of microbes or the very low density of microbial cell production. The growing microbial cells also generally cannot use mixed substrates to derive their nutrition, and the sterilizing costs for these bioreactors are also fairly higher. These problems can be dealt with by optimizing the process parameters like minimizing the microbial cells for the production or by assembling pathways capable of the utilization of mixed substrates. Other measures include minimizing the oxygen demand for the aerobic microbes and reducing the quorum sensing effects, to solve the low density of production problem. In order to tackle the sterilization situation effectively, from an economic point of view, one way out may be to grow strains that are capable of resisting contamination. In conclusion, it may be said that a combination of biological and chemical processes is the way forward for achieving appreciable technology transfer from labs to the industries for biological products, comparable to that of chemical industrial products.

Acknowledgment This chapter is about microbial product commercialization from lab to industry. The authors are grateful to National Institute of Technology (NIT) Rourkela for providing them with quality resources to make this happen. The guidance of Dr. Angana Sarkar, the corresponding author, was constant throughout the writing period of this chapter and it was of great help to the authors. The authors would also like to thank their friends and family who supported and offered deep insight into the study.

References

Bahr A, Wilson AB (2012) The evolution of MHC diversity: evidence of intralocus gene conversion and recombination in a single-locus system. Gene 497(1):52–57

Bell GI, Pictet RL, Rutter WJ, Cordell B, Tischer E, Goodman HM (1980) Sequence of the human insulin gene. Nature 284(5751):26–32

Belwal T, Singh G, Jeandet P, Pandey A, Giri L, Ramola S, Bhatt ID, Venskutonis PR, Georgiev MI, Clément C, Luo Z (2020) Anthocyanins, multi-functional natural products of industrial relevance: recent biotechnological advances. Biotechnol Adv 43:107600

Bengtsson BA (1993) Treatment of adults with growth hormone (GH) deficiency with recombinant human GH. J Clinical Endocrinol Metabol 76(2):309–317

Berlec A, Štrukelj B (2013) Current state and recent advances in biopharmaceutical production in *Escherichia coli*, yeasts, and mammalian cells. J Ind Microbiol Biotechnol 40(3–4):257–274

Bermudez-Aguirre D (2018) Technological hurdles and research pathways on emerging technologies for food preservation. In: Innovative technologies for food preservation. Academic Press, Cambridge, pp 277–303

Bok JW, Hoffmeister D, Maggio-Hall LA, Murillo R, Glasner JD, Keller NP (2006) Genomic mining for Aspergillus natural products. Chem Biol 13(1):31–37

Ciani M, Capece A, Comitini F, Canonico L, Siesto G, Romano P (2016) Yeast interactions in inoculated wine fermentation. Front Microbiol 7:555

Crater JS, Lievense JC (2018) Scale-up of industrial microbial processes. FEMS Microbiol Lett 365 (13):p.fny138
Delihas N (1997) Microbiology. In: Prescott LM, Harley JP, Klein DA (eds) The quarterly review of biology, vol 72. University of Chicago Press, Chicago, pp 472–473
Demain AL (2000) Microbial biotechnology. Trends Biotechnol 18:26–31
Demain AL, Vaishnav P (2011) Natural products for cancer chemotherapy. Microb Biotechnol 4 (6):687–699
Elander RP (2003) Industrial production of β-lactam antibiotics. Appl Microbiol Biotechnol 61(5-6):385–392
Gentschev I, Hess J, Goebel W (1990) Change in the cellular localization of alkaline phosphatase by alteration of its carboxy-terminal sequence. Mol Gen Genet MGG 222:211–216
Goyal R, Singh O, Agrawal A, Samanta C, Sarkar B (2020) Advantages and limitations of catalytic oxidation with hydrogen peroxide: from bulk chemicals to lab scale process. Catal Rev 64(2): 229–285
Gross R, Hanna R, Gambhir A, Heptonstall P, Speirs J (2018) How long does innovation and commercialisation in the energy sectors take? Historical case studies of the timescale from invention to widespread commercialisation in energy supply and end use technology. Energy Policy 123:682–699
Gupta PL, Rajput M, Oza T, Trivedi U, Sanghvi G (2019) Eminence of microbial products in cosmetic industry. Nat Prod Bioprospect 9(4):267–278
Hansen EB, Nielsen DS, LaPointe G (2019) Editorial: microbial food and feed ingredients–reconciling tradition and novelty. FEMS Microbiol Lett 366(11):fnz130
Kenny B, Haigh R, Holland IB (1991) Analysis of the haemolysin transport process through the secretion from Escherichia coli of PCM, CAT or β-galactosidase fused to the Hly C-terminal signal domain. Mol Microbiol 5(10):2557–2568
Meramo S, González-Delgado ÁD, Sukumara S, Fajardo WS, León-Pulido J (2021) Sustainable design approach for modeling bioprocesses from laboratory toward commercialization: optimizing chitosan production. Polymers 14(1):25
Nakano H, Kawakami Y, Nishimura H (1992) Secretion of genetically-engineered dihydrofolate reductase from Escherichia coli using an E. coli α-hemolysin membrane translocation system. Appl Microbiol Biotechnol 37:765–771
Park D, Lee J (2013) Biological conversion of methane to methanol. Korean J Chem Eng 30:977–987
Park JM, Oh BR, Seo JW, Hong WK, Yu A, Sohn JH, Kim CH (2013) Efficient production of ethanol from empty palm fruit bunch fibers by fed-batch simultaneous saccharification and fermentation using Saccharomyces cerevisiae. Appl Biochem Biotechnol 170:1807–1814
Parnell JJ, Berka R, Young HA, Sturino JM, Kang Y, Barnhart DM, DiLeo MV (2016) From the lab to the farm: an industrial perspective of plant beneficial microorganisms. Front Plant Sci 7:1110
Patnaik R (2008) Engineering complex phenotypes in industrial strains. Biotechnol Prog 24(1): 38–47
Rezac S, Kok C, Heermann M, Hutkins R (2018) Fermented foods as a dietary source of live organisms. Front Microbiol 9:1785
Riley R (1988) New directions for agriculture and agricultural research. Food Policy 13(4):405–406
Romain PL (2015) Conflicts of interest in research: looking out for number one means keeping the primary interest front and center. Curr Rev Musculoskelet Med 8(2):122–127
Schwartz W (1962) Michael J. Pelczar Jr. u. Roger D. Reid, Laboratory exercises in microbiology. IX und 173 S. McGraw-Hill Book Co., Inc. New York/Toronto/London 1958. 23 sh 6 d. Z Allg Mikrobiol 2(4):326–327
Shafie MH, Kamal ML, Zulkiflee FF, Hasan S, Uyop NH, Abdullah S, Hussin NAM, Tan YC, Zafarina Z (2022) Application of Carrageenan extract from red seaweed (Rhodophyta) in cosmetic products: a review. J Indian Chem Soc:100613
Sharma A, Kaur J, Lee S, Park YS (2019) Tracking of intentionally inoculated lactic acid bacteria strains in yogurt and probiotic powder. Microorganisms 8(1):5

Shen B (2015) A new golden age of natural products drug discovery. Cell 163(6):1297–1300

Tareq FS, Lee HS, Lee YJ, Lee JS, Shin HJ (2015) Ieodoglucomide C and Ieodoglycolipid, New Glycolipids from a Marine-Derived Bacterium Bacillus licheniformis 09IDYM23. Lipids 50 (5):513–519

Wang X, Wang Y, Zhang A, Duo C, Zhao C, Xie F (2019) Isolation of a highly efficient phenol-degrading fungus and the preparation of an effective microbial inoculum for activated sludge and its enhancement for hydrogen production. Int J Hydrog Energy 44(30):16004–16014

Wynn JP, Hanchar R, Kleff S, Senyk D, Tiedje T (2016) Biobased technology commercialization: the importance of lab to pilot scale-up. In: Van Dien S (ed) Metabolic engineering for bioprocess commercialization. Springer, Cham, pp 101–119

Zucker D (2011) Ethics and technology transfer: patients, patents, and public trust

Chapter 2
Assessment of Microbes and Microbial Products for Future Industrialization

Sabina Khanam

Abstract For life processes man is totally dependent on microbes and on microbial products. Discovery of microbes and their products largely improved the world's health leading to the discovery of antibiotics, vaccines, and many more chemical agents to cure many of these diseases. Consumption of some fermented foods such as fermented dairy products, which contain viable bacteria, widely connected with the reduction of certain diseases such as, cardiovascular diseases, childhood obesity, and type-2 diabetes. These fermented dairy products also improve body composition and weight loss. The quality of food has been improved and protected by many microbial products. Farming practices have been improved by the breakdown of nonedible crops residues for reuse by the new crops. There are so many microorganisms such as nitrogen-fixing bacteria that have been used to inoculate legumes. Biopesticides or microbial insecticides are ecologically good because it is less harmful and has less environmental load.

Keywords Microbes · Microbial products · Health

2.1 Introduction

Study of microorganisms, (bacteria, algae, fungi, viruses, and some parasites like protists and archea), microbial products, and microbial processes has a variety of health benefits. Application of microbial products are very common in the food industry but also very helpful in health interventions, agriculture, and for environment. In food industry, microbial products and microbial cells are used in various food formulations. Microbial fermented products such as fermented honey, juices, and fermented products of animal origin have constituted an essential part of human nutrition and these products have been consumed through naturally microbial fermented products containing viable bacteria. Some dairy products and fermented

S. Khanam (✉)
Department of Biological Sciences, Yobe State University, Damaturu, Nigeria

A. Sarkar, I. A. Ahmed (eds.), *Microbial products for future industrialization*, Interdisciplinary Biotechnological Advances,
https://doi.org/10.1007/978-981-99-1737-2_2

vegetables contain significant amounts of viable microbes (Selhub et al. 2014). Human gut microbiota plays a very important role in the functionality of gut and body homeostasis, which is influenced by some environmental factors, age, type of diet, medical therapies, diseases, and disorders. Microorganisms also identified as efficient modulators of intestinal microbial balance (Gomez-Gallego and Salminen 2016).

Consumption of some fermented foods, such as fermented dairy products, contain viable bacteria widely connected with the reduction of certain diseases such as, cardiovascular diseases, childhood obesity, and type-2 diabetes. These fermented dairy products also improve body composition and weight loss. Manipulation of microbial communities is very beneficial in the treatment of waste water, for the prevention of diseases, treatment of caries, and inflammatory bowel disease and obesity (Werner et al. 2011; Marsh 1994; Maloy and Powrie 2011; Ley et al. 2006; Blekhman et al. 2015; Thorning et al. 2016).

2.2 Microbes and Microbial Products in Plant Health and Nutrition

For plant growth and nutrition, microbes such as nitrogen fixing symbiotic bacteria and mycorrhizal fungi play a very important role by improving the mineral nutrition. In plants roots are metabolically very active and provides continuous food source for the microorganisms present in the soil in the form of cell debris, organic compounds, and sloughed-off dead cells. In the root system there is a region called heterogeneous region or rhizosphere in which microorganisms are very active. Some of the microorganisms are associated with the root but a large fraction of microorganisms develops on the root surface. Many microorganisms can invade root tissue, which is sometimes beneficial or sometimes this invasion is harmful for the plant.

When plants and animals decompose, microorganisms in the soil releases nitrogen, sulfur, carbon, phosphorus, and many trace elements. This process of mineralization is the primary source of atmospheric carbon dioxide. Many bacterial strains such as *Pseudomonas* and *Azospirillum* had been known for having plant growth and promoting effects (Burr et al. 1978; Lin et al. 2015).

There are three main mechanisms by which microbial activity boost plant growth:

1. Repelling or outcompeting pathogenic microbial strains (Mendes et al. 2013)
2. Manipulating the hormonal singling of plants (Verbon and Liberman 2016)
3. By increasing the bioavailability of soil-borne nutrients (van der Heijden et al. 2008)

2.3 Microbial Products as Insect Control Agents

Living organisms are the most commonly used biopesticides, which is pathogenic for the pest. These biopesticides include biofungicides (trichoderma), bioinsecticides (*Bacillus thuringiensis*, *B sphaericus*), and bioherbicides (Phytopthora). There are more than 1500 naturally occurring microorganisms, which are used for the control of insect pests. They induce diseases in target insects and suppress the insect population directly or with the help of chemical insecticides. Use of chemical insecticides leads to several problems because the residues of these insecticides cause toxic effects on some wild animals such as birds and on some insects such as honey bees. Insecticides can also be harmful to humans and contaminate ground water (Lacey and Siegel 2000).

Biopesticides or microbial insecticides are derived from natural materials such as plants, bacteria, animals, and from certain minerals. For example, baking soda has some pesticidal applications and hence is considered a biopesticide. There are so many advantages of using biopesticides in the place of chemicals. These include:

1. Biopesticides or microbial insecticides are ecologically good because it is less harmful and has less environmental load.
2. Biopesticides can contribute more when used as a component of integrated pest management (IPM) programs.
3. It benefits the environment because it decomposes quickly and it is very effective in very small quantities, thereby resulting in lower exposures and avoiding pollution problems.
4. The residues of microbial insecticides are safe for humans and other animals.
5. Microbial insecticides can also be used at the time when crop if ready for harvest.
6. Microbial insecticides do not affect beneficial insects directly in the treated areas because toxic action of microbial insecticides is often specific to a single group of insect species.
7. Microbial insecticides organisms are nonpathogenic and nontoxic to humans, wildlife, and other organisms.
8. Microbial insecticides also enhance the growth of plants and roots by encouraging the beneficial soil microflora (Table 2.1).

2.4 Microbes as Fuel and Energy

A wide variety of microorganisms are used in the process of fermentation for the manufacture of wines, beers, and foods. In several countries such India and Brazil the production of ethanol by the process of fermentation is used for fuel purposes. In the same way the production of biogas is also used as a source of energy in several

Table 2.1 Microbes and their uses

Microbe name	Use
Bacteria	
Bacillus thuringiensis	Toxin of this bacteria has been used as an insecticide to enhance resistance to insect pests in genetically modified crops It is commonly used in organic farming
Bacillus popilliae	It produces spores, which have pesticide active ingredient Spores of this bacterium are used on lawns and ornamental plants around residential areas
Bacillus sphaericus	It is the most commonly used microbial biocontrol agents for medical pets, such as black flies and mosquitoes This bacterium produces spores and these spores contain a protein that damages the gut and paralyzes the gut of mosquito larvae
Fungi	
Beauveria bassiana	It is used to control pests such as whiteflies, termites, and other insects It also infects a huge variety of insects and controls crop infestations by whitefly, aphids, and thrips
Lagenidium giganteum	It infects and kills mosquito larvae. It is widely used as a biological pesticide to control mosquito
Protozoa	
Nosema locustae	It is used to kill caterpillars, crickets, locusts, grasshoppers, and some corn borers
Viruses	
Gypsy moth nuclear polyhedrosis	It is used to control the gypsy moth population
Coding moth granulosis virus	It is used commercially to control insect pest of apple

countries. Microorganisms can generate fuels such as methane, ethanol, hydrogen, butanol, and lipids.

Microalgae and Cyanobacteria possess the potential to reduce the atmospheric CO_2 into biofuels photosynthetically, and methanotrophs can use methane to produce methanol. Bacteria can convert sugar into ethanol. *Geobacter sulfurreducens* and *Shewanella oneidensis* are bacteria that exhibit molecular machinery that helps transfer of electrons from microbial outer membrane to conductive surfaces, this feature can be deployed in bioelectrochemical devices for the generation of bioelectricity and biohydrogen (Liao et al. 2016; Kracke et al. 2015) (Table 2.2).

Table 2.2 Microorganism and their biofuel

Biofuel	Microorganism	References
Ethanol	*Zymomonas mobilis* *Caldicellulosiruptor bescii* *Tricoderma reesei* *Escherichia coli*	Kremer et al. (2015), Chung et al. (2014), Romero-Garcia et al. (2016)
Butanol	*Clostridium acetobutylicum* *Escherichia coli* *Pseudomonas putida*	Lutke-Eversloh and Bahl (2011), Shen et al. (2011), Nielsen et al. (2009)
Isobutanol	*Clostridium thermocellum*	Lin et al. (2015)
Fatty acids	*Saccharomyces cerevisiae* *Yarrowia lipolytica*	Yu et al. (2016), Beopoulos et al. (2009)
Lipids	*Cryptococcus vishniaccii*	Deeba et al. (2016)

References

Beopoulos A, Cescut J, Haddouche R, Uribelarrea JL, Molina-Jouve C, Nicaud JM (2009) *Yarrowia lipolytica* as a model for bio-oil production. Prog Lipid Res 48:375–387. https://doi.org/10.1016/j.plipres.2009.08.005

Blekhman R, Goodrich JK, Huang K, Sun Q, Bukowski R, Bell JT et al (2015) Host genetic variation impacts micro biome composition across human body sites. Genome Biol 16:191. https://doi.org/10.1186/s13059-015-0759-1

Burr TJ, Schroth MN, Suslow T (1978) Increased potato yields by treatment of seed-piece with specific strains of *Pseudomonas fluorescens* and *P. putida*. Phytopathology 68:1377–1383. https://doi.org/10.1094/Phyto-68-1377

Chung D, Cha M, Guss AM, Westpheling J (2014) Direct conversion of plant biomass to ethanol by engineered *Caldicellulosiruptor bescii*. Proc Natl Acad Sci USA 111:8931–8936. https://doi.org/10.1073/pnas.1402210111

Deeba F, Pruthi V, Negi YS (2016) Converting paper mill sludge into neutral lipids by oleaginous yeast *Cryptococcus vishniaccii* for biodiesel production. Bioresour Technol 213:96–102. https://doi.org/10.1016/j.biortech.2016.02.105

Gomez-Gallego C, Salminen S (2016) Novel probiotics and probiotics: how can they help in human gut microbiota dysbiosis? Appl Food Biotechnol 3:72–81. https://doi.org/10.22037/afb.v3i2.11276

Kracke F, Vassilev I, Kromer JO (2015) Microbial electron transport and energy conservation- the foundation for optimising bioelectrochemical systems. Front Microbiol 6:575. https://doi.org/10.3389/fmicb.2015.00575

Kremer TA, LaSarre B, Posto AL, McKinlay JB (2015) N_2 gas is an effective fertilizer for bioethanol production by *Zymomonas mobilis*. Proc Natl Acad Sci U S A 112:2222–2226. https://doi.org/10.1073/pnas.1420663112

Lacey LA, Siegel JP (2000) Safety and ecotoxicology of entomopathogenic bacteria. In: Charles JF, Delecluse A, Nielsen-LeRoux C (eds) Entomopatgenic bacteria: from laboratory to field application. Springer, New York

Ley RE, Turnbaugh PJ, Klein S, Gordon J, I. (2006) Human gut microbes associated with obesity. Nature 444:1022–1023. https://doi.org/10.1038/4441022a

Liao JC, Mi L, Pontrelli S, Luo S (2016) Fuelling the future: microbial engineering for the production of sustainable biofuels. Nat Rev Microbiol 14:288–304. https://doi.org/10.1038/nrmicro.2016.32

Lin PP, Mi L, Morioka AH, Yoshino KM, Konishi S, Xu SC et al (2015) Consolidated bioprocessing of cellulose to isobutanol using *Clostridium thermocellum*. Metab Eng 31:44–52. https://doi.org/10.1016/j.ymben.2015.07.001

Lutke-Eversloh T, Bahl H (2011) Metabolic engineering of *Clostridium acetobutylicum*: recent advances to improve butanoll production. Curr Opin Biotechnol 22:634–647. https://doi.org/10.1016/j.copbio.2011.01.011

Maloy KJ, Powrie F (2011) Intestinal homeostasis and its breakdown in inflammatory bowel disease. Nature 474:298–306. https://doi.org/10.1038/nature10208

Marsh PD (1994) Microbial ecology of dental plaque and its significance in health and disease. Adv Dent Res 8:263–271. https://doi.org/10.1177/08959374940080022001

Mendes R, Garbeva P, Raajimakers JM (2013) The rhizosphere micro biome significance of plant beneficial, plant pathogenic, and human pathogenic microorganisms. FEMS Microbiol Rev 37: 634–663. https://doi.org/10.1111/1574-6976.12028

Nielsen DR, Leonard E, Yoon SH, Tseng HC, Yuan C, Prather KLJ (2009) Engineering alternative butanol production platforms in heterologous bacteria. Metab Eng 11:262–273. https://doi.org/10.1016/j.ymben.2009.05.003

Romero-Garcia JM, Martinez-Pati-o C, Ruiz E, Romero I, Castro E (2016) Ethanol production from olive stone hydrolysates by xylose fermenting microorganisms. Bioethanol 2:51–65. https://doi.org/10.1515/bioeth-2016-0002

Selhub EM, Logan AC, Bested AC (2014) Fermented foods, microbiota, and mental health: ancient practice meets nutritional psychiatry. J Physiol Anthropol 33:2. https://doi.org/10.1186/1880-6805-33-2

Shen CR, Lan EI, Dekishima Y, Baez A, Cho KM, Liao JC (2011) Driving forces enable high titer anaerobic 1-butanol synthesis in *Escherichia coli*. Appl Environ Microbiol 77:2905–2915. https://doi.org/10.1128/AEM.03034-10

Thorning TK, Raben A, Tholstrup T, Soedamah-Muthu SS, Givens I, Astrup A (2016) Milk and dairy products: good or bad for human health? An assessment of the totality of scientific evidence. Food Nutr Res 60:32527. https://doi.org/10.3402/fnr,v60.32527

van der Heijden MGA, Bardgett RD, Van Straalen NM (2008) The unseen majority: soil microbes as drivers of plant diversity and productivity in terrestrial ecosystems. Ecol Lett 11:296–310. https://doi.org/10.1111/j.1461-0248.2007.01139x

Verbon EH, Liberman LM (2016) Beneficial microbes affect endogenous mechanisms controlling root development. Trends Plant Sci 21:218–229. https://doi.org/10.1016/jtplants2016.01.013

Werner JJ, Knights D, Garcia ML, Scalfone NB, Smith S, Yarasheski K et al (2011) Bacterial community structures are unique and resilient in full-scale bioenergy systems. Proc Natl Acad Sci U S A 108:4158–4163. https://doi.org/10.1073/pnas.1015676108

Yu AQ, Juwono NKP, Foo JL, Leong SSJ, Chang MW (2016) Metabolic engineering of *Saccharomyces cerevisiae* for the overproductionn of short branched-chain fatty acid. Metab Eng 34: 36–43. https://doi.org/10.1016/j.ymben.2015.12.005

Chapter 3
Design and Operation of New Microbial Product Bioprocessing System

Ravi Ranjan Kumar and Chitra Bhattacharya

Abstract Microbial products are recognized as the most diverse commercially important chemical structure. Bioprocess offers sustainable techniques for the production of existing and newly developed microbial products. Microbial cell factories and microbial consortia are playing a central role in the bioprocessing system. The advancement of an efficient bioprocessing system is essential for the production of microbial products on an industrial scale. Various components of bioprocessing such as strain improvement, design of bioreactor, media optimization, linked bioreactor system, and downstream processing have been modified to optimize the end product. Recently, numerous approaches have been adopted to accept the challenges in new microbial product development. These advanced approaches include protein engineering, engineering precursor supply, metabolic engineering, pathway engineering, combinatorial biosynthesis, and mutasynthesis for upstream processing. High-throughput, single-use cultivation, optimization by the design of experiments, process modeling, and process systems engineering are advanced methods for mid-stream processing. Integrated continuous downstream processing is an important step for bioprocess economy. Metabolic modification of microbial cell factories is a modern approach for the optimization of microbial products. The application of system biology in metabolic engineering is an advanced technique in optimization of up, mid, and downstream processing. The integrated advanced bioprocess design facilitates manufacturing flexibility, cost-effectiveness, and quality end product. This chapter highlights the design and operation of prevailing and modern approaches to the bioprocessing system for the development of new microbial products.

Keywords Microbial cell factory · Natural product · Metabolic engineering · Upstream processing · Downstream processing

R. R. Kumar · C. Bhattacharya (✉)
Department of Microbiology, Atmiya University, Rajkot, India
e-mail: chitra.bhattacharya@atmiyauni.ac.in

A. Sarkar, I. A. Ahmed (eds.), *Microbial products for future industrialization*, Interdisciplinary Biotechnological Advances,
https://doi.org/10.1007/978-981-99-1737-2_3

3.1 Introduction

Bioprocess is the use of biological sources for valuable product formation from cheaper substrate under optimum conditions. The success of industrial microbiology or biotechnology is observed through new microbial products and their commercialization. Bioprocess technology is a promising tool for fulfilling the demand of microbial products. Bioprocess development initiates with laboratory scale, shifts to pilot scale, and finally scales up to industrial level. Bioprocess engineering require design, direction, and process to innovate, manufacture, and commercialize biological product as per the needs of society. Microbial products are produced by bioprocess engineering. The global market of microbial product is calculated as $186.3 billion, which is expected to increase to $302.4 billion by 2023. Bio-based economy uses scientific and engineering principles for microbial product formation. The microbiology research and their products brought better visions for the development of bioprocess engineering. It is important to find new microbial products and improve the production using various approaches. Microbial-based product has been utilized in health, agriculture, food industry, environment, scientific research, and cosmetics (Sanchez et al. 2012). Microbes were used for the commercial production of antibiotics, organic acids, chemotherapeutic agents, peptides, biosurfactant, polysaccharides, vitamins, enzymes, amino acids, alcohol, and vaccines. However, most of the microbial products are produced in low concentrations during bioprocessing. Therefore, industrially improved strain, genetically engineered microorganisms, novel strategies for bioprocess, and different developmental approaches are used for improvement of microbial products. Strain improvement and product engineering process is an imperative strategy for product development. The unit operation for upstream and downstream processing has been modified by novel approaches. In this chapter, we discussed the upstream process, downstream process, recent approaches used to improvise products, control systems, and high-throughput techniques for microbial product improvement.

3.2 Development of Upstream Process

3.2.1 *Microbial Bioactive Natural Products*

3.2.1.1 Antibiotic

Natural products of microbes are a rich source of antibiotic drug development. They are classified into three categories—non-ribosomal peptides, polyketides, and aminoglycosides. Penicillin produced from *Penicillium notatum* was the first naturally obtained antibiotic; it belongs to the non-ribosomal peptide. Non-ribosomal peptides are expressed from the non-ribosomal peptide synthetase (NRPS). Vancomycin is also an effective non-ribosomal peptide that can inhibit the growth of pathogenic bacteria, such as *Streptococcus pneumoniae*, *Listeria monocytogenes*, *Staphylococcus epidermidis*, etc. (Dasgupta 2012). Polyketides are the most significant secondary metabolite utilized in the pharmaceutical and agricultural field. It is accumulated from polyketide synthases (PKS). Pikromycin obtained from *Streptomyces venezuelae* was the first reported polyketide antibiotic that showed potent activity against multidrug resistant (MDR) respiratory infections (Woo et al. 2014). A broad-spectrum polyketide antibiotic erythromycin A obtained from *S. erythraea* is used widely in clinical applications like whooping cough, syphilis, acne respiratory and gastrointestinal infections for the patients who have an allergy against penicillin (Cobb et al. 2013). Aminoglycosides are one of the classes of antibiotics that inhibit protein synthesis. In 1944, the first aminoglycoside antibiotic, streptomycin reported from *Streptomyces griseus,* was found to be effective against pulmonary tuberculosis. Later, hundreds of antibiotics were discovered and significantly used to treat infectious microorganisms.

3.2.1.2 Antifungal Agents

In 1950, the first antifungal agent nystatin was identified from *Streptomyces noursei*; it showed activity against *Aspergillus* sp. Nystatin plays an important role in the treatment of gastrointestinal, oral, and genital candidiasis (Fjaervik and Zotchev 2005). Amphotericin B is also a polyene antifungal agent formed by *Streptomyces nodosus* especially used as an effective antifungal agent to treat life-threatening fungal infections such as organ transplantation and acquired immunodeficiency (Tevyashova et al. 2013).

3.2.1.3 Anticancer and Antitumor

Actinomycetes are well known for the development of anticancer and antitumor drug. The first anticancer antibiotic, actinomycin, was isolated from *Streptomyces parvulus* and glycopeptide produced by *Streptoalloteichus hindustanus* in the 1940. Actinomycin D, also known as dactinomycin, and bleomycin, which is a glycopeptide, are FDA-approved anticancer drugs used to treat many tumors, such as Ewing's sarcoma, Wilm's tumor, metastatic, squamous cell carcinomas, testicular, ovarian cancer, melanomas, testis tumor, etc. (Demain and Vaishnav 2011). In the 1960s, the FDA-approved daunorubicin and doxorubicin for the treatment of cancer therapy. Daunorubicin is employed in myeloblastic lymphoma whereas doxorubicin is used in breast cancer, solid tumors in children, and soft tissue sarcomas, (Giddings and Newman 2013).

3.2.1.4 Immunosuppressant and Anti-Inflammatory Agents

Immunosuppressive agents overwhelm the immune system and decrease the risk of rejection during organ transplantation. Rapamycin and tacrolimus have the immunosuppressive antigen. Rapamycin blocks the proliferation of interleukin-2 and interleukin-3, platelet, epidermal, insulin growth factor, and reduces acute renal allograft rejection, antitumor, neuroprotective, antineoplastic, and kidney toxicity (Yoo et al. 2017). Tacrolimus is also employed to increase the immunosuppression through the calcineurin inhibition and to break down the T cell activation pathway. Rapamycin and tacrolimus are also used in the treatment of refractory rheumatoid arthritis, a chronic inflammatory disease, expression of inflammatory cytokines, and secondary infection in spinal cord.

3.2.1.5 Antidiabetic, Antiparasitic and Insecticides

Humulin R is the first recombinant biopharmaceutical antidiabetic drug useful for diabetic patients. Biopharmaceutical product insulin glargine is derived from *Escherichia coli*, having the property of insulin analog and the pneumococcal vaccines produced from *S. pneumoniae* (Sanchez et al. 2016). Microbially derived antiparasitic compounds Avermectin B1a, B1b, and their derivatives can significantly reduce the growth of onchocerciasis and lymphatic filariasis (Shen 2015). Spinosad and milbemycin also have insecticidal activity.

3.2.1.6 Biofilm-Inhibitory Agents

Biofilm inhibitory agents obtained from the microorganisms can inhibit the biofilms formation. Some examples of biofilm inhibitory agents are 5-benzylidene-4-oxazolidinones, which inhibit synthesis of biofilm in MRSA and cahuitamycins isolated from marine bacteria *Streptomyces gandocaensis*, which inhibit the development of *Acinetobacter baumannii* biofilms (Park et al. 2016).

3.2.1.7 Biofuels and Bioenergy

Microorganisms have high biomass and potentiality to produce reliable, affordable, and environmentally friendly biofuels, which would replace fossil fuels. In the present scenario, attention was gaining toward the production of microbial oils and biodiesel feedstock. The yeast *Saccharomyces cerevisiae* and the bacterium *Zymomonas mobilis* are employed in the brewing industry to synthesize ethanol from some carbon sources.

3.2.1.8 Microbial Enzymes

Microbial enzymes are biocatalysts used in industries and biopharmaceuticals. Lipstatin produced from the *Streptomyces toxytricini* is one of the popular enzymes that inhibit the pancreatic lipase and is used for the treatment of obesity by breaking down the triglycerides to glycerol (Weibel et al. 1987). Hydrolytic enzymes synthesized from the bacteria and fungi have multipurpose uses such as ripening of fruits, cosmetics, detergent, dairy, etc.

3.2.2 Microbial Cell Factories

For the development of natural products from microorganisms and recombinant protein, bioprocess engineering selection of appropriate host strains are an important aspect of industrialization. Some of the microorganisms or their consortia are listed below that have the capacity to produce natural products using bioprocess technology.

3.2.2.1 Gram-Negative Bacteria

Gram-negative bacteria are well known for the causes of various diseases worldwide because of their high resistivity against antibiotics. On the other hand, some gram-negative bacteria are vastly used in bioprocess industry. *Escherichia coli* is a ability to synthesize natural products and recombinant proteins due to their fast growth, simple culture procedures, cost-effectiveness, high versatility, high productivity, genetic tools can be easily used for the manipulation of their plasmid DNA for laboratory and industrial scales. *E. coli* are successfully used to synthesize natural products such as recombinant humulin, Artemisinin, Erythromycin A, Somatrem, Somatropin, pneumococcal vaccines, human serum albumin, etc.

3.2.2.2 Gram-Positive Bacteria

Gram-positive bacteria are also able to produce natural products because of their flexibility during genetic engineering, production of enzymes, valuable antibiotics production, etc. Some of the gram-positive bacterial species that can produce natural products for industrial application are *Bacillus amyloliquefaciens*, *B. licheniformis,* which are used for high protein production; *Streptomyces* species are able to synthesize antibiotics, antitumors, and *Lactococcus lactis* are used in food, cosmetic, and pharmaceutical industries.

3.2.2.3 Yeast/Fungi

Unicellular eukaryotic single-celled yeast *Saccharomyces cerevisiae* is used to perform post-translational modifications such as phosphorylation, acetylation, O-linked glycosylation, and acylation to express the recombinant proteins such as recombinant human insulin. It is preferably utilized for alcohol production in bio-reactors (Baeshen et al. 2014). Multicellular filamentous fungi are also used for the production of citric acid, gluconic acid, antibiotics, microbial enzymes, and other pharmaceutical products.

3.2.2.4 Microbial Consortia

Microbial consortium and artificial microbial consortium engineering are used to achieve the supreme productivity at industrial level. Microbial consortia are utilized for enhancement of bio-production, optimization of nutrients and physiological factors, and to achieve high productivity (Bhatia et al. 2015).

3.2.3 Strategy for Product Enhancement

3.2.3.1 Strain Improvement

Industries are constantly striving to develop new products to fulfill the demands of consumers. Strain improvement is the process of improvising the production and yield of microorganisms by utilization of certain microbiological, biotechnological, and biochemical processes. There are several ways to improve strains such as whole-genome shuffling, protoplast fusion, ribosomal engineering, mutasynthesis, etc. Whole-genome shuffling is a process for the utilization of multi-parental crossing by shuffling the DNA with a recombinant genome. For example, two strains of *Streptomyces fradiae* are applied for the two rounds of genome shuffling that enhances antibacterial tylosin up to ninefold. Mutasynthesis was used to modify mutant strain *Streptomyces cellulosum* for the production of epothilone by 130-fold. Ribosome engineering is another frequently used technique to increase secondary metabolite production. It was used to mutate rpoB gene, which ultimately increases the transcription of biosynthetic gene clusters in *Streptomyces griseus*, *S. erythraea*, and *S. coelicolor* (Ochi and Hosaka 2013). In another study, production of avilamycin from the recombinant *Streptomyces viridochromogenes* strain has been increased up to 37-fold by ribosome protein engineering (Lv et al. 2013).

3.2.3.2 Protein Engineering

Protein engineering is the technique of developing new enzymes or desirable protein. For the production of enzymes or protein, amino acid sequence is modified by recombinant DNA technology. Protein engineering is highly useful to improve enzyme stability and its efficiency in low water system. Various protein engineering methods are available such as recombinant DNA technology, DNA shuffling, rational designing, site-directed mutagenesis, random mutagenesis, molecular dynamics, peptidomimetics, and X-ray crystallography. The most conventional method in protein engineering is "rational design," which involves "site-directed mutagenesis" of proteins. In the process of site-directed mutagenesis, specific amino acids coding sequence are inserted into a target gene. It is classified into two common methods: overlap extension and whole plasmid single round PCR method. Random mutagenesis process is employed to improvise and enhance the enzyme catalytic activity. Tyrosine hydroxylase produced from yeast are randomly mutated using error-prone PCR to increase the catalytic activity. Similarly, PCR-based random mutagenesis is used to increase enzyme catalytic activity of isopentenyl diphosphate isomerase by 2.53-fold (DeLoache 2015). Protein engineering methods are designed for the development of new enzymes for biotechnological industries such as food and cosmetic industries. These enzymes have extraordinary properties including thermostability, specificity, and catalytic efficiency.

3.2.3.3 Protoplast Fusion

Protoplast is the cell whose outermost layer of the cell wall has been removed. It can be obtained by the action of specific lytic enzymes. Protoplast fusion is the physical phenomenon in which two or more protoplast are conjoining together naturally or in the presence of fusion encouraging agents. Due to the enzymatic action of fusogens, when two or more protoplast are closely attached and adhere together, then, intramembranous protein and glycoproteins will be disturbed. It enhances the fluidity in the membrane and prepared a site for intermixing lipid molecules that allow the coalescence of adjacent membranes. Due to the intramembranous phosphate groups, protoplast carries negative charge particles. When the Ca^{++} ions bind with the protoplast, it causes a reduction reaction in the zeta potential of the plasma membrane so that protoplasts are fused. During protoplast fusion, beneficial genes such as disease resistance, nitrogen fixation, drought, heat and cold, and herbicide resistance genes can transfer from one species to another species. It increases spontaneous growth rate, enhances the production rate, increases the protein quality, frost hardiness, etc. Protoplast fusion also plays a significant role in the development of genetic modification, recombination, and strain improvement in filamentous fungi. Protoplast fusion, protoplast mutagenesis, and transformation are the most abundant techniques for the preparation and regeneration of protoplasts. Intergeneric fusants are used to fuse *Trichoderma reesei* and *Saccharomyces cerevisiae* for bioconversion of cellulosic materials to ethanol. This is one of the best alternative techniques for ethanol production. Protoplast fusion is also helpful in the production of cellulases enzyme by the combination of *Trichoderma reesei* and *Aspergillus niger* (Ahmed and El 2006). *Trichoderma reesei* strain PTr2 showed high CMCase activity with the twofold increment in enzyme activities using protoplast fusion techniques (Prabhavathi et al. 2006).

3.2.3.4 Metabolic Engineering

The term "metabolic engineering" is the combination of two words, "metabolism" and "engineering." Metabolism is the process of enzyme catalytic transformation of organic molecules in cell, and engineering is the technique to develop and achieve the target to make more competent (Ravi and Satish 2011). Metabolic engineering is basically referred to determine alteration of metabolic pathways of organisms for the better understanding and utilization of cellular pathways related to transduction of energy and supramolecular assembly. The desired goal of metabolic engineering is to develop and improvise the product using metabolic pathways, molecular biological techniques, designing, and implementing rational genetic modifications. Various methods are available for metabolic engineering such as genetic circuit design and bacterial metabolism modeling. The major steps involved in metabolic engineering is analysis of metabolic pathway, modification of pathway using in silico model, and improvement of product using various genetic approaches. The popular approaches for metabolic engineering are heterologous expression of entire gene clusters, gene insertion and deletion, engineering regulatory networks, quorum sensing, readdress metabolic pathway, precursors stimulation, and genetic knockout of loci. Microbes are key features for the production and development of many drugs including antibiotics, antitumor agents, immunosuppressant agents, antiviral mediators, antiparasitic agents, and enzyme inhibiting compounds (Gupta 2007). Metabolic engineering is employed to enhance the extension of substrate range, elimination of by-products, and improvement of cellular growth, their yield and productivity of microorganisms. Some of the metabolic engineering techniques maximize the microbial production. Primary metabolic fluxes can also modify by recombinant DNA technology. This technique is also used for continuous expression of antibiotic regulatory proteins (SARP) like actinorhodin from *Streptomyces coelicolor*, transfer of polyhydroxybutyrate (PHB) gene from *Ralstonia eutropha* into *Saccharomyces cerevisiae*, and metabolic expression of folate genes into *Lactobacillus plantarum* WCFS1 for overproduction of folate enzyme (Wegkamp 2010).

3.2.3.5 Recombinant DNA Technology

The new revolution in the field of industrial microbiology has been observed by the discoveries of the double-stranded structure of DNA and the expansion of recombinant DNA technology. Molecular biology techniques were applied with conventional industrial microbiology for the improvised production of recombinant primary and secondary metabolites, industrially important enzymes, alcohol, vitamins, amino acid, and recombinant proteins at an industrial scale.

For the recombination in microorganisms three major processes conjugation, transduction, and transformation are involved. For the rearrangement of internal DNA segments in microorganisms, transposons are considered as one of the major tools to translocate gene segments from one to another site. Modern genetic engineering techniques have been employed in the strain improvement program. It can be utilized to assemble the appropriate characteristics in their genes by in vitro recombinant DNA technology. This technique is particularly valuable in organisms with a complex regulatory system in genetic alteration. Some of the analytical techniques are also involved to combine quantity fluxes and molecular biological methods for the implementation of genetic modification. Various analyzing flux are metabolite balancing method, NMR technique, kinetic modeling, tracer trials, genetic modification analysis, biocatalyst analysis, and control theories (Eggeling et al. 1996). Amino acid was produced by *Corynebacterium glutamicum*, *Brevibacterium*, and *Serratia* strains using RDT. Similarly, various other recombinant strains were used to produce vitamins, organic acids, ethyl alcohol, Lysin, 1,3-propanediol, and carotenoids.

3.2.3.6 Mutasynthesis

Mutasynthesis is an advanced version of the precursor-directed biosynthesis process, which utilizes the mutant microorganisms for the biosynthesis pathways. Mutasynthesis overcomes the drawbacks of precursor-directed biosynthesis such as simplest metabolites production, ease in downstream process, maximum productivity, and amalgamation of various precursor analogues. Selective microbes were initially mutated using random mutagenesis, then, they were screened qualitatively and quantitatively for their improved production. Mutasynthesis is mainly employed for the generation of new analogues for several classes of compounds. Analogue of aminoglycoside-aminocyclitol antibiotic neomycin B2 was the first successful product of mutasynthesis, produced by mutant *Streptomyces fradiae*. Similarly, neomycin analogues, hybrimycin A1 and hybrimycin B2, were produced by double mutant strain of *Streptomyces fradiae*. They were supplemented with aminocyclitols streptamine and epistreptamine for their biosynthesis machinery (Mickel et al. 2004). Mutasynthesis was also successfully implemented for the production of natural products 14- and 16-membered erythromycins and platenolides, balhimycin novomycin, and β-lactam antibiotics. Other analogues such as cahuitamycins D and E having improved antibiofilm property against *A. baumannii*, chloro-substituted nonbenzoquinone analogue with increased therapeutic properties was produced.

3.2.3.7 Precursor Engineering Supply

Precursor engineering supply is used for the enhancement of the primary metabolites, which is involved for the biosynthesis of natural products. Precursor engineering supply can be achieved by manipulation of enzyme pathway, which is linked with precursor. Malonyl-CoA and methylmalonyl-CoA are most significant metabolic precursors, which is obtained from the biosynthesis of polyketides. Methyl oleate is the precursor of FK506 synthesized from the *Streptomyces clavuligerus* CKD1119. Supplementation of methyl oleate increase the internal concentration of methylmalonyl-CoA (Mo et al. 2009). Several precursors including non-ribosomal peptides, flavonoids, polyketides, and alkaloids derived from shikimate pathway are important for the production of aromatic amino acids or other metabolites. *Amycolatopsis* sp. increases production of vancomycin analogue balhimycin using non-ribosomal peptide precursor 3-deoxy-D-arabino-heptulosonate7-phosphate synthase, an enzyme of shikimate pathway (Pham et al. 2019).

3.2.3.8 Pathway Engineering

Pathway engineering can be achieved by the gene expression, gene deletion, and introduction of new genes into the host gene to increase the rate of natural products. Several attempts have been made to increase microbial products. Overexpression of copies of actinorhodin cluster increases 20-fold of actinorhodin in the *Streptomyces coelicolor*. Validomycin A cluster has been engineered to increase validomycin by 34% in *Streptomyces hygroscopicus* (Zhou et al. 2014). Overexpression of Streptomyces antibiotic regulatory protein (SARP) gene, which is a positive regulator of antibiotic from the dosage increase the antibiotic production by several fold. For example, overexpression of chxA (cycloheximide) or mgsA (iso-migrastatin) gene, a family member of SARP increases the antibiotic lactimidomycin production in *Streptomyces amphibiosporus* (Zhang et al. 2016).

3.2.4 Inoculum Development

Inoculum development is an important step in the upstream process to activate and prepare microorganisms for obtaining maximum product during bioprocess. The criteria for inoculum development includes:

- Inoculum must be healthy and in active state that can reduce the lag phase in fermentation process.
- Large volume of optimum size should exist.
- Culture inoculum should have the standard morphological arrangement.
- It must be in sterile and pure form.
- Inoculum must have the significant capacity to product formation.

- The quantity of inoculum used should be between 3% and 10% of the medium volume.
- Inoculum must be used during log phase for ensured activity and optimum production

3.3 Types of Fermentation

There are two types of fermentation processes based on the state of the substrate. If substrate is solid or semisolid, it is called solid-state fermentation whereas submerge fermentation includes liquid media. Most of the fungi and filamentous bacteria grow well and produces product in solid substrate whereas bacteria use submerge condition for optimal growth and product.

3.3.1 Solid State Fermentation (SSF)

Solid-state fermentation is utilizing large- and small-scale applications in bioprocess industries for production of fermented foods such as miso, soy sauce, tempeh, koji, annatto red wine, sake, and tapai. Solid-state fermentation process utilizes solid matrix, moist solid supports, and inert substrate for the cultivation of microorganisms and their metabolic activity in the absence or nonappearance of free water (Thomas et al. 2013). Solid materials have property to absorb water, hence maintaining low water activity without using soluble nutrients. SSF has been practiced since ancient times for the production of primary and secondary metabolites such as antibiotics, biocatalyst-enzymes, biosurfactants, organic acids, and perfumes.

3.3.1.1 Factors Influencing of SSF

The solid-state fermentation process can be affected by the various factors. It can be classified into three major categories: Biological, Physico-chemical, and Mechanical factors.

(a) **Biological factors**: Selection of a suitable microorganism is the most significant step in SSF. The selected organism could have the ability to degrade solid substrate. Some of the examples of microorganisms such ad bacteria, fungi, combination of bacteria, and yeast are the capable organisms for it to decompose the solid substrate and convert it into product. Significant inoculum preparation is important for SSF. Studies suggest that production of antibiotics, enzymes, and acid fluctuate greatly with inoculum size. Substrate selection is a significant step that provides the physical structure, water holding capacity, support to the growth of microorganisms and enhance the biosynthesis of cellular ingredients. Selection of suitable substrate can also be helpful for the enhancement of carbohydrate, lipids, proteins, nucleic acids, and energy for the cell. Due to different particle sizes of food, agro-industrial crops and residues are applied as good source of solid-state fermentation.

(b) **Physico-chemical factors**: Moisture content in the solid substrate has been determined by the water activity. It indicates the amount of free water in the solid substrate and organisms that can grow in SSF. pH has been measured for the growth of microorganisms over a wide range of pH optimum maxima and heterogenicity in the conditions of the system (Behera and Ray 2016). During the solid-state fermentation, temperature arises and reduces microbial activity due to the heat accumulation in the system. So, heat has to be removed from the system to maintain the overheating and disruption of microorganisms and product formation. Oxygen is the main source to grow the microorganisms. By the process of aeration, it can maintain high oxygen levels and low carbon dioxide levels in the solid substrates. Aeration rate directly influences the effect on the growth rate of microorganisms and product formation.

(c) **Mechanical factors**: Agitation plays an important role in SSF as it circulates the air flow more evenly and enhances the growth rate of microorganisms within the entire fermentation bed. Another benefit of agitation is that airflow is more evenly distributed, which improves the conditions for microbial growth within the entire fermented bed.

3.3.2 Submerged Fermentation (SmF)

Submerged fermentation is the process of developing the microorganism's liquid nutrient medium. Liquid substrates such as molasses and broths are used as the free-flowing liquid medium for the conduction of SmF process. SmF is used for the synthesis of biomolecules, enzymes, alcohols, oil, antibiotics, vitamin, amino acid, etc. In the aerobic condition submerged fermentation process is carried out when microorganisms can interact with nutrients in liquid broth to break down the nutrients and synthesized products. Nitrogen and carbon sources like sugars, vegetable juices, molasses, liquid media, fruit, whey, and sulfite waste liquor are used as a substrate of SmF. There are three methods involved in submerged fermentation including batch fermentation, continuous fermentation, and fed-batch fermentation.

3.3.2.1 Batch Fermentation

Batch fermentation is a large-scale microbial culture in which microbes are grown in fixed volume in closed system. Once sterilized media is filled in a bioreactor, microbial inoculum grow in the fixed volume. In this process, operational condition is continuously changed according to time duration. In batch fermentation, addition and removal of media from reactor occurs by the batch-wise process. The intracellular and extracellular materials could be harvested at the end of the process. The main disadvantages of the batch fermentation process are discontinuity of production, limited carbon source utilization, time consumption, and difficulty in optimization processes physical parameter such as temperature and pH (Sindhu et al. 2017).

3.3.2.2 Continuous Fermentation

Continuous fermentation is the open culture system in which continuous movement of culture medium occurs through the bioreactor. In continuous fermentation, one or more feed streams having nutrients such as carbon, nitrogen, micro- and macronutrients are fed continuously and the residues including cell effluents are to be removed incessantly. For the maintenance of steady state, the volume of the flow rate, number of microbes, and production rate are kept constant using chemostat or turbidostat. Continuous fermentation is used for primary metabolite production. Several products have been produced using continuous fermentation such as alcohol, amino acid, vitamin, etc.

3.3.2.3 Fed-Batch Fermentation

Fed-batch fermentation is a semi-open system in which more than one nutrient is fed to the bioreactor without removal of broth. During cultivation, the microbes and product remain present in fermenter until the end of the run. Fed-batch fermentation plays an important role in maintaining the concentration of the fed substrate into the medium at certain level. Fed-batch fermentation technique is used for the maintenance of high cell density culture for the product formation in the industries. The advantages of fed-batch fermentation are the supply of carbon source, nitrogen sources, macronutrients, phosphorus, precursors and inducers, etc., on continuous basis into the culture medium. Therefore, culture volume would be maintained till the time of completion of the process. In the fed-batch fermentation process, fixed volume and variable volume approaches are being applied. During fixed-volume fed-batch fermentation, limiting substrate is added in highly concentrated form, whereas in variable volume fed-batch culture, the volume of limiting substrate has been changed with the time of fermentation owing to the substrate feed. Fed-batch culture has been used for secondary metabolite production.

3.3.3 Design of Bioreactor

Fermenter or bioreactor is considered as the heart of the fermentation process. Bioreactor is the large container that is used to produce optimum product from cheaper substrate using microorganisms at controlled condition such as oxygen transfer, pH, temperature, nutrient, etc. Bioreactor is designed for chemical transformation, determination of metabolic activity, optimum growth rate of microorganisms, enzyme activity, etc. Quality improvement of the product depends on the design of fermenter that can provide good quality products at a low cost. The design of the fermenter and mode of operation relies upon microbial metabolism, microorganism product, product value, condition necessary for the desired product formation, and production scale. Fermenters are designed to produce batch or continuous product at optimum level under sterilized condition. According to the report of Eibl, in the fermentation process three steps, namely, proper mixing of substrate and organisms, uniform shear rates, and mono septic conditions must be maintained (Eibl and Dieter 2008). Spargers are used to provide optimum aeration using direct sparging, indirect and membrane aeration into medium. The characteristic features of good bioreactor design include: robust mechanical design, easy to operate, eco-friendly, cost effective, agitation system, oxygen transporting system, elasticity in the process, aeration system, maintain aseptic conditions, foam control, volume maintenance, heat and pH control system, sampling harbor, sterilization system, regulate the transfer of heat and biomass.

The critical parameter such as oxygen and heat control need special attention for optimum production. Aeration can provide the sufficient amount of oxygen for the growth of microorganisms in the fermenter. For the aeration process the device such as sparger is used in the insertion of oxygen in the fermenter. Different sparger produces various size of bubbles but small-sized bubbles are more preferrable because large bubbles occupy less surface area compared to smaller bubbles. Agitation facilitates vigorous mixing of media with heat and oxygen. Agitation is achieved with the agitator in the fermenter. Agitator consists of motor, shaft, and impellers. Various classes of impellers such as vaned disc, disc turbine, open turbine, and marine propeller are available in the market. Heat can be controlled by heating jacket or heat exchanger.

3.3.4 Linked Bioreactor System

Linked bioreactor system is a novel strategy being used in the continuous production of biological components, innovative products, and biosimilars to strengthen industrial product. In this technique the two-bioreactor system are attached and simulated with each other. Continuous perfusion culture bioreactors play a significant role in the productive modes of bioreactor operation. Perfusion bioreactor maintains the cells proliferation at higher state and it provides the continuous culture inoculum to the second bioreactor, namely, continuous-flow stirred-tank reactor (CSTR). After identification of adjacent optimum system, steady-state parameters are set. Linked bioreactor system is resilient in nature and higher volumetric productivity could be obtained using overwhelming media, laboratory set-up, and long-term operation to obtain higher productivity. By employing this system, culture could also easily replace with fresh and younger cell culture without creation of disturbance in the steady-state system. Sustainable perfusion system can also produce high quality of protein in comparison to traditional bioreactor operation.

3.4 Bioprocess Optimization

Media optimization plays a significant role in the enhancement of product yield. There are two approaches, classical non statistical one-factor-at-a-time (OFAT) and statistical experimental design for the development of bioprocesses. Some of the methods such as factorial design, response surface methodology, and Design-of-Experiment are the various optimization tools that reduce the number of experiments.

Factorial design and response surface methodology: Factorial design and response surface methodology is being used for the evaluation of important experimental variable in bioprocessing in the biotechnological system. It can reduce the biases and required number of experiments. Design-of-experiment (DoE) methodology was used in a favorable direction to find a true optimum value (Roebuck et al. 1995). DoE has been used to increase the performance of upstream and downstream process development.

Screening of factors: Setup of DoE and factorial design is the most efficient approach to initiate the screening of significant variables and has been studied from selected corners in the experimental space (Box et al. 2005). In the factorial designing only a fraction of the possible values are to be investigated. In this modeling, three selected factors are to be screened at two levels of values, and each variable is tested twice at low and high level of significance.

Design of media and mixtures: For the optimization of media and mixture composition, Design-of-Experiment (DoE) plays an important role for the culture media formulation. If one component is dominant in the medium then the effect of this component can also be evaluated by the DoE.

Evaluation of the experimental design: Multiple linear regression (MLR) been used for the compilation of the results of response variables. In MLR, the mathematical model design is used to describe a relationship between one or more independent variables and a response variable.

3.5 Process Measurement

Bioprocess includes several processes such as pH, temperature, oxygen, agitation, aeration, energy requirements, pressure, and flow. All these processes need to be measured using sensor or other devices. pH is measured using electrode or pH sensors. Sensors are of three types- inline, online, and offline sensors. Inline sensors are inserted inside the fermenter to measure the fermenter-linked process. Online sensors are associated with bioreactor but does not insert inside the reactor. Offline sensors measure various parameters outside the bioreactor but are usually associated with product measurement. Besides the sensor, pressure bellow, thermistor, load cell, and flow meters were also used to measure pressure, temperature, volume, and flow of media in bioreactor. Bioprocess measurement is necessary to control all the parameters for optimum growth and production.

3.6 Bioprocess Control

Conventional bioprocess technology contains the various units for the operation to induce the optimal environment of cells growth, cell division, and product manufacture. Measurement is followed by bioprocess control using computer-controlled system. Control loop strategy is frequently used in the system for control of various parameters. On-off controller, proportional-integral-derivative (PID) controller, and advanced controllers are used in bioprocess system. Supervisory set point control and direct digital control system are preferred to control different parameters simultaneously. Bioprocess strategies face unique challenges due to the variability, complexity, and nonlinearity of biotechnological processes. Implementation of contemporary control strategies such as open-loop control, closed loop control, fuzzy logic, and model predictive control are effective to control bioprocess parameters.

Open-Loop Control: It is the oldest and simplest technique applied for the carbon source limiting fermentation process in industries. Open-loop system is utilized to reduce batch to batch variability. It has been vastly used to optimize fed-batch operation condition. Open-loops control strategy has been applied in recombinant lipase B enzyme, ethyl alcohol, poly 3-hydroxybutyrate-co-3-hydroxyvalerate, and production of vaccine (Julia et al. 2019). Besides its efficiency in some product, it has several limitations such as difficulties in mathematical formulation of nonlinear system, incompetency to corrective measurement in the random disturbance at the time of operation, etc.

Closed-Loop Control: Closed-loop system such as proportional Integral (PI) and proportional integral derivative (PID) are designed to overcome the disadvantages of open-loop control system. In this process, feedback term has been added for the controlling action of closed-loop system. The common closed-loop algorithm uses PID controller because of its simplicity, strength, and effortlessness in tuning. A new closeloop control system, SISO nonlinear bioreactor model, is being developed for real-time data collection during fermentation of *Escherichia coli.* It uses internal model control-based PID (IMC-PID) to minimize the absolute errors (Joanofarc et al. 2019).

Fuzzy Logic-Based Control: Fuzzy logic-based control system is designed for the integration of human experience and reasoning. Fuzzy logic-based controllers are used for the development of energy, medicine, economics, and pharmacological sciences. It is a nonlinear controller in bioprocess fermentation industries. In the bioprocess, fuzzy logic-based controller is being designed for the setup of optimum process parameters to achieve the higher growth yield and productivity and tracking to control the bioprocessing technique (Abyad et al. 2017).

Model-Based Control: In this process, modeling is employed for measurement of monitoring, optimization, and control. It has mathematical, statistical, and empirical advancement. Parametric optimization and control (PAROC) is one of the techniques, which is based on model control system and used in biopharmaceutical purification process. For the process of fermentation, various numbers of control strategies have been implemented. For example, DO-stat control strategy and Extremum seeking control strategy, DO stat or pH stat control strategies deals with the concept of indirect feedback control. DO stat control maintains the concentration of dissolved oxygen and pH. Extremum seeking control is working on the basis of process models. It can solve the optimization problem, increase the system capability to eliminate the disturbances and reduce sensitivity. This technique is used for the development of antibiotics (Gomis-Fons et al. 2020).

Challenges for Bioreactor Design and Operation: Several designing issues can affect the fermentation technology. The following are the challenges for the operation of bioreactors:

- Conceptual design of bioreactor has to be systematic and the methodology must be refined and user friendly.
- In the basics of bioreactor designing, fermentation kinetics, mass transfer, and heat transfer mechanisms must be optimized.
- Omics technology should be utilized in the assembly of bioreactor design.
- New cellular system should be generated in the processing of up- and down-scale fermentation.
- New optimization tools should be utilized for the maintenance of large-scale operation at the microfluidics dimension, microbioreactor, parallel process analysis, and manipulation of mass production.
- To reduce the gaps between scales and computational fluid dynamics in bioreactor designing.
- To enhance the production yield, accelerate the media optimization process by using statistical factorial designing methods.
- PAT approaches and multivariate data analysis must be used for the measurement and increment of information flow and modeling.
- Soft sensors can be used to regulate the exploitation of information.

3.7 Recent Progress in Upstream Processing

3.7.1 Quality by Design Approach

Quality by Design (QbD) is a novel concept applied with Design of Experiments (DoE) or high-throughput devices for development of upstream process. In this approach, manufacturing process is integrated with product quality, so that high quality product can be obtained. QbD method has been employed in characterization of various microbial products as well as development of bioprocess. It has been also employed in media optimization to optimize each parameter and process parameters for recombinant protein production at higher scale.

3.7.2 Process Analytical Technology

Optimization of bioprocess parameters can be regulated by real-time monitoring and their control. Process analytical technology (PAT) is applied to ensure quality end product by analyzing, designing, monitoring, and controlling all the parameters periodically. It maintains critical quality attributes (CQAs) for quality product formation. The use of sensor technology and real-time monitoring devices had made it possible to maintain critical quality attributes. It avoids various regulatory issue; hence opens global market to sell products.

3.7.3 Upstream High-Throughput Cultivation Systems

Upstream High-throughput devices (HTPDs) have been developed for laboratory and pilot-scale bioreactors. It has been successfully implemented for screening microorganisms, experiments, process optimization, and control system to save cost of production and time before industrial-scale production. HTPDs were used in stirred mini-tank bioreactors and microplate-based mini-bioreactors for optimization of batch and fed-batch process optimization.

3.8 Downstream Process

Downstream processing (DSP) is a series of operation phase that extract, separate, purify, and formulate the active pharmaceutical ingredient (API) produced during upstream process. Therefore, it requires stepwise separation of impurities according to physical differences of components, minor variations between contaminants and product, as well as specific properties of product. DSP varies from cell to cell and product to product. Microbial downstream processing is broadly divided into four stages, namely, harvesting, clarification, capturing, and polishing.

Microbial products can be intracellular or extracellular. Extracellular products containing medium can be directly separated from microbes by harvesting techniques. Clarification is needed when the product is intracellular; hence it is extracted out by disruption of cell wall, membrane, or both as per their location. Products are clarified by stepwise process such as pH adjustment, flocculation, floatation, or

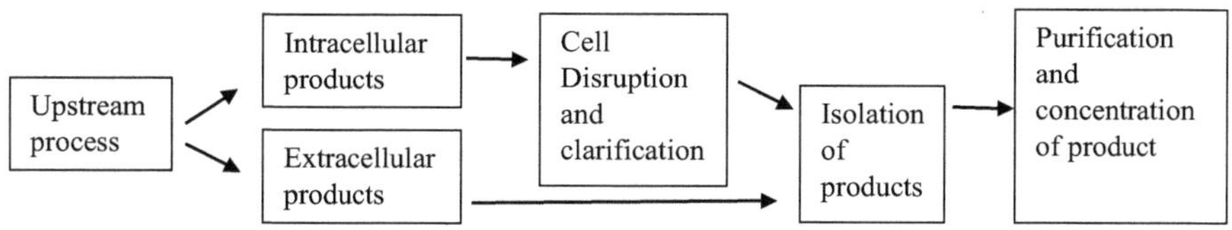

Fig. 3.1 Steps of downstream process

precipitation depending on need. Sometimes intracellular products are extracted and separated by integrated harvesting and clarification. Capturing involves interaction and binding of desired products with stationary phase material while polishing involves capturing or interaction of trace impurities but separation of desired products. Polishing can be further classified as intermediate or final clarification depending on the broad category of contaminants. Capturing and polishing are based on different chromatography principle. Different principles are applied for separation and purification of protein and non-protein microbial products. At the laboratory scale, proteins can be easily precipitated by the salting-out technique. Salt is removed by dialysis and further protein purified by chromatographic principles. Non-protein bioactive products are usually separated by solvent extraction and further purified by chromatography. Large-scale production and purification are challenging steps that require extensive care in purification and removal of contaminants at low cost. The general steps of downstream process have been schematically represented in Fig. 3.1.

3.8.1 Stages of Downstream Process

3.8.1.1 Cell Disruption

Cell disruption is the step of unit operation for intracellular microbial products. Cell disruption techniques are applicable as per location of products such as cytoplasm, membrane, or periplasm of bacteria. Some techniques create pore, some increases membrane permeability, or some disrupt complete cell envelop. Optimizing must be required before cell disruption and protein recovery as they can destroy by harsh treatment or heat generation. Disruption technique can impact product yield, quality of product, process scalability, product stability, cell debris particle size, impurity level, and operating costs. There are four cell disruption methods that differ according to disruption efficiency, scalability, mildness, and cost.

(a) **Mechanical**: Mechanical are considered as low-cost industrially relevant method because of its volumetric output, higher disruption efficiency, ease of handling, and scalability. The principle of disruption rely on cell membrane disruption using solid or liquid shear forces, cavitation, impingement, or bead mill homogenization. The drawback of mechanical methods is heat generation, shear stress, poor selectivity, and micronization of cell debris.

(b) **Physical and Chemical**: Physical method is based on sudden shift of cellular environment such as pH change, raising temperature, and osmotic differences. Chemical disruption is accomplished through chemical additive such as a detergent that disrupts cell membrane. Both physical and chemical methods disrupt cell membrane integrity, increase porosity that allows leakage of cellular components and intracellular proteins. The methods are based on osmotic shock, chelation, thermolysis, freeze-thaw, acidification, and addition of detergent, chaotrope, or solvent. Detergent (cationic, anionic or non-ionic) disrupt inner membrane by micelle formation. Solvent such as glycol and toluene also release soluble protein by changing membrane permeability. Acidification disintegrate outer membrane by reduction of intracellular pH. Metal chelators, mainly Ethylenediaminetetraacetic acid (EDTA), decreases outer membrane stability by pulling divalent atoms (Mg^{2+} or Ca^{2+}) while chaotropic agents solubilizes membrane proteins. Physical methods are mainly used for periplasmic protein whereas chemicals methods are applied for intracellular proteins. Advantages of these methods are simplicity, easy to implement, and low-cost. Disadvantage includes potential product quality, long reaction times, and disposal of additive.

(c) **Enzymatic**: Enzymatic is a gentle treatment method that works in selective manner. Enzymes such as muramidase or lysozyme disrupt peptidoglycan layer of bacteria. Disruption efficiency can be increased by pretreatment of harsh Lipopolysaccharides (LPS) rich layer of gram-negative bacteria with chemical additives, such as Tween. Other enzymes such as DNAse can also use to reduce lysate viscosity resulting from release of DNA during cell disruption. The advantage of enzymatic method is mild treatment, specificity and high rate of product release whereas disadvantage is higher enzyme cost, requirement of large quantity of enzyme and fine cell debris.

(d) **Combined**: Combination of different synergistic membrane treatment method are employed to improve the efficiency, purity and yield of product. Non-mechanical methods are followed by mechanical methods. As an example, LPS layer can be effectively destabilize by EDTA than physical or enzymatic methods could be employ to disrupt cell membrane.

3.8.1.2 Precipitation

Precipitation is the commonly used technique for separation or concentration of fermented product. Various agents are used for precipitation at different stages of downstream process. Polyelectrolytes aggregates cells as well as precipitate a range of compounds. Ammonium and sodium sulfate salts are used in salting-out technique for protein precipitation. The solubility of some molecules is decreased by addition of acid or base. Organic solvents are used for the precipitation of biomolecules. Chilled ethanol precipitate protein by changing dielectric constant whereas dextran is precipitated by methanol. Protein can also be precipitated by protein-binding dye or non-ionic polymers such as polyethylene glycol.

3.8.1.3 Flocculation

Flocculation is a separation process in which chemical additives are used to facilitate interaction between particles, create larger-sized aggregate particles for sedimentation. The suspended small-sized cell debris are agglomerated by the addition of polyelectrolyte or inorganic salts. Polyelectrolytes are water soluble, of high molecular weight organic compounds, such as polystyrene sulfate and polyacrylamide. Flocculation increases the clarification efficiency by removal of cell debris or whole cells from fermentation broth.

3.8.1.4 Sedimentation

Sedimentation is process that allows to settle out particle in suspension under gravity. This is a low-cost technology extensively used for settlement of large flocs (>100 μm diameter). According to Stokes law, the rate of particle sedimentation is directly proportional to the diameter of particles, density of particles, gravitational acceleration but inversely proportional to viscosity.

3.8.1.5 Centrifugation

Centrifugation works on the principle of centrifugal force that increases the solid liquid separation. It acts as additional force beyond the gravity, hence continues to be one of the important techniques for separation of microbial biomass or bioprocessed product. Centrifugal sedimentation rate is directly proportional to density of particle, angular velocity of the centrifuge, distance of the particle from the center of rotation whereas inversely proportional to viscosity. Centrifugation is a highly efficient method that separates particles as small as 0.1 μm. centrifugation can be done in batch or on a continuous basis. Industrial centrifugation is popular in terms of continuous or semicontinuous centrifugation, size of particle to be separated, centrifugal speed, and volume of particle to be separated.

The commonly used industrial centrifuges are tubular, multichambered bowl, disc stack, and screw-decanter centrifuges. Tubular centrifuges possess hollow tubular rotor bowls that allows passes of suspension from bottom to top. Particulate materials are separated from clarified liquid on continuous basis. It operates at centrifugal forces 13,000–17,000 g. Multichambered bowl centrifuges consist of vertically riding interconnecting cylinders that allow fermented broth to pass through the center and collect smaller particles in the outer chambers. The operating speed is 5000–10,000 g. Disc stack also called solid bowl centrifuges comprises a stack of conical discs that supports in separation. It operates at 5000–13,000 g and works semicontinuously having the facility for discharge of the collected material. Screw-decanter centrifuges is advanced centrifuge having high throughput capacity, dewatering, biohazard, and cleanliness facility.

3.8.1.6 Filtration

Filtration is the commonly used technique that separates suspended particles from a gas or liquid by means of filter device. In bioprocess, microbial product present in fluid is passed where as solid particles such as microorganisms, their debris, or biomolecules are left on the filter layer, commonly termed as "filter cake." The filtration efficiency depends on several factors such as size of the organism, porosity of filter, pressure, temperature, and viscosity of the medium. Filtration techniques can be applied for separation of microbial debris as well as purification process. There are two types of "filters—depth and absolute.". The porosity of depth filter is usually more than particle size; hence several folds must be needed for efficient separation. They are autoclavable, composed of asbestos, glass wool, or filter paper. Absolute filters consist of specific pore sizes whose porosity are smaller than the particles size.

Conventional industrial filters are plate and frame filters as well as rotary drum vacuum filters. Plate and frame filters are batch filtration system consisting of alternate stack of filter cloth containing porous plates and hollow frames. Filters are used to harvest microorganisms and for recovery of protein precipitates. Vertical and horizontal pressure leaf filters are another absolute category of filters. Rotary vacuum filters are continuous systems often used for separation of broth having particles size of 0.5–10 μm and 10–40% solids. It consists of hollow perforated drum supports with membrane filter. Drum rotates continuously, accumulate solid on membrane under vacuum, filtrate pass into receiving vessel via drum, and finally solid scrapped out during rotation.

Membrane Filters are absolute filters that encompasses specific pore sizes. Particles or fluid that pass through membrane is called filtrate whereas the remanent is called retentate. Membrane filters are used in both static filtration and crossflow filtration. Static filters work on the principle of static state of slurry during filtration. It leads to accumulation of cake mass, pressure drop, hence decreases filtration efficiency. Crossflow filtration works on the principle of crosswise direction of slurry across the membrane, hence cake nullify the problem of cake mass. Membrane filters are categorized as microfiltration, ultrafiltration, nanofiltration, and reverse osmosis. Microfiltration removes microbial cells and other particles in the range of 10–0.1 μm. Microfiltration is costly but quiet in operation, requires less energy, and is free from bioaerosols. Ultrafiltration is more efficient than microfiltration as it removes ammonium salt from protein, bacterial cell debris, or even virus up to the size of 0.1–0.01 μm. Nanofilters removes particle size 0.01–0.001 μm whereas reverse osmosis works below 0.001 μm.

3.8.1.7 Solvent Extraction

The extraction of non-proteins fermented products can be removed from slurry using solvent extraction techniques. The principle applied for separation is "like dissolves like.". The molar polarization of a compound or organic solvent depends on their dielectric constant. Different fermentation products have different degrees of polarity. A range of solvents are available with varying degrees of polarity. The affinity of products toward the solvent on the basis of polarity is the foundation of solvent extraction. The choice of solvent depends on partition or distribution coefficient K, where K = Concentration of solute in extract/concentration of solute in raffinate. K value decides the ease to extract the compound. If K value will be high, it indicates good separation of the solvent and aqueous phases. Single-stage extraction system is applied for higher K value whereas multistage is applicable for lower K value. Practically, most of the systems have low value of K; therefore, multistage system is popularly used.

Multistage system can be further classified as cocurrent or countercurrent systems. There are n number of mixer/separator vessels in sequence that allows solvent and fermented broth to mix, separate. In a cocurrent system, the solvent and broth flow in same direction, raffinate pass from first to n number of vessels. Countercurrent system is more efficient were solvent flows in opposite direction from n vessel to first vessel. Though, a huge amount of solvent is used, extraction can be achieved at high degree. Industrial-scale Podbielniak centrifugal extractor is based on countercurrent system. Since huge solvents are used and solvents are expensive too, their recovery for reuse is necessary.

3.8.1.8 Solvent Recovery

The recovery of solvent from raffinate can be achieved by distillation process. Distillation is also useful for purification of acetone, alcohol, or other solvents from fermentation broth. Therefore, it is an important industrial process for the purification of beverage products and fuel. The principle of distillation is based on the boiling point of the product, and other components need to separate. There are three stages of distillation:

(a) Evaporation is used for the removal of solvent in the vapor form using evaporator.
(b) Column is used for vapor-liquid separation, to separate more volatile component from less volatile components.
(c) Condenser is used for condensation of vapor for recovery of more volatile solvent fraction

Distillation is further based on batch or continuous basis. In batch distillation, the vapor from evaporator pass through column, condensed and a part of condensate reflux to column. This process continues to repeat to get acceptable recovery of the product. In continuous distillation, initially, total reflux of condensate occurs at intermediate of column. Several columns such as perforated tray column or tower column are popularly used. Supercritical fluid extraction is another technique that utilizes supercritical fluid for extraction of compound. It becomes relatively convenient to extract compound from critical fluids as they can easily convert to gaseous phase at critical temperature and pressure. They are used in the extraction of vegetable oils, hop oils, β-carotene, chemotherapeutic drugs, steroids, vanilla, and caffeine.

3.8.1.9 Salting Out

Salting out is a protein precipitation technique in which salt is used to induce precipitation of protein. With the increasing salt concentration, water molecules get attracted toward salt ions. Salt and protein compete to interact with water, fewer water molecules are left to interact with protein. Protein gets associated with each other and finally start precipitation. Optimum concentration of salt such as ammonium sulfate are required for precipitation of protein. Further, dialysis of protein is required to remove the salt from protein.

3.8.1.10 Chromatography

Chromatography is the most popular technique employed to concentrate and purify metabolic products on the basis of size, polarity, charge, and affinity. The principle of chromatography is based on distribution of molecules between stationary and mobile phase. There are two types of chromatography based on the mode of separation: planar and column. Paper and thin layer are examples of planar chromatography whereas adsorption, ion-exchange, gel permeation, affinity, reverse-phase, high performance liquid, and gas chromatography are examples of column chromatography. In bioprocess, we usually employ analytical or preparative column chromatography. The choice and order of column chromatography technique depends on the nature of products and different chromatography principles applied for purification process.

Adsorption chromatography: Adsorption chromatography comprises separation on the basis of affinity of the solute to the solid matrix by weak hydrogen bond/ van der Waals forces. Inorganic adsorbents such as aluminum oxide, aluminum hydroxide, active carbon, silica, and hydroxyapatite are used to purify nonpolar molecules, whereas organic polystyrene resins are used for organic molecules. Elution of adsorbed solutes are based on increasing ionic strength, frequently by phosphate ion concentration.

Ion exchange chromatography: The principle of ion exchange is based on reversible exchange of ions between a solid phase bearing ion-exchange resin and liquid phase. Cationic resins contain negatively charged group such as carboxylic acid, phosphonic acid, or sulfonic acid whereas anionic resins contain positively charged quaternary ammonium, secondary or quaternary amine. The product can be eluted by changing the ionic strength of the buffer.

Gel permeation chromatography: It separates molecules on the basis of their molecular size. Smaller molecules penetrate the gel pore at greater degree, hence diffuses into gel more quickly than the larger ones that pass through void spaces. Therefore, larger molecules eluted first compared to smaller one. The porous stationary phase consists of crosslinked dextrans, acrylic polymers, cellulose, or agarose (Sepharose).

Affinity chromatography: Affinity chromatography is a selective separation cum purification technique that involves specific interaction between biological materials such as enzyme-substrate, signal-receptor, antigen-antibody, etc. Stationary phase ligands are linked covalently with matrix via spacer arm. Specific molecule separates by specific interaction with ligand, which are further eluted.

Reverse-phase chromatography: Reverse-phase chromatography involves greater polarity of mobile phase as compared to stationary phase. It utilizes modified solid phase containing hydrophobic alkyl chains that separate proteins on the basis of hydrophobicity. Protein with more hydrophobicity bind more strongly with stationary phase hence elute later.

High performance liquid chromatography:

HPLC is a high-resolution technique that uses a pre-fitted column containing small, uniform, and rigid beads. High pressure is used for faster separation and better resolution. A range of chromatographic principle such as ion exchange, size exclusion is used for qualitative and quantitative analysis of samples. Analytes are detected by a vast range of detector. The signals are finally recorded using software in the form of chromatogram.

3.8.2 Product Formulation

Formulation is the final stage of downstream process that transits active substance into a formulated end product suitable for administration, stabilization, storage, and shelf life. The low molecular weight products such as organic acids and solvents are formulated by removal of water to make them concentrate. Proteins are susceptible; therefore, stabilizers such as polymers (polyethylene glycol), polyhydric alcohols (glycerol), salts (sodium chloride) or sugars (sucrose) are added to increase the shelf life of protein. Some molecules are formulated by drying or crystallization to increase their stability and shelf life.

Crystallization—It is based on the arrangement of atoms or molecules into a precise crystal lattice so that an energetic state can be minimized. It is achieved by dissolution of product in a solvent using high temperature followed by cooling or evaporation to make the solution supersaturated. The nuclei will form in supersaturated solution, crystals form and grow simultaneously. Citric acid and cephalosporin C are some examples of crystal product.

Drying—Drying is a technique in which heat transfers to the wet material for removal of water or solvent in vapor form. Different drying techniques are employed as per heat tolerance properties of product to ensure minimum loss of product activity, stability, viability, and nutritional value. Water can be removed by freezing, direct heat contact, indirect heat contact, or radiation.

Freeze-drying, popularly called lyophilization are used for thermolabile products such as protein, vaccines, some enzymes, or cells. In this technique, samples were frozen and finally water present in solid state is removed by sublimation under vacuum. Lyophilizing agents were used to avoid osmotic or thermal shock.

Spray drier is a widely used industrial technique for drying heat-sensitive biological products. The product present in liquid is atomized into small droplets using nozzles and pass-through preheated chamber at faster rate. The chamber is saturated with heated air and the product doesn't contact with the heating surfaces. Rotary drum driers with or without vacuum are used to remove water by heat conduction. Suspension of heat resistant product is applied to the surface of steam heated drum. This is a continuation process in which product is scraped with a knife during rotation.

3.9 Recent Progress in Downstream Processing

3.9.1 High-Throughput Technologies

High-throughput (HT) is an advanced technology, recently used in downstream process to gather more data compared to traditional techniques. It has been used in various methods of cell disruption such as bead mills, sonication, homogenization, etc. In *Escherichia coli*, HT-compatible bead mills and 8-well-sonication device was used. Similarly, 24-well-HT sonication device, microfluidic channels, osmotic shocks, freeze-drying, and thermal treatment was used for cell disruption in bacteria, fungi, and yeasts. High-throughput chromatography devices such as pre-packed PreDictor filter plates, MediaScout RoboColumn, and MediaScout MiniChrom columns were used for monoclonal antibodies. Microfluidics-based methodology was also adopted for multiplexed screening and to increase the speed. Ultra-scale-down device is used for HT depth filtration and simultaneous assessment of multi-layer depth filters. Recently, microfluidic aqueous two-phase extraction systems with inbuilt fluorescence microscopy were used for screening of protein (Sao Pedro et al. 2019).

3.9.2 Single-Use Technologies

Single-Use Technologies is a technique used in downstream process during cell harvest, extraction, and purification of product. Disposable or single-use centrifuge, unifuge, and single-use depth are used in commercial products. Single filtration step replaces primary and secondary filters; hence, lower filtration surface area, cycle time, and buffer requirement. Single-use depth filters have several advantages like consistency, better recovery, ease of scalability, and low-cost. Single-use technology is also used in chromatography for the purification of recombinant proteins (Boedeker et al. 2017).

3.9.3 Downstream Process Analytical Technology

Downstream process analytical technology (PAT) are important tools used for analysis of host cell proteins, protein concentration, protein purity, endotoxin, host cell DNA, misfolding, and process-related impurities in downstream processing. Spectroscopy, circular dichroism, HPLC spectrometry, and other tools were used to screen the critical quality of products. FTIR spectroscopy, fluorescence detectors, and multi-angle light scattering were also used as PAT tools in protein purification.

3.9.4 Modeling Approach

Different Modeling and simulations approaches are used to collect more data while decreasing the number of required experiments. Application of these models can describe the downstream unit operation in detail and make it much convenient. Mechanistic models are based on physicochemical properties of various products whereas empirical models are based on a prior information of product within a definite design. Mechanistic modeling is a process development approach being used to speed up the chromatography process. Online mechanistic model of chromatography is used as a process analytical tool for higher product purity as compared to offline column fractionation. Application of statistical process control and monitoring of protein refolding in *Escherichia coli* showed higher productivity of recombinant therapeutic proteins production (Hebbi et al. 2019).

3.9.5 Continuous Downstream Processing

Continuous process is better than batch process for any kind of product. Though it requires expertise, skill, and initial capital investment, it can reduce overall cost, increases product formation in less time period. The design and techniques for continuous downstream processing for microbial products have been developed and scaled up. Continuous bead milling was used for cellular disruption, whereas continuous centrifugation was used for cell harvesting. Tubular bowl centrifuges and disk stack centrifuge is based on continuous operation used to harvest microorganisms. Aqueous two-phase extraction technique is also updated to operate in continuous mode. It is carried out by spray columns, column contactors, rotating disk contactors, and mixer-settler. Continuous chromatography technique is crucial to accomplish high purity for proteins. Continuous operation could be achieved by operating several chromatography in a cocurrent or countercurrent manner in the subsequent stages. Some examples of continuous chromatography systems are countercurrent chromatography (CCC), continuous annular chromatography (CAC), multicolumn countercurrent, countercurrent tangential (CCT) chromatography, solvent gradient purification chromatography (MCSGP), and simulated moving bed (SMB) chromatography. A countercurrent continuous three-stage single-pass diafiltration membrane was developed for desalting and efficient buffer exchange. Protein formulations by crystallization was also developed for continuous process.

3.9.6 Integrated Continuous Bioprocessing

Competition for optimum production, product stability, cost, scale-up, and quality improvement are some important issues in microbial product bioprocessing system. These problems can be overcome by continuous intensification of upstream and downstream process. Integrated continuous bioprocess system can scale-up methodologies, provide manufacturing flexibility, reduce shortages and variability, simplify overall process, decrease production costs, and enhance product yield. The real-time monitoring using PLC-SCADA can operate the whole continuous process as a unit. The design and operation of an integrated continuous system is used to produce various microbial products such as microbial enzyme, recombinant protein, etc. (Fisher et al. 2018).

3.9.7 Combinatorial Approaches

The integration and optimization of process development approaches for downstream operation could generate deeper understanding of experimental design and produces higher degree product. There are three strategies for process optimization of downstream, namely, expert-, experimental-, and model-based approaches (Nfor et al. 2008). These approaches are extremely convenient for effective development of new microbial products. Combination of these approaches, called combinatorial approaches, provides their full potential for product formation.

(a) **Heuristic approach:** This is the first approach to provide expert knowledge and thumb rules for generation of initial sets of data needed for the high-throughput approach. It could perform complicated separation of impurities from product and also remove destabilizing material.

(b) **Experimental approach:** Experimental approach gets data sets from heuristic and uses high-throughput (HT) experiments, design experiments, and optimal experimental design for production. Since experiments are based on previous experience and data of heuristics, it can be further categorized as universal process design, platform processes, and heuristics for specific processes. It can be used to characterize different processes and create new databases and platforms. Experimental approach is also used to calibrate and validate different models for in silico methods.

(c) **Model-based approach: A** model-based approach is used to reduce experimental effort and is applied in computational optimization. It is based on mechanistic models as well as empirical models like response surface model (RSM). Both models provide quality guideline for upstream and downstream processing of new microbial products. Hybrid approach is a heuristic, experimental and model-based approach. Hybrid approach fasten the process in effective way (Hanke and Ottens 2014).

3.10 Conclusion

Microorganisms are well known for various therapeutical, enzymatic, agricultural, and food products. They have tremendous potential to produce products; therefore, research on finding microbes for novel product has been continuously increasing. Microbial metabolic products are produced in lesser amount; hence, genetic modification, mutasynthesis, protoplast fusion, and various other strategies are used to improve their production. Engineering approach for upstream and downstream has been adopted for product improvement. Statistical tools for media optimization, sensitive measuring devices, and advanced controllers are used. High-throughput techniques in upstream and downstream process, optimization, advanced control system, and integrated continuous approaches improve microbial production to a new height. Continuous process, single-use technology, modeling approaches, integration of upstream with downstream process, and combinatorial approaches are used in downstream processing to increase the production efficiency and decrease production cost. Application of these approaches in bioprocess engineering can further improve new microbial product.

References

Abyad M, Karama A, Khallouq A (2017) Fuzzy Takagi-Sugeno based modelling and control for an alcoholic fermentation process. In: Proceedings of the 2017 international conference on electrical and information technologies (ICEIT), Rabat, Morocco. Institute of Electrical and Electronics Engineers (IEEE), Piscataway, NJ, pp 1–6

Ahmed M, El B (2006) Gene transfer between different Trichoderma species and aspergillus Niger through intergeneric protoplast fusion to convert ground rice straw to citric acid and cellulase. Appl Biochem Biotechnol 135:117–132

Baeshen NA, Baeshen MN, Sheikh A et al (2014) Cell factories for insulin production. Microb Cell Factories 13:141

Behera SS, Ray RC (2016) Solid state fermentation for production of microbial cellulases: recebt advances and improvement strategies. Int J Biol Macromol 86:656–669

Bhatia SK, Yi DH, Kim YH et al (2015) Development of semisynthetic microbial consortia of Streptomyces coelicolor for increased production of biodiesel (fatty acid methyl esters). Fuel 159:189–196

Boedeker B, Goldstein A, Mahajan E (2017) Fully disposable manufacturing concepts for clinical and commercial manufacturing and ballroom concepts. Adv Biochem Eng Biotechnol 165:179–210. https://doi.org/10.1007/10_2017_19

Box GEP, Hunter WG, Hunter JS (2005) Statistics for experimenters, design, innovation, and discovery, 2nd edn. Wiley, New York

Cobb R, Luo E, Freestone YT et al (2013) Drug discovery and development via synthetic biology. In: Zhao H (ed) Synthetic biology. Academic press, Boston, MA, pp 183–206. https://doi.org/10.1016/b978-0-12-394430-6.00010-8

Dasgupta A (2012) Advances in antibiotic measurement. Adv Clin Chem 56:75–104. https://doi.org/10.1016/b978-0-12-394317-0.00013-3

DeLoache WC (2015) An enzyme-coupled biosensor enables (S)-reticuline production in yeast from glucose. Nat Chem Biol 11:465–471

Demain AL, Vaishnav P (2011) Natural products for cancer chemotherapy. Microb Biotechnol 4: 687–699. https://doi.org/10.1111/j.1751-7915.2010.00221.x

Eggeling L, Sahm H, de Graaf AA (1996) Quantifying and directing metabolic flux: application to amino acid overproduction. Adv Biochem Eng Biotechnol 54:1–30

Eibl R, Dieter E (2008) Design of bioreactors suitable for plant cell and tissue cultures. Phytochem Rev 7:593–598

Fisher AC, Kamga MHH, Agarabi C et al (2018) The current scientific and regulatory landscape in advancing integrated continuous biopharmaceutical manufacturing. Trends Biotechnol 37:253–267. https://doi.org/10.1016/j.tibtech.2018.08.008

Fjaervik E, Zotchev SB (2005) Biosynthesis of the polyene macrolide antibiotic nystatin in Streptomyces noursei. Appl Microbiol Biotechnol 67:436–443. https://doi.org/10.1007/s00253-004-1802-4

Giddings LA, Newman DJ (2013) Microbial natural products: molecular blueprints for antitumor drugs. J Ind Microbiol Biotechnol 40:1181–1210. https://doi.org/10.1007/s10295-013-1331-1

Gomis-Fons J, Schwarz H, Zhang L et al (2020) Model based design and control of a small-scale integrated continuous end-to-end mAb platform. Biotechnol Prog 36:e2995

Gupta PK (2007) Metabolic engineering for over production of metabolites. Elem Biotechnol. 458–470

Hanke AT, Ottens M (2014) Purifying biopharmaceuticals: knowledge-based chromatographic process development. Trends Biotechnol 32:210–220

Hebbi V, Thakur G, Rathore AS (2019) Process analytical technology implementation for protein refolding: GCSF as a case study. Biotechnol Bioeng 116:1039–1052. https://doi.org/10.1002/bit.26900

Joanofarc X, Nivedhika D, Patnaik S (2019) Closed-loop performance and analysis of a real-time non-linear bioreactor process. In: Proceedings of the 2019 2nd international conference on power and embedded drive control (ICPEDC), Chennai, India, 21–23 august 2019. Institute of Electrical and Electronics Engineers (IEEE), Piscataway, NJ, pp 406–412

Julia R, Xavier G, Jose LM et al (2019) Continuous operation, a realistic alternative to fed-batch fermentation for the production of recombinant lipase B from Candida Antarctica under the constitutive promoter PGK in Pichia pastoris. Biochem Eng J 147:39–47

Lv XA, Jin YY, Li YD et al (2013) Genome shuffling of Streptomyces viridochromogenes for improved production of avilamycin. Appl Microbiol Biotechnol 97:641–648. https://doi.org/10.1007/s00253-012-4322-7

Mickel SJ, Sedelmeier GH, Niederer D et al (2004) Large-scale synthesis of the anti-cancer marine natural product (+)-discodermolide. part 3: synthesis of fragment C15-21. Org Process Res Dev 8:107–112

Mo S, Ban YH, Park J et al (2009) Enhanced FK506 production in Streptomyces clavuligerus CKD1119 by engineering the supply of methylmalonyl-CoA precursor. J Ind Microbiol Biotechnol 36:1473–1482. https://doi.org/10.1007/s10295-009-0635-7

Nfor BK, Ahamed T, van Dedem GWK et al (2008) Design strategies for integrated protein purification processes: challenges, progress and outlook. J Chem Technol Biotechnol 132: 124–132

Ochi K, Hosaka T (2013) New strategies for drug discovery: activation of silent or weakly expressed microbial gene clusters. Appl Microbiol Biotechnol 97:87–98. https://doi.org/10.1007/s00253-012-4551-9

Park SR, Tripathi A, Wu J et al (2016) Discovery of cahuitamycins as biofilm inhibitors derived from a convergent biosynthetic pathway. Nat Commun 7:107–110. https://doi.org/10.1073/pnas.1716245115

Pham JV, Yilma MA, Feliz A et al (2019) A review of the microbial production of bioactive natural products and biologics. Front Microbiol 10:1404. https://doi.org/10.3389/fmicb.2019.01404

Prabhavathi VR, Mathivanan N, Sagadevan E et al (2006). Intra-strain protoplast fusion enhances carboxymethylcellulase activity in trichoderma reesei. Enzym Microb Technol 38:719–723

Ravi RK, Satish P (2011) Metabolic engineering of bacteria. Indian J Microbiol 51(3):403–409

Roebuck K, Brundin A, Johns M (1995) Response surface optimization of temperature and pH for the growth of *Pachysolen tannophilus*. Enzym Microb Technol 17:75–78

Sanchez B, Ruiz L, Gueimonde M et al (2012) Toward improving technological and functional properties of probiotics in foods. Trends Food Sci Technol 26(1):56–63. https://doi.org/10.1016/j.tifs.2012.02.002

Sanchez GL, Martin L, Mangues R et al (2016) Recombinant pharmaceuticals from microbial cells: a 2015 update. Microb Cell Factories 15:33. https://doi.org/10.1186/s12934-016-0437-3

Sao Pedro MN, Azevedo AM, Aires-Barros MR et al (2019) Minimizing the influence of fluorescent tags on IgG partition in PEG–salt aqueous two-phase systems for rapid screening applications. Biotechnol J 14:1800640. https://doi.org/10.1002/biot.201800640

Shen B (2015) A new golden age of natural products drug discovery. Cell 163:1297–1300. https://doi.org/10.1016/j.cell.2015.11.031

Sindhu R, Pandey A, Binod P (2017) Design and types of bioprocesses. In: Current developments in biotechnology and bioengineering: bioprocesses, bioreactors and controls. Elsevier, Amsterdam, pp 29–43. https://doi.org/10.1016/B978-0-444-63663-8.00002-1

Tevyashova AN, Olsufyeva EN, Solovieva SE et al (2013) Structure-antifungal activity relationships of polyene antibiotics of the amphotericin B group. Antimicrob Agents Chemother 57: 3815–3822. https://doi.org/10.1128/aac.00270-13

Thomas L, Larroche C, Pandey A (2013) Current developments in solid-state fermentation. Biochem Eng J 81:146–161

Wegkamp A (2010) Physiological responses to folate overproduction in *Lactobacillus plantarum* WCFS1. Microb Cell Factor 9:100

Weibel EK, Hadvary P, Hochuli E et al (1987) Lipstatin, an inhibitor of pancreatic lipase, produced by streptomyces toxytricini. I. Producing organism, fermentation, isolation and biological activity. J Antibiot 40:1081–1085. https://doi.org/10.7164/antibiotics.40.1081

Woo MW, Nah HJ, Choi SS et al (2014) Pikromycin production stimulation through antibiotic down-regulatory gene disruption in Streptomyces venezuelae. Biotechnol Bioprocess Eng 19: 973–977

Yoo YJ, Kim H, Park SR et al (2017) An overview of rapamycin: from discovery to future perspectives. J Ind Microbiol Biotechnol 44:537–553. https://doi.org/10.1007/s10295-016-1834-7

Zhang MM, Wang Y, Ang EL et al (2016) Engineering microbial hosts for production of bacterial natural products. Nat Prod Rep 33:963–987. https://doi.org/10.1039/c6np00017g

Zhou TC, Kim BG, Zhong JJ (2014) Enhanced production of validamycin a in Streptomyces hygroscopicus 5008 by engineering validamycin biosynthetic gene cluster. Appl Microbiol Biotechnol 98:7911–7922

Chapter 4
Industrial Aspect of Marine Bioprocessing

Raveena Ann Alex, Joel Augustine, and Jayanthi Abraham

Abstract The marine habitat provides enormous biodiversity and a valuable source of bioactive compounds with therapeutic and biotechnological properties, which are used in various fields. The marine ecosystem consists of largely undiscovered resources that are of potential interest to mankind; only a few of these marine natural products have reached the stage of industrial production. Marine bioprocess engineering focuses on the design and evaluation of processes that manufacture valuable compounds from marine organisms inside an enclosed, controlled environment. Aquatic microorganisms, even though believed to be exceedingly diverse, have not been appreciably utilized by the present bioprocesses. Marine microorganisms have structurally unique bioactive natural products that are not found in terrestrial counterparts. Not only marine bacteria and fungi are rich in biologically active compounds, other organisms from the marine habitat include algae, crustaceans, plants, and fishes. This book chapter deals with the industrial aspect of marine bioprocessing and surveys the scope of bringing together all the undiscovered resources of the marine environment for a better living.

Keywords Marine organisms · Bioprocess · Enzymes · Metabolites · Natural products

4.1 Marine Bioprocessing

Marine bioprocess designing turns out to be a more diverse and dynamic area of science offering numerous possibilities to investigate remarkable marine resources. It is a process that completely utilizes bacterial cells or some of their cellular components to get the desired product. It is used to develop pharmaceuticals, flavors, cosmeceuticals, food supplements, and biomaterials with the help of a biocatalyst. Microorganisms can act as a catalyst in a bioreactor. The harsh physical and

---

R. A. Alex · J. Augustine · J. Abraham (✉)
Microbial Biotechnology Laboratory, School of Biosciences and Technology, Vellore Institute of Technology, Vellore, India

A. Sarkar, I. A. Ahmed (eds.), *Microbial products for future industrialization*, Interdisciplinary Biotechnological Advances,
https://doi.org/10.1007/978-981-99-1737-2_4

chemical conditions of the marine environment have also promoted the production of numerous novel molecules in marine creatures. With respect to terrestrial organisms, the bioactive compounds produced by marine organisms are more diverse in structural and functional features (Kathiresan et al. 2008). They also constitute a reservoir of new bioactive compounds with a significant pharmaceutical potential (Romano et al. 2017).

4.2 Sources of Marine Products

Among terrestrial plants and nonmarine organisms, marine ecosystems provide various natural bioactive compounds. The exploitation of marine environments relies on the development of various techniques from collecting marine samples, to improving spectrometric analysis, separation methods, and processing. In comparison with other marine organisms, algae are one of the predominant producers of new compounds consisting of antitumor and cytotoxic activity. These compounds have been validated with the ability to mediate precise inhibitor activities on a number of key cellular processes, inclusive of apoptosis pathways, angiogenesis, migration, and invasion in both in vitro and in vivo models, revealing their capability to be used as anticancer drugs (Alves et al. 2018).

4.3 Steps Involved in Bioprocessing of Marine Products

Bioprocess can be categorized into three levels: preparation, production, and purification. Preparation of samples always includes nutrient media and equipment sterilization. The keys to the production stage are bioreaction kinetics, oxygen transfer, and operational strategy (Table 4.1). The final step of product purification contains separation operations. Apart from the optimization of these steps, the biocatalyst is able to utilize substrates efficiently and form the desired product. Without this, the bioprocess procedure will be constrained, regardless of the best

Table 4.1 Steps involved in bioprocessing (Rorrer 2015)

Marine bioprocess engineering	
Step 1	Product discovery
Step 2	(a) Cell culture development
	(b) Strain selection/genetic engineering
Step 3	(a) Bioreactor development for biomass production (b) Product elimination strategies
Step 4	Harvesting of cells and extraction of product
Step 5	Purification of product and final product

efforts to maximize the upstream production kinetics and downstream product recovery (Clarke 2013).

4.4 Catalyst Involved in Marine Bioprocessing

There are several enzymes obtained from marine microorganisms, which are used as catalysts in industries. There are several biocatalysts isolated from microbes inhabiting extreme environments. Industrial bioprocessing operates under parameters with high or low pH, temperature, high saline environment, salt concentration, etc. The organism living in extreme conditions is ideal for industrial processes. There are cold-active biocatalysts used in the cosmetic and pharmaceutical industries. It reduces the excess thermal conditions, saves energy, prevents contamination, etc. The enzymes lipase and protease derived from marine environments are used for this process and with base detergent properties. The cold-active enzymes used in beverages are amylase enzymes in amyl syrups, proteases for caviar production, β-galactosidase for milk production, cellulose for coffee production, and so on. Marine enzymes are used in industries to produce secondary metabolites from microorganisms and plants and used to produce food items, pharmaceutical compounds, etc. Some of the enzymes are amylase from *Aureobasidium pullulans* N13d (Pacific Ocean sediment) (Li et al. 2007a, b), temperature tolerant transaminase enzyme group IV Pfam group produced by *Sulfolobus solfataricus* (Sayer et al. 2012). Superoxide dismutase and catalase produced by *Lingulodinium polyedrum* are used in industries to degrade aromatic compounds and are used as biomarkers with antioxidant activity. The bioethanol can be produced from marine algal biomass and the enzyme amylase is extracted from marine *Streptomyces* sp. for this process. It is used as a catalyst, in this reaction amylase undergoes a saccharification step to form ethanol in saline conditions. The usage of these catalysts in industries is cost-effective and increases the yield of the products (Nikolaivits et al. 2017; Trincone 2011).

4.5 Upstreaming and Downstreaming of Bioprocessing

Bioprocessing is regarded to be a very specific process that uses entire living cells or their components to acquire any value products. The living cells consist of microorganisms, enzymes, and similar kinds of living cells. The basis of productivity in bioprocess includes transportation of their energy and mass to several biological and environmental processes. There are two main stages of biotechnological process development responsible for the manufacturing of desired products, which include the upstream process and downstream process. The three aspects of upstream processing consist of the fermentation, the producer microorganism, and the fermentation process. Fermentation media optimization is an essential factor of media

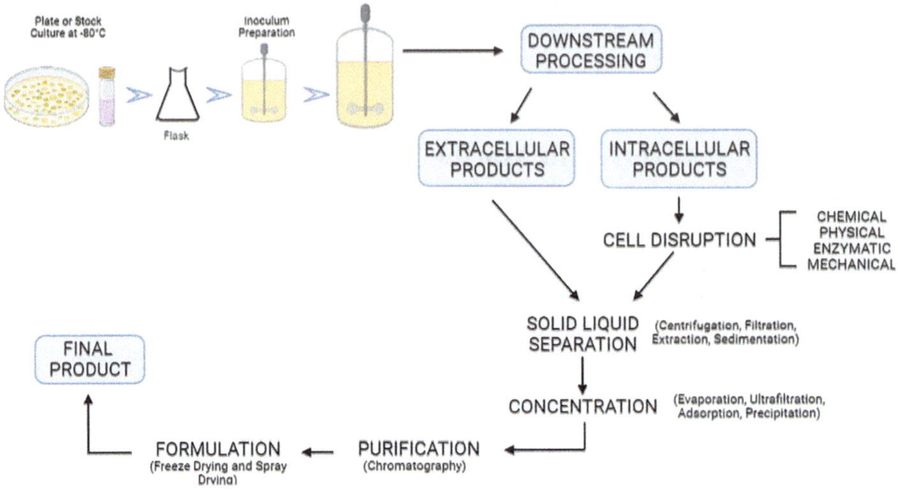

Fig. 4.1 Upstreaming and downstreaming process

improvement and crucial in making maximization of yield and profit. Aspects combined with the producer microorganism encompass the strategy, to begin with obtaining a suitable microorganism, commercial strain improvement to increase productivity and yield, maintenance of strain purity and suitable inoculum preparation. Fermentation is usually executed below rigorously controlled situations optimized for the growth of the microorganism or target microbial product production. The fermentation process is usually conducted in fermentation tanks or bioreactors (Behera and Varma 2017). Besides the upstream process, the second main part of a production process is the downstream process.

Downstream processing contains many steps such as separation, cellular disruption, the capture of the substance, and concentration followed by extraction, purification, polishing, and formulation. In biotechnological process development, the downstream process is underestimated as an ineffective part of a bioprocess step despite being the most expensive process (Silber et al. 2016). For commercialization of the product (different antibacterial compounds from marine sources), efforts have to be made in the downstream process, which are crucial (Fig. 4.1).

4.6 Product Isolation and Purification of the Product

Several natural products have been identified from marine sources; not all of them are being developed further. Marine pharmaceuticals may be the best example of the gap between discovery and commercialization. The majority of the marine products discovered are potential medicines, but they are not available on the market due to their lack of insufficient quantities or the instruments needed for purification, production, and time (Osinga et al. 1999).

The final step in marine bioprocessing is product recovery and is often an experienced step (Osinga et al. 1999). For the production of sufficient products, efficient and well-understood protocols are required.

4.7 Polishing of the Product

In bioprocess, product polishing represents the final processing steps that end with the packaging of the product in a form that is stable, easy to transport, and convenient. Crystallization, desiccation, lyophilization, and spray drying are ordinary unit operations. Depending on the product and its expected use, the polishing process may also include operations to sterilize the product and eliminate or deactivate trace contaminants, which might compromise product safety. Such operations might include the removal of viruses or depyrogenation (Table 4.2).

Table 4.2 Some examples of new microbial species isolated from different marine environment and their industrial use (Poli et al. 2010; Beygmoradi and Homaei 2017)

Sl. no	Marine organism	Marine environment source	Industrial use	References
1.	*Geobacillus* sp. Strain 4004	Sediment in marine hot spring (Italy)	Pharmaceutical application	Nicolaus et al. (2002)
2.	*Pseudoalteromonas* strain 721	Deep-Sea hydrothermal vent	Gelling properties	Rougeaux et al. (1999) Guezennec (2002)
3.	*Hahella chejuensis*	Marine sediment sample (Republic of Korea)	Biosurfactant and detoxification from polluted petrochemical oils	Lee et al. (2001)
4.	*Bacillus licheniformis*	Shallow marine hot spring water (Italy)	Antiviral activity	Maugeri et al. (2002)
5.	*Alteromonas macleodii* subsp. *fijiensis*	Hydrothermal vent of deep sea (North Fijian Basin)	Thickening agent in food processing industry	Loaec et al. (1998)
6.	*Pseudoalteromonas* strain CAM025	Antarctic sea	Cryoprotection	Mancuso et al. (2004)
7.	*Alkalibacillus* sp	Alkaline hypersaline lake (Egypt)	Extracellular glucoamylopullulanase	Mesbah and Wiegel (2014)
8.	*Luteimonas abyssi* XH031	Deep-sea sediment (South Pacific gyre)	Alkalintolerant, high active amylase	Song et al. (2016)
9.	*Halococcus turkmenica*	Saline soil (Turkmenistan)	Alpha-amylase	Santorelli et al. (2016)

4.8 Marine Microorganisms

Marine microorganisms offer significant possibilities for biodiscovery and their potential is not fully explored. The way to achieve this could be to focus on culturable and unculturable microorganisms. The lack of laboratory cultures limits the possibility of growing marine microorganisms for bioprocessing and the discovery of new value-added products. Standard microbiological methods only isolate 0.001–1% of organisms when compared to the huge diversity of marine organisms present in nature (Joint et al. 2010).

4.8.1 Culturable and Nonculturable Organisms

Microorganisms can be of culturable and nonculturable states. For culturing marine bacteria, Zobell marine agar is used, which has a combination of salts that mimic the seawater and helps in the growth of marine microorganisms. Many marine organisms will be in their dormant state as typified by the viable but non-culturable state. In this state, the organisms are metabolically active but are no longer culturable on usual growth media. These non-culturable bacteria may regain their culturability in favorable situations (Zhang et al. 2021). Methods used are fluorescent in situ hybridization (FISH), nucleic acid amplification techniques such as real-time quantitative PCR (RT-QPCR), and reverse transcriptase (RT-PCR). *V. fischeri* mostly seen in the sea is not easily cultivable (Colwell 2009) and *Candidatus Entotheonella* is associated with the marine sponge *Theonella swinhoei* and these bacteria produce free amino acids but it is difficult to cultivate (Mu et al. 2021). Another novel approach to isolating marine bacteria is microencapsulation, which involves micro-droplet encapsulation in an agarose matrix to isolate and analyze cell types independent of their features and activities. Here gel micro-droplets are prepared by dispersing agarose, mixed with water from the natural environment containing bacteria into a nonaqueous phase such as oil to form an emulsion. By immediately cooling the emulsion, the molten agarose solidifies and due to uniform stirring, micro-droplets are formed and a portion of it contains single bacterial cells (Joint et al. 2010). Some bacteria do not grow in the media provided due to the lack of nutrients, for example, exogenous *N*-acetyl muramic acid is required for the growth of *Tannerella forsythia*, L-cysteine is required for *Abiotrophia* and *Granulicatella* (Vartoukian et al. 2010).

4.9 Cosmeceutical

The bioactive compounds produced by marine organisms are used as skincare agents, which are used to develop cosmetics and nutricosmetics. The cosmeceuticals are the combination of "cosmetics" and "pharmaceuticals." These cosmetic items can be used for beautifying, cleansing, altering the appearance, preventing disease, and also used as drugs for certain skincare diseases. The cosmetic products produced are shampoo, face cream, sunscreen, hair gel, ointment, lotion, etc. The nutricosmetics are provided in the form of liquid, pills, functional food, etc. It has the natural capacity to enhance the skin, nails, hair, and so on. These compounds have various advantages and they do not produce any side effects. Cosmeceuticals contain the natural source of enzymes, vitamins, minerals, and phytochemicals in the form of lotions, cream, ointments, and others, which are isolated from different aquatic plants, marine organisms, and microorganisms. Algal resources as cosmeceutical ingredients are underexploited, which can also be used in food diets due to their rich composition of minerals and vitamins. Some marine flora and fauna produce bioactive unique compounds that are not found in terrestrial ecosystems, which can be produced rapidly by modern aquaculture techniques (Kim 2014; Alves et al. 2020).

The active ingredients of marine-derived cosmetics are antiaging, anti-whitening, antioxidant, anti-inflammatory, anti-acne, anti-wrinkling, anti-tyrosinase, cytoprotective, and UV photoprotective. These compounds are extracted from seaweed, sea mud, microalgae, seawater, etc. (Kim 2014). On the basis of pigment composition, seaweeds are classified into red algae (Rhodophyta), green algae (Chlorophyta), and brown algae (Phaeophyta). The anti-inflammatory activity shown by the *Laminaria saccharina* extract is also rich in proteins, vitamins, minerals and polysaccharides and it also regulates the sebaceous gland activity. Algae such as *Undaria pinnatifida, Polysiphonia lanosa,* and *Durvillaea antarctica* produce essential oils, which exhibit anti-inflammatory and antiseptic properties and are used in skin protection and conditioning products. The brown algae produce phlorotannins and polyphenolic phytochemicals, which are used for the production of nutritional supplements, cosmetics, and cosmeceuticals (Karawita et al. 2007). The phlorotannins nourish and beautify skin and it is used in many skincare products. The polysaccharides from the seaweed, such as chitin (exoskeletons of crustaceans), exopolysaccharide (marine bacterial product) fucoidan (brown algae), carrageenan (red algae), ulvans (green algae), alginate, and agar are also used in skin care cosmetics and cosmeceuticals due to its good texture and skin health benefits. The carotenoid pigments extracted from various marine plants, fungi, and bacteria (not from animals) are potential cosmeceutical ingredients, for e.g.: Vitamin A, Fucoxanthin, and Astaxanthin. The *Pelvetia siliquosa* organism (brown algae) produces fucosterol, which has antioxidant and anti-diabetic properties, and it also reduces cholesterol. This brown alga is a good, polyunsaturated fatty acid and amino acid producer, which is also used in pharmaceutical industries (Ruperez 2001).

Microalgae is a small organism that is used in food, pharmaceutical, cosmetic, and medicinal industries due to the abundance of zeaxanthin from *Nannochloropsis oculata* organism, which is used in skin whitening creams. *Arthrospira* and *Chlorella* are the other organisms used in skincare products. Industrial collagen was collected from the bones and skin of cattle and pigs, which are limited now and scientists discovered alternative marine collagen from microalgae, which is safe to use and economical. Due to its antioxidant property, it is used in many skincare products, such as sunscreen. It is also used to treat fish and sponge damage. *Chlorella vulgaris* and *Spirulina maxima* have oil content, which is used to treat dandruff and hair fall and it is used in oils, shampoo, etc.

Seawater is rich in minerals, such as sodium, potassium, magnesium, calcium, chlorine, etc. It helps to hydrate skin, lighten skin, is a deodorizing agent, and it can also be used in hair conditioning. Most skincare cosmetics contain these compounds to improve skin health. The seawater is freeze-dried and used in some skincare products (Mathews 2010). The Dead Sea mud or Dead Sea black mud is used in many skincare products. It nourishes the skin and balances the pH, it has antiaging properties and is used in lotions for skin. It is used to treat various aches, pain, and various health conditions (Table 4.3).

Table 4.3 Typical media used for the cultivation of marine organisms in the laboratory

Sl. no	Media name	Organisms	Components	References
1.	Glucose mineral salts medium (GMSM) (1 L)	Bacteria	Glucose—25 g L^{-1} NH_4NO_3—6 g L^{-1} KH_2PO_4—0.028 g L^{-1} K_2HPO_4—1.6 g L^{-1} $MgSO_4$—0.3 g L^{-1} $CaCl_2.2H_2O$—0.2 g L^{-1} $FeSO_4.7H_2O$	Rangarajan et al. (2012)
2.	Zobell marine agar	Bacteria	Peptone—5 g Yeast extract—1 g $(NH_4)_2SO_4$—1 g $FeSO_4.(NH_4)_2SO_4.6H_2O$—0.1 g Aged Sea water (75%)—1000 mL Agar agar—2% pH—7.4–7.6	Rosenfeld and ZoBell (1947)
3.	Artificial Sea water nutrient broth (ASWNB) (1 L)	Bacteria	Nacl—28.13 g, Kcl—0.77 g, $CaCl_2$—1.60 g, $MgCl_2.6H_20$—4.8 g, $NaHCO_3$—0.11 g, $MgCl_2.7H_20$—3.5 g Peptone—5 g Beef extract—3 g	Mukherjee et al. (2004)
4.	Artificial seawater medium (1 L)	Marine prokaryotes and fungi	Peptone—3.50 g Yeast extract—3.50 g NaCl—23 g $MgCl_2$—5.08 g $MgSO_4$—6.16 g $Fe_2(SO_4)_3$—0.03 g $CaCl_2$—1.47 g KCl—0.75 g Na_2HPO_4—0.89 NH_4Cl—5 g	Lang et al. (2005)

4.10 Therapeutics

The marine environment offers high biodiversity and a number of valuable biological sources that can be used for therapeutic and biotechnological applications. The antiviral, antimicrobial, antioxidant, photoprotective, cytotoxic, genotoxic, anti-inflammatory, and anticancer bioactive compounds of marine organisms have stirred interest in the scientific community for many centuries. The therapeutics derived from the marine organism are being used as anticancer and anticoagulants. Cancer is one of the biggest human health problems in the world and there are several cancer treatment approaches, such as surgical removal, radiotherapy, immunotherapy, and chemotherapy. Cancer therapy causes adverse side effects and reduces patients' quality and in some cases dignity of life. Nowadays, a drug derived from a natural source and its semisynthetic analog structure drugs can be treated against cancer, which are derived from marine organisms. There are some cancer drugs that are isolated from marine organisms, which are currently in use. Cytarabine (Cytosar-U®), which is a derivative from a marine sponge, is used for leukemia; trabectedine (Yondelis®), isolated from a tunicate, is medicinally in use for ovarian cancer and soft tissue sarcoma; eribulin mesylate, Halaven®, used for metastatic breast cancer, is a derivative from a sponge; and brentuximab vedotin (Adcetris®, a conjugated antibody used in Hodgkin's lymphoma and anaplastic large T-cell malignant lymphoma) is derived from a mollusk source (Teixeira et al. 2019). Some fungal metabolite such as terpenoids, isocoumarins, steroids, phenolic compounds, etc., have antiviral, antimicrobial, antioxidant, anti-inflammatory, cytotoxic, genotoxic, anticancer, and kinase inhibition properties which are used as drug targets for number of therapeutic purposes. *Leptosphaeria* sp. fungi and microalgae *Sargassum tortile* cytotoxic metabolites are used as drug against leukemia, HCT-116 colon carcinoma, A549 lung cancer cells, and others (Jeewon et al. 2019; Yanagihara et al. 2005). The organism *S. plicata* and *Halocynthia pyriformis* produce dermatan sulfates which have high heparin cofactors that mediate anticoagulant activity, which is used to treat deep vein thrombosis. Heparin mediates anti-inflammatory and antimetastatic properties. The heparin activity is studied in preclinical rodents with promising results (Pavão 2014). One of the most common age-related diseases is osteoporosis, which is characterized by decreased bone mineral density and microarchitectural deterioration. Marine natural products derived from various marine sources have shown a considerable effect on bone metabolism inhibiting osteoclastogenesis and upregulating osteoblastogenesis through modulating different pathways (RANK/RANKL/OPG) (Chaugule et al. 2017).

4.11 Microalgae As a Source of Biomaterials and Pharmaceuticals

Microalgae are versatile, they thrive in various environments, such as rivers, lakes, soil, and oceans. They have photosynthetic ability and they require light, nutrients pH to grow, some are photoautotrophic and mixotrophic or heterotrophic. They are rich in renewable biological resources, which are used in producing cosmetics, medicines, pharmaceuticals, biomaterials, biofuel, food, etc. The world market for regular drug items is continually developing and algae meet a major portion of this demand. The microalgae biomass contains many bioactive substances, which carry medicinal, antimicrobial, anticancer, and antiviral properties. *Arthrospira*, *Chlorella*, *Dunaliella*, and *Haematococcus* are the microalgae that are used in industry to produce these substances, and their primary and secondary metabolites are used to produce medicines (Orejuela-Escobar et al. 2021). The microalgae extract is used as therapeutics for treating cancer in in vitro studies, which have shown that some microalgae species destroy cancer cell lines like lung carcinoma, human breast, adenocarcinoma, prostate cancer, and colon cancer. *Spirulina platensis* is a microalga that produces a neuroprotective substance that can treat Alzheimer's, Parkinson's, and other neurodegenerative diseases because it slows down or halts diseases affecting the nervous system (Jha et al. 2017). Some anti-allergic bioactive compounds are produced by microalgae, which inhibit histamine and can be used as functional drugs to suppress immunogenic responses and also used for the development of oral vaccines against influenza A, Zika virus, and HIV viruses (Deniz et al. 2017; Bhalamurugan et al. 2018; Tang et al. 2020).

Algal biomass can be converted to biochemical and biomaterials by biorefineries, through chemical and biological conversion. These products mainly are ethanol, isobutanol, levulinic acid, lactic acid, succinic acid, etc., which are produced from edible feedstocks. These can be further used to produce biofuels and other chemicals (Cesário et al. 2018).

4.12 Neuropharmacological Properties of Marine Microorganisms

The terrestrial actinomycetes are used to produce pharmaceutical compounds from their bioactive compounds for a long period of time; however, nowadays, the research is focused on the marine actinomycetes. Marine algae, bacteria, and fungi are other sources of neuropharmacological agents. Neuronal loss and neurodegenerative diseases are caused by the accumulation of free radicals or oxidative stress. An imbalance in the antioxidant causes damage to protein, lipids, and DNA and eventually leads to neurodegenerative diseases. The actinobacteria *Streptomycetes* sp. produce antioxidants, flavonoids, sediments, and other metabolites. The organisms such as *Thermomonosporaceae, Streptosporangiaceae, Micrococcaceae, Nocardiaceae*, and *Streptomycetaceae* that are isolated from the mangroves have potent antioxidant, antitumor, and anti-infectious properties and can be used to treat diabetes and neurodegenerative disease (George et al. 2020). These anti-inflammatory agents protect the neuronal cells against oxidative stress. H_2O_2 is one of the major compounds that damage neuronal cells by causing oxidative stress. It is soluble in water and converts into hydroxyl and superoxide ions, and these superoxide ions damage macromolecules and lead to cell death. Pyrrolopyrazines, isolated from *Streptomyces* sp., has the ability to scavenge free radicals and protect neuronal cells from H_2O_2 (Ser et al. 2016). Alzheimer's disease is caused by the hyperphosphorylation of tau protein and it promotes neurotoxicity, the *Streptomycetes* sp. produce Streptocycline, which inhibits the phosphorylation of tau protein and has anti-inflammatory effects (Alvariño et al. 2018).

The marine brown algae that produce a sterol named fucosterol targets acetyl and butyryl-cholinesterase that are responsible for Alzheimer's disease. It reduces A-β-induced neuronal death, passing through the neuronal membrane (Hannan et al. 2019). The marine algae *Sargassum siliquastrum* produces fucoxanthin, which has antioxidant properties that prevent the H_2O_2-induced DNA damage. *Sargassum hemiphyllum* produces fucoidan, which reduces the hydroxydopamine-induced apoptosis and protects SH-SY5Y from neurotoxicity. The fucoidan increases the production of dopamine and shows neuroprotection properties (Hannan et al. 2020). A pharmacological compound named alkylguanidine produced by *Gymnodinium* sp. and *Gonyaulax* sp. showed a great effect on the central nervous system.

Marine fungi produce some unique diterpenes, which have neuroprotective activities. A unique tetracyclic diterpene named tetraol, which is isolated from *Cephalosporium aphidicola*, is found to inhibit the growth of neuroblastoma in the CNS; these diterpenes inhibit the activity of DNA polymerase (Islam et al. 2016). Cholinesterase is produced by cyanobacteria *Nostoc* sp., which is used to treat Alzheimer's disease. The bacteria *Robiginitalea myxolifaciens* sp. has potent antioxidant activity against L-glutamate toxicity and reactive oxygen species (ROS) scavenging ability (Torregrosa-Crespo et al. 2018).

4.13 Pharmaceutical Aspect of Metabolites from Marine Algae

The biomolecules from marine algae can be used to produce antibiotics, immunosuppressive agents, pigments, anticancer agents, and enzyme inhibitors. Among microalgae and macroalgae, microalgae have high growth rates and short generation periods and can also double their biomass more than once per day within the fastest-growing species. Microalgae contain pigments like chlorophylls, carotenoids, phycobiliproteins, and ketocarotenoids (Zullaikah et al. 2019) and can produce valuable compounds such as antioxidants, enzymes, vitamins, and sterols (Moreno-Garcia et al. 2017; Vignesg and Barik 2019); it is considered as a promising feedstock for synthesis of antimicrobial, antiviral, and anticancer drugs (Rotter et al. 2021). Microalgae *Chaetoceros muelleri* and *Dunaliella salina* biomass extract have antimicrobial properties because of the compounds including cyclocitral, neophytadiene, and phytol. The fungal and bacterial pathogens *Lindra thalassiae* and *Pseudoalteromonas bacteriolytica* can be inhibited by red algae *Peyssonnelia* sp. The diterpenes produced by marine algae is used to make antiviral drugs, bioactive diterpenes isolated from brown algae *Sargassum macrocarpum,* exhibit strong antibacterial activity. *Nostoc flagelliforme* produces an acidic polysaccharide, which has antiviral properties against HSV-1 virus. The fatty acid compounds from marine algae inhibit the growth of several bacteria and viruses. Microalga

Skeletonema costatum has unsaturated and saturated long-chain fatty acids, which inhibit the growth of *Vibrio* sp. The increase in multidrug-resistant bacteria is a serious health problem and commercial antibiotics are failing to inhibit their growth, due to the improper and overuse of drugs. One of the alternatives to overcome this is by searching for novel antibiotics produced by marine algae (Bajpai 2016).

4.14 Bioprocess Engineering Data on Marine Bacteria

Bioprocess engineering comprises the plan, operation, control, and optimization of biochemical processes consisting of different biological pathways involving microorganisms or enzymes under controlled conditions for the proficient biotransformation of unrefined components into a range of products at essential scales.

Bioprocess engineering presents the path from discovery to marketing. Hundreds of bioactive compounds were discovered but only a few products are commercialized due to the limited availability of the metabolites for clinical trials or further alteration by biocatalytic or chemical methods. Despite these uplifting perceptions, it is still consistent that there is an inadequacy of research in bioreactor engineering and designing fermentation protocols in the field of developing marine microorganisms to create high value-added products (Marwick et al. 1999). A major problem faced in marine biotechnology is the fact that less than 5% of all the bacteria observed and isolated are difficult to cultivate in laboratory conditions and this limits the discovery of new marine-derived products.

In industries, marine organisms are used to synthesize their biological compounds for meeting demand. It is done in a controlled environment, where analysis of engineered organisms' manufacturing and processing is focused. Many bioactive compounds cannot be duplicated by chemical synthesis in the laboratory, so the organism is suspended in liquid culture for future use. These processes involve the bioprocess and purification of products, which include the following:

The organism is isolated and characterized

↓

Targeted compound-producing organisms are cultivated in medium with desired compounds in higher concentrations.

↓

The microorganisms are cultivated in a bioreactor in controlled conditions to produce biomass and its compounds.

↓

The cells are harvested by flocculation, de-watering, filtration, centrifugation, etc. After that process, the harvested cells are mixed with appropriate organic solvents.

↓

Biomass extraction is done by using the chromatographic technique, which is fractionated and purified subsequently (Rorrer 2015).

The compounds include pigments, exopolysaccharides, fatty acids, enzymes, and other bioactive compounds. Some of the compounds are not sufficient after the purification, thus are difficult to commercialize (Zaborsky 1999). The cultivable microorganisms are easy to bioprocess compared to other organisms and most of the bacteria used for bioprocessing are Gram-negative bacteria. Marine fungi grow on wood, soil, corals, etc. But the bacteria can grow in natural conditions in selective media. Depending upon the bacteria, suitable pressure and temperature have to be provided by optimizing their culture parameter. For metabolite production, batch cultivation is used, and around 3–1000 L amount of culture is required to produce some metabolites such as thiomarinol, hydroxyakalone, macrolactins, and chalcomycin B by the organisms, *Alteromonas* sp., *Agrobacterium aurantiacum, Bacillus* sp., and *Streptomyces* sp. In a study conducted by Lang et al. (2005), production and purification of bioactive compounds were performed using submerged fermentation followed by solvent extraction (Table 4.4).

Table 4.4 Marine natural products and analogs or derivatives produced by marine organisms in therapeutic agents

Organism		Compound	Property	References
Marine sponge		Cytarabine	Anticancer agent	Pavão (2014)
Tunicate		Trabectedine	Anticancer agent	Pavão (2014)
Microalgae	*Sargassum tortile*	Cytotoxic metabolite	Anticancer agent	Pavão (2014)
Fungus	*Fusarium subglutinans*	Subglutinol A and B	Immunosuppressive	Jeewon et al. (2019)
	Ascomycetous 222	Balticolid	Antiviral	Teixeira et al. (2019)
	Leptosphaeria sp.	Cytotoxic metabolite	Anticancer agent	Pavão (2014)
	Aspergillus ostianus	Circundatin B	Antibacterial	Jeewon et al. (2019)
	Penicillium sp.	Cladosporin, epiepotormin, phyllistine	Antibacterial	Teixeira et al. (2019)
	Fusarium sp.	Fusarielin E	Antifungal	Jeewon et al. (2019)
	Zopfiella latipes	Zopfiellamide A and B	Antifungal	Jeewon et al. (2019)
	Cephalosporium sp.	*Graphislactone A*	Free radical scavenger	Teixeira et al. (2019)
Bacteria	*Brevibacillus laterosporus*	Tauramamide	Antibacterial	Debbab et al. (2010)
	Marinispora (NPS008920)	Lipoxazolidinone	Antimicrobial	Debbab et al. (2010)

4.15 Metabolite Production from Marine Microorganisms

Marine microorganisms produce various metabolites, which have properties including antibacterial, antifungal, antiviral, various exotoxins, hormones, immunomodulatory, pigments, enzyme inhibitors, etc. Many of them are further used in industries, to make cosmetics, pharmacology, food, and agriculture. Microorganisms such as bacteria, fungi, and cyanobacteria produce bioactive compounds for survival and existence. These metabolites are structurally diverse as well as low-molecular-weight complex bioactive compounds.

4.15.1 Marine Bacteria

The marine bacteria in the genera of, *Actinomyces*, *Bacillus*, *Micrococcus*, and *Serratia*, show antibacterial activity against Gram-positive bacteria. Tauramamide is a lipopeptide produced by *Brevibacillus laterosporus* bacteria with methyl and ethyl ester. It can inhibit Gram-positive bacteria, such as *Enterococcus* sp. The *Actinomyces* isolated from marine sediments produce a metabolite named chromopyrrolic acid, which inhibits bacteria, like *S.aureus, S.epidermidis*, and *Enterococcus faecalis*. And the *Actinomycete* strain CNQ-418 and its crude extract have a strong antibiotic property (Debbab et al. 2010; Andryukov et al. 2019). *Flavobacterium* sp. produces proteorhodopsin that reduces organic sulfur compounds for energy production (De Carvalho and Fernandes 2010). The bacteria genera, *Streptomyces* anthracycline daunomycin and its derivative, doxorubicin, the polyketides aclarubicin and mithramycin, the glycopeptide bleomycins A2 and B2, the chromopeptide actinomycin D and mitomycin C are used to treat cancer and its cytotoxic effect, which target the cancer cells. The *Halichondria okadai* bacterial metabolite halichondrin and its synthetic analog, eribulin mesylate are used to treat breast cancer. *Bacillus subtilis* 109GGC020 has linear lipopeptides gageopeptides A-D and gageotetrins A–C has antimicrobial and cytotoxic activity and is used to treat lung cancer, and gageopeptides A, B, C, and D has antifungal and cytotoxic activity against human myeloid leukemia. *Streptomyces koyangensis* SCSIO 5802 bacterial derivative, neoabyssomicin D, shows antiviral activity against vesicular stomatitis virus. These metabolites of marine microorganisms can be used in the pharmaceutical and therapeutic industries (Santos et al. 2020).

Microorganisms produce compounds for their defense against pathogenic microorganisms. These compounds are exploited in therapeutics, cosmetics industries, etc. *Pseudomonas* sp. produces many bioactive compounds: *P. borbori* produce benzaldehyde, quinoline, quinolone, phenanthren, phthalate, andrimid, moiramides, zafrin, pyrroles, pseudopeptide pyrrolidinedione, phloroglucinol, phenazine, and bushrin; these act as antimicrobial agents. *P. phenolica* produces brominated biphenyl compounds, which inhibit *S. aureus*. *Pseudoalteromonas tunicata* is renowned for its antifungal activity, which produces an alkaloid compound consisting of

4-methoxypyrrole. Exopolysaccharides of *Geobacillus thermodenitrificans* has immunomodulatory and antiviral effects against immunocompetent cells, the EPS is easy to use in industries due to its extreme temperature resistant, pH resistant, and salinity resistant properties. It is used to produce bioemulsifier and biosurfactants (Bhatnagar and Kim 2010).

4.15.2 Cyanobacteria

Cyanobacteria has the ability to produce a wide variety of bioactive compounds such as antitumor, antibiotic, antiviral, antimalarial, immunosuppressive agents, etc. *Fischerella* and *Nostoc* sp. produce an alkaloid named hapalindole, which has antibacterial activity against *Mycobacterium tuberculosis, Staphylococcus aureus, Salmonella typhi, Pseudomonas aeruginosa, Escherichia coli,* and *Enterobacter aerogenes*. Noscomin is an antibacterial compound produced by *Nostoc commune,* which can inhibit *Bacillus cereus, Staphylococcus epidermidis*, and *Escherichia coli*. γ-linolenic acid produced by *Spirulina* sp. can be used to prevent cardiovascular disease. Dolastatin 10 and dolastatin 12, which are anticancer peptides produced by *Symploca* sp. and *Leptolyngbya* sp. are used to treat cancer. Phycocyanobilin pigment from cyanobacteria has anti-inflammatory and antioxidant activity (Augustine et al. 2021).

4.15.3 Fungus

Since the disclosure of new substances in the marine environment, new sources are being researched constantly. From the last decade, the attention toward marine fungi by scientists has increased due to the numerous bioactive compounds produced by them in relation to its occurrence, habitat, symbiotic association, etc. The fungal strains are known to produce drugs that were previously recorded only among marine plants and animals. Fungal strains such as *Aspergillus, Penicillium, Talaromyces*, and *Trichoderma* are prevalent in both terrestrial and marine environments, which have the ability to produce various bioactive compounds (Nicoletti and Vinale 2018). *Aspergillus terreus* SCSGAF0162, isolated from Gorgonian corals produces butyrolaceton and lactone derivatives and it exhibits inhibitory activity against acetylcholinesterase and also shows antiviral activity. *Penicillium oxalicum* fungi synthesis dihydrothiophene-condensed chromones and oxalicumons, which have cytotoxicity against several carcinoma cell lines. *Nigrospora oryzae,* isolated from the south Chinese sea has two compounds named citrinins and nigroporins B, which have antifungal activity against *Aspergillus versicolor*. Endophytic fungi, *Neosartorya pseudofischeri,* isolated from starfish named *Acanthaster,* produce a novel glycotoxin in glucose-peptone-yeast extract medium, which inhibit drug-resistant bacteria *S. aureus* and *E.coli* and has cytotoxic property against human cell lining and colon cancer. The *Penicillium* sp. isolated from sea urchin, *Brisaster latifrons* have tanzawaic acid derivatives, which have an anti-inflammatory effect. Two cerebrosides, penicillosides A and B, are isolated from *Penicillium* sp. from the red sea, among which penicillosides A exhibits antifungal activity against *Candida albicans* and penicillosides B exhibits antibacterial activity against *S. aureus*. *Aspergillus* sp. from the mangrove forest of Hainan Island, China produces isocoumarin derivatives and aspergifuranone, which exhibit α-glucosidase activity. The fungi, *Pestalotiopsis vaccinia,* produces a naphthalene derivative that exhibits anti-enterovirus and COX-2 inhibitory activity. *Aspergillus versicolor,* isolated from sponges, produces a compound named 5-methoxydihydrosterigmatocystin, which has potent activity against *Bacillus subtilis* and *S. aureus*. The fast-growing ability and easy availability of *Aspergillus* and *Penicillium* are mostly used to produce therapeutic compounds from their bioactive compounds (Jin et al. 2016). Tambjamine is an alkaloid-like compound, isolated from *Pseudoalteromonas tunicate*, which has antifungal property.

4.15.4 Actinomycetes

Studies suggest that rare *Streptomyces* sp. from marine sources are potential sources of bioactive compounds. This includes 70–80% of total derived bioactive compounds. Some of the bioactive metabolites (antifungal) from *Streptomyces* class are nystatin (*Streptomyces noursei*), natamycin (*Streptomyces natalensis*), and amphotericin B (*Streptomyces nodosus*). Some bioactive compounds, which show antimicrobial activity are vancomycin (*Streptomyces orientalis*), thienamycin (*Streptomyces cattleya*), tetracycline (*Streptomyces rimosus*), etc. (Nair and Abraham 2020).

4.16 Biopolymer Production

Biopolymers derived from marine organisms have a variety of unique material characteristics and commercial applications. There are numerous industrial applications for capsular, complex extracellular polymeric substances (EPS). *Shewanella colwelliana* produces an excess of EPS of which the complex structure helps in biofilm composition and EPS can function as adhesins, which can be used as underwater surface coatings and bioadhesives. EPS has simple α-1-4-linked dextrans to highly complex branched heteropolysaccharides, attached with repeating xanthan and colanic acid. The bacteria that are isolated from the colonized fish produce EPS, which has drag-reduction properties it is applicable in drilling, ship efficiencies, oil cleaning, and viscosity reduction (Brmadsky and Rosenberg 1992). Some marine and nonmarine bacteria and seaweeds produce EPS, which has the property of surfactant alginate, and is used to produce different types of processed food, paper, adhesives, textiles, paints, ceramics and is also used in chemical industries, etc. The *Halomonas* sp. has the ability to produce poly-β-hydroxy-alkanoates (PHAs), and its thermoplasticity property forms the base material of the biodegradable plastics. It is similar to polypropylene and is used to make 200 products including packing materials, pipes, domestic materials, etc. It is produced by using fermentation by engineered organisms and production is time-consuming and costly. Many marine bacteria produce an excess number of pigments and are used in industries to make different products due to their cheaper fermentation and purification (Weiner 1997; Simon-Colin et al. 2008). The extract of bacterial exopolysaccharide (EPS) and its derivatives are rich in bioactive compounds: *V. diabolicus* has EPS with high amounts of glucuronic acid, N-acetyl-glucosamine, N-acetylgalactosamine, etc., which are used as bone healing materials. The recently identified *Rhodovulum* sp. PS88 can be used in wastewater treatment plants because of its better growth and high flocculation properties and these cells exhibit good settling characteristics (Watanabe et al. 1999).

4.17 Enzymes and Other Proteins Production

Marine organisms can produce enzymes, which can offer novel biocatalysts with salt tolerance, high thermostability, pressure, and adaptation to cold. Countless marine microorganisms furnish new approaches and new varieties of enzymes (Sekar and Kim 2020). Marine enzymes are unique protein molecules with novel properties from an organism depending upon its natural habitat. Marine organisms such as bacteria, fungi, sponges, plants, fishes, prawns, crabs, and algae remain unexploited and are identified as unexplored enzyme sources. Amylases (*Bacillus licheniformis*), proteases (*Bacillus, Pseudomonas, Clostridium, Rhizopus, Penicillium, Aspergillus*), *lipases* (*Penicillium oxalicum, Aspergillus flavus, and Streptomyces*), chitinases, keratinases, peroxidase, pyrophosphatase, DNA polymerases, and xylanases are some of the marine-derived enzymes. These enzymes are widely used in pharmaceuticals, beverages, leather, food processing, candy factories, and also in wastewater treatment plants. Some reports of the marine organism producing enzymes are proteases from marine bacterium *Vibrio harveyi* (Estrada-Badillo and Márquez-Rocha 2003) and alkalophilic and salt-tolerant fungus *Engyodontium album* (Sarkar et al. 2010), purified thermostable DNA polymerase derived from eubacterium *Thermotoga maritima* (Gelfand et al. 1996), Extracellular amylase from marine yeast *Aureobasidium pullulans* (Li et al. 2007a, b), pyrophosphatase from *Thermus thermophilus* (Bolchakova et al. 2004), and Lignin peroxidase from marine fungus *Caldariomyces fumago* (Irvine and Venkatadri 1994).

4.18 Pigments Produced by Marine Microorganisms

Marine microorganisms have a great ability to produce a wide range of pigments and bioactive compounds because of their adaptation to extreme conditions. Marine bacteria, fungi, and algae produce biologically active pigments, Prodiginines (red pigment), violacein (violet pigment), carotenes, etc. Many marine bacteria produce an excess number of pigments and are used in industries to make different products due to their cheaper fermentation and purification (Weiner 1997). The microalga *Dunaliella bardawil* and fungus *Blakeslea* produce β-carotene, which is used in food coloring and also used to produce vitamin A synthesis. Fucoxanthin produced by brown algae has anti-inflammatory and antioxidant properties and fucoxanthin does not exhibit toxicity. The rare carotenoids produced by *Nostoc* and *Anabaena* are used to produce hydrogen and polyhydroxyalkanoates. *Robiginitalea myxolifaciens* sp. produce myxol from γ carotenoids, which have potent antioxidant activity and neuroprotective properties. Algae producing chlorophyll is used for food coloring in the pharmaceutical industry because it possesses antimutagenic, anticarcinogenic, and antioxidant activity. The fungus *Monascus* sp. is rich in pigments and it is used as a food coloring agent (Dewapriya and Kim 2014). Some organisms produce melanin pigments to protect their DNA from UV rays, while biofilm protects from protease activity etc. Melanin is a chromophoric compound and it is used in the manufacturing of dyes, sunscreens, colorings, etc. A study on red pigment extracted from coral reefs is also capable of showing cytotoxicity effects against various cell lines (Karuppiah et al. 2013; Abraham and Chauhan 2018).

4.19 Biosurfactants

Several microbes exhibit bioactive compounds ,which have amphipathic properties, surface and emulsifying activity, and are classified as biosurfactants. Biosurfactants have a hydrophilic and hydrophobic region. It has major applications in petrochemical, food, cosmetic industries, etc. The marine bacteria *Alcaligenes* sp. Produces glucose lipids that contain four β-hydroxydecanoic acids linked by ester bonds. It is used in oil-polluted marine environments. In a study conducted by Mukherjee and Shivapathasekaran (Mukherjee et al. 2009), biosurfactant production by *Bacillus circulans* strain showed enhanced surface and antimicrobial activity. One of the common characteristics of biosurfactants is to relax or decrease surface tension, which increases solubility so that biosurfactants can interact on the interfaces and affect the adhesion and detachment of bacteria and all these properties together make the biosurfactant confer against antibacterial, antifungal, and antiviral activities (Floris et al. 2018; Satpute et al. 2010).

References

Abraham J, Chauhan R (2018) Profiling of red pigment produced by Streptomyces sp. JAR6 and its bioactivity. 3 Biotech 8(1):1–9

Alvariño R, Alonso E, Lacret R, Oves-Costales D, Genilloud O, Reyes F, Botana LM (2018) Streptocyclinones A and B ameliorate Alzheimer's disease pathological processes in vitro. Neuropharmacology 141:283–295

Alves C, Silva J, Pinteus S, Gaspar H, Alpoim MC, Botana LM, Pedrosa R (2018) From marine origin to therapeutics: the antitumor potential of marine algae-derived compounds. Front Pharmacol 9:777

Alves A, Sousa E, Kijjoa A, Pinto M (2020) Marine-derived compounds with potential use as cosmeceuticals and nutricosmetics. Molecules 25(11):2536

Andryukov B, Mikhailov V, Besednova N (2019) The biotechnological potential of secondary metabolites from marine bacteria. J Mar Sci Eng 7(6):176

Augustine J, Suresh A, Abraham J (2021) Harnessing the use of cyanobacteria for sustainable agricultural farming practices. In: Gorawala P, Mandhatri S (eds) Agricultural research update. Nova Publisher, Hauppauge, pp 89–119

Bajpai VK (2016) Antimicrobial bioactive compounds from marine algae: a mini review. Indian J Mar Sci 45(9):1076–1085

Behera BK, Varma A (2017) Upstream processes. In: Microbial biomass process technologies and management. Springer, Cham, pp 45–107

Beygmoradi A, Homaei A (2017) Marine microbes as a valuable resource for brand new industrial biocatalysts. Biocatal Agric Biotechnol 11:131–152

Bhalamurugan GL, Valerie O, Mark L (2018) Valuable bioproducts obtained from microalgal biomass and their commercial applications: a review. Environ Eng Res 23:229–241

Bhatnagar I, Kim SK (2010) Immense essence of excellence: marine microbial bioactive compounds. Mar Drugs 8(10):2673–2701

Bolchakova E, Novikov A, Slater M, Huang F, Hartnett J, & Velikodvorskaya G (2004) U.S. patent application no. 10/706,229

Brmadsky G, Rosenberg E (1992) Mmbioi Emi 24:63–76

Cesário MT, da Fonseca MMR, Marques MM, de Almeida MCM (2018) Marine algal carbohydrates as carbon sources for the production of biochemicals and biomaterials. Biotechnol Adv 36(3):798–817

Chaugule SR, Indap MM, Chiplunkar SV (2017) Marine natural products: new avenue in treatment of osteoporosis. Front Mar Sci 4:384

Clarke KG (2013) Bioprocess engineering: an introductory engineering and life science approach. Elsevier, Amsterdam

De Carvalho CC, Fernandes P (2010) Production of metabolites as bacterial responses to the marine environment. Mar Drugs 8(3):705–727

Debbab A, Aly AH, Lin WH, Proksch P (2010) Bioactive compounds from marine bacteria and fungi. Microb Biotechnol 3(5):544–563

Deniz I, García-Vaquero M, Imamoglu E (2017) Trends in red biotechnology: microalgae for pharmaceutical applications. In: Microalgae-based biofuels and bioproducts, 1st edn. Woodhead Publishing, Cambridge, MA, pp 429–460. ISBN 978-0-08-101023-5

Dewapriya P, Kim SK (2014) Marine microorganisms: an emerging avenue in modern nutraceuticals and functional foods. Food Res Int 56:115–125

Estrada-Badillo C, Márquez-Rocha FJ (2003) Effect of agitation rate on biomass and protease production by a marine bacterium Vibrio harveyi cultured in a fermentor. World J Microbiol Biotechnol 19(2):129–133

Floris R, Rizzo C, Giudice AL (2018) Biosurfactants from marine microorganisms. In: Metabolomics-New Insights into Biology and Medicine. IntechOpen, London

Gelfand DH, Lawyer FC, Stoffel S (1996) Mutated thermostable nucleic acid polymerase enzyme from thermotoga maritima. Biotechnol Adv 14(3):340–340

George B, Varathan P, Suchithra TV (2020) Meta-analysis on big data of bioactive compounds from mangrove ecosystem to treat neurodegenerative disease. Scientometrics 122(3): 1539–1561

Guezennec J (2002) Deep-sea hydrothermal vents: a new source of innovative bacterial exopolysaccharides of biotechnological interest? J Ind Microbiol Biotechnol 29(4):204–208

Hannan M, Dash R, Sohag AAM, Moon IS (2019) Deciphering molecular mechanism of the neuropharmacological action of fucosterol through integrated system pharmacology and in silico analysis. Mar Drugs 17(11):639

Hannan M, Dash R, Haque M, Mohibbullah M, Sohag AAM, Rahman M et al (2020) Neuroprotective potentials of marine algae and their bioactive metabolites: pharmacological insights and therapeutic advances. Mar Drugs 18(7):347

Irvine Robert L, Rajagopalan Venkatadri (1994) Method of producing extracellular products from aerobic microorganisms. U.S. Patent No. 5,342,765

Islam MT, da Silva CB, de Alencar MVOB, Paz MFCJ, Almeida FRDC, Melo-Cavalcante AADC (2016) Diterpenes: advances in neurobiological drug research. Phytother Res 30(6):915–928

Jeewon R, Luckhun AB, Bhoyroo V, Sadeer NB, Mahomoodally FM, Rampadarath S, Puchooa D, Sarma V, Kumar Durairajan SS, Hyde KD (2019) Pharmaceutical potential of marine fungal endophytes. In: Jha S (ed) Endophytes and secondary metabolites. Reference series in phytochemistry. Springer, Cham, pp 1–23

Jha D, Jain V, Sharma B, Kant A, Garlapati VK (2017) Microalgae-based pharmaceuticals and nutraceuticals: an emerging field with immense market potential. Chem Bio Eng Rev 4:257–272

Jin L, Quan C, Hou X, Fan S (2016) Potential pharmacological resources: natural bioactive compounds from marine-derived fungi. Mar Drugs 14(4):76

Joint I, Mühling M, Querellou J (2010) Culturing marine bacteria–an essential prerequisite for biodiscovery. Microb Biotechnol 3(5):564–575

Karawita R, Senevirathne M, Athukorala Y et al (2007) Protective effect of enzymatic extracts from microalgae against DNA damage induced by H2O2. Mar Biotechnol 9:479–490

Karuppiah V, Aarthi C, Sivakumar K, Kannan L (2013) Statistical optimization and anticancer activity of a red pigment isolated from Streptomyces sp. PM4. Asian Pac J Trop Biomed 3(8): 650–656

Kathiresan K, Nabeel MA, Manivannan S (2008) Bioprospecting of marine organisms for novel bioactive compounds. Sci Trans Env Technovation 1:107–120

Kim SK (2014) Marine cosmeceuticals. J Cosmet Dermatol 13(1):56–67

Lang S, Hüners M, Lurtz V (2005) Bioprocess engineering data on the cultivation of marine prokaryotes and fungi. In: Ulber R, Le Gal Y (eds) Mar Biotechnol II. Springer, Berlin, pp 29–62

Lee HK, Chun J, Moon EY, Ko SH, Lee DS, Lee HS, Bae KS (2001) Hahella chejuensis gen. Nov., sp. nov., an extracellular-polysaccharide-producing marine bacterium. Int J Syst Evol Microbiol 51(2):661–666

Li H, Chi Z, Duan X, Wang L, Sheng J, Wu L (2007a) Glucoamylase production by the marine yeast Aureobasidium pullulans N13d and hydrolysis of potato starch granules by the enzyme. Process Biochem 42(3):462–465

Li H, Chi Z, Wang X, Ma C (2007b) Amylase production by the marine yeast Aureobasidium pullulans N13d. J Ocean Univ China 6(1):60–65

Loaec M, Olier R, Guezennec J (1998) Chelating properties of bacterial exopolysaccharides from deep-sea hydrothermal vents. Carbohydr Polym 35(1–2):65–70

Mancuso Nichols CA, Garon S, Bowman JP, Raguénès G, Guezennec J (2004) Production of exopolysaccharides by Antarctic marine bacterial isolates. J Appl Microbiol 96(5):1057–1066

Marwick JD, Wright PC, Burgess JG (1999) Bioprocess intensification for production of novel marine bacterial antibiotics through bioreactor operation and design. Mar Biotechnol 1(5): 495–507

Mathews I. (2010) Marine based ingredients. http://wwwspecialchem4cosmeticscom/

Maugeri TL, Gugliandolo C, Caccamo D, Panico A, Lama L, Gambacorta A, Nicolaus B (2002) A halophilic thermotolerant bacillus isolated from a marine hot spring able to produce a new exopolysaccharide. Biotechnol Lett 24(7):515–519
Mesbah NM, Wiegel J (2014) Purification and biochemical characterization of halophilic, alkalithermophilic protease AbCP from Alkalibacillus sp. NM-Fa4. J Mol Catal B Enzym 105:74–81
Mukherjee A, Raichur AM, Modak JM, Natarajan KA (2004) Bioprocessing of polymetallic Indian Ocean nodules using a marine isolate. Hydrometallurgy 73(3–4):205–213
Mukherjee S, Das P, Sivapathasekaran C, Sen R (2009) Antimicrobial biosurfactants from marine Bacillus circulans: extracellular synthesis and purification. Lett Appl Microbiol 48(3):281–288
Nair S, Abraham J (2020) Natural products from actinobacteria for drug discovery. Advances in pharmaceutical biotechnology recent progress and future applications. Springer, Singapore, pp 333–364
Nicolaus B, Lama L, Panico A, Moriello VS, Romano I, Gambacorta A (2002) Production and characterization of exopolysaccharides excreted by thermophilic bacteria from shallow, marine hydrothermal vents of Flegrean Ares (Italy). Syst Appl Microbiol 25(3):319–325
Nicoletti R, Vinale F (2018) Bioactive compounds from marine-derived aspergillus, penicillium, talaromyces and trichoderma species. Mar Drugs 16(11):408. https://doi.org/10.3390/md16110408
Nikolaivits E, Dimarogona M, Fokialakis N, Topakas E (2017) Marine-derived biocatalysts: importance, accessing, and application in aromatic pollutant bioremediation. Front Microbiol 8:265
Orejuela-Escobar L, Gualle A, Ochoa-Herrera V, Philippidis GP (2021) Prospects of microalgae for biomaterial production and environmental applications at biorefineries. Sustainability 13(6): 3063
Osinga R, Tramper J, Wijffels RH (1999) Marine bioprocess engineering: from ocean to industry. Aquat Ecosyst Health Manag 17(8):303–304
Pavão MS (2014) Glycosaminoglycans analogs from marine invertebrates: structure, biological effects, and potential as new therapeutics. Front Cell Infect Microbiol 4:123
Poli A, Anzelmo G, Nicolaus B (2010) Bacterial exopolysaccharides from extreme marine habitats: production, characterization and biological activities. Mar Drugs 8(6):1779–1802
Rangarajan V, Dhanarajan G, Kumar R, Sen R, Mandal M (2012) Time-dependent dosing of Fe2+ for improved lipopeptide production by marine bacillus megaterium. J Chem Technol Biotechnol 87(12):1661–1669
Romano G, Costantini M, Sansone C, Lauritano C, Ruocco N, Ianora A (2017) Marine microorganisms as a promising and sustainable source of bioactive molecules. Mar Environ Res 128: 58–69
Rorrer GL (2015) Bioprocess engineering of phototrophic marine organisms. In: Kim S-K (ed) Springer handbook of marine biotechnology. Springer, Berlin, Heidelberg, pp 257–294
Rosenfeld WD, ZoBell CE (1947) Antibiotic production by marine microorganisms. J Bacteriol 54(3):393–398
Rotter A, Barbier M, Bertoni F, Bones AM, Cancela ML, Carlsson J et al (2021) The essentials of marine biotechnology. Front Mar Sci 8:158
Rougeaux H, Guezennec J, Carlson RW, Kervarec N, Pichon R, Talaga P (1999) Structural determination of the exopolysaccharide of Pseudoalteromonas strain HYD 721 isolated from a deep-sea hydrothermal vent. Carbohydr Res 315(3–4):273–285
Ruperez P (2001) Antioxidant activity of sulphated polysaccharides from the Spanish marine seaweed nori. In: Proceedings of the COST 916 European conference on bioactive compounds in plant foods. Health effects and perspectives for the food industry. European Cooperation in the Field of Scientific and Technical Research, Tenerife, Canary Islands, p 114
Santorelli M, Maurelli L, Pocsfalvi G, Fiume I, Squillaci G, La Cara F et al (2016) Isolation and characterisation of a novel alpha-amylase from the extreme haloarchaeon Haloterrigena turkmenica. Int J Biol Macromol 92:174–184

Santos JD, Vitorino I, Reyes F, Vicente F, Lage OM (2020) From ocean to medicine: pharmaceutical applications of metabolites from marine bacteria. Antibiotics 9(8):455

Sarkar S, Pramanik A, Mitra A, Mukherjee J (2010) Bioprocessing data for the production of marine enzymes. Mar Drugs 8(4):1323–1372

Satpute SK, Banat IM, Dhakephalkar PK, Banpurkar AG, Chopade BA (2010) Biosurfactants, bioemulsifiers and exopolysaccharides from marine microorganisms. Biotechnol Adv 28(4): 436–450

Sayer C, Bommer M, Isupov M, Ward J, Littlechild J (2012) Crystal structure and substrate specificity of the thermophilic serine: pyruvate aminotransferase from Sulfolobus solfataricus. Acta Crystallogr D Biol Crystallogr 68(7):763–772

Sekar A, Kim K (2020) Production of industrial important enzymes from marine isolates. In: Kim S-K (ed) Encyclopedia of marine biotechnology, vol 4. Wiley, Hoboken, pp 2323–2330

Ser HL, Tan LTH, Palanisamy UD, Abd Malek SN, Yin WF, Chan KG et al (2016) Streptomyces antioxidans sp. nov., a novel mangrove soil actinobacterium with antioxidative and neuroprotective potentials. Front Microbiol 7:899

Silber J, Kramer A, Labes A, Tasdemir D (2016) From discovery to production: biotechnology of marine fungi for the production of new antibiotics. Mar Drugs 14(7):137

Simon-Colin C, Raguénès G, Cozien J, Guezennec JG (2008) Halomonas profundus sp. nov., a new PHA-producing bacterium isolated from a deep-sea hydrothermal vent shrimp. J Appl Microbiol 104(5):1425–1432

Song Q, Wang Y, Yin C, Zhang XH (2016) LaaA, a novel high-active alkalophilic alpha-amylase from deep-sea bacterium Luteimonas abyssi XH031T. Enzym Microb Technol 90:83–92

Tang DYY, Khoo KS, Chew KW, Tao Y, Ho SH, Show PL (2020) Potential utilization of bioproducts from microalgae for the quality enhancement of natural products. Bioresour Technol 304:122997

Teixeira TR, Santos GSD, Armstrong L, Colepicolo P, Debonsi HM (2019) Antitumor potential of seaweed derived-endophytic fungi. Antibiotics 8(4):205

Torregrosa-Crespo J, Montero Z, Fuentes JL, Reig García-Galbis M, Garbayo I, Vílchez C, Martínez-Espinosa RM (2018) Exploring the valuable carotenoids for the large-scale production by marine microorganisms. Mar Drugs 16(6):203

Trincone A (2011) Marine biocatalysts: enzymatic features and applications. Mar Drugs 9(4): 478–499

Vartoukian SR, Palmer RM, Wade WG (2010) Strategies for culture of 'unculturable' bacteria. FEMS Microbiol Lett 309(1):1–7

Watanabe M, Suzuki Y, Sasaki K, Nakashimada Y, Nishio N (1999) Flocculating property of extracellular polymeric substance derived from a marine photosynthetic bacterium, Rhodovulum sp. J Biosci Bioeng 87(5):625–629

Weiner R (1997) Biopolymers from marine prokaryotes. Trends Biotechnol 15(10):390–394. https://doi.org/10.1016/s0167-7799(97)01099-8

Yanagihara M, Sasaki-Takahashi N, Sugahara T, Yamamoto S, Shinomi M, Yamashita I, Hayashida M, Yamanoha B, Numata A, Yamori T et al (2005) Leptosins isolated from marine fungus Leptoshaeria species inhibit DNA topoisomerases I and/or II and induce apoptosis by inactivation of Akt/protein kinase B. Cancer Sci 96:816–824

Zaborsky OR (1999) Marine bioprocess engineering: the missing link commercialization. In: Progress in industrial microbiology, vol 35. Elsevier, Amsterdam, pp 403–408

Zhang XH, Ahmad W, Zhu XY, Chen J, Austin B (2021) Viable but nonculturable bacteria and their resuscitation: implications for cultivating uncultured marine microorganisms. Mar Life Sci Technol 3:189–203

Chapter 5
Application of Cutting-Edge Molecular Biotechnological Tools in Microbial Bioprocessing

Madhumita Priyadarsini, Kailash Pati Pandey, Jeetesh Kushwaha, and Abhishek S. Dhoble

Abstract Microbes are cosmopolitan in nature, and due to their molecular simplicity, they can be modified by using different biotechnological tools to produce bio-economically significant compounds in the field of pharmaceutics, agriculture, wastewater treatment, biofuels, and so on, cost-effectively. Using a wild variety of microbes for the industrial production of bioactive compounds is economically not viable. Therefore, molecular biotechnological techniques are being used to engineer these "wild strains" of microorganisms to make microbial bioprocessing economically and industrially sustainable. The methods such as molecular cloning, gene delivery, and genome editing are used to make microbial processing more efficient. In the recent era, recombinant DNA technology is one of the most advanced and widely used methods, through which the targeted gene can be inserted and expressed inside a particular organism by using a vector. Among all other techniques, CRISPR is the most advanced technique widely used for genome editing because of its high specificity and rapid modification of DNA at the genomic level. Thus, with the help of biotechnological methods, the process of microbial bioprocessing can be made more productive. This chapter describes the brief history of biotechnological development while emphasizing the cutting-edge biomolecular techniques and their applications in microbial bioprocessing.

Keywords Microbial bioprocessing · Molecular biotechnological tools · Recombinant DNA technology · Genetic engineering · CRISPR

M. Priyadarsini · K. P. Pandey · J. Kushwaha · A. S. Dhoble (✉)
School of Biochemical Engineering, Indian Institute of Technology (BHU), Varanasi, India
e-mail: asdhoble.bce@iitbhu.ac.in

A. Sarkar, I. A. Ahmed (eds.), *Microbial products for future industrialization*, Interdisciplinary Biotechnological Advances,
https://doi.org/10.1007/978-981-99-1737-2_5

5.1 Introduction

Microorganisms are being used in human civilization since ancient times. In ancient times microbes were used to produce fermented foods such as beer, yogurt, pickles, bread, etc., even before the discovery of microbes. The first evidence of the production of beer through Fermentations by yeast is around 7000 B.C. in Babylonia. Egyptians produced leavened bread in 4000 B.C. With the recent advancements in microbiology and biotechnology, the area in which microbes are used has expanded. Using microbes for the production of the desired product is called microbial bioprocessing. Though these products can also be produced by other means such as chemical processing, microbial bioprocessing is preferred over other means because of convenience and economic viability.

It all began in the seventeenth century when Antonie Van Leeuwenhoek invented the microscope and noticed the microorganisms for the first time. This discovery opened a whole new era of microbiology, which eventually led to the foundation of microbial bioprocessing. The process of microbial bioprocessing speeded up in the nineteenth century after the discovery of enzymes. In the late nineteenth century, an experiment was performed by Edward Buchner in which he used yeast cell lysate to produce ethanol from sucrose. At the start of the twentieth century, microbial production of amino acids, organic acids, and vitamins started. Microbial bioprocessing has provided us with products that have changed our lives and life spans completely. These products include industrial, agricultural products, food, and pharmaceutical products.

5.1.1 The Emergence of Molecular Biotechnological Tools in Microbial Bioprocessing

The process of microbial bioprocessing was not always as efficient as it looks today. In most of the cases, the yield was relatively low, making the whole operation economically unviable. Before the emergence of molecular biotechnological tools in microbial bioprocessing, wild variety (strains of microbes found naturally) was used for product formation. Still, they have a poor growth rate and low production are genetically unstable and are susceptible to product inhibition. New mutant strains are being developed with the help of molecular biotechnological techniques; they are genetically stable, usually have a higher product conversion rate, higher tolerance to product inhibition, and the capability of using a wide variety of substrates.

Moreover, many microbial products, such as antibiotics, are secondary metabolites, which are generally produced by microbes in stress conditions and that too in minimal quantities. Thus, genetic manipulation of these microbes is essential for the optimum yield of the desired product. The drastic difference in yield from wild and mutant strains can be understood through the example of Penicillin production by *Penicillium spp.* Industrial strain *P. chrysogenum P2* produced by UV and X-ray mutagenesis produces 85 times more penicillin than the original penicillium strain (Ziemons et al. BMC Biotechnology (2017)).

Similarly, removing undesirable properties can also be done through strain improvement, e.g., removing antibiotic resistance from *Lactobacillus fermentum* (Derkx et al. Microbial Cell Factories 2014). In addition to increasing yield and stability or removing specific undesirable properties through modification of existing genes, new genes can also be inserted in suitable microbes to produce desirable compounds, e.g., the production of insulin through *Escherichia coli* (Biochemistry: Goeddel et al. 1979).

Multiple ways can be used for strain development. Mutagens are used for strain development, and these mutagens can be in the form of radiation such as UV-rays, X-rays, or chemicals like acridine orange, ethidium bromide, etc. Protoplast fusion is another method used for the purpose of strain development. With the recent advancement in the field of biotechnology, new methods have been developed through which genes can be added, removed, or modified with specificity, and even expression of the desired gene can be monitored. These methods include rDNA technology, MALDI-TOF, CRISPR, RT-PCR, RP-HPLC, etc.

5.1.2 *The Historical Aspect of Molecular Biotechnological Techniques and Some Important Events*

With the development of several molecular biotechnological tools, drastic changes came in this field. This journey is from using yeast "unknowingly" to the editing gene with high specificity. Some important events and techniques developed are listed in Table 5.1.

5.2 Molecular Biotechnological Tools

5.2.1 *Molecular Techniques: Recombinant DNA Technology*

Recombinant DNA Technology includes techniques that allow the implementation of classical gene cloning and recombinant methods for identifying and determining genes, their functions, and their further manipulation. In this set of experiments, the gene is first identified, then the DNA containing the gene of interest is extracted and amplified. Further, it is ligated to a suitable vector and transferred to an appropriate host. In the host, it will integrate into the host genome and express itself. In this way, one can obtain a recombinant and better yielding microorganism. We require specific tools for this set of experiments, such as restriction endonucleases, vectors, host microorganisms, etc.

Table 5.1 Timeline of molecular biotechnological developments in microbial bioprocessing

Year	Advancement
1911	Development of microinjection technique by Marshall A. Barber
1931	Development of electrophoresis by Arne Tiselius
1952	Construction of first continuous cancer cell line HeLa by George Otto Gey
1953	Discovery of a double-stranded helix of DNA by Watson and Crick
1958	Synthesis of DNA in a test tube by Arthur Kornberg
1960	Foundation of Bioinformatics
1962	Green fluorescent protein was discovered in jellyfish *Aequorea victoria by* Osamu Shimomura
1963	Vaccine for measles by Samuel Katz and John F Enders
1964	The technique of polyacrylamide gel electrophoresis developed
1970	Discovery of restriction enzyme Synthesis of the first artificial gene by Har Govind Khorana
1973	First successful rDNA experiment by Stanley Cohen and Herbert Boyer Development of southern blotting by sir Edwin Mellor Southern Development of protoplast fusion by Keller and Melcher
1975	Hybridoma technology developed by Georges Kohler and Cesar Milstein Discovery of viral vectors by J. Michael Bishop and Harold Varmus
1976	Expression of yeast gene in *E. coli*
1977	Development of northern blotting by James Alwine, David Kemp and George Stank
1978	Production of human insulin by *E. coli*
1979	Development of western blotting
1980	Development for MALDI-TOF
1983	Discovery of polymerase chain reaction (PCR) by Kary Mullis
1984	Discovery of DNA fingerprinting technique
1987	Discovery of yeast artificial chromosome (YAC)
1996	The first mammal "Dolly" cloned
1998	Discovery of antisense RNA by Andrew fire and Craig Mello
2011	Development of TALEN
2012	Role of CRISPR-Cas9 in gene-editing

5.2.1.1 Restriction Enzymes

These are the set of enzymes that cut the DNA. Usually, it cuts the DNA at specific sites. These sites are present anywhere in the DNA sequence, and they can also cut the DNA "in-between." So, these are also regarded as restriction endonucleases. Restriction enzymes are naturally produced by some microorganisms (e.g., bacteria) for their defense against different viruses. In the late 1960s, restriction endonucleases were first discovered in *E. coli*, which led to remarkable changes in this field (Gupta et al. 2016).

The nomenclature of restriction endonucleases contains four parts generally. The first capital alphabet denotes the name of the genus, and the small letters represent the species name. After that, the following letter indicates the strain of the microorganism. Lastly, the roman number signifies the order of identification of the enzyme in the microorganisms, e.g., *EcoRI,* where *Eco* stands for *Escherichia coli, R* denotes the strain of *E. coli*, i.e., *RY13*, and the last *I* stand for the order of identification (Oyewusi et al. 2021).

Restriction sites are very specific for each restriction enzyme. Generally, restriction enzymes recognize four, six, or eight palindromic base-pair sequences. Among all these, 4 and 6 base pair sequences are most common, as they might be found after every 44 or 46 (i.e., 256 or 4096) base-pairs, respectively, in a sequence. Some of the examples are listed in the Table 5.2. Certain restriction enzymes cut at eight base-pair sequences, such as *NotI.* Its recognition sequence is GCGGCCGC. These sequences occur once in 48, i.e., 65 kbp approximately, which generates huge DNA fragments. As the recognition sites are not located frequently and generate much larger fragments than others, researchers generally avoid it being used in rDNA techniques. Thus, they are regarded as "rare cutters."

The endonucleases cut their specific sites and generate either cohesive or blunt ends. Then the blunt end, the cohesive end, is mostly used as it has flanking regions, which are helpful during ligation. Restriction enzymes can be categorized into isoschizomers and neoschizomers. Isoschizomers are enzymes that recognize and cut at identical sites. For example, *SphI* and *BbuI*, whose recognition site is 5′ CGTAC↓G 3′. Whereas those enzymes that recognize identical sequences but cut at different locations are called neoschizomers, e.g., *SmaI* and *XmaI* both recognize CCCGGG but cut it differently, as shown in Table 5.2.

5.2.1.2 Vectors

After isolating the gene of interest (GOI), its transfer into the host will be necessary to produce our desired product. Certain transporting vehicles are needed for the transfer and maintenance of our gene of interest. These transporting vehicles are termed vectors. An exemplary vector should contain the origin of replication, which gives the ability to replicate inside the host. It should have multiple cloning sites where the gene of interest will be inserted. Selectable markers are also necessary for the selection of recombinant vectors after insertion of GOI. Apart from all this, a vector should be of low molecular weight. The low molecular weight helps in easy handling and causes minor damage due to shearing. They might present in multiple numbers, which facilitate their easy isolation.

Vector can be of two types: cloning vector and expression vector. Cloning vectors are generally used to transfer GOI into the host, whereas expression vectors are used for the expression of the targeted gene of interest. At first, it will be inserted into the host genome, and then it will be transcribed to produce the RNA, and further, the RNA will be translated to produce the desired protein. The cloning vector contains the origin of replication sequence, drug resistance marker sequence, multiple cloning sites, etc. The expression vector includes regulators, promoters, etc., along with the gene sequences of cloning vectors.

Table 5.2 Some frequently used restriction endonucleases and their restriction digestion sites

Restriction enzymes	Recognition sites	End generated (Blunt or sticky)
AluI	AG↓CT TC↑GA	Blunt
TaqI	T↓CGA AGC↑T	Sticky
HaeIII	GG↓CC CC↑GG	Blunt
Sau3AI	↓GATC CTAG↑	Sticky
MboI	↓GATC CTAG↑	Sticky
HpaII	C↓CGG GGC↑C	Sticky
EcoRI	G↓AATTC CTTAA↑G	Sticky
BamHI	G↓GATCC CCTAG↑G	Sticky
PvuI	CGAT↓CG GC↑TAGC	Sticky
HindII	GTpy↓puAC CApu↑pyTG	Blunt
HindIII	A↓AGCTT TTCGA↑A	Sticky
ClaI	AT↓CGAT TAGC↑TA	Sticky
BglII	A↓GATCT TCTAG↑A	Sticky
KpnI	GGTAC↓C C↑CATGG	Sticky
PstI	CTGCA↓G G↑ACGTC	Sticky
SalI	G↓TCGAC CAGCT↑G	Sticky
SmaI	CCC↓GGG GGG↑CCC	Blunt
XmaI	C↓CCGGG GGGCC↑C	Sticky
NotI	GC↓GGCCGC CGCCGG↑CG	Sticky
SbfI	CCTGCA↓GG GG↑ACGTCC	Sticky

According to the size of our gene of interest, we can choose a suitable vector for its transfer into the host. It can be plasmids, phagemids, cosmids, λ phage, P1 phage, BACs, YACs, etc. (Gupta et al. 2016).

5.2.1.2.1 Host Selection

It is crucial to select the host organism that will allow insertion of the vector, integration of the gene of interest to its own genome, and also allow its expression. *E. coli* is the common host that is generally used. In some cases, other microorganisms like yeast and bacteria can also be used.

5.2.1.2.2 RDT Procedure

At first, the GOI is identified. Then it is extracted using various DNA extraction methods. PCR will be done to amplify the desired DNA. A suitable vector is selected (Khan et al. 2016). Both the gene of interest and the vector will go through restriction digestion using the same restriction endonucleases. The desired gene is inserted into the vector. The host is then transformed with a recombinant vector. The gene of interest will get integrated into the host genome, and further, it will produce our desired product (Fig. 5.1). Thus, we can get the product from our altered and edited genome.

5.2.1.2.3 Application

Recombinant DNA technology is used to make the modified strain of the organism, which produces more products than the wild strain. The alternation of genes and their product helps us a lot to produce many enzymes, hormones, vitamins, and other valuable products in adequate quantities.

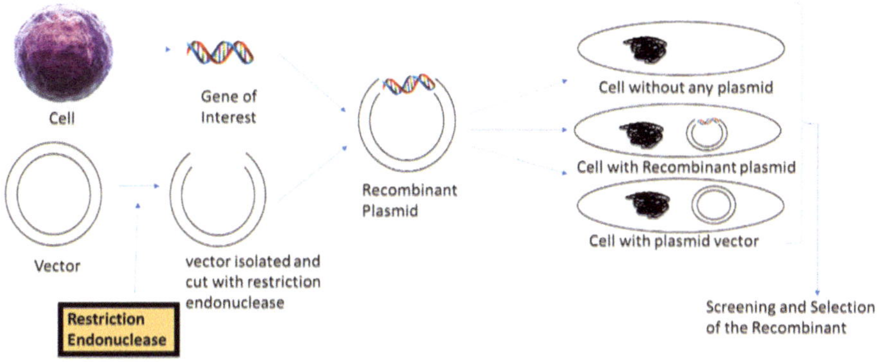

Fig. 5.1 Steps of recombinant DNA technology

5.2.2 Dynamic Single-Cell Analysis Techniques: Flow Cytometry

The term "cyto" means cells, and "metry" means measurement. It is a technique in which we measure the different properties of cells when they flow sequentially through the flow cytometer device (Álvarez-Barrientos et al. 2000). We can detect the cells' shape, size, granulation, and internal composition using this technique. The total number of cells and their diversity can also be diagnosed. Optics-based diagnosis is made by flow cytometer, in which cells are counted and sorted using a laser beam. Scattering plays a significant role here. The forward and side scattering help to detect the cellular configuration, different granulations, and heterogeneity of the experimental sample (Koch et al. 2014). Within seconds a flow cytometer can interrogate more than a thousand cells for many diverse parameters. Different cellular mechanisms such as cell division, cell death, necrosis, apoptosis, etc., can also be diagnosed. Considering its real-time monitoring while maintaining the viability of the cell, it can be regarded as a more effective tool than general microscopy.

The key steps to perform flow cytometry includes the labeling of our sample. Labeling can be done using fluorescent dyes (such as DAPI, SYBR Green) (Guo et al. 2020). Then the samples first go through the fluidic system, which focuses the cells in a single file before their exposure to the laser beam. It is also called as hydrodynamic focusing of cells. The cells are then exposed to the laser beam and forward (narrow-angle) and side scattering (90° angle) of the laser beam observed (Fig. 5.2). The detection system then records these optical responses. Then the optical signal is transformed into an electrical signal by the photodiodes. Subsequently, the signal is amplified by amplifiers. From the analog signal, the digital signal is generated by using Analog-to-Digital Converter (ADC). The generated data

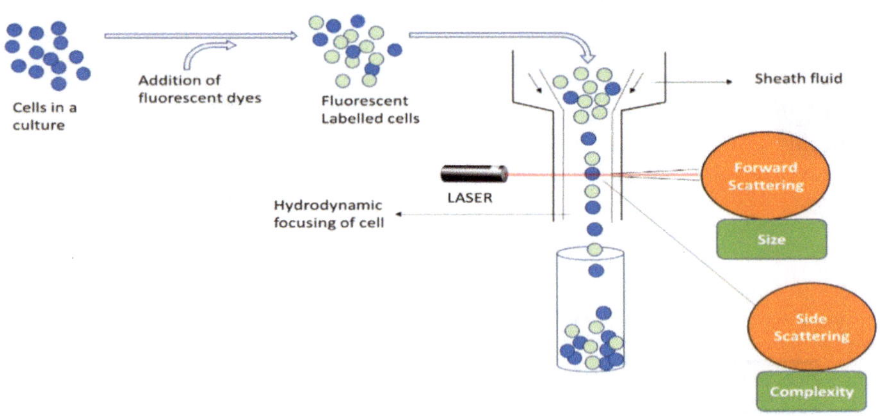

Fig. 5.2 Showing forward and side scattering through flow cytometry

is then processed by the computer software system, which gives the final output of the flow cytometry.

Flow cytometry is used chiefly due to its significant applications in the different fields of immunology (immunophenotyping, measurement of antigen-specific responses, intracellular cytokinin analysis, etc.), cell biology (cell cycle analysis, cellular function analysis, signal transduction pathway, etc.), molecular biology (fluorescent protein analysis, DNA analysis, RNA expression analysis, etc.), cancer biology (cell proliferation analysis, apoptotic analysis), microbiology(for the detection of a wide range of microbes like bacteria, virus, fungus (Cao et al. 2021), etc.), etc.

5.2.3 *Other Techniques*

5.2.3.1 Fish

Fluorescence in situ hybridization is a macromolecular cytogenic identification technique based on sequence complementarity of either DNA or DNA-RNA hybrids (Cui et al. 2016). In this technique, probes are used, which are fluorescent labeled and bind to specific fluorophore coupled parts of the nucleic acid. To examine particular binding, fluorescence microscopy techniques are used. Thus, a specific chromosome can be checked for the presence or absence of a specific sequence of nucleotides. It can be used for genetic diagnostic of various diseases. In the biomedical field, it plays a significant role as it is used for detection of different tumor/cancerous cells, pre-detection of various inherited diseases at the fetal level, identification of the diverse pathogen, detection of the evolutionary relationship between two species, etc. (In et al. 2008).

5.2.3.2 DNA Microarray

It is a genetic detection technique in which gene expression can be detected. In other words, it is well known as a biochip or DNA chip. The main principle behind this technique is the complementarity of the sequence of nucleic acids. The whole procedure of microarray can be divided into mainly two parts. One is probe designing and cDNA production. At first, the probe is designed and attached to a surface (generally glass, silicon surface) where a complete set of the experiment is carried out. Then the DNA is isolated and fragmented using restriction endonucleases. The DNA cut pieces and their complementary DNA sequences are allowed to bind to the DNA chip probe (Heller 2002). The washing procedure is followed to remove the nonspecific DNA fragments. Thus, only highly complementary DNA hybridization results. The fluorescent detection techniques can identify the DNA or specific gene sequence by exposing the sample to the laser beam, and the fluorescence emission is estimated using computer software. This computer software later shows the final result of DNA detection/gene expression based on fluorescence emission data. This emerging technology is popular because of its high-speed, specific detection of particular genotypes. It is widely used in the detection of mutations, structural and genetic variations. In the diagnosis of cancer (Russo et al. 2003), tumors, tumor suppressor genes, cancer biomarkers, pre-detection of cancer (Govindarajan et al. 2012), etc., microarray plays a vital role. Nowadays, it is modified and evolved to analyze cellular and proteomic samples.

5.2.3.3 16s rRNA Sequencing Technique

It is a robust technique for the identification of the evolutionary relationship between bacterial species. It is mainly a gene sequencing method that relies on the 16s rRNA gene, which is one of the most conserved gene sequences. It also contains some variable gene sequences, which is due to random changes over time. Thus, this technique is used worldwide by researchers to identify bacterial species.

In this technique, the bacteria species are first grown, and their DNA is extracted using different isolation methods. After the isolation of desired DNA, it is amplified using PCR techniques. The amplicons are then subjected to agarose gel electrophoresis, which differentiates them according to their respective lengths. The DNA fragment having the required appropriate dimensions are selected and put through cycle sequencing. Sequencing of both the stands should be performed to get more accurate results. The sequencing product is again electrophoresed for purification purposes. The sequencing data is then analyzed using computer software and compared to known databases such as GenBank. Likewise, phylogenetic, taxonomical identification is done using this technique. Although the conventional microbiology techniques can be used to identify bacterial species (Oyewusi et al. 2021), that might consume more time as some bacteria take more time to produce viable colonies (Patel 2001). For those microorganisms, it is the best technique.

5.2.3.4 Polymerase Chain Reaction

It is one of the nucleic acid-based genotype amplification methods that allow us to either amplify certain regions of the genome or the entire genome. This revolutionary technology helps us to get multiple copies of the desired sequences in less time. So, it has a significant role in molecular biology and different diagnostic techniques (Guest et al. 2017).

PCR comprises three basic steps. The first step is to denature the double-stranded DNA using high temperatures. After denaturation, the temperature is lowered, and primers are allowed to anneal to the single-stranded DNA. Extension steps followed the annealing procedure by which the primers extend with the help of the polymerase using the dNTPs and yield the double-stranded DNA (Kalendar et al. 2017). In this way, the amplification is carried out. The extension temperature is a little bit higher than the annealing temperature. It is done to check the nonspecific binding of primers to the DNA strand. Generally, the temperature for denaturation, annealing, and extension is calculated according to the composition of nucleotides in the DNA strands. The melting temperature (Tm) is calculated for certain nucleotide sequences according to the numbers of purines and pyrimidines present in the DNA strand. In general, the denaturation temperature is set at 92–95 °C, which breaks all the hydrogen bonds between the DNA fragments. The temperature is then lowered to 55–65 °C (Oyewusi et al. 2021) so that the primer will anneal to the DNA strands. The annealing temperature varies from the melting temperature by 3–5°.

Two types of primers are used basically. One is the forward primer, and the other is the reverse primer. For elongation, the temperature is increased to 72–75 °C, in which the primers extend in a $5' \rightarrow 3'$ direction and form the complete DNA strand. The DNA polymerase used in PCR is Taq DNA polymerase, extracted from thermophile *Thermus aquaticus* that can stand high temperatures. In this way, two stands are formed from a single strand. The PCR is run for 35 cycles, and nearly a billion copies can be created (Fig. 5.3). When Kary Mullis invented PCR in 1983 (Shampo and Kyle 2002), the temperature was maintained using different water baths, but nowadays, due to advancements in technology, automated PCR machines are available in which the total temperature cycle can be programmed before the onset PCR reaction.

According to the necessity of mankind, PCR can be of many types, such as nested PCR, RT-PCR, real-time PCR, multiplex PCR, quantitative real-time PCR, hot-start PCR, etc. (Adzitey et al. 2013). RT-PCR stands for Reverse Transcriptase- Polymerase Chain Reaction. In this type of PCR, using reverse transcriptase the RNA strands are first transformed into cDNA and then amplified using PCR machinery. In the recent Covid-19 outbreak, the RT-PCR is broadly used to detect the presence of the Coronavirus (Han et al. 2021; Zhang et al. 2021).

5.2.3.5 DGGE

Denaturing Gradient Gel Electrophoresis denoted by DGGE. It is a type of PCR-based genetic fingerprinting technique. As the name suggests in this technique, at first denaturation step is performed. This technique generally uses chemical-based denaturation. Certain polyacrylamide gels are prepared, which contain the denaturing agents in a linear gradient. DGGE technique utilizes the stability of GC-rich nucleotide sequences over AT-rich sequences (Hori et al. 2006). According to the composition of the different DNA sequences, they melt at different concentrations of denaturing agents on the gel and form different bands on the gel matrix. According to the band formation, the samples can be analyzed. It is a useful technique for detecting Single Nucleotide Polymorphisms (SNPs), point mutations at a particular gene (Muyzer and Smalla 1998; Strathdee and Free 2013). Thus, it can be utilized for the detection of many diseases. It also has importance in microbial ecology (Muyzer 1999), as using these molecular fingerprinting techniques, microbial diversity can be interrogated in a complex mixed population of microorganisms. Clones can also be screened using DGGE.

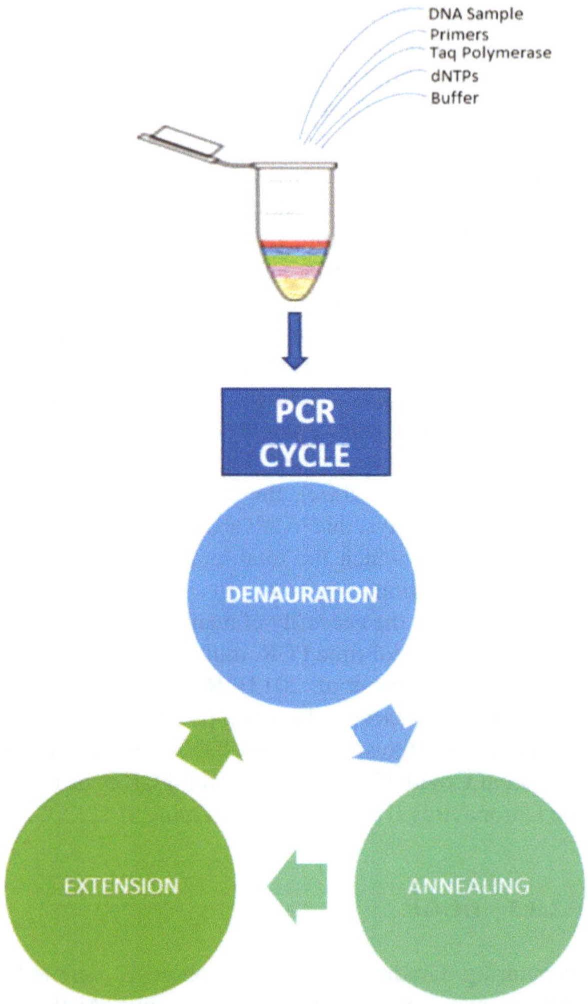

Fig. 5.3 Steps of polymerase chain reaction

5.2.3.6 TGGE

TGGE stands for Temperature Gradient Gel Electrophoresis. Like DGGE, it also uses DNA denaturation methods before fingerprinting but uses a different temperature gradient (Muyzer and Smalla 1998). So, TGGE can be considered as the refined form of DGGE. The main working principle behind the technique is the reduction in electrophoretic mobility of the dsDNA pieces due to denaturation. This technique exploits the temperature sensitivity of nucleic acids. The gel is provided with a temperature gradient; as the nucleic acid sample run on that gel, they melt according to their nucleotide composition and form bands. Unlike DGGE, heteroduplex analysis can be done using this technique (Muyzer 1999). It is independent of the size of the nucleic acid but depends upon its composition. It is also used to determine the point mutations, screen clones, and determine the microbial diversity in a population (Muyzer and Smalla 1998).

5.2.3.7 SSCP

Single-strand conformational polymorphism (SSCP) analysis is a PCR-based electrophoresis technique widely used in genotyping and detection of mutation. Using this technique, DNA polymorphism can also be studied. The total procedure includes four basic steps. First, the desired DNA is isolated and amplified using PCR (Dong and Zhu 2005). Single-stranded DNA is formed by denaturing the double-stranded DNA. Using low temperature, the single-stranded DNA is allowed to form specific three-dimensional folding. Then electrophoresis is performed, and the difference in mobility of the two single strands is observed. The two ssDNA migrate differently according to their sequences. Hence, the difference in their sequence can be deciphered. Wild-type and mutant strains of microorganisms can be differentiated using this technique.

5.2.3.8 RISA

Ribosomal intergenic spacer analysis (RISA) is an rRNA-based genetic analysis. In this technique, the intergenic spacer region analysis is done between different ribosomal RNA genes. At first, primers are produced using the conserved sequences of rRNA, and PCR amplification is carried out (Fisher and Triplett 1999). The complex product of PCR amplification is separated using the gel electrophoresis technique. Like DGGE, the polyacrylamide gel is stained with suitable dye and then observed under a UV transilluminator. The resulting banding patterns are analyzed to detect the genotyping. According to the difference in the bacterial population, the banding pattern also differs from species to species. So, RISA techniques can be used to detect the taxonomy and variation between different microbial colonies. It is a simple, robust, and cost-effective technique.

5.2.3.9 RFLP

In the Restriction Fragment Length Polymorphism (RFLP) technique, a vital role is played by the restriction enzymes. After the isolation of DNA, it is digested using different restriction endonucleases. The DNA contains specific recognition sites where the restriction endonucleases will cut and generate fragments of different lengths (Mittal et al. 2011). Electrophoresis follows restriction digestion, which visualizes the variation in homologous DNA fragments (Fig. 5.4). Different DNA strands can be distinguished according to their length using this technique (Sibley et al. 2012). Due to mutation, the restriction sites might get altered, and the restriction enzyme will not recognize it. So, the number of fragments generated by

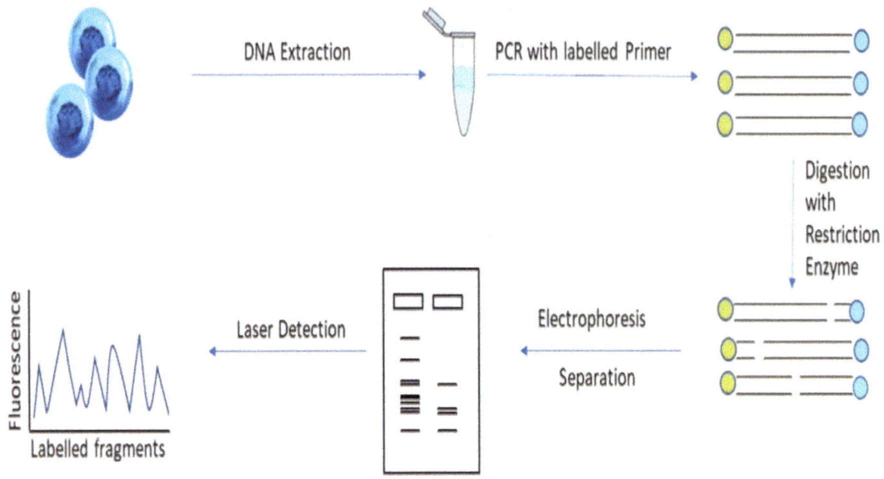

Fig. 5.4 Representation of RFLP

such mutated polymorphic DNA will vary from the wild type. Hence, it is considered a valuable technique to study the mutation or any variation in the DNA sequences (Bombach et al. 2011).

5.2.3.10 GCMS

Gas Chromatography-Mass Spectroscopy is an analytical technique that is a fusion of chromatography with spectroscopy. The basic methodology includes extraction of the particular sample, its hydrolysis followed by derivatization. In recent years, the diagnosis consists of a selected-ion-monitoring method (SIM). This technique is convenient for detecting and quantifying trace elements and the small amounts of products produced in a bioreactor. Volatile and semi-volatile substances such as alcohols, metabolic steroids (Krone et al. 2010), hormones, etc., can be analyzed using GCMS.

5.2.3.11 LCMS

LCMS is the abbreviation of Liquid Chromatography-Mass Spectroscopy. Non-volatile, polar samples can be quantified using this analytical technique. This technique amalgamates the capabilities of Mass Spectrometry and Liquid Chromatography. The identification of components of complex compounds in proteomics and metabolomics can be done using LCMS. It can also be used for structural identification, lipidomic profiling, bio affinity detection, metabolite screening, impurity diagnosis, toxicology testing in Drug discovery (Lynch 2017), and molecular biology experiments.

5.3 The Measure Breakthrough in These Techniques

5.3.1 CRISPR-CAS Technique

Recently, different programmable editing molecular tools have evolved like TALEN (Transcription activator-like effector nucleases) and zinc finger nuclease, which have remarkably changed genome modification efficiency, but they have some limitations. Atsuo Nakata and colleagues discovered a region in the *E. coli* K12 strain in 1987 that included five identical 29-nucleotide sequences spaced by 32 nucleotides downstream of the *iap* gene. The remarkable characteristic of repeating spacers and direct repeats gave attention to CRISPRs (Das and Paudel 2021). CRISPR (clustered regularly interspaced short palindromic repeats)/Cas9 shows important upgradation over a few years. It allows site-specific DNA genome editing. This technique is based on prokaryotic microorganisms' defense mechanism from viral predators; Cas9 nuclease is guided by non-coding RNA to cause site-specific DNA cleavage (Das and Paudel 2021). The NHEJ (non-homologous-end joining) or HDR (homologous-directed repair) pathways are used by cellular DNA repair mechanisms to engage in the repairing process. In Molecular Research, CRISPR is becoming a robust tool for molecular genome editing of DNA in mammalian cells. The Cas9 enzyme had a revolutionary impact in different fields of science, like medicine, agriculture, and biotechnology.

5.3.1.1 CRISPR/Cas9 Structure

This nuclease system is made up of two parts: a Cas enzyme that cuts the target sequences and a single guide RNA (sgRNA) that binds to the 20-base pair target sequence (bp). The complementary target sequence is bound by the sgRNA, which is followed by two cytosine nucleotides (Jiang and Doudna 2017). sgRNA prefers to bind to the opposing DNA strand, followed by nucleotides containing two guanines (NGG). The sequence is known as a PAM (Protospacer Adjacent Motif) sequence. The sequence differs based on Cas9's origin.

5.3.1.2 Principle

The crRNA sequence unique to the target DNA and the tracrRNA sequence that interacts with the Cas9 enzyme makes up a single guide RNA (sgRNA). It interacts with the Cas9 recombinant protein, which exhibits DNA endonuclease activity. Then, the cleavage of the target-specific double-strand DNA takes place in the resulting complex. Non-homologous end-joining is used to repair this cleavage. The DNA repair process is a high-risk approach that can result in insertions/deletions (INDELs) or cause the gene's function to be disrupted.

5.3.1.3 Mechanism of Action

CRISPR/Cas9 induces DNA repair mechanisms by breaking a particular double-strand at the targeted locus. Constitutive Knockouts (through insertion/deletion) by non-homologous end-joining and Knockin (via HDR) (Jiang and Doudna 2017) through homologous recombination are the two types of genome modification employed for the repair (Fig. 5.5).

5.3.1.4 CRISPR/Cas9 Knockout and Knockin

Cas9 is a nuclease that targets the specific genomic sequence guided by a guide RNA molecule, where part of guide RNA contains the homologous sequences to the targeted genomic sequence. After a while, guide RNA finds out the target sequence that matches it and if that sequence is downstream of a PAM sequence. Then, the Cas9 nuclease cleaves both strands of the targeted spot of the genome (Das and Paudel 2021). In response to cleavage, cells activate their repairing mechanism using a non-homologous end-joining repair (NHEJ) pathway. As a result, base pairs are inserted or deleted, resulting in indel or gene knockouts. Their applications include modifying promoter sequence or gene, inserting a reporter gene, and creating a clinically relevant SNP for a disease model. It is observed that Knockin efficiency is lower than knockout efficiency. Moreover, knockin at a single copy of the gene/sequence is generally followed by a knockout/indel of the second copy. The Cas9/sgRNA complex uses the non-homologous end-joining (NHEJ) pathway to cleave both copies, which is then repaired imprecisely, resulting in an indel.

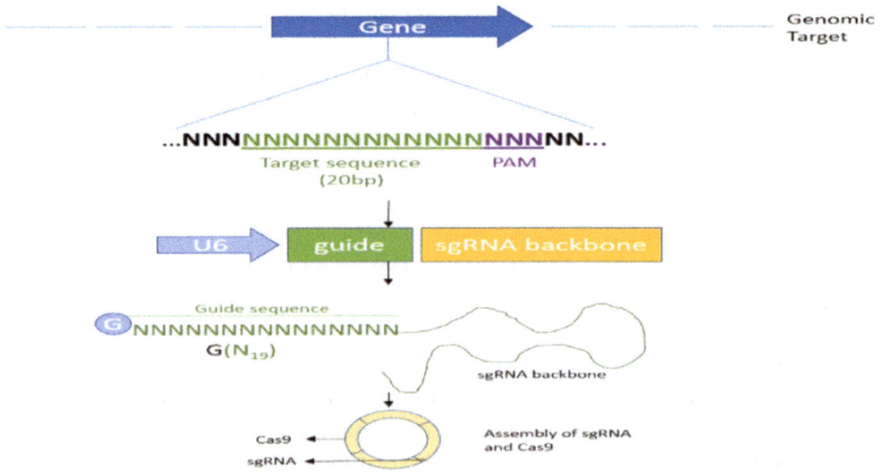

Fig. 5.5 CRISER Cas9 mechanism

5.3.2 *TALEN: Transcription Activator-Like Effector Nuclease Technique*

Working with artificial restriction enzymes created by coupling the TAL effector DNA-binding domain with the DNA cleavage domain is the basis of the Talen approach. Restriction enzymes cleave a certain sequence of DNA strands (Bedell et al. 2012). It can be genetically modified to bind practically any DNA sequence. Transcription activator-like effectors (TALEs) could conveniently attach any useful sequence of DNA. In the field of molecular genome editing, it's becoming popular (Gaj et al. 2013).

During plant infection, *Xanthomonas* bacteria release the TAL effectors proteins. In the DNA-binding domain, 33–34 amino acid sequences are almost always conserved, which include different 12th and 13th amino acids. RVD (Repeat Variable Di Residue) refers to these two sites, which are extremely variable with a strong link with recognition of particular nucleotide. This allows alteration of precise DNA-binding domains by determining the correct RVD combination.

The nonspecific DNA cleavage domain of the End of *Fokl* endonuclease is exploited to produce hybrid nucleases that are functional in various cell types (Gaj et al. 2013). *Fokl* domain is a dimer that requires two constructs with perfect DNA-binding domains that are oriented and spaced correctly for specific genomic sites. The count of bases between two different TALEN-binding sites and the presence of amino acids in the center of TALE DNA-binding domain and the *Fokl* cleavage domain demonstrate crucial characteristics for obtaining numerous activities.

The interaction between TALE-binding domain and amino acid sequence permits perfect modifications in proteins. Following the assembly of TALEN constructs, they are inserted into plasmids and transfected into chosen cells after the expression of gene products that penetrate the nucleus to access the genome. Aside from that, TALEN constructions might be carried to cells as mRNAs, eliminating the possibility of genomic interaction with TALEN expressing the protein (Bedell et al. 2012). During gene editing, the chances of introgression and the level of homology-directed repair (HDR) via mRNA vector increase significantly.

Double-strand breaks (DSB) are introduced into genomes using TALEN technology, which causes cells to engage in their repair mechanisms. Non-homologous end-joining is used to reassemble the sides of double-strand breaks (NHEJ). Because of that repair mechanism occurrence of many errors takes place in the genome via chromosomal rearrangement, insertion, or deletion; such types of errors make the chances of rendering the product of genes coded at a particular location nonfunctional (Bedell et al. 2012). Because the activity of these enzymes varies depending on the species, target gene, cell type, and nuclease used, developing new systems should be done under the supervision of experts. However, in the presence of double-stranded DNA fragments, NHEJ could be used to induce DNA into a genome. The foreign DNA is then inserted at the DSB through a homology-directed repair pathway, with the repair enzymes being templates from transfected double-stranded sequences.

5.3.3 CASFISH Technique

Cas9-mediated fluorescence in situ hybridization (CASFISH) is a technology that uses CRISPR-associated caspase 9 (Cas9) complexes as probes to fluorescently mark sequence-specific genomic loci without causing global DNA denaturation (Dassi 2016). The approach of using in vitro constituted nuclease deficient clusters consistently in interspaced short palindromic repeats comes from the Researchers of Howard Hughes Medical Institute, University of California (Berkeley), and Albert Einstein College of Medicine.

This approach demonstrates the labeling of repetitive DNA elements in centromeres, coding gene loci, G-rich telomeres, and pericentromeric DNA elements powerfully and quickly. Because of the interaction of dCas9 with an array of sgRNAs that surface arbitrary target loci, it can visualize non-repetitive genomic sequences. Simultaneous probing of multiple targets takes place with the help of a stable dCas9/sgRNA binary complex, which binds with high affinity to its target DNA. Different colored dCas9/sgRNA complexes are used in this approach, resulting in multicolor labeling of target loci in cells. Under ideal conditions, the CASFISH assay is effective in identifying primary tissue sections. It's a quick, powerful, cost-effective, and trouble-free method that's proven to be useful in basic research and genomic-based diagnosis.

5.3.4 SAGE Technique

Serial analysis of gene expression (SAGE) is a technique to quantify gene expression levels. It's a process that involves isolating specific sequence tags (9–10 bp in length) from specific messenger RNAs and applying a uniform sequence of tags to lengthy DNA molecules for complete sequencing. This technique is utilized in research to investigate nearly any form of molecular event that occurs as a result of cellular transcription changes, and it offers the complete gene expression profile of a specific cell or tissue. SAGE is highly efficient at precise cellular conditions to detect the set of particular genes and estimate it with the profiles developed for a set of cells stored in varying conditions.

5.3.5 MALDI-TOF

MALDI-TOF is an analytical technique that ionizes the sample into charged molecules and converts them into the measurable ratio of mass-to-charge (m/z). The ionization source in this spectroscopy is matrix-assisted laser desorption/ionization, and the mass analyzer is Time-Of-Flight (TOF). It's a soft ionization, which participates in laser exposure to the matrix of small sized molecules to convert them into analyte molecules in the gaseous phase. Large biomolecules get converted into fragments by heating and other traditional techniques like peptides, lipids, organic macromolecules, and saccharides (Fig. 5.6).

During the flight, different ions m/z are dispersed in time over a field-free drift path of known length. The lighter ones will arrive at the detector first, followed by the heavier ones. MALDI-TOF is accurate, fast, and cost-effective for clinical Microbiology; it can be used to identify different microorganisms (viruses, fungi, and bacteria), biochemical characteristics of bacterial and fungal species.

5.4 Significance of Molecular Biotechnological Tools in Microbial Bioprocessing

Before the development of biotechnological tools, microbial bioprocessing was not effective, and yield was low in quantity. It was limited to the production of the compounds that microbes produce naturally. With the help of biotechnological tools now, the yield has been increased, but the incorporation of new genes can also produce compounds that microbes do not produce in nature. With the advancement in biotechnology, microbial bioprocessing has evolved significantly and, and its applications have extended.

5.4.1 Application in the Pharmaceutical Sector

Biotechnology plays an essential role in producing pharmaceutical compounds, as most of them are produced either in a low quantity or not produced by microbes naturally. Antibiotics are immensely important since they inhibit microbial development and are essential for a person suffering from a bacterial illness. These compounds are produced by microbes in the stress conditions only and that too in minimal quantities. Mutant strains are being created with the help of microbial biotechnological tools, which have a significantly higher yield in comparison to wild type, e.g., *Penicillium chrysogenum* (mutant strain) produces around 56 times higher yield than the wild type strain.

Production of insulin by *E. coli* is only possible because of biotechnological tools. Gene responsible for insulin production is incorporated in E. coli, and the produced

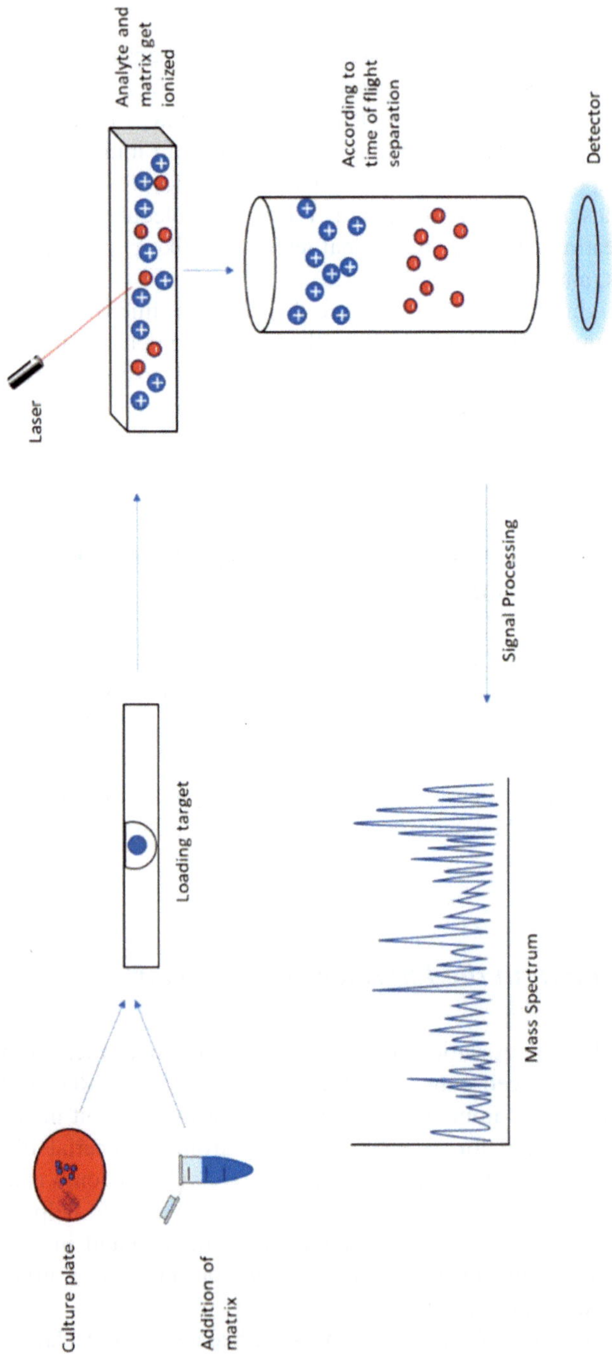

Fig. 5.6 Schematic representation of MALDI-TOF

insulin is isolated and purified. Similarly, many amino acids, hormones, antibiotics, and multiple other compounds of pharmaceutical importance are produced.

5.4.2 Application in Food Production and Processing

Biotechnology has evolved the food processing sector by improving the microorganisms for yield improvement, process control, and efficiency, as well as the safety, quality, and consistency of bio-processed food products. During fermentation, the growth of microbes and their metabolism result in the production of a diverse range of metabolites. Texture-forming substances like xanthan gum, amino acids, organic acids, and flavor components like aldehydes are among these metabolites. These metabolites are processed through downstream processing for purification and isolation.

Microorganisms are also used for the production of alcoholic beverages. Cereal grains such as barley, wheat, and rice are used to produce beer and ale. Similarly, wine is produced from grapes. These processes generally involve different species of *Saccharomyces*. In addition to their role in fermentation as agents of physical and biological change, microbes can directly be used as a food source. Many species of bacteria, yeasts, and some other fungi are being used as food sources. Mushrooms (*Pleurotus eryngii, Agaricus bisporus*) are important fungi used as a food source directly. Microorganisms can also be used to supplement other foods called single-cell protein (SCP), such as *Spirulina*. It is available in stores too. Moreover, microbes are also used as prebiotics. The most commonly used microbes as probiotics are *Lactobacillus* and *Bifidobacterium*.

5.4.3 Application in Waste Management/Role in Bio-Remediation

Microbes are an important part of our ecosystem as they degrade the organic matter and make it available for recycling in nature. With increasing urbanization, waste management is a major challenge throughout the world. Discharge of industrial and urban waste continuously and uncontrollably into the environment has become a global issue. Several toxins and other harmful compounds are released into the environment with untreated waste. Bioremediation does not only remove pollutants from the environment but is also an eco-friendly and more effective process. By the use of microorganisms, these pollutants can be removed from the environment. Different methods are being used for in situ and ex situ bioremediation of waste material.

The contamination of soil through heavy metals is one of the major environmental problems. Heavy metal contamination has been cleaned up using microorganisms such as bacteria, algae, and fungi. Cadmium (Cd) and lead (Pb) are being removed using mercury-resistant bacteria such as *Alcaligenes faecalis*, *Pseudomonas aeruginosa*, and *Brevibacterium iodinium* (Jaysankar De, N. Ramaiah & L. Vardanyan). *Pseudomonas putida* is used for the degradation of various petroleum products. These species have been genetically engineered to degrade petroleum (camphor, octane, xylene, and naphthalene) by plasmid transfer. Research is in progress to develop other species capable of degrading other environmental hazards (Piyush Pandey and Naveen Kumar Arora).

5.4.4 Application in Biofuel Production

Fossil fuel is limited and rapidly depleting with time; the depletion rate increases day by day. This has undoubtedly drawn the interest of researchers in other renewable sources of commercial fuel. Biotechnological methods would be extremely useful for manipulating microbes to economically create biofuel from lignocellulosic biomass via metabolic pathways. Bioconversion of lignocellulosic biomass feedstock to biofuel is becoming increasingly popular around the world.

5.5 Future Aspects

With the ever-increasing dataset from the above-mentioned cutting-edge biotechnological tools, there would be demand for data analytics and automation techniques to find their practical applications in microbial bioprocessing. For instance, a high-throughput technique like flow cytometry can analyze roughly 100,000 cells within a minute. From each such cell, various scattering signals could be collected. The current standard is 19 such parameters. In the future, this is projected to rise to 100 parameters – all from a single cell. The complexity of data (100,000 cells multiplied by 100 parameters from each cell) would require techniques like machine learning to make sense in practical applications.

Since microbial processes are at the heart of any bioprocess operations, next-generation control system would integrate online analysis of microbial signals and adaptive control strategies based on bioinformatic information. Tools like artificial intelligence (AI) might play a crucial role in this regard. Systems analysis and automation are another two disruptive technologies on the horizon. The sooner bioprocess professionals start adapting to these changing technologies and techniques the better it would be for this domain area moving forward.

5.6 Conclusion

Microbes are an integral part of our lives for centuries and they may continue to play a crucial role in industrial bioprocesses. In the last few decades, tremendous growth has been witnessed in molecular biology tools that need to be integrated into industry-scale bioprocess operations. This biotechnological development along with futuristic tools like machine learning and artificial intelligence has a huge potential to play in various applications related to microbial bioprocessing.

References

Adzitey F, Huda N, Rusul G, Ali R (2013) Molecular techniques for detecting and typing of bacteria, advantages and application to foodborne pathogens isolated from ducks. 3 Biotech 3: 97–107. https://doi.org/10.1007/s13205-012-0074-4

Álvarez-Barrientos A, Arroyo J, Cantón R, Nombela C, Sánchez-Pérez M (2000) Applications of flow cytometry to clinical microbiology. Clin Microbiol Rev 13(2):167–195. https://doi.org/10.1128/CMR.13.2.167-195.2000

Bedell VM et al (2012) In vivo genome editing using high efficieny TALEN system. Nature 491(7422):114–118. https://doi.org/10.1038/nature11537.In

Bombach P et al (2011) Resolution of natural microbial community dynamics by community fingerprinting, flow cytometry, and trend interpretation analysis. Adv Biochem Eng Biotechnol 124:151–181. https://doi.org/10.1007/10_2010_82

Cao R et al (2021) Evaluation of the vital viability and their application in fungal spores' disinfection with flow cytometry. Chemosphere 269:128700. https://doi.org/10.1016/j.chemosphere.2020.128700

Cui C, Shu W, Li P (2016) Fluorescence in situ hybridization: cell-based genetic diagnostic and research applications. Front Cell Dev Biol 4(September):1–11. https://doi.org/10.3389/fcell.2016.00089

Das BD, Paudel N (2021) A review on reliability and validity of CRISPR/Cas9 Technology for gene editing. Jordan J Biol Sci 14(3):503–511

Dassi E (2016) Preface. Methods Mol Biol 1358:v–viii. https://doi.org/10.1007/978-1-4939-3067-8

Dong Y, Zhu H (2005) SCP analysis, single-strand conformational polymorphism analysis. Methods Mol Med 108:149–157

Fisher MM, Triplett EW (1999) Automated approach for ribosomal intergenic spacer analysis of microbial diversity and its application to freshwater bacterial communities. Appl Environ Microbiol 65(10):4630–4636

Gaj T, Gersbach CA, Barbas CF (2013) ZFN, TALEN, and CRISPR/Cas-based methods for genome engineering. Trends Biotechnol 31(7):397–405. https://doi.org/10.1016/j.tibtech.2013.04.004

Govindarajan R, Duraiyan J, Kaliyappan K (2012) Microarray and its applications. J Pharm Bioallied Sci 4:310–313. https://doi.org/10.4103/0975-7406.100283

Guest S, Section E, Al ET, O. Of (2017) Identification of microorganisms by modern analytical techniques. J AOAC Int 1607(34):31–35. https://doi.org/10.5740/jaoacint.17-0207

Guo Y, Rosa LFM, Müller S, Harnisch F (2020) Monitoring stratification of anode biofilms in bioelectrochemical laminar flow reactors using flow cytometry. Environ Sci Ecotechnol 4: 100062. https://doi.org/10.1016/J.ESE.2020.100062

Gupta V, Sengupta M, Prakash J, Tripathy BC (2016) Basic and applied aspects of biotechnology. Basic and Applied Aspects of Biotechnology. Springer, Singapore. pp 1–520. https://doi.org/10.1007/978-981-10-0875-7

Han MS, Byun JH, Cho Y, Rim JH (2021) RT-PCR for SARS-CoV-2: quantitative versus qualitative. Lancet Infect Dis 21(2):165. https://doi.org/10.1016/S1473-3099(20)30424-2

Heller MJ (2002) DNA microarray technology : devices, systems, and applications introduction. Annu Rev Biomed Eng 4:129–153. https://doi.org/10.1146/annurev.bioeng.4.020702.153438

Hori T, Haruta S, Ueno Y, Ishii M, Igarashi Y (2006) Direct comparison of single-strand conformation polymorphism (SSCP) and denaturing gradient gel electrophoresis (DGGE) to characterize a microbial community on the basis of 16S rRNA gene fragments. J Microbiol Methods 66:165–169. https://doi.org/10.1016/j.mimet.2005.11.007

In F et al (2008) Hybridization Is Used to Localize DNA Sequences on Chromosomes Using FISH to Identify the Positions of Genes Using Collections of FISH Probes to " Paint " Entire Chromosomes, 1–5.

Kalendar R, Muterko A, Shamekova M (2017) In silico PCR tools for a fast primer, probe, and advanced searching. In: Domingues L (ed) PCR: methods and protocols, methods in molecular biology, vol 1620. Springer, New York

Jiang F, Doudna JA (2017) CRISPR—Cas9 structures and mechanisms. Annu Rev Biophys 46: 505–531

Khan S et al (2016) Role of recombinant DNA technology to improve life. Int J Genomics 2016: 2314–436X

Koch C, Harnisch F, Schröder U, Müller S (2014) Cytometric fingerprints: evaluation of new tools for analyzing microbial community dynamics. Front Microbiol 5:273. https://doi.org/10.3389/FMICB.2014.00273

Krone N, Hughes BA, Lavery GG, Stewart PM, Arlt W, Shackleton CHL (2010) Gas chromatography/mass spectrometry (GC/MS) remains a pre-eminent discovery tool in clinical steroid investigations even in the era of fast liquid chromatography tandem mass spectrometry (LC/MS/MS). J Steroid Biochem Mol Biol 121(3–5):496–504. https://doi.org/10.1016/j.jsbmb.2010.04.010

Lynch KL (2017) Toxicology: liquid chromatography mass spectrometry. Elsevier, Amsterdam, pp 109–130

B Mittal, P Chaturvedi, and S Tulsyan (2011) Restriction fragment length polymorphism. doi: https://doi.org/10.1016/B978-0-12-374984-0.01314-0

Muyzer G (1999) DGGE/TGGE a method natural ecosystems for identifying genes from. Curr Opin Microbiol 2:317–322

Muyzer G, Smalla K (1998) Application of denaturing gradient gel electrophoresis (DGGE) and temperature gradient gel electrophoresis (TGGE) in microbial ecology. Antonie Van Leeuwenhoek 73:127–141

Oyewusi HA et al (2021) Functional profiling of bacterial communities in Lake Tuz using 16S rRNA gene sequences gene sequences. Biotechnol Biotechnol Equip 35(1):1–10. https://doi.org/10.1080/13102818.2020.1840437

Patel JB (2001) 16S rRNA Gene Sequencing for Bacterial Pathogen Identification in the Clinical Laboratory. Mol Diagn 6(4):313–321

Russo G, Zegar C, Giordano A (2003) Advantages and limitations of microarray technology in human cancer. Oncogene 22:6497–6507. https://doi.org/10.1038/sj.onc.1206865

Shampo MA, Kyle RA (2002) Kary B. Mullis–Nobel Laureate for procedure to replicate DNA. Mayo Clin Proc 77(7):606. https://doi.org/10.4065/77.7.606

Sibley CD, Peirano G, Church DL (2012) Molecular methods for pathogen and microbial community detection and characterization: current and potential application in diagnostic microbiology. Infect Genet Evol 12(3):505–521. https://doi.org/10.1016/j.meegid.2012.01.011

Strathdee F, Free A (2013) Denaturing Gradient Gel Electrophoresis (DGGE). Methods Mol Biol 1054:145–157. https://doi.org/10.1007/978-1-62703-565-1

Zhang Z et al (2021) Insight into the practical performance of RT-PCR testing for SARS-CoV-2 using serological data: a cohort study. Lancet 2(2):e79–e87. https://doi.org/10.1016/S2666-5247(20)30200-7

Chapter 6
Engineering Strategies for the Biovalorization of Hemicellulosic Fraction into Value-Added Products: An Approach Toward Biorefinery Concept

Teena Chandna, Sai Susmita Gorantla, T. Chandukishore, R. Satish Babu, and Ashish A. Prabhu

Abstract Dwindling petroleum reservoirs and an exponential increase in chemical commodities demands have ushered the need for the development of a sustainable process. To mitigate the dependence of petroleum-based feedstock there is a paradigm shift toward biorefinery concept. Biorefinery allows us to recirculate the unusable end products into a new production cycle as substrate, thereby reducing the waste and at the same time producing value-added products out of it. Lignocellulosic Biomass (LCB) is an attractive feedstock for the production of value-added products due to its ample availability and cost-effective nature. However, most of the studies are focused on the utilization of cellulosic fraction of LCB, while the hemicellulosic fraction, which is the second most abundant material is discarded as waste. The foremost bottleneck is majority of the microbial cell factories lack pentose utilization pathway. However, for the development of efficient biorefinery, it is imperative that both cellulosic and hemicellulosic fraction of the LCB should be effectively utilized. In this chapter, we are showcasing the advantages of natural xylose utilizing microbes in the production of biochemicals and also we will review the metabolic engineering approach to improve the xylose metabolizing capability of microbial cell factories and to produce chemicals such as ethanol, 2,3 butanediol xylitol, succinic acid, itaconic acid, etc.

Keywords Biorefinery · Lignocellulosic biomass · Xylose · Metabolic engineering · Biochemicals

---

Teena Chandna, Sai Susmita Gorantla and T. Chandukishore contributed equally with all other contributors.

T. Chandna · S. S. Gorantla · T. Chandukishore · R. Satish Babu (✉) · A. A. Prabhu (✉)
Department of Biotechnology, National Institute of Technology Warangal, Warangal, India
e-mail: satishbabu@nitw.ac.in; ashishp@nitw.ac.in

A. Sarkar, I. A. Ahmed (eds.), *Microbial products for future industrialization*, Interdisciplinary Biotechnological Advances,
https://doi.org/10.1007/978-981-99-1737-2_6

6.1 Background

Over the few years, drastic increase in the population worldwide resulted in the exponential usage of petrochemicals-related products, which is ultimately leading to the depletion of fossil resources. There is an urgent need to search for a sustainable replacement for fossil fuels (Chintagunta et al. 2021). In recent years, much attention is focused on utilizing renewable resources derived from agro-industrial waste for the production of biochemicals and biofuels, which are economically viable and aid in improving bioeconomy. Among the renewable resources, lignocellulosic biomass (LCB) has been extensively studied due to its characteristics such as low cost, availability, and high carbon content. LCB is the major organic waste produced from agriculture and composed mainly of cellulose (40–50% (w/w)), hemicellulose (20–30%), and lignin (15–20%) (Banerjee et al. 2017). Though LCB is abundantly available, reducing it to various compounds through degradation and hydrolysis is a great challenge. Cellulose is the most abundant polymer on earth and is majorly composed of D-Glucose units linked by beta 1-4 glycosidic linkages (Banerjee et al. 2017; Himmel et al. 2007). There is a great scientific interest to find a sustainable pathway to break those high tensile branched chains into ready-to-use glucose units for the production of commercially important biofuels and biochemical. The second most important component of lignocellulosic biomass is hemicellulose. Hemicellulose is majorly composed of xylose, galactose, mannose, and arabinose, which are basically C5 and C6 sugars (Chandel et al. 2020; Cheng and Wang 2013). The properties of hemicellulose vary according to the biomass sources, carbohydrate percentage and branching pattern (Kumar et al. 2020; Pérez et al. 2002). Due to the complex composition of hemicellulose, exploiting it exclusively in biofuel and biochemical production is at the research and pilot stage only. Lignin, hemicellulose, and cellulose form the high tensile structure, which makes them highly resistant to various normal chemical and physical treatments (Banerjee et al. 2019a, b) (Fig. 6.1).

6.2 Classification of Lignocellulosic Biomass

To date most of the industries rely on petroleum-based crude resources for the manufacturing of chemicals, consequently there is paradigm shift from chemical to biological renewable sources such as LCB. In the context of the search for better organic sustainable replacement biomass is been divided into basically three types (Clauser et al. 2021).

First-generation Biomass: In this class biomass basically originates from starchy food products, edible seeds oil, and animal fats. Though they have the best energy and carbon replacement for fossil fuels, they are not economical because these products are also important raw materials for various other commercially important products and also leads to food versus feed debates (Clauser et al. 2021).

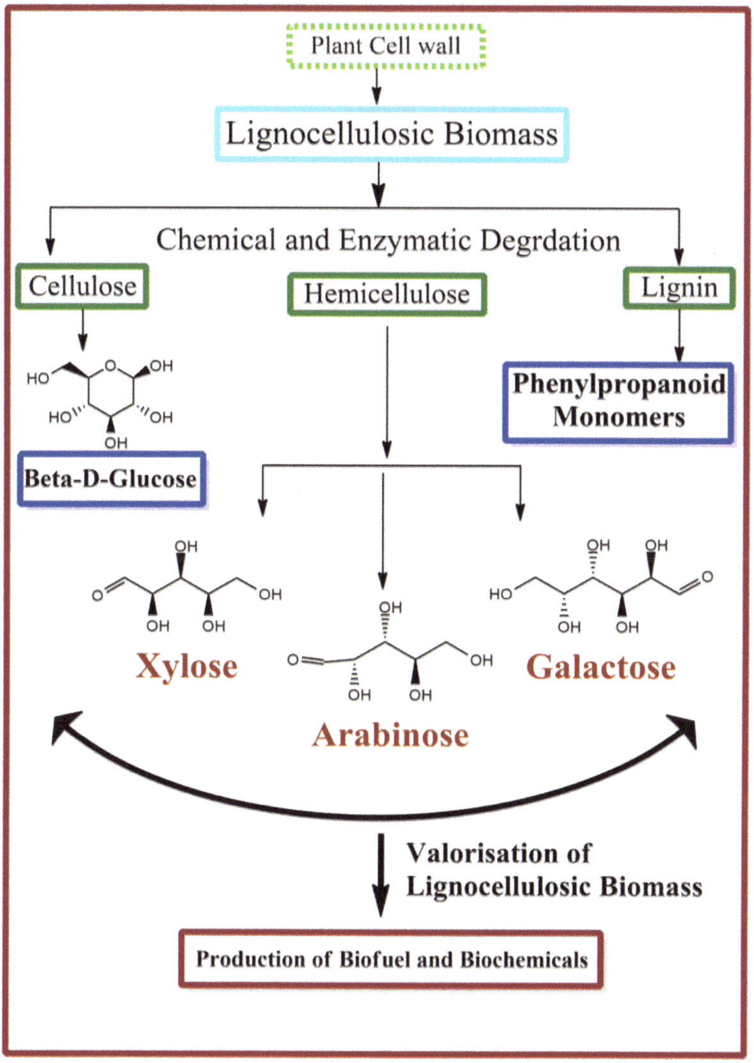

Fig. 6.1 Sugar content of LCB fractions utilized for the production of biochemicals and biofuels

Second-Generation Biomass: This class of biomass is the buzz in the present scenario; due to various properties such as abundance, they are being generated as waste by-products from numerous processes. Biomass in these types of generation includes lignocellulosic waste from agriculture, grass from energy crops, virgin wood, and forest residues. On incorporating various modern molecular biology techniques for strain improvement and applying it to fermentation technology, this class could be a good future biomass for biofuel production (Clauser et al. 2021).

Third-Generation Biomass: This class could be called as a combination of both first-generation and second-generation, which includes algae and seaweeds that are in various research stages brought into use as a replacement for petroleum products for biofuel production (Clauser et al. 2021).

6.3 Classification of Biorefneries Based on the Substrate and Product Formed

Phase 1: In this class of biorefinery generally they involve only single type of biomass and concentrates on only a single product, which is produced by a single process. The example includes the production of biofuel from starchy grains such as corn grains (Kamm and Kamm 2004; Lanigan et al. 2011).

Phase 2: Here though a single type of biomass is used multiple products are produced and multiple processes are involved. An example includes multiple products produced such as carbohydrate derivatives using starch-based products (Kamm and Kamm 2004; Lanigan et al. 2011).

Phase 3: This class of biorefinery is the most advanced type where it not only produce multiple products but also involves multiple types of biomasses and multiple processes. There is strong scientific research going on all over the world by the scientific fraternity to incorporate this lignocellulosic biomass in the phase 3 type of biorefinery (Kamm and Kamm 2004; Lanigan et al. 2011).

Phase 3 biorefinery is further classified into (a) Lignocellulosic-based biomass, (b) Green refinery, (c) Whole crop-based biomass, and (d) Two platform concept biomass. (Biorefinery 2019).

6.4 Status of Production of Bio-Refinery Using Lignocellulosic Waste

The advent of biorefinery has mitigated the dependence on fossil fuel-based chemical production and also helps to overcome adverse effects of Green House gases on environment (Brethauer and Studer 2015; Chandel et al. 2020). Some countries like Brazil depend on first-generation biomass for the production of biofuel; this could be only possible because of low-cost labor. The most important reason other than labor is readily available resources such as sugar and starchy-based biomass for which per liter bioethanol was costing around 0.2USD/L (Clauser et al. 2021). But the application of the same concept to the Sub-continental and populated Asian countries such as India is quite challenging. Most of the first-generation biomass is from edible food crops and which severely affects the food supply chain, consequently, the second- and third-generation biomass which are basically by-products of agricultural waste and algal waste are abundant in nature, utilizing this biomass will be

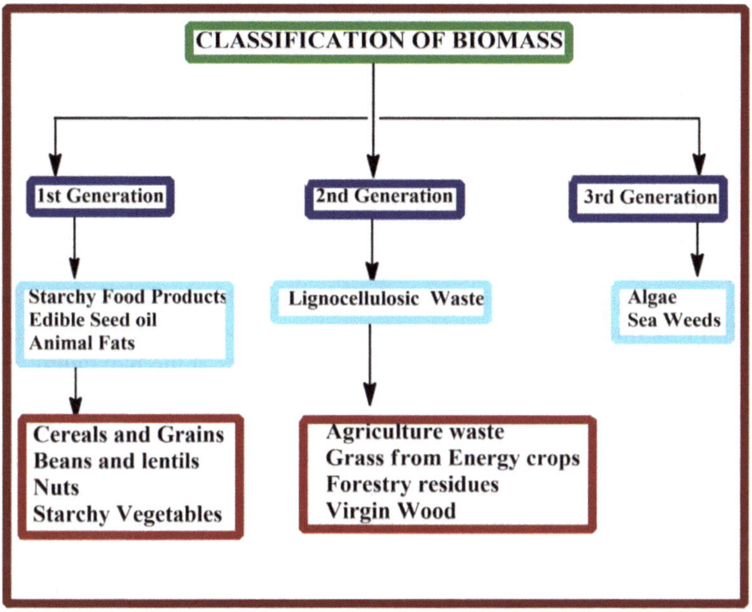

Fig. 6.2 Classification of biomass and their respective sources

sustainable and environmentally friendly (Cheng and Wang 2013). Figure 6.2 shows the schematic of the various value-added products produced from biomass.

6.5 Lignocellulosic Biomass Conversion to Biofuels and Biochemicals

Lignocellulosic biomass can be converted into biofuels and valued biochemical in two ways one is a biochemical process and the other is a thermochemical process. In the case of biochemical process fermentation of sugar-based biomass is involved, whereas in the case of thermochemical process gasification of biomass is done (Lanigan et al. 2011). The biochemical process basically consists of hydrolysis, extraction, transesterification, and fermentation. There is a need to find a solution for the so-called "biomass recalcitrance" problem posed by the composite and complex structure of lignocellulosic biomass (McCann and Carpita 2015). Lignocellulosic biomass when undergoing acid or enzymatic hydrolysis give rise to C6 and C5 sugars such as glucose, galactose, arabinose, xylose, and phenylpropanoid compound. Though there are numerous bacteria and fungi groups that produce cellulases and hemicellulases, using them in biofuel and biochemical production is not an economical choice due to their yield constraints and demand overall in the world. There is a need to search for microorganisms that can produce cellulosomes

(Brethauer and Studer 2015). Cellulosomes are composite enzymatic structures attached to the cell wall which is basically comprised of hydrolytic enzymes, such as endogluconases, cellobiohyrolases, and beta-glucossidases (Patel A and Shah A 2021). Lignin, which accounts for 15–20% of total composition of lignocellulosic biomass, when they undergo hydrolytic treatment and pyrolysis, produces various inhibitory compounds such as 5-Hyroxymethylfurfural, furfural, acetols, and phenolic compounds (Zeng et al. 2014). These compounds inhibit cellular replication of ethanologenic microorganisms, and in turn decrease the yield of rate-limiting products. In the case of the thermochemical process, lignocellulosic biomass undergoes combustion, pyrolysis, and gasification to majorly yield syngas or bio-oil (Shen et al. 2015). Both thermochemical and biochemical processes face technical, economic, and social challenges to use the lignocellulosic biomass at the maximum.

The above-mentioned various technical and economic challenges could be solved by the incorporation of engineered microbes in pretreatment and fermentation stages of conversion of lignocellulosic biomass to value-added chemicals. As already discussed above lignocellulosic biomass pretreatment is a challenge due to the complex nature posed by lignin and hemicellulosic structures. Finding microbes which can exclusively deconstruct the lignocellulosic biomass for bioethanol production and produce less inhibitory compounds such as furfural, 5-hydromethylfurfural and phenyl compounds is of great scientific interest. There are various studies which are under progress to understand the lignin composition and yield of lignin by controlling lignin biosynthesis genes. Lignin biosynthesis involves various enzymes and pathways, such as phenylalanine ammonia to cinnamyl alcohol. Downregulation of transcription factors, oxidative enzymes, or target gene control, modification of various interdependent pathways can reduce lignin production (Fu et al. 2011).

Apart from utilizing cellulosic components most of the microbe cannot swiftly utilize the C5 sugars present in LCB as they lack the metabolic pathway for pentose sugar utilization. In order to make LCB bio-economical in nature, the simultaneous usage of C5 and C6 sugars is inevitable. Over the last few years tremendous amount of research is being carried out to modify various cell factories using synthetic and systems biology approach in order to make them suitable for integrated biorefinery (Lugani et al. 2020; Prabhu et al. 2020) (Fig. 6.3).

6.6 Structural and Functional Properties of Hemicellulose: Possibilities of Using Hemicellulose in Bio-Refinery

An integrated biorefinery is a facility that uses a zero-waste strategy to convert fuels and high-value platform chemicals into carbohydrates, lignocellulosic biomass, and oils. Biorefineries involve dry and paper mills, the pulp that creates several products from biomass. Biorefineries, in comparison to petroleum refineries, use a wider range of feedstocks and processing processes (Isakov et al. 1990). During the last

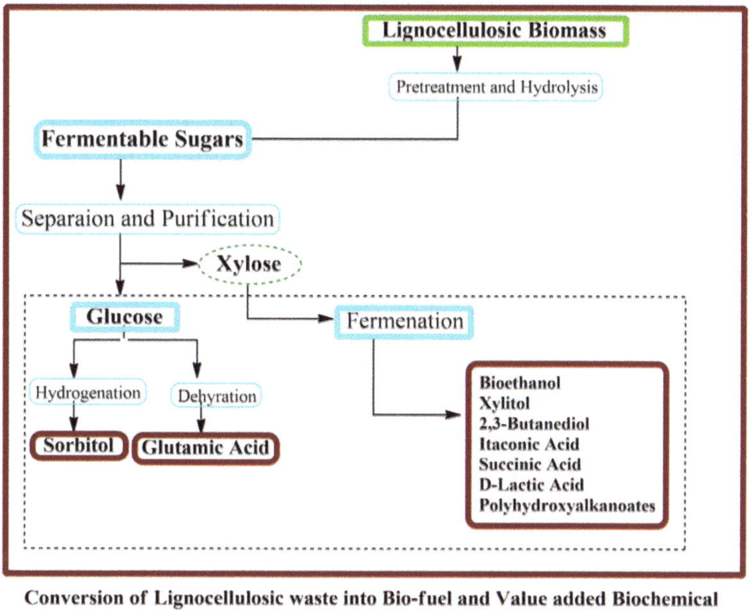

Fig. 6.3 Conversion of lignocellulosic waste into value-added biochemicals and biomass

decade, there has been a noticeable increase in the development of refineries using a renewable source, which is the keen research of interest for scientists and industrial enterprises. The main driving causes for these efforts are the scarcity of natural sources and the replacement of petroleum-formed products that are linked to demands for solutions to global environmental concerns. Because only a small proportion (typically cellulose or hemicellulose) of the annual harvest of plant products produced by land and forest are used, converting hemicelluloses to usable goods could be a key solution to the challenges stated above. Previously (Ebringerová 2005), research in the subject of hemicellulose was primarily focused on converting plant biomass into fuel like bioethanol, chemicals, and sugars in form of heat energy. Hemicelluloses, on the other hand, are appealing as biopolymers that can be used in a variety of applications, including food and non-food uses, in their original or modified forms. The purpose of this chapter is to provide an overview of hemicellulose structural and functional variety and the production of value-added chemicals.

6.7 Hemicellulose

Hemicellulose is a heterogeneous polysaccharide that contains xylans, mannans, arabinoxylans, galactans, glucomannans, β-(1→3,1→4)-glucans, arabinans, glucuronoxylans, and xyloglucans, thus it is hard to separate hemicellulose into its pure form of sugar monomer (Sista Kameshwar and Qin 2018). It has a symmetrical configuration with β-(1–4) linkage. It is made up of different sugar units and in variable proportions with substituents. Therefore, it can also be called as hexose and pentose branched polymers. Hemicelluloses are not water soluble; however, they can be made soluble by aqueous alkali. Chemical alteration can improve hemicellulose solubility. This characteristic is frequently used to distinguish between traditional gums and hemicelluloses. The Golgi membrane of plants has glycosyl transferases, which synthesize hemicellulose using nucleotide sugars. Different types of glycoside hydrolases are secreted by microorganisms for the breakdown of hemicellulose. Heteropolysaccharide cell walls are made up of hemicellulose; it has some similarities with polymers. It is nontoxic and biodegradable and easily available like any other polysaccharides, having high potential to be used as pharmaceutical polymers. Different species have different hemicellulose structural details and various cell types based on various plants. The most important role is to tether micro-fibrils, which strengthens the plant cell wall. Xyloglucans dominate in dicots and arabinoxylans in monocots. Food and feed are the main components of hemicellulose, which form the main part of the lignocellulosic biomass.

6.8 Various Sources of Hemicellulose

Olorunsola et al. (2018) explained that plant cell walls contain hemicelluloses in addition to cellulose and lignin. Hemicellulose is a branching polymer of hexose and pentose sugars, β-(1–4)-D-glucose polymer forms cellulose, pectin is a galacturonic acid polymer. Hemicellulose makes up 25–30% of the cell wall of the plant and is required for maintaining the cell wall's structure. The stiffness is provided by cellulose, which is embedded in the hemicellulose, and the entire system is held together by lignin. As a result, the plant cell wall binds these three polymers together. The plant's middle lamella is predominated by the hemicellulose. They are found in large amounts in leaves, seeds, kernels, reeds, and wood.

6.9 Different Classes of Hemicellulose Biopolymer

Hemicelluloses are typically classified into four structurally distinct types of polysaccharide; Figure 6.4 depicts the classification of hemicellulose biopolymer present in LCB.

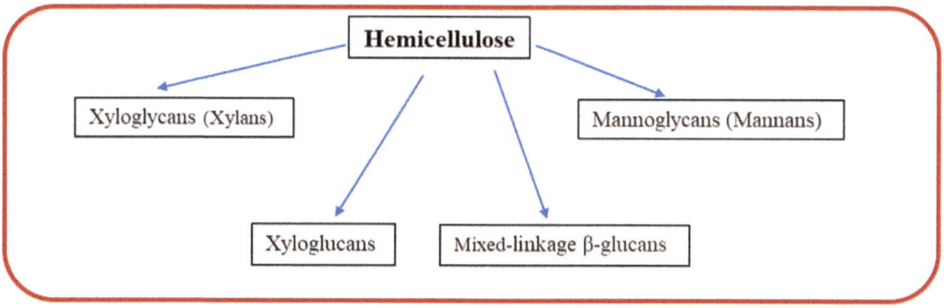

Fig. 6.4 Classification of Hemicellulose based on the composition and linkages

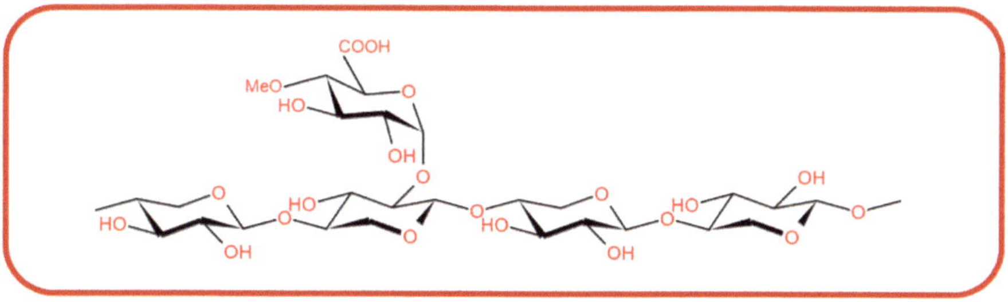

Fig. 6.5 Example of xyloglycans (O-Methyl-D-Glucurono-D-Xylan)

They have a variety of structural differences, such as localization, side-chain types, distribution, and types of glycoside linkage and main macromolecular chain distribution.

Xyloglycans: D-xylose-β -(1, 4) is the backbone present in the heteropolymer xylan. Xylans are the most abundant hemicelluloses in hardwood. They make up around 30% of all lignocellulosic materials. Xylans can be found in a variety of plant materials, including sorghum stalks, cane stalks, and cobs. They are derived from starch manufacturing hulls, hardwoods, husks, and softwoods. Based on the side chain, xylan is divided into three categories, i.e., homoxylan, glucuronoxylan, and arabinoxylan. Backbone and the side chain of homoxylans are primarily comprised of xylose. The major sugar unit in the backbone and side chain is the same pentose sugar. The backbone of glucuronoxylans is a xylose sugar and the side chain are mostly made of glucuronic acid. β-(1–4) linked D-xylose residues form the backbone of arabinoxylans, whereas the α-L-arabinose is made up of short side chains (Fig. 6.5) (Ebringerová 2005)).

Mannoglycans: Mannans are heteropolymers and have a backbone of β-(1,4)-D-Mannopyranose. On the basis of the side chain, they are divided into three subtypes. homomannan, glucomannan, and galactomannan are the three types. Mannose makes up the backbone of homomannans, and mannose also makes up the majority of the side chain. Homomannan is a relatively rare hemicellulose. Mannose-rich backbone forms glucomannans and it has a glucose-rich side chain. A secondary cell wall is made up of them. A mannose-rich backbone is also present in

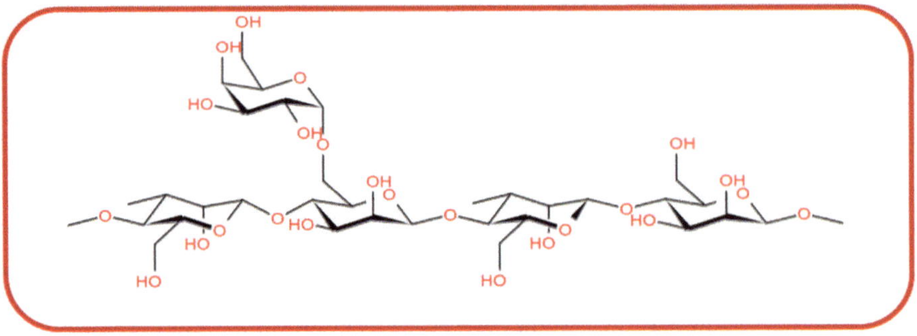

Fig. 6.6 Example of mannoglycans (D-Galacto-D-Mannan)

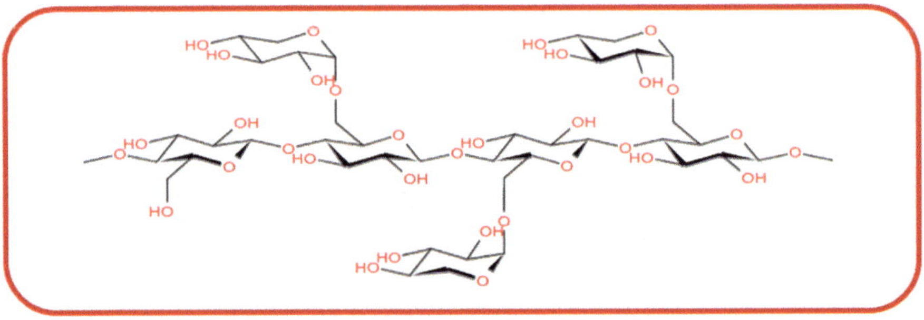

Fig. 6.7 Example of Xyloglucans (D-Xylo-D-Glucan)

galactomannans and it has a short galactose-rich side chain too. They're commonly in the cell walls of storage tissues, especially in leguminous seeds. The amount of galactose residue affects the hemicellulose solubility and viscosity (Fig. 6.6) (Olorunsola et al. 2018).

Xyloglucans: Xyloglucans made up of β-(1–4) linked glucose unit forms the backbone of xyloglucans. The backbone has a cellulose-like structure. At position 6, it is adorned with -D-xylopyranose. The cellulose microfibrils are tightly linked to this form of hemicellulose. As a result, extracting the xyloglucans is tough. Leguminous seeds like tamarind and afzelia contain xyloglucan hemicellulose (Isakov et al. 1990).

Xylogalactans: The backbone of xylogalactans is made up of galactose units. α-Dxylopyranose residues are used to embellish the backbone (Fig. 6.7). Leguminous seeds, such as Prosopis africana, contain xylogalactans (Al-Rudainy et al. 2019).

6.10 Functional Properties

Hemicellulose, after cellulose, is the second most prevalent polysaccharide of plant, with amorphous and highly branched short sugar chains. In comparison to cellulose, it has low molecular weight and is degradable and nontoxic. Hemicellulose is a bio-polymer made from renewable sources like polysaccharides and protein. Because of its low cost and biodegradability, it has a lot of potentials to replace non-biodegradable petroleum-based polymers. Weerasooriya et al. (2020) stated that its strong water resistance, superior flexibility, and low gas permeability makes hemicellulose the most promising material for film applications. Furthermore, hemicellulose is highly brittle, has low mechanical characteristics with some hygroscopic nature, and has humidity sensitivity with semi-crystalline structure but the high hydroxyl groups limit its use as a raw material.

S. Banerjee et al. (2019a, b) explained that to overcome the resistant character of associated cellulose and preserve retrieved components for further valorization, several physical, chemical, and biological preprocessing techniques can be used in a biorefinery to extract the hemicellulose component of the biomass. The hemicellulose extraction (91–96%) from cobs with 16% alkali has been reported for utilizing hydrothermal-assisted alkali pretreatment. The traditional hemicellulose extraction using the alkali method, which is from Canut husks, yields the highest after 16 h, but steam-assisted alkali pretreatment cuts the extraction time in half (8 h of incubation). Microwave irradiation has been demonstrated to boost the accessibility of hemicellulose in recent research. Coma proposed blending hemicellulose with other available commercially natural polymers and biopolymers as a viable solution for overcoming the aforementioned constraint in 2013. Hemicellulose's amorphous and highly branching nature makes it simple to convert into a variety of value-added chemicals. Because it's more unstable than other lignocellulosic biomass, it requires less rigorous pretreatment to degrade. The use of the hemicellulose fraction in the synthesis of biofuels and biochemicals from lignocellulose helps to justify the overall economics of the process.

6.11 Pretreatment Methods

Unlike cellulose, hemicellulose is a heteropolymer that comprises approximately 40% of the cell walls of lignocellulosic biomass. Hemicellulose comprises of a fraction of hexose sugars, consisting mainly of glucose, mannose, galactose, and pentose sugars such as xylose and arabinose. The hemicellulose due to their complex heteropolymeric nature, have an amorphous structure. They lack strength and can be hydrolyzed relatively easily by various pretreatment methods. The schematic of the various pretreatment process for the hydrolysis of hemicellulosic fraction of LCB is shown in Table 6.1.

Table 6.1 List of various pretreatment methods to hydrolyze hemicellulose

Method	Effect on feed	Reagent/force used	References
Milling	Disintegrates the structural framework	Inert balls; Shear force	Lin et al. (2010)
Pyrolysis	Thermal degradation, degrading cellulose into carbon compounds and H_2	Heat; temperature higher than 300 °C	Xiao et al. (2010)
Steam explosion	Degrades hemicellulose fraction, and effects cellulose crystalline structure	Pressurized steam	Auxenfans et al. (2017)
Dilute acid treatment	Transformation in lignin structure, hemicellulose hydrolysis	Sulfuric acid, hydrochloric acid	Nguyen et al. (2000)
Alkali treatment	Removes lignin fractions, and solubilization of hemicellulose	Sodium hydroxide, potassium hydroxide, lime	McIntosh and Vancov (2010)
CO_2 explosion	Allows penetration in small pores and hydrolyze hemicellulose	3.45–13.8 MPa pressured gas, at 200 °C	Zheng et al. (1995)
Ozonolysis	Oxidation on lignin faction	O_3 gas is sparged in the reactor	Travaini et al. (2016)
Wet oxidation	Targets hemicellulose and lignin linkage in moist biomass. Maximum recovery of individual components	Pressured oxygen gas at about 200 °C	Arvaniti et al. (2012)
Organosolv	Partial hydrolysis and enhanced accessibility of cellulose to cellulases	Formic acid, Peracetic acid with catalyst	Zhang et al. (2014)
Ionic liquids	Breaks non-covalent interactions. Effective for cellulose recovery	NaCl	Zhang et al. (2018)
Biological	Enzymatic reactions altering the LC structures at specific locations.	Microbial cultures	–

6.12 Biological Pretreatment

Biological pretreatment is very specific and occasionally coupled with other pretreatment methods for better yields. The pretreatment methods enhance the availability of the woody biomass for the enzymatic treatment. For hydrolysis of LB various cellulases and hemicellulases act upon the biomass. For complete hydrolysis of hemicellulosic faction into its constituent monomers, six different hemicellulases are required, including acetylxylan esterase (AXE), xylanase, xylosidase, alpha-glucuronidase, alpha-arabinofuranosidase, and ferulic acid esterase. Biological pretreatment has a low energy requirement, is an easy, manageable, and eco-friendly mode of operation. To make the process easy and more feasible, research is going on for the reduction of pretreatment steps and making strains richer in hydrolases. *Trichoderma reesei*, *Xylaria*, and *Daldinia* spp. are capable of fully degrading the components of hemicellulose (Lundell et al. 2010) that makes their genome a center of attention for this purpose. In a separate study *T. orientalis* EU7–22 was modified through homologs insertion using binary vector pUR5750 vector to knockout creA gene, which acts as a carbon catabolite repression transcription factor in the strain. The study reported that the knockout of creA strain recorded an increase in xylanase and β-Glucosidase by 2.75, 2.51-fold in inducing media and 9.24 and 11.79-folds in glucose repressing media (Long et al. 2018). A study showed gene manipulation of *Trichoderma reesei* in order to modulate the secreted enzyme mixture, by overexpressing xylanase regulator 1 XYR1 gene, which is a major regulator of hemicellulase expression genes. The strain was also engineered to lack four major cellulases encoding genes. The resulting secreted enzymatic mixture included cellulases EGIII and BGLI and several hemicellulases with proteins associated with non-hydrolytic cellulose degradation. The testing on untreated corn fiber showed considerable amount of glucose and hydrolytic release of xylose, arabinose, and mannose (Zhang et al. 2021a). Despite its high market potential due to the expression of these enzymes the problem of carbon catabolite repression (CCR) has been persistent, hampering the quantity of enzyme production. Shibata N and colleges disrupted α-tubulin *tubB*. The strain demonstrated upregulation of both cellulases and hemicellulases genes and cellobiose and other sugar transporter genes were upregulated. The strain was observed to simultaneously uptake glucose and cellobiose suggesting release from CCR (Shibata et al. 2021). In a different study deletion of xyloglucanase gene *cel74a* of *Trichoderma reesei* impacted the enzymatic expression. Cellulases *cel7a*, *cel7b*, and *cel6a* were expressed in lesser quantities while the expression of hemicellulases *xyn1* and *xyn2* was enhanced in presence of sugarcane bagasse (Lopes et al. 2021).

6.13 Metabolism of Xylose to Produce Various Commercially Important Products

The result of the pretreatment step leaves behind monomers or oligomers that are uptake and processed by enzymes of microorganism systems in the culture media. Such organized process aims to concentrate the desired metabolite with significant market value for the commercial or pharmaceutical scale. Xylose is metabolized following four major pathways. Eukaryotic species metabolize xylose via oxido-reductase pathway. While in prokaryotes xylose is metabolized via the isomrease pathway, and non-phosphorylative pathway (Dahms and Weimberg pathway) (Halmschlag et al. 2020; Harcus et al. 2013). Native xylose uptaking eukaryotic species (mycelial fungi and yeasts) like *Barnettozyma* species, *Candida tropicalis*, *Candida shehatae* undergo oxido-reductase pathway. It is a two-step pathway catalyzed by D-xylose reductase and xylitol dehydrogenase to obtain D-xylulose with xylitol as intermediate (Fig. 6.8). In bacteria such as *Pseudomonas fragi* the pathway is known as isomerase pathway. The conversion is a single step and is catalyzed by xylose isomerase converting d-xylose to D-xylulose. In both eukaryote and prokaryote species D-xylulose is further phosphorylated by xylulose kinase to get d-xylulose-5-phosphate. This compound enters the organism's native metabolism via pentose phosphate pathway. Apart from these, prokaryotes have additional

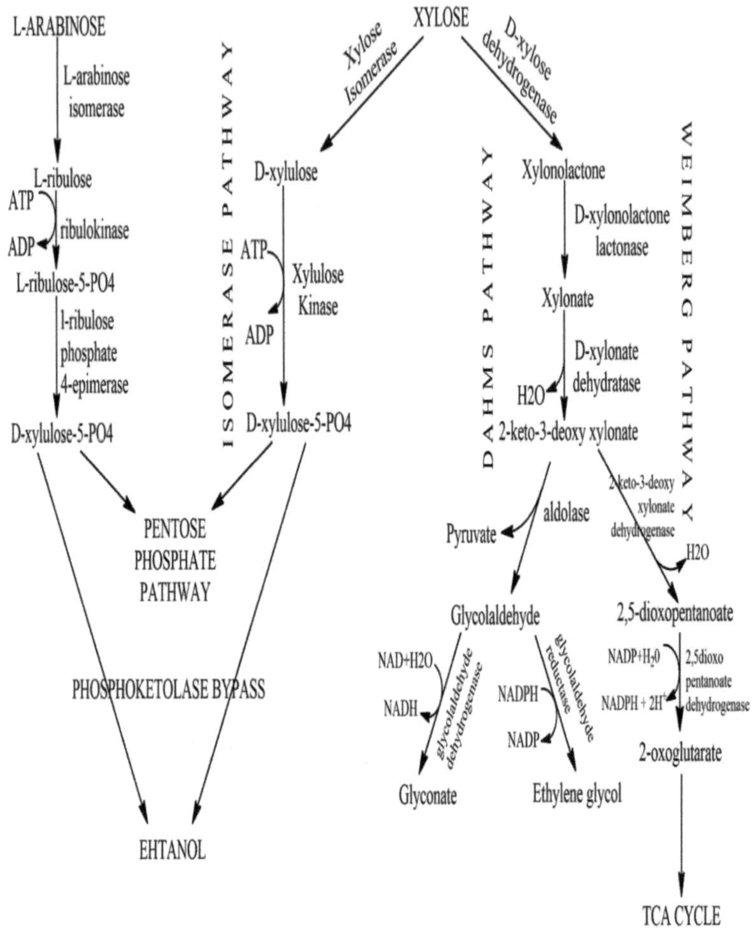

Fig. 6.8 Bacterial metabolic pathways for utilizing L-arabinose and xylose

oxidative pathways, Weimberg pathway (Fuente-Hernandez et al. 2013) and Dahms pathway (de Fátima Rodrigues de Souza et al. 2018) (found in *Pseudomonas fragi*).

Both have a common intermediate 2-keto-3-hydroxy-xylonate. Weimberg pathway produces 2-ketoglutarate and assimilates xylose via TCA cycle, while dahms pathway produces pyruvate and glycoaldehyde and directs the xylose carbon flux into various pathways with pyruvate and glycoaldehyde as intermediates (Meléndez-Hevia et al. 2009). Other monomer obtained from hemicellulose hydrolysis is l-arabinose. L-arabinose metabolism follows separate pathways in bacteria and yeast.The end product of both pathways is D-xylulose-5-phosphate(Verho et al. 2004). Fungal pathway converts L-arabinose, which is catabolized by reactions catalyzed by aldose reductase and L-arabinitol-4-dehydrogenase to produce L-xylulose and further converted to D-xylulose-5-phosphate by sequential oxido-reductase pathway and single step phosphorylation by xylulose kinase (covered in

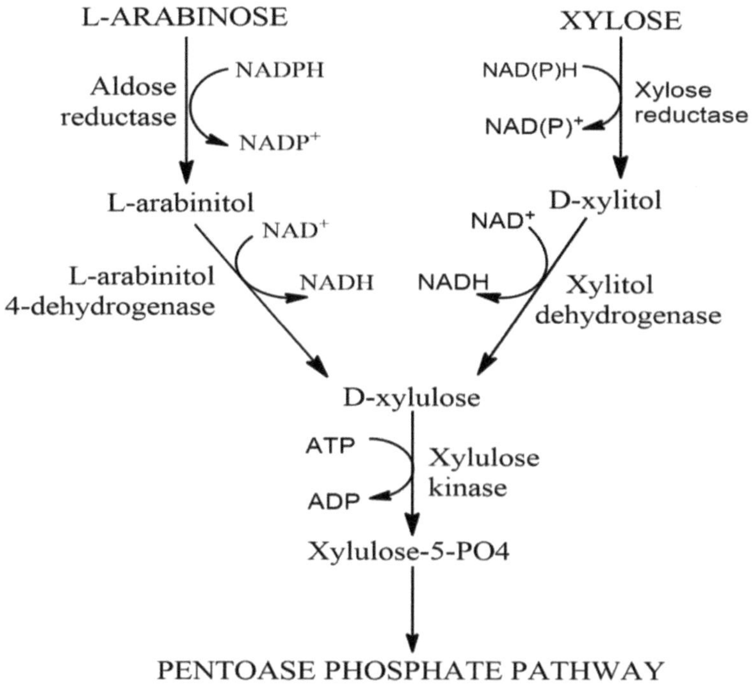

Fig. 6.9 Yeast metabolic pathways for utilizing the L-arabinose and xylose

xylose metabolism). In bacteria arabinose is converted to L-ribulose, L-ribulose-5-phosphate, and D-xylulose-5-phosphate, by reaction catalyzed by L-arabinose isomerase, Ribulokinase, and L-Ribulose phosphate 4-epimerase (Fig. 6.9).

6.14 Production of Commercially Important Biochemicals from Xylose Using Engineered Microorganisms

Plant biomass is humanity's sole anticipated sustainable supply of energy and minerals. Cellulosic materials are particularly appealing in this setting because of being budget-friendly and due to the plentiful supply of the material. The synthesis of hemicellulolytic biomass, enzymes, biomass hydrolysis, and zymolysis of the resultant sugars to produce products in one process step via a hemicellulolytic microbe is a promising technique for overcoming this barrier. If recombinant microorganisms with the requisite combination of substrate consumption and product production qualities can be produced, "consolidated bioprocessing" (CBP) can offer significant cost savings (Ajao et al. 2018). Genetically modified microorganisms for the production of value-added products using lignocellulosic biomass have been the upcoming subject of interest. The valorization of lignocellulosic biomass

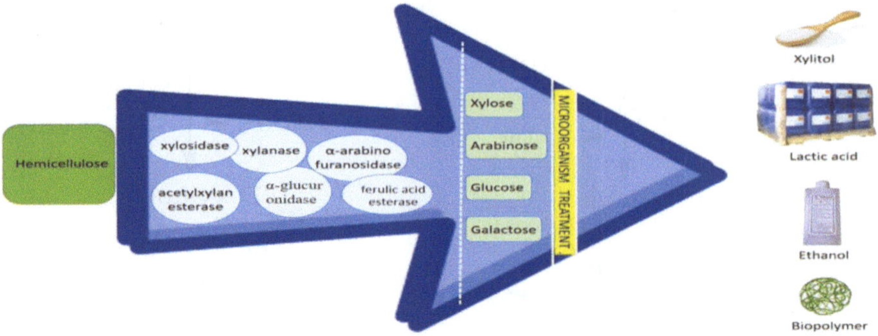

Fig. 6.10 Diagrammatic explanation of various commercially important products produced using hemicellulose

releases cellulose and hemicellulose as the main byproducts, which are further used in the production of value-added biochemicals (Fig. 6.10).

6.15 Bioethanol Production

After the pretreatment of lignocellulosic biomass under extreme carbon dioxide conditions, and after the pretreatment of molasses green solvents, lignin & polysaccharides are separated from cellulose. The product obtained from this process can be directly used in bioethanol production (Olorunsola et al. 2018), explained that enzymatic hydrolysis can be done from the obtained cellulose. The most prevalent procedures for bioethanol production are SSF, i.e., simultaneous hydrolysis and fermentation and SHF, i.e., separated hydrolysis and fermentation. Lignin depolymerization on lignin derivatives and mineralization activity has been shown by genus *Streptomyces* bacteria like *Streptomyces flavovirens, Streptomyces badius ATCC 39117, Amycolatopsis sp. ATCC 39116/75iv2*, and *Streptomyces cyaneus CECT 3335*. Many fungal strains are also used for lignocellulosic biomass pretreatment to produce bioethanol that includes *Pleurotus ostreatus, Ceriporiopsis subvermispora, Pycnoporus cinnabarinus Ceriporia lacerate, Cyathus stercoreus, etc.* (Abe et al. 2021).

Bioethanol can be produced by immobilized microorganisms; immobilization seems to be a leading trend in the field of enzymology. Wirawan et al. (2020) studied that immobilized *Z.mobilis ATCC 29191* can produce ethanol from pretreated alkaline molasses using continuous co-fermentation compared to *P.stipitis 21,775* and they concluded that the two-stage process that includes high yield ethanol production from glucose by immobilized *Z. mobilis* and production of ethanol by *P.stipitis* using xylose produced the yield of 70.60% in SHF and 81.08% in SCF (Fig. 6.11).

Fig. 6.11 Biofuel production from lignocellulosic biomass

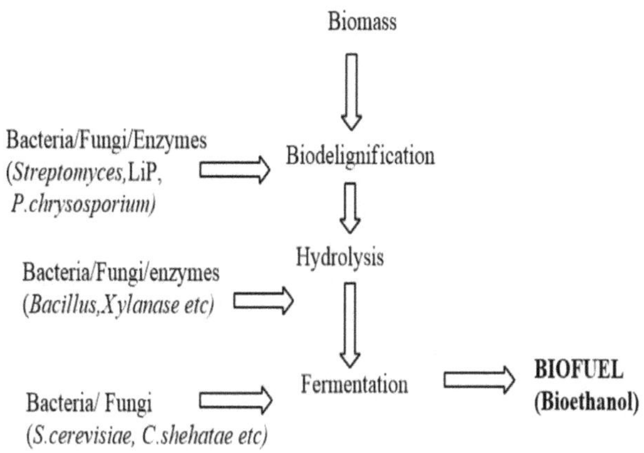

6.16 Ethanol Production from Xylose Fermenting *S. Cerevisiae*

Xylose is abundant in hemicellulose but few yeasts have the ability to acquire and absorb xylose. Many yeasts have xylose coding genes, i.e., XR (xylose reductase), XDH (xylose dehydrogenase), and XK (xylose kinase) but still, some yeasts cannot grow on the xylose nor can show any xylose-fermenting ability among themselves. It indicates another possible factor that yeast can metabolize xylose by regulating xylose pathway expression or characterizing the enzymes used in the metabolism. *Scheffersomyces stipites* is one the most studied xylose fermenting yeasts.

Xylose pathway expressing *Sc. stipitis* genes have been most exploited than almost any eukaryotic xylose-pathway for heterologous xylose production in the eukaryotic host. *Spathaspora passalidarum*, beetle-associated yeast has recently been found as the native yeast, which is xylose-fermenting and can produce ethanol. When compared to *Sc. stipitis* and the *Sp. passalidarum* it has an NADH-preferred XR that allows it to eat xylose and make ethanol more efficiently under both anaerobic and aerobic conditions. Their sensitivities to ethanol and the presence of inhibitors in lignocellulosic hydrolysates, like hydroxymethyfurfural, and acids are low, and fermentation of xylose by native yeast strains is highly dependent on culture conditions. Due to its high yield and resistance to environmental stressors, involving tolerance to alcoholic products, *S. cerevisiae* is a vital microbial cell factory in today's biopharmaceutical and food sectors, particularly as an ethanol producer. *S. cerevisiae*, on the contrary, can't consume xylose but may use xylulose via the endogenous xylulokinase (XKS1) and pentose pathway. To generate effective *S. cerevisiae* xylose-fermenting strains, metabolic engineering techniques for adding xylose consumption pathways and optimizing its metabolisms were used (Kwak and Jin 2017).

6.17 D-Lactic Acid Production

Lactic acid is a versatile chemical of industrial importance. It has applications in the pharmaceutical industry, cosmetic industry, packaging industry, and food industry. Lactic acid serves as a monomer to the biopolymer polylactic acid (PLA) that serves as the biological alternative to polyethylene and polystyrene polymers (Datta et al. 1995). Lactate is produced by the action of lactate dehydrogenase (LDH) enzyme on pyruvate, while in bacterial species the pyruvate enters the phosphoketolase pathway to produce lactate (Zaunmüller et al. 2006). The essential factor for these species is not the presence of the enzyme encoding gene, but the significant carbon flux flow in the pyruvate to lactate conversion node of metabolism, and the presence of all the required transporter molecules for the effective flow of metabolites favoring the process. Lactic acid can be produced from a wide range of organisms but utilization of xylose as the carbon source is restricted (Zhang et al. 2021b). The most commonly used species for lactic acid production are *Rhizopus oryzae, Lactobacillus pentosus*, and *Bacillus coagulans*. In a study by Maas et al., *Rhizopus oryzae* with xylose as sole carbon source produced 21.1 g/L lactic acid. When cultured in a media containing mixture of sugars (glucose and xylose), the strain (pregrown on xylose) produced lactic acid as the major fermentation product (33.4 g/L). But in actual lignocellulosic hydrolyzate media, the titer was 3.1 g/L (Maas et al. 2006). Another independent study screened 56 strains of *Rhizopus oryzae* for lactic acid production using untreated wheat straw powder. The yield was roughly 0.23 g/g (Saito et al. 2012) indicating the need for engineered strains for better bioconversion. To make the process a more cost-effective process we require a cheaper substrate. *Lactobacillus pentosus* was grown on hydrolyzed hemicellulose biomass giving the titer of 42.5 g/L. The same study with multiple carbon sources in the form of hemicellulose hydrolysate and partially digested cellulignin were used resulting in 65 g/L with 0.93 g/g yield (Wischral et al. 2019). A study produced a maximum concentration of 75 g/L (Ourique et al. 2020) using *Bacillus coagulans NL01*. Another study for *Bacillus coagulans* C106 reported 83.6 g/L in batch culture, and 215.7 g/L in the fed-batch mode of operation (Ye et al. 2013). A techno-economic study conducted for *E. coli* metabolizing corn stover hydrolysate produced 40 g/L of lactic acid.

According to Zhang et al. (2018) when the fermentation of lignocellulosic hydrolysate was done using recombinant strains of *E.coli*, D-Lactic acid was produced, which has a high yield in the minerals containing medium and can consume pentose sugar together, which is an appealing characteristic. But, D-lactic acid productivity is limited, and low Ph conditions are difficult to avoid. Moreover, there is presently no information on the efficiency of modified *Escherichia coli* on lignocellulose hydrolysate, and the investigation of lactic acid synthesis from lignocellulosic biomass using recombinant *E. coli* is under inception. *E. coli* can metabolize a variety of carbon sources, which is one of the benefits of employing it as a D-lactic acid-producing host. However, because glucose is usually digested first, pentose sugars are used late or incompletely. A strategy to prevent glucose effect in recombinant strain has been proposed, which includes replacement of L-LDH gene (ldhL) with another gene, i.e., D-LDH gene (ldhA) and by the deletion of ptsG gene, which codes for glucose phosphotransferase, catabolite repression was eliminated. To improve the cell growth for glucose utilization adaptive evolution was applied, which resulted in co-utilization of the substrates, glucose and xylose, in strain *E.coli JH15,* yielding 83 g/L of d-lactic acid in 96 h.

6.18 Xylitol Production

Xylitol is an alditol, sugar alcohol. It is used as a sugar substitute as it helps to overcome various health-related problems, especially dental problems as we saw above; bacteria do not produce xylitol, and also do not metabolize it, which also provides it an antibacterial property, and makes it a sugar substitute for all the diseases that are associated with high consumption of sucrose or glucose. It helps in the remineralization of dental enamel due to its ability to bind with calcium. Xylitol can be produced via chemical methods and by fermentation of hemicellulose sugar d-xylose (Borges and Pereira 2011). This section covers some recent studies of the production of xylitol from various hemicellulosic sources.

Yeasts are the preferable sources for xylitol production due to the presence of native xylose reductase gene. Yeast from genera *Pichia, Kluyveromyces*, and *Candida* are well known for xylitol production. A study by Saha, B.C. et al. screened 99 yeasts for xylitol production. The strain *Barnettozyma populi* Y-12728, which is not popularly known for xylitol production, was observed to produce xylitol in hydrolyzed hemicellulose from corn stover. The strain was chosen for its inability to utilize arabinose as a carbon source. Culture on xylose-rich complex media produced a maximum xylitol titer of 33.8 g/L with the yield of 0.69 at a dry cell weight of 1.43 g (Saha and Kennedy 2021). To make the xylitol production more economically viable, a study optimized the utilization of sulfated zirconia hydrolyzed corn cob by *Candida tropicalis* and obtained a xylitol yield of 0.76 g/g (Wan et al. 2021) and in a fed-batch culture it was reported to be 58 g/L with 0.73 g/g yield (Ling et al. 2011). Narisetty V et al. conducted xylitol production in using xylose-rich olive pits and sugarcane bagasse hydrolysates that end up in waste streams from olive oil and sugar production plants. The study reported xylitol accumulation in fed-batch culture was found to be 71.9 and 86.6 g/L with a conversion yield of 0.74 and 0.75 g/g by *Pichia fermentans* (Narisetty et al. 2021). Another interesting and uncommon hemicellulosic source will be cocoa pod husk, which has only a few reported studies for xylitol production. In a study, *Candida boidinii* XM02G consumed hydrolyzed cocoa pod husk to produce 11.34 g/L of xylitol with 0.52 g/g bioconversion (Santana et al. 2018).

The alcohol and acid derivatives of xylose, xylitol, and xylonic acid are chemicals with numerous uses in the food and pharmaceutical sectors. (Zhang et al. (2021a) produced a recombinant *Candida tropicalis,* mutant strain, which is deficient in uracil and produced from xylose mother liquor for xylitol production. Xylitol assimilation is blocked by gene deletion and a recombinant strain created from

Yarrowia lipolytica by heterologous expression of dehydrogenase gene and it yielded 97.10 g/L xylitol from 300 g/L of mother liquor with a fermentation time of 120 h (Dhar et al. 2016). The potent industrial system, *Corynebacterium glutamicum* is used for producing different platform chemicals. Dhar et al. (2016) experimented that the production of xylitol from sugars of lignocellulose using recombinant *C. glutamicum.* Sorghum stover converted into xylitol using reductase xylose gene along with expression of three poly-cistronic genes from various microbial sources. The strain produced 27 ± 0.3 g/L xylitol as compared to the synthetic source.

6.19 Itaconic Acid Production

One of the most significant organic acids derived from lignocellulose biomass is itaconic acid (IA). IA, also known as 2-methylenebutanedioic acid, or propylene dicarboxylic acid, is an unsaturated dicarboxylic acid (C5H6O4). IA is readily water and alcohols soluble, stable at room temperature, and stable in middle-basic, neutral, and acidic environments because of its mild acidity. It has a white crystalline powder appearance and has no odor (Teleky and Vodnar 2019).

Kautola et al. (1985) reported the first itaconic acid producer from xylose by immobilizing *Aspergillus terreus NRRL 1960* spores in gel beads of calcium alginate in reactors. The highest yield obtained was 54.5% submerged with initial glucose of 55 g/L and with the productivity of 0.32 g/L/h, 44.8% from xylose, which has productivity of 0.20 g/L/h. Kautola also juggled about the production of itaconic acid from xylose in continuous and batch reactors using *A. terreus TKK 200*; the strain responded badly. Recently, Klement showed that *U. maydis* can grow on xylose by slowly utilizing it. But *U. maydis* is not a good xylose or glucose-producing strain. Li et al. (2011) reported the maximum itaconic acid production strain with 2.5 g/L production after expressing gene CAD and two genes, which code for acid transporters (mfsA, mttA) from *Aspergillus terreus* into *Aspergillus niger* and deleting oahA gene, i.e., oxaloacetate hydrolyze gene and parameters optimizing. Furthermore, the hemoglobin domain overexpressed and replaced with basal level production of citric acid N_{201} and with low oxygen supply in culture conditions that have shown to increase the production of itaconic acid level in *Aspergillus niger.*

6.20 2,3-Butanediol

2,3-Butanediol finds multiple applications in several industries such as food, medicine, polymers and fabrics, cosmetics, energy industry, as antifreeze agents, etc., and is traditionally extracted from petroleum products. A variety of carbon sources such as pentose and hexose sugars, sucrose, lactose, and other disaccharides, etc., can be used for the production of 2,3-butanediol. 2,3-butanediol (BDO) production in microbes is carried out by three enzymes: acetolactate synthase (ALS), acetolactate decarboxylase (ALDC), and acetoin reductase (AR) acting on pyruvate in anaerobic conditions. Microbes with the potential to produce BDO by native pathway include *Clostridium* species, *Paenibacillus* strains, *Pyrococcus furiosus, Klebsiella pneumoniae*, and *Serratia marcescens.* The major setback for the utilization of hydrolyzed hemicellulose resources is that the feed composition cannot be strictly controlled. *Pantoea agglomerans* and *Klebsiella pneumoniae* have the potential of a versatile system for BDO production (Lee et al. 2019). It reported production of BDO from soybean hull hydrolysate with the simultaneous metabolization of glucose, arabinose, and xylose. Maximum production obtained was 21.9 g/L and 20.5 g/L for *Klebsiella pneumoniae* and *Pantoea agglomerans,* respectively (Cortivo et al. 2019). Relatively higher concentrations have been reported for BDO production using fed-batch. 28.923 g/L production of BDO using corncob through simultaneous saccharification and fermentation by strain *Enterobacter cloacae* UV4 (Zhang et al. 2018). Biomass processing is a tedious, time-consuming, and expensive process. In this next study, *Klebsiella pneumoniae* PM2, a well-known xylose utilizing strain was grown on a whole slurry of pretreated oil palm biomass under optimized conditions resulting in 75.03 $\pm$ 3.17 g/L of BDO (Rehman et al. 2021). To achieve the highest economic efficiency there is a constant need to formulate the processing and the utilization of waste products so that they could be used to bio-convert them into value-added products. Corncob molasses, which is a leftover waste, post xylitol production, has been utilized for BDO production. Corncob molasses has a high concentration of various sugars; on fermentation with *Klebsiella pneumoniae SDM* produced a maximum concentration of 78.9 g/L (Wang et al. 2013).

6.21 Succinic Acid

Succinic acid is an organic acid and an economically important platform chemical. Succinic acid production from xylose as carbon source is carried out through xylose assimilation via Weimberg pathway that directs the xylose carbon flux to TCA cycle, from there succinic acid is obtained. Hemicellulosic acid hydrolysate of sugarcane bagasse is a viable carbon source for succinic acid production. As per the study (Xi et al. 2013), fermentation using *Actinobacillus succinogenes* produced a maximum of 19.7 g of acid per liter succinic acid concentration (65.7% yield), 30 g/L of sugar in detoxified hydrolysate. Non-detoxified hydrolysate containing 30 g/L of sugar (22.4 g of xylose, 3.9 g of arabinose, and 3.6 g of glucose) yielded 23.7 g/L of succinic acid (79% yield). Borges and Pereira (Borges and Pereira 2011) has also reported the production of succinic acid using hemicellulosic hydrolysate of sugarcane bagasse using *Actinobacillus succinogenes* with similar results.

6.22 Polyhydroxyalkanoates (PHA/PHB)

Polyhydroxyalkanoates are polyesters found in the cytoplasm of different microorganisms as insoluble compounds in discrete granular form for the storage of various substances. The polyesters being biodegradable and biocompatible are promising alternatives to conventional plastics. Further, they provide the added advantage of being produced from renewable resources such as lignocellulosic biomass instead of nonrenewable petroleum extracts. The most popular among the different polyhydroxyalkanoates are polyhydroxybutyrate (PHB). It is characterized by properties that are very similar to the properties of polypropylene. These properties include processability, high melting point, hydrophobicity, optical purity, and inertness making it a potential alternative for the replacement of conventional petrochemical products. Silva et al. (2004) reported the production of polyhydroxybutyrate from *Bacillus cepacia* IPT048 and *Bacillus sacchari* IPT101 using detoxified hydrolysate of sugarcane bagasse in which it was observed that the use of hydrolysate as the carbon source provided the maximum yield (0.29 and 0.39 g of polyhydroxybutyrate per gram of hydrolysate for IPT 048 and ITP 101, respectively) as compared to the use of a combination of pure glucose and pure xylose (which showed yields of 0.19 and 0.22 g of polyhydroxybutyrate per gram of the carbon source for IPT 048 and ITP 101, respectively). Similarly, Cesário et al. (2014) reported the use of hydrolyzed wheat straw for polyhydroxybutyrate production by *Burkholderia sacchari* DSM 17165 strain in which 0.19 g of polyhydroxybutyrate was obtained for per gram of hydrolysate with 60% grams of polyhydroxybutyrate produced per gram of cellular dry weight. *Burkholderia sacchari* is an important strain for biorefinery application; polymer yield of *B. saccharia* from xylose is 0.38 g/g. It has potential to reach 72% PHA of the complete dry cell weight (Oliveira-Filho et al. 2021).

6.23 Conclusion

Hemicellulose and lignin fractions of lignocellulosic biomass are essential components of plants and trees providing them a robust structure that makes them highly resistant to microbial degradation. But at the same time due to the robust structure, this biomass does not fit well in the biochemical production market. This biomass degrades gradually in its wild habitat by wild microbial species and the enormous amounts of unutilized lignocellulosic biomass in nature make it the ideal feed target for biorefineries. Hemicellulosic biorefineries are industrial facilities that are designed for target-specific bioconversion of the biomass into biochemicals, energy, and other value-added products. But it comes with its challenges including the high transportation cost, variable feed composition, and heterogeneous nature of hemicellulose, moderate to low product yields, tedious processing of product feed, and high capital investment. Studies from cutting-edge research continue to come forward to develop a better understanding of physical, chemical, and biological characteristics of the biomass and resolve or ease the standing issues to make these biorefineries more efficient, including hybrid pretreatment methods, new and engineered microorganisms, and economic downstream processing techniques. As future aspects, biorefineries can be operational at locations that connect industries producing agro waste (ex: wood industries, food industries, sugarcane mills, oil mills) and the industries that process or utilize the respective biorefinery product as raw material, in order to make both raw materials and finished goods more accessible. Biorefineries have the potential to lead to economic growth parallel to waste management and have the potential of partially replacing the petrochemical resources thus fighting global warming and climate change and thus the future of sustainable growth.

References

Abe MM, Branciforti MC, Brienzo M (2021) Biodegradation of hemicellulose-cellulose-starch-based bioplastics and microbial polyesters. Recycling 6:22. https://doi.org/10.3390/RECYCLING6010022

Ajao O, Marinova M, Savadogo O, Paris J (2018) Hemicellulose based integrated forest biorefineries: implementation strategies. Ind Crop Prod 126:250–260. https://doi.org/10.1016/j.indcrop.2018.10.025

Al-Rudainy B, Galbe M, Hernandez MA, Jannasch P, Wallberg O (2019) Impact of lignin content on the properties of hemicellulose hydrogels. Polymers 11:35. https://doi.org/10.3390/polym11010035

Arvaniti E, Bjerre AB, Schmidt JE (2012) Wet oxidation pretreatment of rape straw for ethanol production. Biomass Bioenergy 39:94–105. https://doi.org/10.1016/j.biombioe.2011.12.040

Auxenfans T, Crônier D, Chabbert B, Paës G (2017) Understanding the structural and chemical changes of plant biomass following steam explosion pretreatment. Biotechnol Biofuels 10:1–16. https://doi.org/10.1186/s13068-017-0718-z

Banerjee R, Chintagunta AD, Ray S (2017) A cleaner and eco-friendly bioprocess for enhancing reducing sugar production from pineapple leaf waste. J Clean Prod 149:387–395. https://doi.org/10.1016/j.jclepro.2017.02.088

Banerjee R, Chintagunta AD, Ray S (2019a) Laccase mediated delignification of pineapple leaf waste: an ecofriendly sustainable attempt towards valorization. BMC Chem 13:1–11. https://doi.org/10.1186/s13065-019-0576-9

Banerjee S, Patti AF, Ranganathan V, Arora A (2019b) Hemicellulose based biorefinery from pineapple peel waste: xylan extraction and its conversion into xylooligosaccharides. Food Bioprod Process 117:38–50. https://doi.org/10.1016/j.fbp.2019.06.012

Biorefinery (2019) EPA public access 1–30. https://doi.org/10.1007/s10098-018-1568-5

Borges ER, Pereira N (2011) Succinic acid production from sugarcane bagasse hemicellulose hydrolysate by Actinobacillus succinogenes. J Ind Microbiol Biotechnol 38:1001–1011. https://doi.org/10.1007/s10295-010-0874-7

Brethauer S, Studer MH (2015) Biochemical conversion processes of lignocellulosic biomass to fuels and chemicals – a review. Chimia (Aarau) 69:572–581. https://doi.org/10.2533/chimia.2015.572

Cesário MT, Raposo RS, de Almeida MCMD, van Keulen F, Ferreira BS, da Fonseca MMR (2014) Enhanced bioproduction of poly-3-hydroxybutyrate from wheat straw lignocellulosic hydrolysates. New Biotechnol 31:104–113. https://doi.org/10.1016/j.nbt.2013.10.004

Chandel AK, Garlapati VK, Jeevan Kumar SP, Hans M, Singh AK, Kumar S (2020) The role of renewable chemicals and biofuels in building a bioeconomy. Biofuels. Bioprod Biorefining 14: 830–844. https://doi.org/10.1002/bbb.2104

Cheng H, Wang L (2013) Lignocelluloses feedstock biorefinery as petrorefinery substitutes. In: Biomass Now. IntechOpen, London. https://doi.org/10.5772/51491

Chintagunta AD, Zuccaro G, Kumar M, Kumar SPJ, Garlapati VK, Postemsky PD, Kumar NSS, Chandel AK, Simal-Gandara J (2021) Biodiesel production from lignocellulosic biomass using oleaginous microbes: prospects for integrated biofuel production. Front Microbiol 12:1–23. https://doi.org/10.3389/fmicb.2021.658284

Clauser NM, González G, Mendieta CM, Kruyeniski J, Area MC, Vallejos ME (2021) Biomass waste as sustainable raw material for energy and fuels. Sustainability 13:1–21. https://doi.org/10.3390/su13020794

Cortivo PRD, Machado J, Hickert LR, Rossi DM, Ayub MAZ (2019) Production of 2,3-butanediol by Klebsiella pneumoniae BLh-1 and Pantoea agglomerans BL1 cultivated in acid and enzymatic hydrolysates of soybean hull. Biotechnol Prog 35:1–8. https://doi.org/10.1002/btpr.2793

Datta R, Tsai S, Bonsignore P, Moon S, Frank JR (1995) Technological and economic potential of poly(lactic acid) and lactic acid derivatives. FEMS Microbiology Reviews 16:221 231

de Fátima Rodrigues de Souza R, Dutra ED, Leite FCB, Cadete RM, Rosa CA, Stambuk BU, Stamford TLM, de Morais MA (2018) Production of ethanol fuel from enzyme-treated sugarcane bagasse hydrolysate using d-xylose-fermenting wild yeast isolated from Brazilian biomes. 3 Biotech 8:1–9. https://doi.org/10.1007/s13205-018-1340-x

Dhar KS, Wendisch VF, Nampoothiri KM (2016) Engineering of Corynebacterium glutamicum for xylitol production from lignocellulosic pentose sugars. J Biotechnol 230:63–71. https://doi.org/10.1016/j.jbiotec.2016.05.011

Ebringerová A (2005) Structural diversity and application potential of hemicelluloses. Macromol Symp 232:1–12. https://doi.org/10.1002/masy.200551401

Fu C, Mielenz JR, Xiao X, Ge Y, Hamilton CY, Rodriguez M, Chen F, Foston M, Ragauskas A, Bouton J, Dixon RA, Wang ZY (2011) Genetic manipulation of lignin reduces recalcitrance and improves ethanol production from switchgrass. Proc Natl Acad Sci U S A 108:3803–3808. https://doi.org/10.1073/pnas.1100310108

Fuente-Hernandez A, Corcos PO, Beauchet R, Lavoie J-M (2013) Biofuels and co-products out of hemicelluloses. In: Liquid, Gaseous Solid Biofuels. Intech Open, London, pp 3–46. https://doi.org/10.5772/52645

Halmschlag B, Hoffmann K, Hanke R, Putri SP, Fukusaki E, Büchs J, Blank LM (2020) Comparison of isomerase and Weimberg pathway for γ-PGA production from xylose by engineered Bacillus subtilis. Front Bioeng Biotechnol 7:1–17. https://doi.org/10.3389/fbioe.2019.00476

Harcus D, Dignard D, Lépine G, Askew C, Raymond M, Whiteway M, Wu C (2013) Comparative xylose metabolism among the ascomycetes C. albicans, S. stipitis and S. cerevisiae. PLoS One 8:1–10. https://doi.org/10.1371/journal.pone.0080733

Himmel ME, Ding SY, Johnson DK, Adney WS, Nimlos MR, Brady JW, Foust TD (2007) Biomass recalcitrance: engineering plants and enzymes for biofuels production. Science 315:804–807. https://doi.org/10.1126/science.1137016

Kamm B, Kamm M (2004) Principles of biorefineries. Appl Microbiol Biotechnol 64:137–145. https://doi.org/10.1007/s00253-003-1537-7

Isakov IF, Stepanov EA, Dronov AF, Shumov ND, Blinnikov OI, Archibong A (1990) Diagnostika i lechenie spaechnoĭ kishechnoĭ neprokhodimosti u deteĭ [Diagnosis and treatment of adhesive intestinal obstruction in children]. Khirurgiia (Mosk) 8:3–7. Russian

Kautola H, Vahvaselkä M, Linko Y-Y, Linko P (1985) Itaconic acid production by immobilizedAspergillus terreus from xylose and glucose. Biotechnol Lett 7:167–172. https://doi.org/10.1007/BF01027812

Kumar B, Bhardwaj N, Agrawal K, Chaturvedi V, Verma P (2020) Current perspective on pretreatment technologies using lignocellulosic biomass: An emerging biorefinery concept. Fuel Process Technol 199:106244. https://doi.org/10.1016/j.fuproc.2019.106244

Kwak S, Jin YS (2017) Production of fuels and chemicals from xylose by engineered Saccharomyces cerevisiae: a review and perspective. Microb Cell Factories 16:1–15. https://doi.org/10.1186/s12934-017-0694-9

Lanigan BA, Clark PJ, Deswarte DF (2011) Microwave processing of lignocellulosic biomass for production of fuels. University of York, London, p 258

Lee GB, Kim YJ, Lim JK, Kim TW, Kang SG, Lee HS, Lee JH (2019) A simple biosynthetic pathway for 2,3-butanediol production in Thermococcus onnurineus NA1. Appl Microbiol Biotechnol 15:3477–3485. https://doi.org/10.1007/s00253-019-09724-z

Li A, van Luijk N, ter Beek M, Caspers M, Punt P, van der Werf M (2011) A clone-based transcriptomics approach for the identification of genes relevant for itaconic acid production in aspergillus. Fungal Genet Biol 48:602–611. https://doi.org/10.1016/j.fgb.2011.01.013

Lin Z, Huang H, Zhang H, Zhang L, Yan L, Chen J (2010) Ball milling pretreatment of corn Stover for enhancing the efficiency of enzymatic hydrolysis. Appl Biochem Biotechnol 162:1872–1880. https://doi.org/10.1007/s12010-010-8965-5

Ling H, Cheng K, Ge J, Ping W (2011) Statistical optimization of xylitol production from corncob hemicellulose hydrolysate by Candida tropicalis HDY-02. New Biotechnol 28:673–678. https://doi.org/10.1016/j.nbt.2010.05.004

Long C, Cheng Y, Cui J, Liu J, Gan L, Zeng B, Long M (2018) Enhancing Cellulase and Hemicellulase production in Trichoderma orientalis EU7-22 via knockout of the creA. Mol Biotechnol 60:55–61. https://doi.org/10.1007/s12033-017-0046-3

Lopes DCB, Carraro CB, Silva RN, de Paula RG (2021) Molecular characterization of xyloglucanase cel74a from trichoderma reesei. Int J Mol Sci 22:4545. https://doi.org/10.3390/ijms22094545

Lugani Y, Rai R, Prabhu AA, Maan P, Hans M, Kumar V, Kumar S, Chandel AK, Sengar RS (2020) Recent advances in bioethanol production from lignocelluloses: a comprehensive review with a focus on enzyme engineering and designer biocatalysts. Biofuel Res J 7:1267–1295. https://doi.org/10.18331/BRJ2020.7.4.5

Lundell TK, Mäkelä MR, Hildén K (2010) Lignin-modifying enzymes in filamentous basidiomycetes–ecological, functional and phylogenetic review. J Basic Microbiol 50:5–20. https://doi.org/10.1002/jobm.200900338

Maas RHW, Bakker RR, Eggink G, Weusthuis RA (2006) Lactic acid production from xylose by the fungus Rhizopus oryzae. Appl Microbiol Biotechnol 72:861–868. https://doi.org/10.1007/s00253-006-0379-5

McCann MC, Carpita NC (2015) Biomass recalcitrance: a multi-scale, multi-factor, and conversion-specific property. J Exp Bot 66:4109–4118. https://doi.org/10.1093/jxb/erv267

McIntosh S, Vancov T (2010) Enhanced enzyme saccharification of Sorghum bicolor straw using dilute alkali pretreatment. Bioresour Technol 101:6718–6727. https://doi.org/10.1016/j.biortech.2010.03.116

Meléndez-Hevia E, De Paz-Lugo P, Cornish-Bowden A, Cárdenas ML (2009) A weak link in metabolism: the metabolic capacity for glycine biosynthesis does not satisfy the need for collagen synthesis. J Biosci 34:853–872. https://doi.org/10.1007/s12038-009-0100-9

Narisetty V, Castro E, Durgapal S, Coulon F, Jacob S, Kumar D, Kumar Awasthi M, Kishore Pant K, Parameswaran B, Kumar V (2021) High level xylitol production by Pichia fermentans using non-detoxified xylose-rich sugarcane bagasse and olive pits hydrolysates. Bioresour Technol 342:126005. https://doi.org/10.1016/j.biortech.2021.126005

Nguyen QA, Tucker MP, Keller FA, Eddy FP (2000) Two-stage dilute-acid pretreatment of softwoods. Appl Biochem Biotechnol 84–86:561–576. https://doi.org/10.1385/ABAB:84-86:1-9:561

Oliveira-Filho ER, Gomez JGC, Taciro MK, Silva LF (2021) Burkholderia sacchari (synonym Paraburkholderia sacchari): an industrial and versatile bacterial chassis for sustainable biosynthesis of polyhydroxyalkanoates and other bioproducts. Bioresour Technol 337:125472. https://doi.org/10.1016/j.biortech.2021.125472

Olorunsola O, Akpabio EI, Akpabio I, Adedokun MO, Olorunsola EO (2018) Emulsifying Properties Properties of of Hemicelluloses. In: Karakuş S (ed) Science and technology behind nanoemulsions. IntechOpen, London. https://doi.org/10.5772/intechopen.74473
Ourique LJ, Rocha CC, Gomes RCD, Rossi DM, Ayub MAZ (2020) Bioreactor production of 2,3-butanediol by Pantoea agglomerans using soybean hull acid hydrolysate as substrate. Bioprocess Biosyst Eng 43:1689–1701. https://doi.org/10.1007/s00449-020-02362-0
Patel A, Shah AR (2021) Integrated lignocellulosic biorefinery: gateway for production of second generation ethanol and value added products. J Bioresour Bioprod 6:108–128.
Pérez J, Muñoz-Dorado J, De La Rubia T, Martínez J (2002) Biodegradation and biological treatments of cellulose, hemicellulose and lignin: an overview. Int Microbiol 5:53–63. https://doi.org/10.1007/s10123-002-0062-3
Prabhu AA, Ledesma-Amaro R, Lin CSK, Coulon F, Thakur VK, Kumar V (2020) Bioproduction of succinic acid from xylose by engineered Yarrowia lipolytica without pH control. Biotechnol Biofuels 13:113. https://doi.org/10.1186/s13068-020-01747-3
Rehman S, Khairul Islam M, Khalid Khanzada N, Kyoungjin An A, Chaiprapat S, Leu SY (2021) Whole sugar 2,3-butanediol fermentation for oil palm empty fruit bunches biorefinery by a newly isolated Klebsiella pneumoniae PM2. Bioresour Technol 333:125206. https://doi.org/10.1016/j.biortech.2021.125206
Saha BC, Kennedy GJ (2021) Optimization of xylitol production from xylose by a novel arabitol limited co-producing Barnettozyma populi NRRL Y-12728. Prep Biochem Biotechnol 51:761–768. https://doi.org/10.1080/10826068.2020.1855443
Saito K, Hasa Y, Abe H (2012) Production of lactic acid from xylose and wheat straw by Rhizopus oryzae. J Biosci Bioeng 114:166–169. https://doi.org/10.1016/j.jbiosc.2012.03.007
Santana NB, Teixeira Dias JC, Rezende RP, Franco M, Silva Oliveira LK, Souza LO (2018) Production of xylitol and bio-detoxification of cocoa pod husk hemicellulose hydrolysate by Candida boidinii XM02G. PLoS One 13:1–15. https://doi.org/10.1371/journal.pone.0195206
Shen Y, Jarboe L, Brown R, Wen Z (2015) A thermochemical-biochemical hybrid processing of lignocellulosic biomass for producing fuels and chemicals. Biotechnol Adv 33:1799–1813. https://doi.org/10.1016/j.biotechadv.2015.10.006
Shibata N, Kakeshita H, Igarashi K, Takimura Y, Shida Y, Ogasawara W, Koda T, Hasunuma T, Kondo A (2021) Disruption of alpha-tubulin releases carbon catabolite repression and enhances enzyme production in Trichoderma reesei even in the presence of glucose. Biotechnol Biofuels 14:1–16. https://doi.org/10.1186/s13068-021-01887-0
Silva LF, Taciro MK, Michelin Ramos ME, Carter JM, Pradella JGC, Gomez JGC (2004) Poly-3-hydroxybutyrate (P3HB) production by bacteria from xylose, glucose and sugarcane bagasse hydrolysate. J Ind Microbiol Biotechnol 31:245–254. https://doi.org/10.1007/s10295-004-0136-7
Sista Kameshwar AK, Qin W (2018) Understanding the structural and functional properties of carbohydrate esterases with a special focus on hemicellulose deacetylating acetyl xylan esterases. Mycology 9:273–295. https://doi.org/10.1080/21501203.2018.1492979
Teleky B-E, Vodnar DC (2019) Biomass-derived production of Itaconic acid as a building block in specialty polymers. Polymers (Basel). 11:1035. https://doi.org/10.3390/polym11061035
Travaini R, Martín-Juárez J, Lorenzo-Hernando A, Bolado-Rodríguez S (2016) Ozonolysis: an advantageous pretreatment for lignocellulosic biomass revisited. Bioresour Technol 199:2–12. https://doi.org/10.1016/j.biortech.2015.08.143
Verho R, Putkonen M, Londesborough J, Penttilä M, Richard P (2004) A novel NADH-linked L-Xylulose reductase in the L-arabinose catabolic pathway of yeast. J Biol Chem 279:14746–14751. https://doi.org/10.1074/jbc.M312533200
Wan L, Gao Z, Wu B, Cao F, Jiang M, Wei P, Jia H (2021) Hydrolysis of corncob hemicellulose by solid acid sulfated zirconia and its evaluation in xylitol production. Appl Biochem Biotechnol 193:205–217. https://doi.org/10.1007/s12010-020-03412-9

Wang R, Li L, Zhang B, Gao X, Wang D, Hong J (2013) Improved xylose fermentation of Kluyveromyces marxianus at elevated temperature through construction of a xylose isomerase pathway. J Ind Microbiol Biotechnol 40:841–854. https://doi.org/10.1007/s10295-013-1282-6
Weerasooriya PRD, Nadhilah R, Owolabi FAT, Hashim R, Khalil HPSA, Syahariza ZA, Hussin MH, Hiziroglu S, Haa MKM (2020) Exploring the properties of hemicellulose based carboxymethyl cellulose fi lm as a potential green packaging. Current Research in Green and Sustainable Chemistry 2:20–28. https://doi.org/10.1016/j.crgsc.2020.05.001
Wirawan F, Cheng C-L, Lo Y-C, Chen C-Y, Chang J-S, Leu S-Y, Lee D-J (2020) Continuous cellulosic bioethanol co-fermentation by immobilized Zymomonas mobilis and suspended Pichia stipitis in a two-stage process. Appl Energy 266:114871. https://doi.org/10.1016/j.apenergy.2020.114871
Wischral D, Arias JM, Modesto LF, de França Passos D, Pereira N (2019) Lactic acid production from sugarcane bagasse hydrolysates by Lactobacillus pentosus: integrating xylose and glucose fermentation. Biotechnol Prog 35:e2718. https://doi.org/10.1002/btpr.2718
Xi YL, Chen KQ, Dai WY, Ma JF, Zhang M, Jiang M, Wei P, Ouyang PK (2013) Succinic acid production by Actinobacillus succinogenes NJ113 using corn steep liquor powder as nitrogen source. Bioresour Technol 136:775–779. https://doi.org/10.1016/j.biortech.2013.03.107
Xiao R, Chen X, Wang F, Yu G (2010) Pyrolysis pretreatment of biomass for entrained-flow gasification. Appl Energy 87:149–155. https://doi.org/10.1016/j.apenergy.2009.06.025
Ye L, Zhou X, Hudari MSB, Li Z, Wu JC (2013) Highly efficient production of L-lactic acid from xylose by newly isolated Bacillus coagulans C106. Bioresour Technol 132:38–44. https://doi.org/10.1016/j.biortech.2013.01.011
Zaunmüller T, Eichert M, Richter H, Unden G (2006) Variations in the energy metabolism of biotechnologically relevant heterofermentative lactic acid bacteria during growth on sugars and organic acids. Appl Microbiol Biotechnol 72:421–429. https://doi.org/10.1007/s00253-006-0514-3
Zeng Y, Zhao S, Yang S, Ding SY (2014) Lignin plays a negative role in the biochemical process for producing lignocellulosic biofuels. Curr Opin Biotechnol 27:98–45. https://doi.org/10.1016/j.copbio.2013.09.008
Zhang CY, Peng XP, Li W, Guo XW, Xiao DG (2014) Optimization of 2,3-butanediol production by Enterobacter cloacae in simultaneous saccharification and fermentation of corncob residue. Biotechnol Appl Biochem 61:501–509. https://doi.org/10.1002/bab.1198
Zhang Y, Yoshida M, Vadlani PV (2018) Biosynthesis of d-lactic acid from lignocellulosic biomass. Biotechnol Lett 40:1167–1179. https://doi.org/10.1007/s10529-018-2588-2
Zhang L, Chen Z, Wang J, Shen W, Li Q, Chen X (2021a) Stepwise metabolic engineering of Candida tropicalis for efficient xylitol production from xylose mother liquor. Microb Cell Factories 20:105. https://doi.org/10.1186/s12934-021-01596-1
Zhang W, Guo J, Wu X, Ren Y, Li C, Meng X, Liu W (2021b) Reformulating the hydrolytic enzyme cocktail of trichoderma reesei by combining XYR1 overexpression and elimination of four major Cellulases to improve Saccharification of corn fiber. J Agric Food Chem 70:211–222. https://doi.org/10.1021/acs.jafc.1c05946
Zheng Y, Lin HM, Wen J, Cao N, Yu X, Tsao GT (1995) Supercritical carbon dioxide explosion as a pretreatment for cellulose hydrolysis. Biotechnol Lett 17:845–850. https://doi.org/10.1007/BF00129015

Chapter 7
Emerging Microbial Enzymes for Future Industrialization

Taniya Sengupta Rathore

Abstract Continuous research is being carried out for several years to identify microorganisms having the potential to produce biocatalyst to be used commercially by different industries to oppose toxic chemicals for ecofriendly processing. The special characteristics of these enzymes are being exploited for their industrial applications and commercialization. The basic characteristic includes the stability and activity of the enzymes in a wide range of pH and temperature. The enzymes have proven to be utilized by different biological-based industries such as pharmaceutical, food, leather, textile, animal feed, and waste management. The chapter highlights the present involvement and status of microbial enzymes in the Indian as well as global market. It also deals with the emerging prospects of microbial enzymes for the growth of future industrialization.

Keywords Biocatalyst · Microbial enzyme · Enzyme market · Industrialization

According to a Biospectrum Article by Narayan Kulkarni (2021), the Indian enzyme market in 2019–20 has touched Rs. 2596 crore with a growth rate of 10%. There are 25 major Indian industries that are either involved in the marketing or formulation of enzymes. There are also companies that are directly involved in the production of enzymes both at the multinational and local levels. These manufactured enzymes are used in different industrial sectors such as food processing, pharmaceuticals, detergent, leather, textile, and paper and pulp industries.

The emerging enzyme production in India has reduced the import of industrial used enzymes from the exporter by 5%. The major exporters of Indian enzymes include China, Netherland, Japan, the USA, Denmark, Germany, Singapore, Italy, Austria, and Spain. Except for China, Netherland, and Japan, the rest of the countries saw a fall in enzyme export to India.

T. S. Rathore (✉)
Department of Biotechnology, O P Jindal University, Raigarh (C.G), India
e-mail: taniya.rathore@opju.ac.in

A. Sarkar, I. A. Ahmed (eds.), *Microbial products for future industrialization*, Interdisciplinary Biotechnological Advances,
https://doi.org/10.1007/978-981-99-1737-2_7

On the other hand, the Indian enzyme market is not only restricted to the local market, but is exporting to different countries such as the USA, China, Denmark, Belgium, Iran, and Bangladesh, which is worth more than 10 crores in the year 2019–20. Apart from these major importers, in the same year a surge in Indian enzyme demand has been observed by the Indian market from the countries such as Algeria, Baharain, Burundi, Costa Rica, Chile, Ghana, Kuwait, Morocco, Malawi, Sudan, and Tunisia.

According to the report by TechSci Research (June, 2020), these new markets will open opportunities for Indian enzymes and thus the Indian Enzyme Market is likely to undergo robust growth by 2025, with its main application in dairy industries for the production of curd, cheese, yogurt, etc.

Biocon and Novozymes were the leading enzyme industries in India, which contributed more than 50% of the total enzyme revenue in the year 2005–06 (Biospectrum 2006). The list of other leading manufacturers with their bioproduct has been presented in Table 7.1.

The Enzyme market has been segmented on the basis of the following criteria:

Table 7.1 Industrial enzyme, manufacturer, and application of Indian market (Singh et al. 2016)

Industrial enzyme	Manufacturer	Established	Applications
Protease, xylanase, α-amylase, glucoamylase	Novozymes India	1983	Food and beverages, baking beverages, textile, oil and fats, alcohols, etc.
Protease, xylanase, amylase, phytase, β-mannanase, α-galactosidase	Advanced Enzymes Technologies Ltd.	1989	Pharma. Food and beverages, animal feed, textile, detergent, biofuel, etc.
Protease, xylanase, amylase, glucoamylase, cellulase, phytase, pectinase, lipase, catalase, etc.	MAPS Enzyme Ltd.	1975	Leather, textile, feed, etc.
Protease, amylase, cellulase, phytase, pectinase, laccase, catalase, mannanase	Rossari Biotech	1997	Pharma. Food and beverages, animal feed, textile, detergent, chemical, etc.
Amylase, cellulase, protease, alkaline pectinase, catalase, xylanase, β-glucanase, etc.	Lumisbiotech		Food and beverages, textile, feed, etc.
Protease, amylase, pectinase, xylanase, catalase, amyloglugosidase, etc.	Anthem Cellutions India Ltd.	2007	Pharma, food and beverages, textile, detergent, feed, paper, etc.
Papain, lysozyme, penicillin amidase, lipase etc.	Aumgene Biosciences	2004	Pharma, food and beverages, textile, detergent, feed, paper, etc.
Amylase, protease, cellulase, lipase, phytase, nattokinase	Zytex India Pvt. Ltd.	1947	Neutraceuticals, textile, animal feed, etc.

1. Types of the enzymes
2. Sources of the enzymes
3. Industrial application of the enzymes

7.1 Types of Enzymes

All biochemical reactions carried out as a part of metabolic processes of the cells require a specific biocatalyst known as enzymes. Enzymes show high specificity in terms of their substrates and product; thus, a specific biochemical reaction requires a specific enzyme. Enzymes are generally proteinous in nature or may possess a nonproteinous prosthetic group.

Enzymes are macromolecules made up of long chains of amino acids bonded with amide bonds. By molecular mass, it ranges from kilodalton to megadalton. The specificity of the substrate is determined by the catalytic site of the enzyme, which is buried within the hydrophobic region of the macromolecule. Its specificity classifies enzymes and provides its nomenclature. IUBMB (International Union of Biochemistry and Molecular Biology) in association with IUPAC (International Union for Pure and Applied Chemistry) established an Enzyme Commission (EC) for the systematic classification and nomenclature of enzymes (Liese et al. 2006). The Enzymes have been classified into six main classes according to the reaction they participate as a catalyst, shown in Table 7.2.

According to the types of enzymes, the market has been fragmented into carbohydrase, protease, lipase, nuclease, polymerase, etc. Among these, carbohydrase shows the largest market share in the year 2019–20 and is predicted to be highest in future financial years. This is due to the fact that carbohydrate is one of the most prominent molecules in pharmaceutical and food industries for which carbohydrase is extensively used. Followed by carbohydrase, the next revenue generator is

Table 7.2 Enzyme classes

S. no.	Class	Reaction involved	Enzymes
1	Oxidoreductase	Transfer of oxygen or hydrogen between the molecules	Dehydrogenase, oxidase, oxygenase, peroxidase
2	Transferase	Transfer of atom or group of atoms between the molecules	Transketolase, acyltransferase, transaminase, fructosyltransferase
3	Hydrolase	Hydrolytic cleavage of bond	Protease, amylase, lipase, cutinase, phosphatase
4	Lyases	Non-hydrolytic cleavage by addition or elimination of group(s)	Pectase, lyase, hydratase, succinase, fumarase
5	Isomerase	Change in the position of group within same or different molecules	Isomerase, epimerase, racemase
6	Ligase	Utilizing ATP/GTP for covalent bonding	Synthetase, ligase

protease that is required for protein processing. Polymerase and nuclease are also considered to be potential enzymes in the future market.

7.2 Sources of Enzymes

Enzymes are very difficult to chemically synthesize, so are obtained easily from living organisms. On the basis of sources from which enzymes are obtained, the global enzyme market is categorized into microbes, plants, and animals. Among the three sources, microorganisms are the major segmented source due to their several characteristics such as low production cost, easy availability, low processing time, nontoxicity, eco-friendliness, and low energy input. Due to the large-scale production of microbial enzymes, it dominated the market. Advance techniques of genetic engineering and its vast application also make them top the global enzyme market. As compared to plant enzymes, microbial enzyme purification is more tedious. The advanced technique of purification has made it easier with a little high costlier. But its high production and source availability made it cheaper compared to other sources. It is expected to be a lucrative segment by the year 2027.

Recombinant DNA technology can be easily used for the genetic manipulation of microbes for the production of stable enzymes in elevated quantities (Illanes et al. 2012). Enzymes isolated from different microbes show high performance under are a wide range of physical and chemical conditions. Apart from industrial use, microbial enzymes are also used as therapeutical agents in different health disorders caused genetically such as sacrosidase enzyme is prescribed orally for the patient suffering from inherited congenital sucrose-isomalate deficiency, unable to digest sucrose (Treem et al. 1999). Similarly, phenylalanine ammonia lyase is given to persons with genetically phenylketonuria disorder for degradation of phenylalanine (Sarkissian et al. 1999).

Industrial production of microbial enzymes with advanced genetic and protein engineering is done as per the demand in the market. Figure 7.1 shows the flow chart for the industrial production of enzymes.

7.3 Industrial Applications of Enzymes

Uses of enzymes especially microbial enzymes are increasing rapidly over conventional methods in different sectors such as food, pharmaceuticals, house hold products, textiles, leather, paper and pulp, etc., due to their various characteristics such as high efficiency, environment friendliness, high quality, etc. (Gurung et al. 2013a, b). The following are the industries where microbial enzymes are tremendously used.

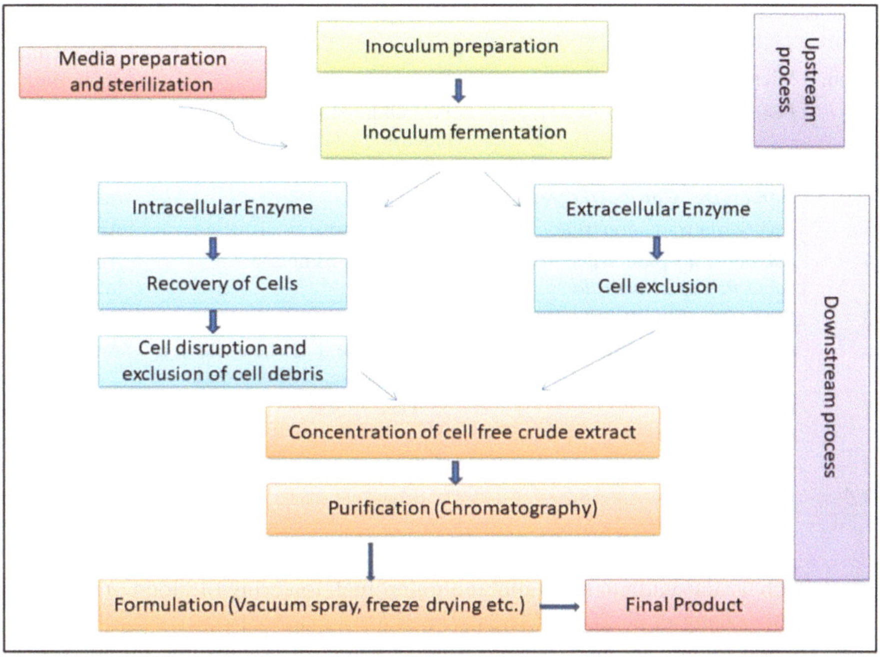

Fig. 7.1 Flowchart for Industrial production of microbial enzyme (Singh et al. 2016)

7.3.1 *Pharmaceutical Industries*

Enzymes are used as therapeutic agents against enzyme deficiency and digestive disorders by the pharmaceutical industries. Apart from providing therapies against disorders they are vigorously used in diagnosis purposes such as manufacturing of ELISA, diabetic kits, etc.

Microbial enzymes are largely acceptable in the formulation of medicines for various diseases. Microbial enzyme Nattokinase, considered as the fibrinolytic enzyme, is used as a prominent therapeutic agent against thrombosis disorders. It is used for skin repairing from burns and clots as well as useful in removing dead skins (Cho et al. 2010). Microbial Lipase is used to synthesize optically active organic compounds such as alcohols, esters, acids, etc. Phenol oxidase enzymes such as tyrosinase play important an role in the synthesis of melanin pigment, additionally it is a potent enzyme for synthesis of intermediate called L-DOPA, which when injected converted to dopamine, a potent drug which control myocardium neurogenic injuries in Parkinson's disease (Zaidi et al. 2014). Acid protease is used for the treatment of alimentary dyspepsia. Rhodanase and dextranase are, respectively, used to treat cyanide poisoning and decay of teeth (Okafor 2007). Chitosan is hydrolyzed into chitosan oligosaccharide by microbial enzyme chitosanase. This active chitosan oligosaccharide is used as several therapeutic

Table 7.3 Therapeutic applications of microbial enzymes (Kaur and Sekhon 2012)

S. no.	Microorganisms	Enzymes	Therapeutic application
1	*Penicillium* sp.	Penicillin oxidase, Rifamycin B oxidase	Antibiotic synthesis
2	*Klebsiella pneumonia, Serratia marcescens, Citrobacter freundii*	β-Lactamase	Antibiotic resistance
3	*Saccharomyces cerevisiae*	Ribonuclease	Antiviral
4	*Lactobacillus plantarum, Mycobacterium* sp., *Nocardia* sp.	Superoxide dismutase	Anti-inflammatory
5	*Streptococci* sp., *Bacillus subtilis*	Streptokinase	Anticoagulant
6	*Escherichia coli, Pseudomonas acidovorans, Acinetobacter*	L-Tyrosinase, L-Galactosidase, L-Asparaginase, L-Glutaminase	Antitumor
7	*Bacillus* sp. *Aspergillus oryzae*	Amylase, lipase	Digestive disorder
8	*Clostridium perfringens*	Collagenase	Skin ulcers
9	*Pseudomonas aeroginosa*	Rhodanase, laccase	Detoxification

agents such as antioxidant, antitumor, antimicrobial, lowering of blood pressure, and cholesterol (Zhang et al. 2012; Thadathil and Velappan 2014).

In clinical diagnosis, enzymes are used to detect their substrate qualitatively as well as quantitatively. Some of the enzymes are also used for the detection and conversion of their harmful substrates into a neutral compound, for example, cholesterol is detected and converted by the enzyme cholesterol oxidase. Glucose oxidase is used in a biosensor to detect the glucose level in blood. Urease and urate oxidase are used for the detection of urea and uric acid, respectively. Lipase, glucokinase, and carboxyl esterase are for determining the presence of triglycerides. Creatinase is used for predicting creatinine (Dordick 2013). Similarly, putrescine oxidase is used to detect food spoilage by detecting putrescine (Le Roes-Hill and Prins 2016).

In biotechnological application, microbial enzymes as intensively used in research and developments in gene manipulation such as restricted endonuclease obtained from several microorganisms. Different microbial enzymes also played a vital role in PCR and RT PCR diagnosis of the recent pandemic viruses and also help to predict future mutated strains. Restriction endonuclease is used for site-specific cleavage of DNA, polymerase enzymes are responsible for polymerization in PCR techniques (IAEA. 2020). Table 7.3 illustrates the therapeutical applications of some microbial enzymes.

7.3.2 Food and Beverage Industries

With an increase in the world population, there will also be an increase in the demand for food. Various microbial enzymes are used in food industries. The enzymes are used efficiently in enhancing the production of food, nutritional value, and other characteristics such as aroma, texture, flavor, and color. As an emerging technology, enzymes are also used in fat substitution and sweetener production (Li et al. 2012).

Industrial enzymes are also used in brewing industries to produce high-quality products. In fruit and vegetable juice industries, microbial enzymes are used as a part of processing ingredient to increase the efficiency of products such as peeling, juicing, extraction, and clarification. During juice processing enzymes such as amylase, pectinase, and cellulase are used for obtaining a good quality product with cost-effectiveness (Garg et al. 2016). These enzymes digest the starch, cellulose, pectin, and protein rapidly and thus improve the processing time and quality. On the other way, amylase is used for clearing cloudy juice and cellulase and pectinase are used for enhancing the stability and texture of the juice (Sivaramakrishnan et al. 2006). Pectin esterase and lyase are used to maintain the pectin level in juice required for the thickness of the juice for the best quality. For citrus fruit, the bitter components are separated by using microbial enzymes named naringinase and limoninase (Li et al. 2012).

For the production of distilled alcohol, microbial amylase is used for the hydrolysis of starch, which is further fermented. Amylase also reduces the turbidity of the liquor. It is also used in beer production for converting starch of barley into sugar.

7.3.3 Dairy Industries

Dairy enzymes such as protease, lipase, esterase, lactase, lysozyme, lactoperoxidase, aminopeptidase, catalase, etc., are important part of the food industry enzyme used to enhance the quality and shelf life of milk and milk products. Dairy enzymes are used to produce curd, yogurt, cheese, etc. (Qureshi et al. 2015).

Microbial rennet containing papain and chymosin is used for cheese production, which crosses the revenue of approximately 33% globally. Lipolysis of milk fat by lipase is done to enhance the flavor and reaction of processing time. Transglutaminase and various microbial protease are used to polymerize and inhibit allergic reactions of milk protein, respectively (Kieliszek and Misiewicz 2014).

The lactase enzyme is used to break lactose from the milk and also enhance sweetness and solubility. It is helpful for persons having lactose intolerance disorder (Soares et al. 2012).

7.3.4 Baking Industries

Different microbial enzymes are used in baking for improving the quality and prolonging its freshness. These enzymes are used for flour enhancing, dough stabilizing, softening, and improving color, texture, and volume. The use of enzyme increased the efficiency of the bread-making process. Amylase in bread manufacturing causes softening, increasing moisture content, and also improves the shelf life by acquiring freshness for prolonged period. Transglutaminase is used to improve flour quality. Lipase and xylanase are used for increasing the stability of the dough whereas lipoxygenase and glucose oxidase causes the whitening and strengthening of the dough (Adrio and Demain 2014).

7.3.5 Textile Industries

Conventional chemicals used in textile industries for dyeing, bleaching, etc., are toxic to the environment and cause hazardous impacts. Alternatively, enzymes are used in advanced technology for fiber processing, thus improving the quality of ecofriendliness (Choi et al. 2015). Hydrolase and oxidoreductase are used for cotton pretreatment. Enzymes such as amylase, cellulase, cutinase, pectinase, lipase, and esterase are used as bio polishing agents for different fabrics. They are used for cotton softening, wool finishing, and synthetic fiber modification (Chen et al. 2013). Laccase, catalase, peroxidase, and ligninase are used for bio bleaching and dye decolorization (Mojsov 2012). Applications of different microbial enzymes in textile industries are shown in Tables 7.4 and 7.5.

7.3.6 Animal Feed Industry

With the increase in the demand for milk and meat, the animal feed enzyme market is growing continuously. It has been estimated that there is an increase of 7.3% in market revenue for feed enzymes from the year 2015–2020 (Singh et al. 2016). Feed enzymes are used in the formulation and production of animal feed, which not only enhances the protein quality but also helps to remove toxic materials from the animal body. In poultry, microbial enzymes used as feed are phytase, protease, amylase, galactosidase, glucanase, xylanase, etc., which increases the nutritional quality basically the protein quality of feeds (Adrio and Demain 2014).

Phytase is the largely used feed enzyme used to bind natural phosphorus in feeds based on cereals (Frias et al. 2003). Xylanase and glucanase are used to degrade cellulose and hemicellulose for easy digestion of cellulosic feeds. Protease is also used similarly for breaking complex proteins into amino acids. Feed enzymes help to make the feed cost-effective by enhancing the nutritional value (Adrio and Demain 2014). Table 7.6 illustrates the sources and applications of different feed enzymes.

Table 7.4 The applications of microbial enzymes in different segments of food industries

S. no.	Segments	Microorganisms	Enzymes	Applications
1	Beverages	*Bacillus* sp., *Aspergillus* sp., *Streptomyces, Rhizopus*	Amylase	Starch hydrolysis
		Bacillus sp., *Klebsiella* sp.	Pullulanase	Starch saccharification
		Aspergillus oryzae, Penicillium sp.	Pectinase	Depectinization
		Aspergillus niger, Trichoderma atroviride	Cellulase	Fruit liquefaction
		Bacillus subtilis, Aspergillus niger	B-Glucanase, protease	Inhibit haze formation
		Aspergillus sp.	Naringinase, Limoninase	Remove bitterness
		Aspergillus niger	Glucose oxidase	Oxygen removal from liquors
2	Dairy	*Aspergillus* sp.	Proteinase (acid)	Milk coagulation
		Aspergillus niger, Bacillus subtillis, Lactobacillus sp.	Proteinase, lipase, aminopeptidase	Fast cheese ripening
		Aspergillus niger	Catalase	Cheese processing
		Escherichia coli	Lactase	Lactose reduction
		Streptomyces sp.	Transglutaminase	Protein crosslinking
3	Baking	*Bacillus* sp., *Aspergillus* sp.	Amylase	Flour adjustment
		Aspergillus niger	Lipase, xylanase	Dough conditioning
		Aspergillus niger, Penicillium chrysogenum	Glucose oxidase. Transglutaminase	Dough strengthening
		Bacillus stearothermophilus	α-Amylase	Enhancing bread shelf life

7.3.7 Paper and Pulp Industries

Environmental awareness has increased the use of microbial enzymes in this sector. The enzymes not only control the use of toxic chemicals but also improved the processing strategy by reducing processing time and energy consumption (Srivastava and Singh 2015). For starch coating, deinking and paper cleaning amylase enzyme is used. Xylanase and ligninase are used for removing hemicellulose and lignin (Kuhad et al. 2011). Laccase is used as an alternative to chlorine used in pulp processing. Additionally, mannase is used to enhance the brightness of the paper (Clarke et al. 2000). Table 7.7 details the applications and sources of enzymes used in this sector.

Table 7.5 Textile application of microbial enzymes (Singh et al. 2016)

S. no.	Microorganisms	Enzymes	Applications
1	*Aspergilus niger, Penicilliun funiculosum*	Cellulose	Cotton softening
2	*Bacillus licheniformis*	Amylase	Desizing
3	*Bacillus subtilis*	Laccase	Bleaching
4	*Aspergillus sp.*	Catalase	Bleach termination
5	*Aspergillus niger* *Bacillus subtilis*	Protease	Degumming of silk
6	*Candida antartica*	Lipase	Removal of lubricants
7	*Clostridium histolyticum*	Ligninase, Collagenase	Wool finishing
8	*Pseudomonas mendocina, Fusarium solanipisi*	Cutinase, Lyase	Cotton scouring

Table 7.6 Sources and applications of feed enzymes

S. no.	Microorganisms	Enzymes	Applications
1	*Aspergillus niger*	Phytase	Utilization of natural phosphorus
2	*Aspergillus* sp., *Bacillus* sp.	Xylanase	Cellulose breakdown
3	*Aspergillus niger*	β-Glucanase	Digestive aid

Table 7.7 Paper and pulp industries' enzyme: sources and applications

S. no.	Microorganisms	Enzymes	Applications
1	*Bacillus* sp., *Aspergillus* sp.	Amylase, cellulase	Deinking
2	*Trichoderma ressie, Thermomyces lanuginosus*	Xylanase	Hemicellulose removal
3	*Bacillus subtilis*	Laccase	Bleaching
4	*Bacillus subtillis*	Protease	Biofilm removal
5	*Candida antarctica*	Lipase	Pitch control

7.3.8 Leather Industry

Leather processing causes severe health and environmental hazards. A bulk amount of amylase, protease, and lipase are used as biodegradable alternatives to amine, sulfide, and lime in leather processing. They are used in dehairing process. Enzymes are used at different levels of leather processing for improving quality. They are applied for curing, soaking, dehairing, bating, liming, degreasing, and tanning. Details of enzymes are shown in Table 7.8.

Table 7.8 Applications of microbial enzymes in leather industries

S. no.	Microorganism	Enzyme	Applications
1	*Alcaligenes faecalis, Aspergillus niger, Bacillus subtilis*	Protease	Dehairing, bating, soaking
2	*Apergillus* sp.	Lipase	Degreasing
3	*Aspergillus* sp., *Bacillus sutillis*	Amylase	Fiber splitting

Table 7.9 Detergent enzymes

S.no	Microorganism	Enzyme	Applications
1	*Aspergillus* sp. *Bacillus subtilis*	Amylase	Remove carbohydrate strain
2	*Aspergillus oryzae, Bacillus subtillis*	Protease	Remove protein strain
3	*Aspergillus* sp.	Lipase	Remove fat strain
4	*Aspergillus niger, Bacillus* sp.	Cellulase	Clarification of color
5	*Fusarium solani*	Cutinase	Removal of triglycerides
6	*Bacillus* sp.	Mannanase	Removal of mannan

7.3.9 Detergent

According to the global enzyme market, 25–30% of the total enzyme market depends upon detergent. The additive enzymes in detergent are used to remove the stains of oil, fat, starch, and proteins. Additionally, they are used to retain the quality, texture, and color of the fabrics. Commonly used enzymes are amylase, protease, pectinase, cellulase, and lipase. Protease is used to remove protein stains such as blood, egg, sweat, etc., cutinase is used in dish washing and laundry detergents. Protease and amylase are also used in dish wash for removing starchy and protein food particles (Keshwani et al. 2015). Details are illustrated in Table 7.9.

7.4 Emerging Microbial Enzymes

Growing concern toward environmental safety and human health has led to the use of enzymes in different industrial sectors. Various research and development units have been set in different sectors for utilizing enzymes by exploiting an enormous source that is microorganisms. Advance techniques of genetic and protein engineering also made it a successful approach and it became a basic need for future industrializations. Apart from the above-mentioned applications of microbial enzymes as industrial enzyme, there are also several aspects such as the advancement of enzyme technology that has reduced the use of harmful radioactive elements in several immunoassays for the detection of proteins and hormones (Mane and Tale 2015).

The synthesis of biopolymers with vast applications is an emerging industrial sector. The utilization of microbial enzymes for the synthesis of biopolymers has several advantages over conventional chemical methods related to environmental issues. Biopolymers are highly demanding nowadays in medical, packaging, and electronic industries. In in vitro conditions, biopolymers could be synthesized by using lipase, peroxidase, and transglutaminase (Gurung et al. 2013a, b).

The use of enzymes in cosmetics is tremendously increasing. People are shifting toward biological products. In cosmetics such as sun screens, hair gels, dyes, mouth washes, and tooth pastes enzymes are used as free radical scavengers. To treat skin damages from pollutions super oxide dismutase is commonly used. Protease enzyme is also used to remove dead skins. Papain and glycosidase are used in tooth paste for treating tooth and gum decay. Laccase, peroxidase, and papain are used in hair dying and waving (Li et al. 2012). Enzymes are also applicable in synthesizing contact lens cleaner.

Synthesis of organic compounds by using enzymes are practically more significant as they are eco-freindly in nature. Lipase is the highest used enzyme in the synthesis of chemicals such as alcohols, acids, and esters. The production of acrylamide using nitrile hydratase is used by the Nitto chemical company in Japan (Singh et al. 2016). The Glucose isomerase is used to synthesize corn syrup in millions of tons every year commercially and is also used as an alternative sweetener.

Apart from involving in industrial production and processing, enzymes show a high prospect in waste treatments released from the industries. Industrial effluents contain a large quantity of hazardous toxic chemicals. Microbial enzymes are used and new strains of microbes are in search from the environment itself to degrade these toxic pollutants. Genetic manipulations and protein engineering are also evolved as essential tool for producing more stable enzymes with diverse applications. Enzyme-based biosensors are in progress to detect and remediate toxic pollutants from the environment.

7.5 Conclusion

The prospect of microbial enzymes in industrial use is increasing rapidly as they have significant potential for many different industries by which the demand of the growing population can be met. Microbial enzymes not only show their potential in production and processing in different industries but rather could be used as an alternative to food supply. They are actively used in waste management. The enzymes are the biocatalysts that reduce the activation energy of a particular reaction thus accelerating the rate of reaction causing cost- and time-effective high-quality products.

References

Adrio JL, Demain AL (2014) Microbial enzymes: tools for biotechnological processes. Biomol Ther 4(1):117–139

Biospectrum (2006). https://www.biospectrumindia.com/news/73/2915/industrial-enzymes.html. Accessed Feb 2022

Biospectrum (2021). https://www.biospectrumindia.com/views/17/19331/indian-industrial-enzymes-market-touches-rs-2596-cr-with-10-growth-rate.html. Accessed Feb 2022

Chen S, Su L, Chen J, Wu J (2013) Cutinase: characteristics, preparation, and application. Biotechnol Adv 31(8):1754–1767

Cho YH, Song JY, Kim KM (2010) Production of nattokinase by batch and fed-batch culture of *Bacillus subtilis*. New Biotechnol 27(4):341–346. https://doi.org/10.1016/j.nbt.2010.06.003

Choi JM, Han SS, Kim HS (2015) Industrial applications of enzyme biocatalysis: current status and future aspect. Biotechnol Adv 33:1443–1454

Clarke JH, Davidson K, Rixon JE (2000) A comparison of enzyme aided bleaching of softwood pulp using a combination of xylanase, mannanase and a-galactosidase. Appl Microbiol Biotechnol 53:661–667

Dordick JS (2013) Biocatalysts for industry. ISBN 9781475745979 encyclopedia of occupational health and safety (beverage industry). http://www.ilocis.org/documents/chpt65e.html. Accessed Feb 2022

Frias J, Doblado R, Antezana JR (2003) Inositol phosphate degradation by the action of phytase enzyme in legume seeds. Food Chem 81(2):233–239

Garg G, Singh A, Kaur A (2016) Microbial pectinases: an ecofriendly tool of nature for industries. 3 Biotech 6(1):47–59

Gurung N, Ray S, Bose S, Rai V (2013a) A broader view: microbial enzymes and their relevance in industries, medicine and beyond. Biomed Res Int 2013:1–18

Gurung N, Ray S, Bose S, Rai V (2013b) A broader view: microbial enzymes and their relevance in industries, medicine and beyond. Biomed Res Int 2013:329121. https://doi.org/10.1155/2013/329121

IAEA (2020). https://www.iaea.org/newscenter/news/how-is-the-covid-19-virus-detected-using-real-time-rt-pcr. Accessed Feb 2022

Illanes A, Cauerhff A, Wilson L (2012) Recent trends in biocatalysis engineering. Bioresour Technol 115:48–57

Kaur R, Sekhon BS (2012) Enzymes as drugs: an overview. J Pharm Edu Res 3(2):29–41

Keshwani A, Malhotra B, Kharkwal H (2015) Natural polymer based detergents for stain removal. World J Pharm Sci 4(4):490–508

Kieliszek M, Misiewicz A (2014) Microbial transglutaminase and its application in the food industry. A review folia. Microbiology 59:241–250

Kuhad RC, Gupta R, Singh A (2011) Microbial cellulases and their industrial applications. Enz Res 2011:1–10. https://doi.org/10.4061/2011/280696

Le Roes-Hill M, Prins A (2016) Biotechnological potential of oxidative enzymes from actinobacteria. In: Dhanasekaran D, Jiang Y (eds) Actinobacteria. Intech Open, Lodndon. https://doi.org/10.5772/61321

Li S, Yang X, Yang S (2012) Technology prospecting on enzymes: application, marketing and engineering. Comp Struc Biotechnol J 2:1–11

Liese A, Seelbach K, Wandrey C (2006) Industrial bio transformations, 2nd edn. Wiley, Weinheim, pp 273–447

Mane P, Tale V (2015) Overview of microbial therapeutic enzymes. Int J Curr Microbiol App Sci 4(4):17–26

Mojsov K (2012) Microbial alpha-amylases and their industrial applications: a review. Int J Manage IT Eng 2(10):583–609

Okafor N (2007) Biocatalysis: immobilized enzymes and immobilized cells. In: Modern industrial microbiology and biotechnology, p 398

Qureshi MA, Khare AK, Pervez A (2015) Enzymes used in dairy industries. Int J Appl Res 1(10):523–527
Sarkissian CN, Shao Z, Blain F (1999) A different approach to treatment of phenylketonuria: phenylalanine degradation with recombinant phenylalanine ammonia lyase. Proc Natl Acad Sci U S A 96:2339–2344
Singh R, Kumar M, Mittal A, Mehta PK (2016) Microbial enzymes: industrial progress in 21st century. Biotech 6:174
Sivaramakrishnan S, Gangadharan D, Nampoothiri KM (2006) α-Amylases from microbial sources: an overview on recent developments. Food Technol Biotechnol 44(2):173–184
Soares I, Tavora Z, Patera R (2012) Microorganism-produced enzymes in the food industry. In: Valdez DB (ed) Food industry, scientific, health and social aspects of the food industry, pp 83–94. https://doi.org/10.5772/31256
Srivastava N, Singh P (2015) Degradation of toxic pollutants from pulp & paper mill effluent. Discovery 40(183):221–227
Thadathil N, Velappan SP (2014) Recent developments in chitosanase research and its biotechnological applications: a review. Food Chem 150:392–399. https://doi.org/10.1016/j.foodchem.2013.10.083
Treem WR, McAdams L, Stanford L (1999) Sacrosidase therapy for congenital sucrase-isomaltase deficiency. J Pediatr Gastroenterol Nutr 28:137–142
Zaidi KU, Ali AS, Ali SA (2014) Microbial tyrosinases: promising enzymes for pharmaceutical, food bioprocessing, and environmental industry. Biochem Res Int 2014:1–16. https://doi.org/10.1155/2014/854687
Zhang H, Sang Q, Zhang W (2012) Statistical optimization of chitosanase production by *Aspergillus* sp. QD-2 in submerged fermentation. Ann Microbiol 62(1):193–201

Chapter 8
Bioethanol Production from Microbial Fermentation of Prospecting Biomass

Debapriya Sarkar, Renupama Bhoi, and Angana Sarkar

Abstract Among the many other difficulties of modern societies, energy scarcity, and depletion of fossil fuels are one of the most concerning ones. Due to the exponential growth of population and the advancement of science and technologies, there is no option that the rise of energy and fuel demand would become static. This crisis led to the investigation of an easy and sustainable source of energy that can last along with human society. Bioethanol has been identified as one of the most attractive alternatives or conjugation of conventional fossil fuels as it can be produced by a wide range of processes. Bioethanol fermentation is widely practiced all over the world but the substrate for the conventional processes involves food crops like corn, sugarcane, etc. This makes the process not so sustainable as it competes with the food industry, which itself is facing day to day increase in demand. Hence the alternative biomasses, which can refer as prospecting biomass, are crucial for making the process of bioethanol production more sustainable and economical. It has been observed that in the last decade over 39 countries' research (1753 research articles) has been conducted to find out a potential substrate for bioethanol fermentation as suggested by the search in the Web of Science database by the keyword "Bioethanol AND Substrate." The keyword analysis of this search result by Citespace software revealed that over 32 different types of biomass had been reported with their potential to become a sustainable substrate for bioethanol fermentation. The studied biomass was further characterized into four different groups, viz., agro waste, forest waste, fruit waste, and algal biomass. In this current chapter, detailed studies on these biomasses had been discussed.

Keywords Bioethanol fermentation · Substrate · Biomass · Agro waste · Algal biomass

---

D. Sarkar · R. Bhoi · A. Sarkar (✉)
Department of Biotechology and Medical Engineering, National Institute of Technology, Rourkela, India
e-mail: sarkara@nitrkl.ac.in

A. Sarkar, I. A. Ahmed (eds.), *Microbial products for future industrialization*, Interdisciplinary Biotechnological Advances,
https://doi.org/10.1007/978-981-99-1737-2_8

8.1 Introduction

The utilization of renewable energy such as solar, wind, bioenergy, and geothermal energy has become important as resource depletion and pollution are elevating to a concerning level. In contrast to other renewable resources, biofuel, particularly bioethanol, has many distinct benefits, including its usage as a liquid fuel that can be instantly used in existing automotive engines, distribution through the existing fossil fuel system, and promotion of rural economies (S. Morteza Raeisi et al. 2016). The growing demand for bioethanol has resulted in an overabundance of food and agricultural land being used in its production. Food prices have risen as a result, and the bioethanol business has seen its growth inhibited. Bioethanol is one of the exemplary biofuels that are often produced from renewable resources such as trees and plants and can now be used as an alternative energy source for gasoline engines to improve global energy security. Bioethanol is used as a gasoline substitute to reduce carbon monoxide (CO) and unburnt hydrocarbon (UHC) emissions. However, due to the high cost of manufacturing and the limited amount of available feedstock, full reliance on this type of biofuel has not yet been effectively established (Tyagi et al. 2019). The construction of an alternative method of producing bioethanol would be a possible solution to the current complications. Due to infrastructure, carbohydrate constituents like cellulose, lignin, hemicellulose, and carbon-based materials are regarded as a viable feedstock for bioethanol synthesis. Wheat straws, aquatic weeds, bagasse, fruits, vegetable waste, and mustard plant residue are only a few examples of cellulosic materials used in biorefineries. Agricultural industry wastes, livestock wastes, vegetable wastes, fruit, and crop leftovers are the four categories of agricultural by-products. Pretreatment, hydrolysis, fermentation, and purification are all part of the bioethanol manufacturing operation from agricultural waste biomasses.

Bioethanol synthesis from lignocellulosic materials is expected to be one of the cost-effective options. Even though hypothetical ethanol yields from starch and sugar are greater than lignocellulose, such traditional sources are inadequate for the global production of bioethanol. Crop residues are seasonal, economical, and easily available in this regard. Feedstocks do not necessitate the use of distinct ground, water, and energy resources also, these do not have any nutritional value. Many complications must be addressed to make bioethanol manufacturing economically viable. The feedstock, conversion technique, hydrolysis process, and fermentation arrangement are the four essential factors for the transformation of macromolecules into simpler forms for efficient bacterial fermentation (Zhu et al. 2015). Price, availability, processing, and management are the most significant feedstock challenges. Biomass modification as well as appropriate and cost-effective treatment technology to release cellulose and hemicellulose from its complex with lignin are sticking points for conversion technology. Lignocellulose is made up of cellulose, hemicellulose, and lignin. Cellulose is a linear homopolymer made up of repeated glucose units linked by 1,4 glycosidic bonds. Hemicellulose is a short branched heteropolymer made up of D-galactose, D-xylose, D-fructose, D-glucose, and D-mannose (Nakanishi et al. 2017). Lignin is a water-insoluble carbohydrate molecule that is strongly linked to these two carbohydrate polymers. Pretreatment for depolymerization is required to free cellulose and hemicellulose preceding the breakdown of the feedstock. Hydroxylation of cellulose and hemicellulose is required to yield microbial fermentable sugars such as glucose, arabinose, xylose, mannose, and galactose, and fermentation of reducing sugars. Enzymatic hydrolysis is perhaps the most effective alternative procedure for saccharification of complicated polymers in this regard. The issue with enzymatic hydrolysis would be to create an effective procedure for depolymerizing cellulose and hemicellulose to yield fermentable monomers with a higher percentage. Fruit and vegetable processing waste include higher content of soluble sugars, pectin, proteins, fiber, and lignin. Citrus processing waste can also be a useful resource of limonene that can be separated as a by-product in high yields (Tsukamoto et al. 2013).

Algae are photoautotrophic microorganisms that possess chlorophyll, which captures photons and can grow in a photoautotrophic or heterotrophic environment. Phototrophic algae use CO_2 from the atmosphere to produce nutrition like carbohydrates. Heterotrophic algae, on the other hand, keep growing by consuming carbon sources (Möllers et al. 2014). Algae can grow in all seasons in all types of environments, including saline waters, freshwater, lakes, and deserts. They are rich in fat and oil content and easy to culture and harvest (Huang et al. 2013). Microalgae and macroalgae are two types of algae. Microalgae are photosynthetic microorganisms that are either prokaryotic or eukaryotic. With their single-celled or basic colonies formations, they may persist in harsh environments. They could generate a lot of fat, protein, and carbohydrate in a short period since they are photosynthetic organisms. Aside from bioethanol and biodiesel, microalgae produce a range of advanced products and byproducts like methane, biogas, acetone, livestock feed, medicines, and cosmetics. The elemental composition of microalgae varies depending on the method of culture and the environment under which it is grown (Chow et al. 2015). They have a high content of proteins, fat, and carbohydrates (Kumar et al. 2016). Microalgae have long been suspected to be a potential biofuel resource due to their advantages over conventional crops like maximum biomass production efficiency per unit of surface and duration, non-competitiveness for territory, or food market with crops. They can also be grown on nonarable land, have better mineralization and nutrient accumulation, and promotes economic growth through appropriate reprocessing and contain the ability to use manufacturing or assembly as a source of reasonably priced fertilizer containing a high quantity of organic carbon, nitrogen, and phosphorus (Rizza et al. 2017).

8.2 Study on Worldwide Research Prospect of Different Biomass As a Substrate for Bioethanol Fermentation

Bioethanol fermentation is widely practiced and popularized all over the world where the major substrates are either corn or sugarcane (Mohanty and Swain 2019; Rossi et al. 2021). The selection of the substrates for bioethanol production makes the process a competition of the global food market (Susmozas et al. 2020). Raising eyebrows of different communities, bioethanol production developed a debate on food versus fuel (Esmaeili et al. 2020). To establish more sustainable and flawless energy production, the scientific community is searching for alternative substrates for bioethanol production. Waste management and utilization of non-accomplished resources are being investigated for their potential to be a superb substrate for bioethanol production. The study on 1753 different research articles indexed in the Web of Science database from 2011 to 2021 had revealed that in 39 different countries all over the world this search for an alternative substrate for bioethanol production is going on (Fig. 8.1). The search was conducted using the keywords "Bioethanol AND Substrate." Among these countries, the Republic of China has produced 300 articles related to this topic in the last decade as per the Web of Science database. India is the second-highest cited country with 229 citations for the study of the same topic as per the Web of Science database. Other notable countries are Brazil (169 citations), the United States of America (131 citations), Spain (76 citations), and South Korea (75 citations). The worldwide interest in this particular topic implied the importance of an alternative, sustainable, and capable substrate for bioethanol production.

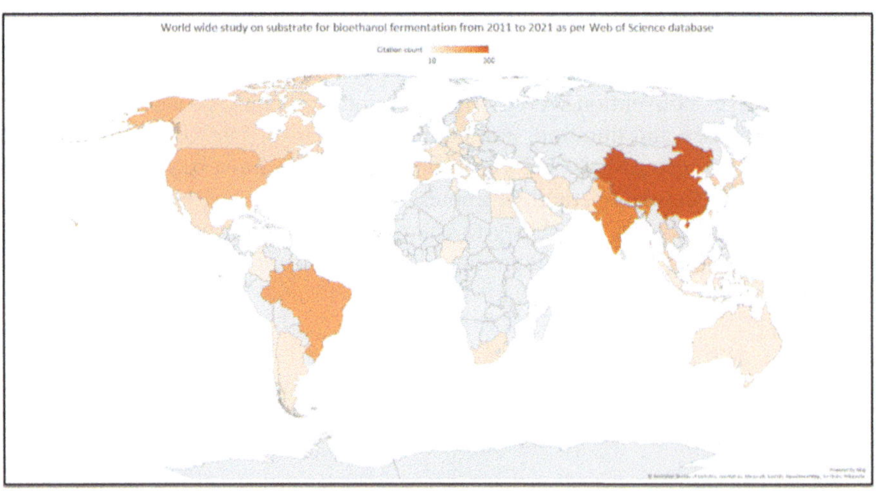

Fig. 8.1 Worldwide study on substrate for bioethanol fermentation from 2011 to 2021 as per Web of Science database

8.3 Explored Biomass for Bioethanol Production

The search results from the Web of Science database with the keyword "Bioethanol and Substrate" for the timeline of 2011–2021 were analyzed based on keywords in Citespace software (Synnestvedt et al. 2005), which revealed there was a total of 32 different types of biomass which were investigated for its potential to be a superb substrate for bioethanol fermentation (Fig. 8.2). Among these biomasses, the most dominant one was lignocellulosic biomass, which includes a wide range of plant-based biomass with high cellulose content. Apart from that, wheat straw, sugarcane bagasse, rice straw, corn stover, and straw are notable mentions with high citations. On the contrary, non-accomplished resources like cassava stem, switchgrass, corn-cob, elephant grass, grape pomace, etc., are also less investigated. Emerging substrates like microalgae, which are very recently studied as a potential substrate for bioethanol production are also being studied to a notable extent. Among this diverse range of studied biomass, many have shown their potential to replace the conventional and costly substrates to make the process more sustainable and accomplished.

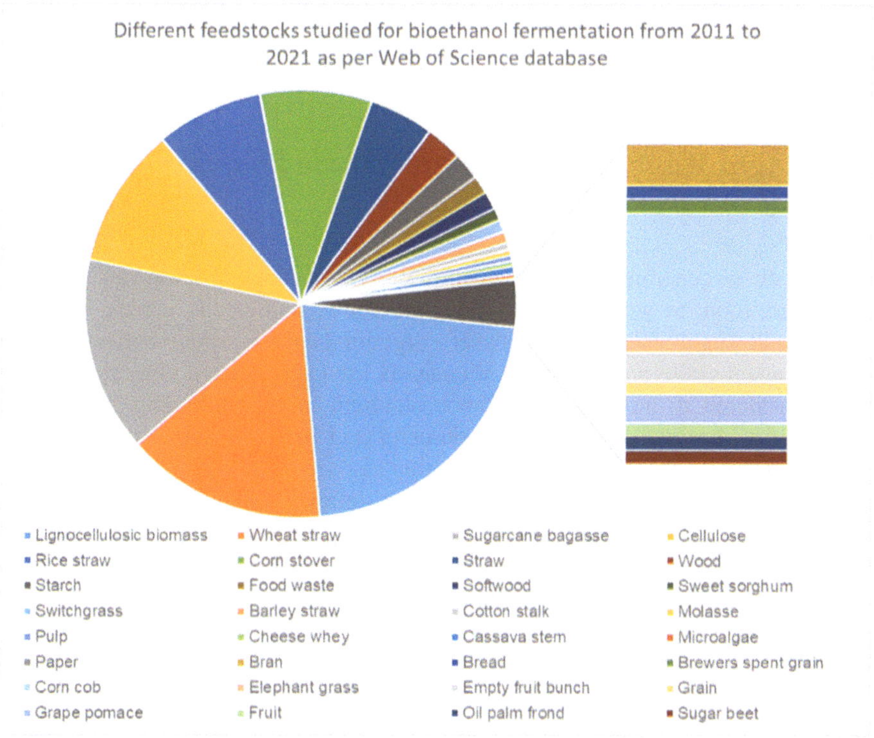

Fig. 8.2 Different feedstocks that were studied for bioethanol fermentation from 2011 to 2021 as per the Web of Science database

8.4 Classifications of Biomass for Bioethanol Production

To study the different biomass used for bioethanol production, it would be much more convenient to classify the wide range of biomass into four major categories, viz., agro waste, fruit waste, forest waste, and algal biomass (Table 8.1). Agro waste includes the waste byproducts of crops after collecting the edible or valuable parts, which includes straw, stem, cob, pulp, bran, etc. This is one of the most popular and well-studied substrates for bioethanol fermentation (Kumar et al. 2020a, b). Fruit wastes include the nonedible part of fruits, which includes peels, brunches, pomace, seeds, etc. This type of wastes is considered to be the most resourceful and drawn interest of biorefinery prospects (Vy et al. 2020). Forest wastes include dead leaves, softened woods, sawmill dust, etc., and other forestry by-products. This type of waste is characterized by a high content of cellulose, hemicellulose, and lignin, which could be easily converted into fermentable sugars that will act as a substrate for bioethanol fermentation (Hossain et al. 2021; Megersa 2020). Algal biomass has opened up a completely different approach to bioethanol production which is identified as fourth-generation bioethanol production. Easy cultivation of algal biomass and the simple extraction process of fermentable components from it with high consistency and reproducibility has drawn the significant interest of the scientific community (Ungureanu et al. 2020; Kumar et al. 2020a, b). Later in this chapter, a detailed study on these different types of biomass has been conducted (Table 8.1).

8.4.1 Agro Waste As Raw Material for the Production of Bioethanol

Agro waste is generated in huge quantities by a variety of industries including agriculture, food, as well as dairy industries. Straw, stem, stalk, leaves, bran, cob, strover, and sugarcane bagasse, sweet sorghum bagasse, spent coffee grounds, brewer's spent grains are among the abundantly used agro wastes. Once these wastes were considered to have no importance although in recent findings it has been considered to be one of the most resourceful and practical substrates for energy production due to the high content of usable resources. Many attempts have been undertaken to transform these lignocellulosic leftovers into useful products like biofuels, chemicals, and animal feed, with varying degrees of success. Due to its high carbohydrate and crude protein content, agro waste can be used as a substrate for ethanol production due to its abundant availability. The lignocellulosic agro waste is pretreated to transform into elementary carbohydrates. The rice straw was pretreated with grinding and 1% H_2SO_4 or mild NaOH before being fermented by *Candida shehatae* NCL-3501 (Raeisi et al. 2016). Different microbes like *Saccharomyces cerevisiae* MTCC 174, *Scheffersomyces stipites,* and *Spathaspora passalidarum* were reported to be used for the fermentation of sugarcane bagasse pretreated with 1% H_2SO_4 at 60 °C or 1.5% NaOH at 130 °C (Tyagi et al. 2019;

Table 8.1 Classification of biomass studied for bioethanol production and their pros and cons

Sl. no	Classes of biomass	Name of biomass	Pretreatment	Microorganism	Cons	Pros	Productivity of ethanol	References
1	Agro waste	Rice straw	Chopping and 1% H_2SO_4 or by mild NaOH	*Candida shehatae NCL-3501*	The inhibition of the final product lowers the ethanol yield. Contamination most likely developed as a result of the lengthy step procedure. It requires the use of an efficient enzymatic hydrolysis process for better ethanol yield	It is easy to cultivate and does not compete with food. It's a fuel generation method that's both environmentally friendly and cost-effective	10–40 g/L	Raeisi et al. (2016), Zhu et al. (2015)
		Sugarcane bagasse	1% H_2SO_4 at 60° or 1.5% NaOH at 130 °C	*S.cerevisiae MTCC 174* or *Scheffersomyces stipitis and Spathaspora passalidarum*			0.35–15.41 g/L	Tyagi et al. (2019), Nakanishi et al. (2017)
		Wheat straw	Grinding/milling or dilute 0.75% H_2S0_4 at 160 °C	*Clostridium acetobutylicum and Escherichia coli FBR5*			0.43–12 .41 g/L	Qi et al. (2019), Saha et al. (2015)
		Corn cobs	Acid hydrolysis and 12.70% $Na_3PO_4{\cdot}12H_2O$ at 121 °C	*Malbranchea cinnamomea, Scytalidium, thermophilium, and s.cerevisie*			10.3–36 g/L	Arumugam et al. (2020), Nakanishi et al. (2017)
2	Fruit waste	Orange	Acid hydrolysis and	*Candida parapsilosis IFM 48375 and Candida parapsilosis NRRL Y-12969*	It has the problem of microbial inhibition occurring during fermentation if essential oils are not extracted from biomass, resulting in a low bioethanol yield	Bioresources such as essential oil and nanocellulose can be found in citrus processing biomass. It is a low-cost way of producing bioethanol from a	0.07–7.53 g/L	Tsukamoto et al. (2013), Yang et al. (2018)
		Apple pomace	1% H_2SO_4 at 120°	*Pichia stipitis*			8.75–14.36 g/L	Kut et al. (2020), Evcan and Tari (2015)

(continued)

Table 8.1 (continued)

Sl. no	Classes of biomass	Name of biomass	Pretreatment	Microorganism	Cons	Pros	Productivity of ethanol	References
		Banana peel	Acid catalysis	*Saccharomyces cerevisiae*		five-carbon molecule. It is a suitable biomass for commercial scale scaling up of the process	5–38 g/L	Palacios et al. (2017)
		Pineapple	Ultrasonic pretreatment and NaOH alkali treatment	*Saccharomyces cerevisiae and Trichoderma harzianum*			4–6 g/L	Casabar et al. (2020)
	Forest waste	*Prosopis juliflora*	3% nitric acid with solid to liquid ratio 1:10	*Kluyveromyces marxianus*	Incorrect substrate assimilation by microorganisms during fermentation might lower ethanol yield. Pretreatment has a high cost, and inhibitor development happens after pretreatment	Its benefits include the ability to conduct each stage under perfect conditions and the low cost of saccharification and fermentation process	1.3–21.45 g/L	Sivarathnakumar et al. (2019), Nanssou et al. (2016)
		Wood chips	Steam explosion	*Saccharomyces cerevisiae*			10.60 g/100 g of dry weight	Cotana et al. (2014)
	Algae	*Chlorella vulgaris FSP-E*	1% H_2SO_4 121 °C for 20 min	*Z. mobilis*	The end-product inhibition reduces the amount of bioethanol. Contamination is a possibility as a result of the lengthy procedure	The acid hydrolysis procedure produces more effective ethanol and can also be used to produce huge amounts of bioethanol	11–20 g/L	Huang et al. (2013), Kumar et al. (2016)
		Cyanobacterium Synechococcus sp	Sonication, lysozyme and a-glucanase	*S. cerevisiae*			7–30.0 g/L	Möllers et al. (2014), Chow et al. (2015)
		Desmodesmus sp.	10% dry w/v with 2% (v/v) H_2SO_4 at 120 °C for 30 min and it followed by lyophilization	*S. cerevisiae*			40–61.2 g/L	Rizza et al. (2017)

Nakanishi et al. 2017). Pretreatment of wheat straw included grinding/milling or treating with 0.75% H_2SO_4 at 160 °C, followed by fermentation with *Clostridium acetobutylicum* and *Escherichia coli* FBR5 (Qi et al. 2019). Corn straw was pretreated with acid hydrolysis and 12.70% $Na_3PO_4.12H_2O$ at 121 °C, followed by fermentation and saccharification with *Malbranchea cinnamomea, Scytalidium, thermophilium*, and *S. cerevisiae* (Arumugam et al. 2020). Product inhibition affects ethanol yield, which is one of the drawbacks of bioethanol production from agricultural waste biomass, which might be the result of a multistep process. It necessitates the use of a high-efficiency hydrolysis method. The use of enzymes in hydrolysis of such wastes faces difficulties as the operational temperature would differ from the optimal temperature for that particular enzyme.

8.4.2 Fruit Waste Biomass in Bioethanol Production

Fruit waste accounts for a large portion of the resourceful anthropogenic organic waste generated globally. It contains the highest amount of free sugars compared to other lignocellulosic wastes. Fruit wastes such as banana peels, pineapple peels, apple pomace, and orange peel are investigated for their ability to produce total reducing sugars, pentose sugars, and bioethanol. The processing of fruit waste using acid degradation and fermentation with *Candida parapsilosis* IFM 48375 and *Candida parapsilosis* NRRL Y-12969 has drawbacks, such as inhibition of microbial growth would be inevitable if essential oils are not extracted from biomass, leading to a lower bioethanol yield. Pretreatment of apple pomace with 1% H_2SO_4 at 120 °C with fermenting organisms *Pichia stipitis* yields 8.75–14.36 g/L of ethanol (Kut et al. 2020; Evcan and Tari 2015). Acid hydrolysis pretreatment of the banana peel followed by fermentation with *Saccharomyces cerevisiae* yields 5–38 g/L of bioethanol (Palacios et al. 2017). *Saccharomyces cerevisiae* and *Trichoderma harzianum* aid in the fermentation process, yielding 4–6 g/L of bioethanol from pineapple peel. Samples hydrolyzed with 2% NaOH and treated with 30 min of sonication, were used for ethanol fermentation to examine their potential for bioethanol production. This was conducted in an inert atmosphere, with the hydrolysis samples being inoculated aseptically with *S. cerevisiae* at optimum pH (Casabar et al. 2020). Bioethanol synthesis from fruit waste has several downsides. Percent higher loading of biomass stimulates the production of inhibitory molecules, which could inhibit fermentation (Yang et al. 2018). Due to the high and diverse content of phytoactive compounds microbial growth would be affected in multiple ways making the process unpredictable and complex. Different instrumentation is needed, and the technique isn't yet well established for rapid quantification of such diverse phytoactive compounds contributing to the nonviability of the process (Evcan and Tari 2015).

8.4.3 Forest Waste As Raw Material in Bioethanol Production

Organic material from forest residue has the most adequate amount of lignocellulosic material, which is used to produce bioenergy by utilizing cellulose and hemicellulosic constituents. It is also a second-generation feedstock comprising hardwood, softwood, and its byproducts like wood chips and sawdust. The advantages of forestry waste feedstock are worldwide availability and low cost. Forest residues are collected from a forestry region that does not require any agricultural land; therefore forest feed-stock eliminates the competition for arable land. In India, the biennial deciduous species *Prosopis juliflora* is widely distributed. The constituents of *P. juliflora* were estimated to contain 25–30% cellulose, 40–45% lignin, 11–28% extractives, and 3–15% residues (Sivarathnakumar et al. 2019). The dust-free and air-dried hardwood collected from the forest was pretreated with nitric acid and pasteurized for 1 h at 121 °C. The batch fermentation dynamics of biomass development, substrate consumption, and bioethanol generation using wooden stems of *P. juliflora* were studied using *Kluyveromyces marxianus*. Sequential enzymatic hydrolysis and fermentation were performed on the pretreated substrates at pH 4.9 and 41 °C, and the maximum yield of bioethanol was determined to be 21.45 g/L (Sivarathnakumar et al. 2019). The difficulties of producing bioethanol from forest waste are inconsistent substrate assimilation by microorganisms throughout the fermentation affecting the final ethanol yield, the high cost of pretreatment, and microbial growth inhibition of the pretreated broth.

8.4.4 Algal Biomass in Bioethanol Production

Algae have recently drawn attention as the most potential source of biofuels as it eliminates the requirement for large arable lands and related resources. According to the Department of Energy, algae have the potential to produce more energy than the current biofuel crops grown on land. Algae are also good at recycling carbon from the air. Algae absorb and fix up atmospheric carbon dioxide converting it to organic carbon. When compared to petroleum fuel, algae-based biofuel produced through fatty acid transesterification is estimated to lower greenhouse emissions. Photosynthesizing biomass has the potential to generate biofuel and many other important bio-based products. Cyanobacteria and other eukaryotic microalgae, such as green algae, red algae, and diatoms, are the most prevalent photosynthetic organisms in the environment. *Synechococcus* cells after treatment with freezing, lysozyme, and 2-α-glucanases facilitated the breakdown and mobilization of monomeric sugar for fermentation by *S. cerevisiae*. Photosynthetic organisms' propagation under nitrate restriction produced starch biomass from cyanobacteria. Basic enzymatic treatment transformed this biomass into a feedstock appropriate for fermenting by *S. cerevisiae*. To achieve the highest ultimate ethanol concentration, no harmful chemical procedures or segregation of carbohydrate-containing components of biomass are required to yield 30 g/L of ethanol (Möllers et al. 2014). Similarly, the *Desmodesmus* sp. was pretreated using 10% dry w/v H_2SO_4 and 2% (v/v) H_2SO_4 at 120 °C for 30 min before freeze-drying (Rizza et al. 2017).

8.5 Conclusion

The study of different biomass as a substrate for bioethanol fermentation is largely practiced all over the world as the sustainability of bioethanol production largely depends on substrate selection. In the last decade, a wide range of different biomass had been investigated for its potential use. Biomass are grouped into four different categories: agro waste, fruit waste, forest waste, and algal biomass. All of these four types of biomass have their pros and cons. Despite that, some of these had exhibited breakthrough results to establish themselves as the sustainable substrate for bioethanol fermentation. Regardless of the improvement of the potential substrate, the advancement of the process technologies would also be required for the development of proper sustainable economic and advanced bioethanol production.

References

Arumugam A, Vishal Malolan V, Ponnusami V (2020) Contemporary pretreatment strategies for bioethanol production from corncobs: a comprehensive review. Waste and Biomass Valorization 12:1–36

Casabar JT, Ramaraj R, Tipnee S, Unpaprom Y (2020) Enhancement of hydrolysis with Trichoderma harzianum for bioethanol production of sonicated pineapple fruit peel. Fuel 279: 118437

Chow T-J, Hsiang-Yen S, Tsai T-Y, Chou H-H, Lee T-M, Chang J-S (2015) Using recombinant cyanobacterium (Synechococcus elongatus) with increased carbohydrate productivity as feedstock for bioethanol production via separate hydrolysis and fermentation process. Bioresour Technol 184:33–41

Cotana F, Cavalaglio G, Gelosia M, Nicolini A, Coccia V, Petrozzi A (2014) Production of bioethanol in a second generation prototype from pine wood chips. Energy Procedia 45:42–51

Esmaeili SA, Haji AS, Szmerekovsky J, Dybing A, Pourhashem G (2020) First-generation- vs. second-generation: a market incentives analysis for bioethanol supply chains with carbon policies. Appl Energy 277:115606

Evcan E, Tari C (2015) Production of bioethanol from apple pomace by using cocultures: conversion of agro-industrial waste to value added product. Energy 88:775–782

Hossain N, Mahlia TMI, Miskat MI, Chowdhury T, Barua P, Chowdhury H, Nizamuddin S et al (2021) Bioethanol production from forest residues and life cycle cost analysis of bioethanol-gasoline blend on transportation sector. J Environ Chem Eng 9(4):105542

Huang NL, Huang MD, Chen TLL, Huang AH (2013) Oleosin of subcellular lipid droplets evolved in green algae. Plant Physiol 161(4), pp. 1862–1874.

Kumar VB, Pulidindi IN, Kinel-Tahan Y, Yehoshua Y, Gedanken A (2016) Evaluation of the potential of Chlorella vulgaris for bioethanol production. Energy Fuel 30(4):3161–3166

Kumar SPJ, Sampath Kumar NS, Chintagunta AD (2020a) Bioethanol production from cereal crops and lignocelluloses rich agro-residues: prospects and challenges. SN Appl Sci 2(10):1–11

Kumar AN, Yoon J-J, Kumar G, Kim S-H (2020b) Biotechnological valorization of algal biomass: an overview. Syst Microbiol Biomanufactur 1:1–11

Kut A, Demiray E, Karatay SE, Dönmez G (2020) Second generation bioethanol production from hemicellulolytic hydrolyzate of apple pomace by Pichia stipitis. Energ Sources Part A Recovery Utilization Environ Effects 44:1–12

Megersa S (2020) Application of wood rot wild mushrooms in bioethanol production from sawdust of sawmills of Oromia Forest and wildlife Enterprise, Ethiopia. World News Nat Sci 29(3)

Mohanty SK, Swain MR (2019) Bioethanol production from corn and wheat: food, fuel, and future. In: Bioethanol production from food crops. Academic, New York, pp 45–59

Möllers KB, Cannella D, Jørgensen H, Frigaard N-U (2014) Cyanobacterial biomass as carbohydrate and nutrient feedstock for bioethanol production by yeast fermentation. Biotechnol Biofuels 7(1):1–11

Nakanishi SC, Soares LB, Biazi LE, Nascimento VM, Costa AC, Rocha GJM, Ienczak JL (2017) Fermentation strategy for second generation ethanol production from sugarcane bagasse hydrolyzate by Spathaspora passalidarum and Scheffersomyces stipitis. Biotechnol Bioeng 114(10): 2211–2221

Nanssou PA, Kouteu YJ, Nono, and César Kapseu. (2016) Pretreatment of cassava stems and peelings by thermohydrolysis to enhance hydrolysis yield of cellulose in bioethanol production process. Renew Energy 97:252–265

Palacios S, Ruiz HA, Ramos-Gonzalez R, Martínez J, Segura E, Aguilar M, Aguilera A, Michelena G, Aguilar C, Ilyina A (2017) Comparison of physicochemical pretreatments of banana peels for bioethanol production. Food Sci Biotechnol 26(4):993–1001

Qi G, Xiong L, Li H, Huang Q, Luo M, Tian L, Chen X, Huang C, Chen X (2019) Hydrotropic pretreatment on wheat straw for efficient biobutanol production. Biomass Bioenergy 122:76–83

Raeisi SM, Tabatabaei M, Ayati B, Ghafari A, Mood SH (2016) A novel combined pretreatment method for rice straw using optimized EMIM [Ac] and mild NaOH. Waste Biomass Valorization 7(1):97–107

Rizza LS, Smachetti MES, Do Nascimento M, Salerno GL, Curatti L (2017) Bioprospecting for native microalgae as an alternative source of sugars for the production of bioethanol. Algal Res 22:140–147

Rossi LM, Jean MR, Gallo LHC, Mattoso MS, Buckeridge PL, Allen DT (2021) Ethanol from sugarcane and the Brazilian biomass-based energy and chemicals sector. ACS Sustain Chem Eng 9(12):4293–4295

Saha BC, Nichols NN, Qureshi N, Kennedy GJ, Iten LB, Cotta MA (2015) Pilot scale conversion of wheat straw to ethanol via simultaneous saccharification and fermentation. Bioresour Technol 175:17–22

Sivarathnakumar S, Jayamuthunagai J, Baskar G, Praveenkumar R, Selvakumari IAE, Bharathiraja B (2019) Bioethanol production from woody stem Prosopis juliflora using thermo tolerant yeast Kluyveromyces marxianus and its kinetics studies. Bioresour Technol 293:122060

Susmozas A, Martín-Sampedro R, Ibarra D, Eugenio ME, Iglesias R, Manzanares P, Moreno AD (2020) Process strategies for the transition of 1G to advanced bioethanol production. PRO 8(10): 1310

Synnestvedt MB, Chen C, Holmes JH (2005) CiteSpace II: visualization and knowledge discovery in bibliographic databases. In: AMIA annual symposium proceedings, vol 2005. American Medical Informatics Association, p 724

Tsukamoto J, Durán N, Tasic L (2013) Nanocellulose and bioethanol production from orange waste using isolated microorganisms. J Braz Chem Soc 24:1537–1543

Tyagi S, Lee K-J, Mulla SI, Garg N, Chae J-C (2019) Production of bioethanol from sugarcane bagasse: current approaches and perspectives. In: Applied microbiology and bioengineering. Academic, New York, pp 21–42

Ungureanu Nicoleta, Valentin Vladut, and Sorin-Stefan Biris (2020) Capitalization of wastewater-grown algae in bioethanol production. Engineering for rural development: Jelgava, Latvia. 1859–1864

Vy NT, Thuy YU, Manmai N, Whangchai K, Ramaraj R (2020) Impact and significance of pretreatment on the fermentable sugar production from low-grade longan fruit wastes for bioethanol production. Biomass Conversion Biorefin 12:1–13

Yang P, Yun W, Zheng Z, Cao L, Zhu X, Dongdong M, Jiang S (2018) CRISPR-Cas9 approach constructing cellulase sestc-engineered Saccharomyces cerevisiae for the production of orange peel ethanol. Front Microbiol 9:2436

Zhu S, Huang W, Huang W, Wang K, Chen Q, Yuanxin W (2015) Pretreatment of rice straw for ethanol production by a two-step process using dilute sulfuric acid and sulfomethylation reagent. Appl Energy 154:190–196

Chapter 9
Microbial Biodiesel for Future Commercialization

P. Kavya, R. C. Theijeswini, and M. Gayathri

Abstract In spite of the increased energy requirement in the mechanical globe and the contamination issues induced by extensive utilization of renewable energies, the necessity for developing sustainable and eco-friendly energy sources is rising. Biodiesels are a significant alternative for petroleum-derived fuel with the advantages like renewability, eco-friendliness, lubricity, and flexibility. Microbes are highly efficient feed stocks for production of biodiesel owing to their least process requirements, rapid production rates, and ease of ramp up for technical development. The production of sustainable biodiesel using microbes has gained appreciable advances in commercialization. Recently, biodiesels derived using *Rhodosporidium* species, fungus, lignocellulose, and microalgae are commercialized. Sustainable biodiesel has been commercialized in different countries mainly Brazil, South Africa, the USA, the European Union, Indonesia, Malaysia, and Germany. The recent advances in microbial biodiesel commercialization and global biodiesel commercialization are described in this chapter. Different microbial sources and techniques for biofuel production are also summarized.

Keywords Sustainable · Eco-friendly · Energy sources · Biodiesel · Microbial · Commercialization

9.1 Introduction

In the contemporary timeline, the anticipation of dwindling fossil fuel reserves has become an inevitable reality. This crisis has rather led to a situation that imperatively demands the high-scale development of alternative energy resources, which are ought to be economically feasible, and sustainable along with being eco-friendly. Presently the most superior resources for energy supply around the world are

P. Kavya · R. C. Theijeswini · M. Gayathri (✉)
Department of Bio-Medical Sciences, School of Bio Sciences and Technology, Vellore Institute of Technology, Vellore, India
e-mail: gayathrigopinath@vit.ac.in

A. Sarkar, I. A. Ahmed (eds.), *Microbial products for future industrialization*, Interdisciplinary Biotechnological Advances,
https://doi.org/10.1007/978-981-99-1737-2_9

petroleum, natural gas, and coal. Nevertheless, the deficient and very finite amount of these reserves has pushed the consideration of switching to alternative fuels from renewable sources to the edge. Most of the renewable sources pose dominance over the existing fossil fuels as they possess various benefits. These include reduced greenhouse gases and pollution. Biodiesel can be soundly acknowledged as one such alternative form of energy. Among the ample amount of choices available, biodiesel has been getting greater prominence due to certain undeniable characteristics they possess including biodiesel can be efficiently produced from an extensive array of resources like oily sludge dispensed from factories, unproductive animal fat, waste cooking oil and even microbes; numerous technological options are available for the production of biodiesel formulated on the raw material's quality, providing probable alternatives to reduce the expense of the gross production (Gebremariam and Marchetti 2018); biodiesel completely lacks sulfur in comparison with contemporary petroleum diesel fuel; carbon monoxide production is minimized along with the reduction in fumes, hydrocarbons, and particulate matters (Hasan and Rahman 2017), and the presence of an additional amount of free oxygen in contrast with petroleum diesel fuel results in absolute combustion and lessened emission (Fazal et al. 2011). Other notable features include: a greater flashpoint, built-in lubricity, and biodegradability (Knothe 2008).

This chapter predominantly focuses on biodiesel derived from microbes, the production aspects and commercialization around the globe. Microbial oils that are synthesized via a copious amount of microorganisms, which includes both the broad categories, prokaryotes (bacteria), and eukaryotes (filamentous fungi, yeast and microalgae), which tend to build up intracellular oils in volumes up to 20% of dry cell weight, are now gaining a greater emphasis as renewable raw material substitutes for the production of biodiesel. There is a wide variety of microbial sources, which are used in the production of biodiesel including microbial lipids/microbial oils, microbial lipases, filamentous fungi, microalgae, yeast, bacteria, isoprenoids, etc. (Yuzbasheva et al. 2014). Microorganisms have got numerous benefits in contrast to plants, which comprise a smaller life-cycle; do not require a big piece of land area; despite the climatic and seasonal variations, there is a constant accumulation of biomass; a small footprint and lipid content is high (Wahlen et al. 2012).

Eukaryotic-derived microbial oil synthesizers are observed to accumulate biomass with greater intensity and have shown exceeding yields in comparison with the prokaryotic yield of the target product (Liang and Jiang 2013). In the present scenario, prokaryotes are more efficiently used as model objects rather than eukaryotes as they cannot aid in certain sectors associated with the production of biodiesel. The microbial oils that are produced comprise neutral lipids, for instance, free fatty acids, triacylglycerol, and stearin esters. They also play a role as the cell's energy reserve (Yuzbasheva et al. 2014). Triacylglycerol serves as the majority in the neutral lipid composition. This composition varies with each microbe.

Despite the various advantages, there are a number of other non-beneficial factors that hamper the production rate and industrial-scale commercialization of microbial biodiesel which comprises of: (1) severe microbial culture conditions; (2) highly expensive and complex procedure of: extraction of microbial oil and microbial harvesting; (3) aside from the raw materials needed, the transesterification process requires a highly efficient and economical catalysts. Therefore, homogeneous catalysts were developed for the same (NaOH, H_2SO_4) (Bušić et al. 2018). Nonetheless, retrieval and reuse of the soluble catalysts from the final transesterification products appeared to be very difficult. However, the heterogeneous catalysts that were developed to combat the disadvantages and associated inefficiency of homogeneous catalysts, too exhibit finite drawbacks like: intricacy in the preparation process, inconsistent activity, severe reaction conditions, and so on (Bohlouli and Mahdavian 2021).

In the present arena, there are diverse countries that are commercializing microbial biodiesel worldwide including Brazil, the European Union, Germany, the USA, Malaysia, Indonesia, Taiwan, India, the Philippines, and China.

9.2 Microbial Sources for Biodiesel

In the necessity to discover a potential alternative to the environmentally unfriendly and nonrenewable, conventional petroleum gas, biodiesel might be the new potential alternative in the field of science and technology. With biodiesel broadening the idea of sustainable, eco-friendly and renewable fuel, microbes are tending to gain the priority in this aspect. Microbial biodiesel can be soundly approved as a significant scientific breakthrough as various researches carry on manifesting yeast, fungi, bacteria, and microalgae as a propitious capable raw material for the production of biodiesel. The different microbial sources for biodiesel production are listed in Fig. 9.1.

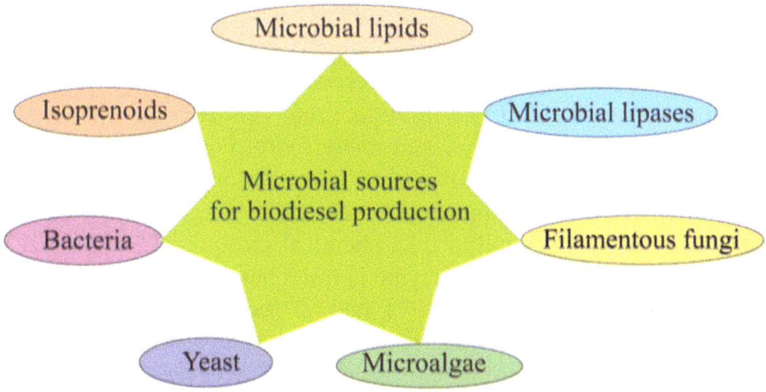

Fig. 9.1 Microbial sources for biodiesel production

9.2.1 Microbial Lipids

Those microbes that can build up lipid in addition to 20% of their biomass such as yeasts, algae, molds and bacteria (i.e., bacillus) are collectively known as "oleaginous microorganisms." With respect to type and construction, microbial lipids share a fair similarity with plant's oil (Ram and Tyagi 2019). Accumulation of lipids occurs in oleaginous microbes when the cell is deprived of nitrogen and/or other nutrients excluding carbon. It includes magnesium, phosphorous, iron, and zinc. Due to the high similarity of single-cell oil with plant oil's framework and fatty acid make-up, single cell oil seem promising for application in the production of biodiesel (Madani et al. 2017).

The three general methods for producing microbial oil and using it as a substrate for biodiesel manufacturing and other products can be listed as oleaginous microorganisms having a high potential for lipid synthesis are screened; raising lipid content through environmental parameter optimization and mutation and the use of by-products in the manufacture of lipids. Microbial lipids are certainly preferable because the life cycle of microbes is short, the requirement of labor is minimal, less affected by climate and season and scaling up is easier (Ochsenreither et al. 2016).

In the case of depletion of a major nutrient, the cell enters a lipid storage phase, in which surplus carbon is converted to storage lipid. Oil reserves might be mobilized and converted into cellular components if the cells return to an environment where nitrogen could be accessible (Ratledge 2002; Ratledge and Wynn 2002). The surplus carbon is converted to lipid stores during the second phase. Cells cannot grow and replicate due to the lack of nitrogen. Isocitrate dehydrogenase is a crucial enzyme in the citric acid cycle that becomes inactive in the absence of nitrogen. As a result of the oleaginous yeasts inability to metabolize isocitrate, isocitrate and citric acid accumulate in mitochondria and are quickly transported to the cytoplasm (Madani et al. 2017).

9.2.2 Microbial Lipases

Lipase enzyme has been synthesized by copious amounts of fungus and bacteria, which include *Pseudomonas aeruginosa*, *Aspergillus niger*, *Staphylococcus epidermidis*, *Burkholderia cepacia*, and many more. For the production of crude lipase, fungi broth culture and solid-state fermentation might be viable options (Bhan and Singh 2020). Since lipases are good catalysts, their manufacturing and utilization could be a better option than chemical catalysts (Li et al. 2012). Even though the main source of lipases enzyme is microbes thriving in the environment, lipase is also produced with the aid of natural foods (i.e., plants and animals) (Toldrá-Reig et al. 2020; Tae Hwang et al. 2014).

Esterification and transesterification reactions are catalyzed by a lipase enzyme which is a biocatalyst. An enzyme catalyst like lipase enzyme has several benefits in comparison with a chemical catalyst including the requirement of less temperature and energy, less time consuming, absence of formation of soap, environmentally friendly, and minimal amounts of water required for washing (Bhan and Singh 2020).

Long chains of triacylglycerol's are hydrolyzed by the lipase enzyme, which acts as a catalyst. The transesterification reaction catalyzed by lipase enzyme occurs in two steps. This can be explained as: (1) Hydrolyzing of the fatty acid ester bond aids in the conversion of triglycerides to monoglycerides; (2) Ester is formed by alcohol, which plays the role of an acyl acceptor. The presence of H_2O is a major determinant of lipase activity. Water aids in retaining the lipase activity; the oil–water interface is formed as a result of a combination of water and oil, which aids in enzyme activation and active site reformation by conformational change (Al-Zuhair 2007).

Microbial lipases are found to be predominantly extracellular in nature; also, lipolytic genes in several organisms can be modulated by lipids. In the presence of an aqueous medium, an active and highly specific 3-D structure is exhibited by the lipase enzyme, which contains both polar and nonpolar groups. The role of lipase as a catalyzing enzyme can be determined by certain parameters including the area of the interface, nature of the interface, and interfacial properties. The lipase enzyme is activated by an adsorption mechanism, and the interface aids in the opening of the catalytic active site of lipase.

9.2.3 Fungi

Fungi are one of the oleaginous microbes, which can aid in the synthesis of biodiesel. The fungus produces and builds up large quantities of triacylglycerides inside their cell, almost up to 70% of their biomass weight. They share significant similarities with the plant's oil in terms of construction and energy value (Lunin et al. 2013). Fungal single celled oil poses certain advantages such as very small life cycle and production remains consistent irrespective of seasonal variations.

Microbial lipids were commercially synthesized for the first time from *Mucor circinelloides*. Its lipid accumulation capacity was researched widely, and its TAG has been put forward as a feedstock for synthesizing biodiesel through the direct transformation of lipids (Vicente et al. 2009). *Umbelopsis isabellina*, a phylogenetically related species, has turned out as a potential species for the conversion of biomass into biodiesel (Rossi et al. 2011).

Between eukaryotic organisms, the metabolic pathway of lipid synthesis does not exhibit a definite difference. Therefore, no difference can be observed between oleaginous fungi and non-oleaginous fungi. The parameters that affect the amount of significant lipid accumulation include metabolic pathway regulation, provision of precursors (acetyl co-enzyme A, G-3-P (glycerol-3-phosphate), malonyl coenzyme A) and supply of the cofactor, NADPH (Rossi et al. 2011).

9.2.4 Microalgae

Photosynthetic microbes that have the ability to convert sunlight, water (H_2O), and carbon-dioxide (CO_2) into algal biomass are widely known as microalgae (Metting and Pyne 1986). Most of them are quite promising in the production of biodiesel because they are exceptionally abundant in oil. This prime feature can be utilized in biodiesel production with the help of conventional technology (Mallick et al. 2016). Currently, microalgae are gaining immense importance as it is expected to be the sole source of biodiesel production to replace fossil diesel. Contrary to the other kinds of oil crops present, microalgae are quite advantageous. This is because they grow exceedingly faster and are immensely rich in oil. A surprising fact is that, just within 24 h they can produce twice the biomass.

Microalgae could have oil concentration approximately 80 percentages by weight of dry biomass. For the potential and maximum production of biodiesel, microalgae that own immense oil productivities are chosen. Varieties of hydrocarbons, complexes, and lipids are produced with respect to the species of the microalgae. This indicates that not all algal oil is a potential candidate.

Even though heterotrophic microbes that are grown on a carbon source (organic and natural), for instance, sugar are potential oil carriers, their efficiency is lower in contrast with photosynthetic algae (Gong and Jiang 2011). This can be reasoned as heterotrophic microbes require natural carbon sources for their growth, which are in turn synthesized by photosynthesis in crop plants. The more the quantity of microalgal biomass, the more algal oil can be produced (Song et al. 2008). Microalgal biodiesel would have to meet existing requirements in order to gain user acceptability. Most of the microalgal oils are less likely to meet "European biodiesel criteria" due to their composition, anyways this does not account for a serious limitation. Partially catalytic hydrogenation of microalgal oil, which is quite similar to the process used for making margarine from vegetable oil, might easily, lessen the amount of unsaturation and content of the fatty acids with greater than 4 double bonds. The cost for producing microalgal based biodiesel can be significantly reduced by following certain strategies including bringing on a production ideology based on biorefinery; genetic engineering can be used as a key to improvise the microalgae's competence, to bring on advancements and improvements in the field of photo bioreactor engineering (Chisti 2007).

9.2.5 Yeast

Oleaginous yeasts can be soundly recognized as efficient candidates for biodiesel production due to the potential features they own, which comprises of maximal unicellular growth rate, very short duration of life, scale-up can be done with ease, and very fast lipid gathering capability in the detached lipid bodies (Phukan et al. 2019). To add on, they also possess the capability of utilizing cheap fermentation media, which comprises agricultural and industrial nutritional remains (Ratledge 1993).

Yarrowia lipolytica, *Rhodosporidium toruloides*, *Cryptococcus curvatus*, *Lipomyces starkeyi* build up greater than 20% of lipids of the dry cell weight they own (Patel et al. 2016). On a per-cell basis, oleaginous yeast varieties produce lipid yields of up to 65% dry weight along with biomass yields of around 10–100 g/L in 3–7 days. Furthermore, most of the yeast lipids are long-chain fatty acid molecules, which possess high similarities with common vegetable oils utilized in the production of biodiesel. This is the main reason why it is considered one of the potential options for the adoption of carbs-based microbial oil.

The intracellular product produced by oleaginous yeasts by the utilization of numerous carbon substrates along with the conversion of various lipids to triacylglycerol form is known as bio-oil. By the process of transesterification process, such a second-generation fuel could be applied in biodiesel production. Intracellular lipids of yeasts primarily consist of triacylglycerides, which are analyzed as efficient candidates for durable biodiesel precursors. However, synthesizing biodiesel from yeast on its own is not economically attainable in the contemporary timeline (Kongruang et al. 2020). Usage of nutrient sources with low value and implementation of economical downstream procedures in yeast cultivation lowers costs and makes the yield lipids highly competitive with the other oils in the production of biodiesel (Ratledge and Cohen 2008).

9.2.6 Bacteria

Some kind of bacteria serves as reasonable candidates in the production of biodiesel due to their oil accumulative properties (Koppolu and Vasigala 2016). Non-photosynthetic bacteria have been reported to build up storage material in the form of substantial quantities of wax esters and triacylglycerol (Rojo 2008). Oleaginous bacteria act as a great source of triacylglycerides but still, their use as a potential candidate for biodiesel production is very limited in comparison with yeast and microalgae (Kumar et al. 2020).

Certain bacterial species like *Rhodococcus, Arthrobacter, Acinetobacter*, and *Gordonia* have definite oil synthesizing ability. However, *Rhodococcus* sp. has gained much significance as researches have proved that they have a strong capability for growing on various substrates (Patel et al. 2020). Cyanobacteria own a photosynthetic membrane that ultimately serves as a huge reservoir where the diacylglycerols along with membrane lipids are accumulated. This in turn helps in harnessing the synthesis of biodiesel by the process of transesterification followed by methanol and then catalysts (Liu et al. 2011).

9.2.7 *Isoprenoid*

In recent times, the field of agriculture has contributed immensely toward the constant supply of lignocellulose (i.e., plant biomass), which can be efficiently used as a sustainable mode of feedstock, which results in carbon-neutral transportation fuels. In an attempt to extract this potential, researchers have created a new class of biodiesel fuels based on isoprenoids. To identify which compounds are compatible with the existing engine technology, researchers chemically produced and ran tests on several biodiesel compounds based on isoprenoids. *Escherichia coli* is being utilized to develop biosynthetic or metabolic pathways for the manufacture of these attested isoprenoid-based biodiesel choices. For the generation of target biodiesels, some of these methodologies have been built and evaluated. The development of new pathways genes and the optimization of the process of production are now being investigated (Gupta and Phulara 2015).

9.3 Processes for Microbial Biodiesel Production

Although biodiesel is a significant alternative over petroleum diesel in different aspects, it is imperiled due to the high expense of raw materials and lack of industrially and economically feasible technology for its adequate production from any kind of raw material. There are various processes with varying commercial viability employed for microbial biodiesel production, based on the availability of raw materials, production expenditure, and related challenges. There are five major routes by which microbial biodiesel can be produced including transesterification, metabolic engineering, pyrolysis, enzyme catalysis, and microalgal torrefaction. The different processes for microbial biodiesel production are listed in Fig. 9.2.

9.3.1 *Transesterification*

Transesterification is a simple and less expensive process while the choice of raw material is crucial as it possesses about 70% of the production rate. Transesterification involves the production of biodiesel from fat and oil raw materials. Transesterification reaction involves the catalytical conversion of fat/oil into esters or biodiesel (Rezania et al. 2019) (Fig. 9.3).

Different factors such as type of catalyst, type of reactor, nature of raw material, stirring speed, temperature, reaction time, type of solvent, and degree of solvent to oil have been demonstrated to be significant to get maximum biodiesel production through transesterification process. The type of catalyst and raw material are the significant factors for the efficient production of biodiesel, as these parameters can affect the expenditure of biodiesel production. The transesterification reaction is

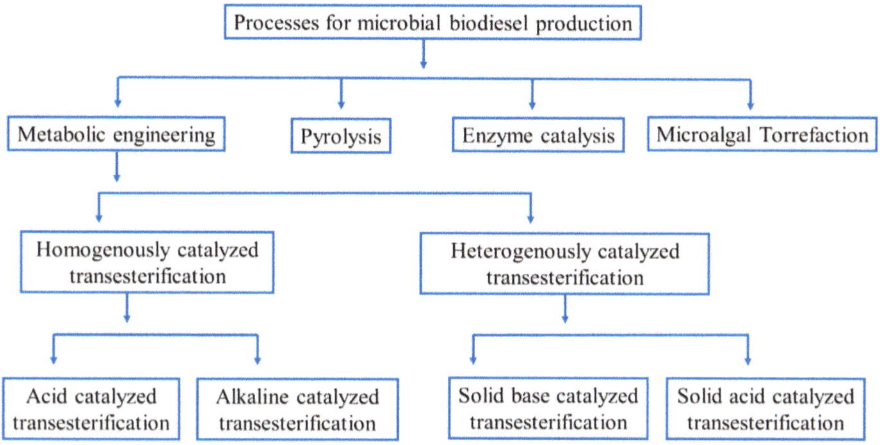

Fig. 9.2 Processes for microbial biodiesel production

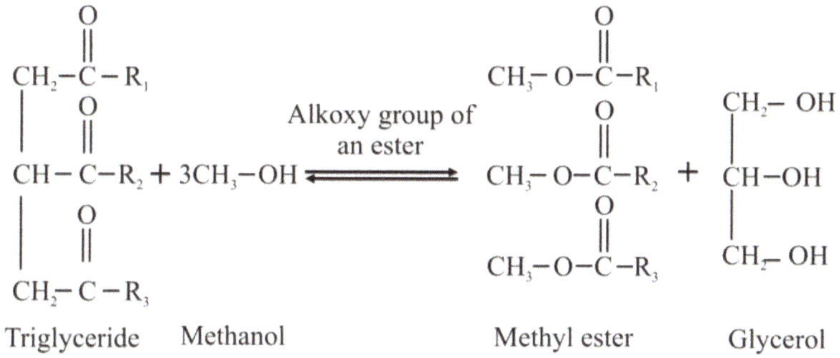

Fig. 9.3 Transesterification reaction

performed in the presence of enzymatic, acidic, or alkaline catalysts. It can be homogeneously or heterogeneously catalyzed, in accordance with the solubility of catalyst in the reactants (Athar and Zaidi 2021).

9.3.1.1 Homogeneously Catalyzed Transesterification

Homogeneous catalysts can be alkaline or acidic and the choice largely relies on the free fatty acid and water content of the oil. Homogeneous catalysts for transesterification also possess many drawbacks including great energy demand, soap formation as a by-product, laborious and expensive removal of catalyst, and huge quantity of waste water production.

9.3.1.1.1 Acid Catalyzed Homogeneous Transesterification

Acid catalyzed transesterification involves triglyceride reacting with an alcohol in the presence of an acid catalyst to make esters or biodiesel. This method is favorable and economically feasible in producing biodiesel from triglyceride sources. A few remarkable advantages of acid catalysts are less sensitivity for the free fatty acid content of the oil and esterification and transesterification reactions taking place simultaneously.

Esterification involves free fatty acid reacts with an alcohol in the presence of acid catalyst to make ester (biodiesel) and water as products. Generally, sulfuric acid, hydrochloric acid, phosphoric acid, ferric sulfate, and organic sulfonic acids are employed as acid catalysts for esterification reaction (Fig. 9.4).

Nevertheless, acid catalyzed reactions are time consuming and need an elevated reaction temperature, bigger quantity of alcohol, and low activity of catalyst compared to the alkali catalyzed reactions (Talha and Sulaiman 2016).

9.3.1.1.2 Alkaline Catalyzed Homogeneous Transesterification

Homogeneous transesterification in the presence of an alkaline catalyst is the less expensive and fastest method in comparison with other enzymatic methods. It reaches an increased production of biodiesel with increased purity. The typical homogeneous alkaline catalysts that have been strongly employed as biodiesel catalysts are sodium hydroxide (NaOH) and potassium hydroxide (KOH). Even though biodiesel production in the presence of sodium and potassium methoxides as catalysts is improved, they are not commonly employed due to high cost (Atabani et al. 2012).

9.3.1.2 Heterogeneously Catalyzed Transesterification

Catalysts that are immiscible with the reaction blend are known as heterogeneous catalysts. It may be acidic or alkaline kind and is more easily removed from the products. Heterogeneous transesterification is carried out by solid catalysts as a functional substitute for homogeneous catalysts. Solid catalysts can be reused and facilitate both esterification and transesterification reactions to take place concurrently; hence it leads to less expensive production of biodiesel. Heterogeneous

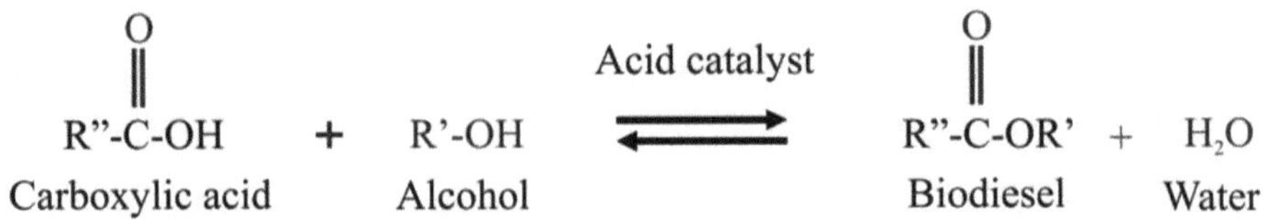

Fig. 9.4 Esterification reaction

catalysts are synthesized from renewable substances and are more efficient than homogeneous catalysts as they overcome the drawbacks of homogeneous catalysts (Wang and Yang 2007).

9.3.1.2.1 Heterogeneous Solid Base Catalytic Transesterification

Solid base catalysts are highly dynamic compared to solid acid catalysts. The solid base performs the heterogeneous catalytic process, which contributes to less expensive biodiesel production. If calcium oxide and magnesium oxide are proved to be dynamic and stable, it would be advantageous for commercial biodiesel production, as the particular metal oxides are less expensive and easily available (Abbaszaadeh et al. 2012).

9.3.1.2.2 Heterogeneous Solid Acid Catalytic Transesterification

Even though heterogeneous solid acid catalysts have lesser activity, it has been employed in different industrial processes since they contain diverse acid sites with distinct strength of Lewis acidity, in comparison with homogeneous acid catalysts. Solid acid catalysts have different benefits including simultaneous esterification and transesterification, removal of the purification process of biodiesel, a simple process of separation of the catalyst from the products, and minimizing corrosion (Dalai et al. 2006).

9.3.2 Metabolic Engineering

Metabolic engineering is an advanced approach for creating extremely functional microbial cells and a chief component for biodiesel production. It has been widely employed to produce needed products in different engineered microbial hosts. A broad spectrum of novel compounds has been synthesized through metabolic engineering (Adegboye et al. 2021). It is also a favorable approach proposed to overcome economic challenges (Naghshbandi et al. 2020).

Isoprenoids are a potential raw material owing to their distinct properties indistinguishable from that of fossil fuels. Large-scale research has been performed to employ metabolic engineering approaches in microbes essentially, to overcome the restrictions related to their production and to develop industrially efficient microbial strain for isoprenoid-derived biodiesel production (Gupta and Phulara 2015). Isoprenoid-derived fuels have indistinguishable energy content, combustion properties, and storage and transportation properties with fossil fuels. Microorganisms synthesize isoprenoids through the natural pathway, but natural titer is very less for industrialization. Metabolic engineering approaches have the potentiality to redirect the metabolic pathways and hence increase the titer for microbial-based advanced biofuel synthesis (Das et al. 2020).

Metabolic engineering approaches are demonstrating benefits for the production of biodiesel and biodiesel raw materials in various microbes including yeasts, microalgae, blue-green algae, and bacteria. In comparison with conventional random mutagenesis for microbial strain advancement, metabolic engineering approach is sensible, fast, and immensely powerful. Metabolic and genetic engineering along with synthetic biology and systems biology tools is the key to synthesizing greatly efficient microbial cell factories for biodiesel production (Majidian et al. 2018).

9.3.3 Pyrolysis

Pyrolysis is the process of converting one organic matter into another by applying heat in the absence of air or by heating and using a catalyst resulting in rupture of chemical bond and formation of diverse small molecules (Abbaszaadeh et al. 2012). The process of pyrolysis takes place at temperatures ranging from 400 to 600 °C. Conversion of triglycerides employing thermal breakage produces aromatic compounds, alkanes, alkenes, alkadienes, and carboxylic acids, which are fuel constituents for the diesel engine and it suggests a promising technology for biodiesel production (Maher and Bressler 2007). Nevertheless, a few unwanted compounds are also formed owing to the absence of oxygen during the process. It is a comparably slow chemical reaction to convert renewable energy sources into biofuels such as biodiesel by applying heat (Saidur et al. 2011). This method is simple, functional, eco-friendly, and lacks side products. Partial volatility, vulnerability, and minimum heating value of the products are a few undesired characteristics that generally restrict the utilization of biodiesel produced via pyrolysis. Pyrolysis generates gases, bio-oil, and char based on the degree of heating. According to the working conditions, pyrolysis is divided into three subdivisions: slow pyrolysis, fast pyrolysis, and ultra-fast pyrolysis. Slow pyrolysis takes place at 10 °C and the main products include gas, char, and tar. Fast pyrolysis takes place at 100 °C and the main products include thinner bio-oil, gas, and char. Ultra-fast pyrolysis takes place at 10 ° C and the main products include bio-oil and gas.

9.3.4 Enzyme Catalysis

Enzymes are biological catalysts and it has become very significant in biofuel production recently (Atabani et al. 2012). Biological catalysts are lipases that are demonstrated to be capable of performing transesterification reactions that produce biodiesel. Lipase catalyzed transesterification involves triglyceride reacting with an alcohol in the presence of lipase as a catalyst to produce biodiesel and glycerol. Lipase catalysis has many advantages compared to chemical catalysis including lesser energy demand, purified biodiesel and glycerol, increased yield of products, simple product recovery, and no waste water as by-product. Biodiesel production employing enzymes is generally suggested since it doesn't include washing, purification, neutralization, and saponification steps. Enzyme catalyzed transesterification process possesses numerous advantages compared to chemical catalyzed transesterification including easy to remove products, lack of side products, needs only moderate process conditions, and reusability of the catalysts. Enzymatic catalysts are generally employed in the immobilized form to confirm their solidity and recycling. This helps in less costly production of biodiesel and leads to commercialization (Rakchai and Kittikun 2016).

9.3.5 Microalgal Torrefaction

Microalgae are demonstrated as favorable raw materials for biodiesel production and have been widely analyzed. Carbohydrate, lipid, and protein are the three chief components of microalgal cells; these are synthesized and degenerated during microalgal growth. These changes may notably affect the properties of biodiesel during thermochemical conversion. Torrefaction has a major role in microalgal solid biodiesel production and has gained increasing interest. Torrefaction is a less strong pyrolysis for producing solid biofuel by applying 200–300 °C of temperature for 60 min.

On the basis of heating mode and carrier gas, torrefaction can be subdivided into conventional torrefaction, microwave torrefaction, wet torrefaction, oxidative torrefaction, etc. In conventional torrefaction, raw material is heated at 200–300 °C for 10–60 min in nitrogen containing atmosphere. An elevated temperature in association with a lesser time reduces expenditure and hence leads to huge energy utilization. Microwave torrefaction involves the employment of microwave irradiation, which pyrolyzes biomass from the inner portion of the material and achieves a higher heating efficiency with lesser time and easy scale up. Wet torrefaction takes place at 180–260 °C for 10–240 min. The process takes place under increased pressure and is similar to hydrothermal liquefaction. It is suitable for moisture containing raw materials, since it doesn't require dewatering process. Wet torrefaction produces solid biofuel with higher quality compared to conventional torrefaction. It can also reduce the ash content of solid biofuel, hence making it appropriate for huge ash containing microalgal upgrading. Oxidative torrefaction takes place at 200–300 °C for 10–60 min in the presence of air. It includes oxidation and devolatilization of raw material. This method is less time consuming, less expensive, and needs lesser energy input. Torrefaction is necessary for microalgal biomass amelioration, as torrefied products commonly have higher energy density. After torrefaction, hydrophobicity, calorific capacity, grindability, storage, and transportation properties are appreciably improved. Microalgal solid biodiesel is assessed by regulating its cellular level constituents. As a result of variations among microalgal species and cultivation conditions, the quantities of cellular constituents in microalgae can vary to a great extent. This may alter the function and properties of microalgal biodiesels. Synthetic biology in combination with significant torrefaction technologies should be employed for the adequate production of microalgal solid biofuel (Ho et al. 2020).

9.4 Advantages and Disadvantages of Microbial Biodiesel

9.4.1 Advantages of Microbial Biodiesel

Biodiesel is an ideal fuel for densely contaminated cities as it is renewable, noncombustible, inexhaustible, safe, transportable, sustainable, environment friendly, and lacks or contains a low quantity of ash, aromatic compounds, sulfur, carbon, nitrogen, particulate matter, and trace elements. It reduces greenhouse gas emissions and protects the ozone layer. It minimizes risks for cancer and neonatal imperfections by 90% owing to its lesser contaminating combustion. Biodiesel aids rural advancement to reinstate deteriorated lands. Furthermore, it has adequate potential for rural employment opportunities. Biodiesel possesses a higher cetane number compared to petroleum diesel, which minimizes the delay for combustion. The production process is easy, less time-consuming, and is a cheap resource. Biodiesel possesses higher lubricity properties. This enhances lubrication in different parts of the vehicle including fuel pumps and injector units, which reduces engine wear and tear and enhances engine functionality. Biodiesel possesses a higher flash point; hence it is safe for transportation, operating, dispersal, employment, and storage (Saidur et al. 2011; Kamm 2004; Atabani et al. 2012).

9.4.2 Disadvantages of Microbial Biodiesel

Biodiesel has a lower energy density; this causes increased fuel consumption. It has lower volatility that leads to the generation of deposits in the engine owing to partial combustion properties. It has a higher viscosity and lower volatility, hence requiring higher injector pressure. Oxidation stability of biodiesel is lesser; hence it oxidizes into fatty acids in the presence of oxygen and leads to corrosion of the vehicle parts including injector, fuel tank, and pipe. Sources of biomass are dispersed and may not be obtainable in adequate amounts to use as a national energy source (Saidur et al. 2011; Atabani et al. 2012).

Some of the advantages and disadvantages of microbial biodiesel are listed in Table 9.1.

Table 9.1 Major advantages and disadvantages of microbial biodiesel

Advantages of microbial biodiesel	Disadvantages of microbial biodiesel
Renewable, noncombustible, inexhaustible, safe, transportable, sustainable, easily obtainable, environment-friendly, biodegradable, and highly lubricant	Lower energy density
Reduces greenhouse gas emissions and protects the ozone layer	Increase in fuel consumption
Generally low quantity of ash, aromatic compounds, sulfur, carbon, nitrogen, particulate matter, trace elements	Higher viscosity and lower volatility
Biodiesel possesses a higher cetane number and flash point	Causes corrosion in vehicle parts
Production process is easy, cheap resource and less time consuming	Sources of biomass are totally dispersed and may not be obtainable in adequate amounts to use as a national energy source
Increases rural employment opportunities	Lower oxidation stability

9.5 Recent Advances Toward Microbial Biodiesel Commercialization

9.5.1 Rhodosporidium *Species-Derived Biodiesel*

Microbial lipids are a promising source of lipids that could be utilized in biodiesel production (Maza et al. 2021). Recently *Rhodosporidium* sp. has gained remarkable significance in biodiesel production owing to its capability to collect high lipid content. Furthermore, it can flourish on a broad range of substrates, has more inhibitor tolerance, and rapid growth rate. With the execution of needed properties employing genetic engineering techniques, *Rhodosporidium* sp. is anticipated to be highly utilized in the future concerning lipid accumulation and biodiesel production (Saini et al. 2020).

9.5.2 Fungal-Derived Biodiesel

Reduction in the number of nonrenewable fuel reservoirs, environmental issues, and low yields from plant and animal sources cause the necessity to develop microbes for sustainable and renewable energy production. Fungal hosts exhibit commercial advantages compared to other microbes, owing to their strong tolerance for industrial fermentation. Even though fungi have been employed to produce a wide spectrum of chemicals, presently fungal hosts are not the main commercial organism (Adegboye et al. 2021; Ekas et al. 2019).

Microbial lipids (single cell oils) are alluring raw materials for biodiesel production owing to their quick production rates, least work necessities, and ease of scale-up for commercial production. Among the microbial lipid producers, the barely investigated molds show beneficial properties such as hydrolyzing different organic wastes that favor downstream processing. Even though different oleaginous molds have been investigated for microbial lipid production, excessive production expenses and technological problems cause the process less feasible in comparison with typical lipid sources for biodiesel production (Mhlongo et al. 2021).

9.5.3 Lignocellulose-Derived Biodiesel

Lignocellulosic biodiesel has been recognized worldwide as a source of sustainable energy that provides ecological benefits while reducing competition between feed and fuel supplies. Since it contains polysaccharides, cellulose, and hemicellulose, it can be converted into biodiesel. Even though lignocellulose is a sustainable feedstock for the production of biodiesel, less expensive conversion mechanics are required to overcome its intrinsic recalcitrance that hinders the extensive use of lignocellulosic biodiesel. Hence the development of an uncomplicated, sustainable, and cost-efficient pretreatment facility will speed up the lignocellulosic biodiesel toward commercialization. Developing solid microbial conversion facilities that can convert a broad range of carbon sources obtained from biomass raw material to biodiesel is of significance in the field of biorefinery research (Ko et al. 2020).

The increasing interest in biodiesel from lignocellulose raw material in the USA and Europe can provide a way ahead to replace petroleum with sustainable biodiesel, which is likely to reduce greenhouse gas emissions. The bioconversion of raw materials into biodiesel plays a critical role in the commercialization of next-generation biodiesel (Usmani et al. 2021).

9.5.4 Microalgae-Derived Biodiesel

Microalgae are a potent raw material with favorable properties (Song et al. 2008). Biodiesel production using microalgae as a sustainable energy source is a substitute for the usage of fossil fuels. Still, the commercialization of microalgae is controversial due to different factors related to the life cycle appraisal and feasibility of microalgae-derived biodiesel. With the favorable characteristics of huge volume of raw material and being noncompetitive with crops, microalgae are the potential for bioenergy supply. In order to commercialize microalgae-derived biodiesel on a large scale, a flawless downstream process is needed to reduce the production expense (Song et al. 2008; Khoo et al. 2020). Successful development of microalgae-derived biodiesel industry needs the ideal combination of technological innovations, associated with less expenditure in the practical accomplishment and consolidated scale up for production and commercialization (Gendy and El-Temtamy 2013).

Microalgal lipids are potential raw materials to produce biodiesel owing to sustainability, greater energy content, and greater cetane number. Nevertheless, currently, lipid productivity is very less for commercial purposes. Even though stress manipulation is a potential alternative to improve the microalgal lipid content, the production process is time consuming and lacks the data regarding mechanisms of triacylglycerides production as a consequence to stress. The development of promising higher lipid-producing microalgal strains is encouraging owing to advanced metabolic engineering and synthetic biology tools (Shahid 2020).

9.6 World Microbial Biodiesel Commercialization Policies

9.6.1 Biodiesel in Brazil

Biodiesel is attaining importance as a higher level alternative of energy source to substitute fossil fuel in transportation. Nowadays, biodiesel commercialization is constantly rising as the biofuel has been brought in to over 60 countries globally. Many countries' governments have encouraged commercial production of liquid biofuels such as biodiesel and sponsored it with financial grant. Governments also brought many biofuel policies to promote the sustainable biofuel development in the countries. A few biofuel projects have been commercialized with government support. Beneficial regulatory and cost-effective mechanisms have played a crucial role in the production, employment, and commercialization of biofuels in those countries.

The government of Brazil launched a biodiesel program in 1980, which was unsuccessful due to high production expenses. The government restarted the biodiesel program in 2004 with many objectives including promotion of energy security and diversification, reduction of oil import, and generation of rural employment opportunities. Brazil encourages biodiesel development through legal and financial policies to assure the execution of the plan. The Brazilian Government also grants the tax alliance for the biodiesel producers and sellers. These actions significantly accelerated biodiesel development in Brazil (Su et al. 2015).

9.6.2 Biodiesel in the European Union

The main motives for the development of biodiesel in the European Union are reduction of greenhouse gas release, energy security, and development and advancement of agriculture. Government sets renewable energy goals, supports industrial developments, and funds scientific research for biodiesel development. European Union council issued the Bill on the Promotion and Use of Energy from Renewable Sources in 2009 to promote the sustainable development of biodiesel and reduce greenhouse gas emissions (Su et al. 2015).

9.6.3 Biodiesel in Germany

Among the states under Europe, Germany has the highest degree of bioenergy development. The European federal government has brought the goals of 2020 for the contribution of renewable energies in overall energy consumption, for expediting the renewable energy development in Germany. The goals are to increase the contribution of renewable energies in electricity production and to replace petroleum with biodiesel to reduce greenhouse gas emissions (Noguera 2013).

9.6.4 Biodiesel in the USA

The US biodiesel industry is going through rapid development. The major impetuses for the development of different biomasses in the USA are: to ensure oil security, to generate more job opportunities and advance agriculture, to encourage the development of a sustainable economy, and to investigate new industrial technologies in order to create different renewable energy sources for biodiesel production. In order to encourage the development of biodiesel, the USA has accredited laws and regulations associated with renewable energy sources recently, such as tax policies and economic support and loan assurance for scientific investigation and development of biodiesel (Su et al. 2015).

9.6.5 Biodiesel in India

The Indian National Policy on biofuels needs biodiesel production to utilize nonedible oil seeds to be cultivated on wastelands. A minimum support price for biodiesel raw material and minimum purchase price for biodiesel production has been approved by the government. The government also excluded the central excise tax to encourage biodiesel production. The biodiesel commercialization in India is in the developing stage and a commercially viable approach is necessary for developing a sustainable biodiesel market. The third generation raw material such as microalgae can be implemented as a potential energy source for future biodiesel production (Dewangan et al. 2018).

9.6.6 Biodiesel in Malaysia

Malaysian government brought forward the National Biofuel Policy in 2006 to encourage the development and utilization of biodiesel. The development of Malaysian biodiesel policy is mainly focused at minimizing the dependence on petroleum-derived fuels, encouraging the need for palm oil moreover making its prices affordable (Yatim et al. 2016). Following National Biofuel Policy, the Malaysian Biofuel Act of 2007 regulated the obligatory utilization of biodiesel and authorizing of activities associated with its production, storage, and marketing (Yusoff et al. 2021).

9.6.7 Biodiesel in Indonesia

The objectives of the energy policy of Indonesia include the development of renewable energy sources to decrease the use of fossil fuels and logical energy pricing by removing nonessential subsidies (Hsu et al. 2020). After the approval of the National Biofuel Policy, biodiesel production in Indonesia raised significantly over the years (Sorda et al. 2010).

9.6.8 Biodiesel in Thailand

The objectives of Thailand's energy policy include the development of sustainable energy, promotion of bioenergy, and ensuring proper price limits for energy. Since biodiesel commercialization was not successful even after Thailand's national energy policy was passed, the government put forward a policy that mandates the production of biodiesel. Following that many biodiesel producing companies have been established and biodiesel production has been increased (Kumar et al. 2013).

9.6.9 Biodiesel in the Philippines

The Biofuels Act 2006 was passed by the Philippines government primarily to develop and employ renewable energy sources, alleviate greenhouse gas emissions, develop rural employment opportunities and incomes, and make certain the availability of renewable energy. Biodiesel production was sufficient in the Philippines over the years, particularly after the approval of the Biofuel Act (Montefrio 2011).

9.6.10 Biodiesel in China

China is the largest energy-consuming, largest CO_2 emitting as well as largest oil importer country in the world. The major motives for the development of biomass in China are to assure oil security, minimize oil imports, and reduce greenhouse gas emissions. China has created policies and regulations to encourage the sustainable development of biomass and has granted subsidies for the tax to encourage its utilization (Guerrero-Lemus et al. 2019).

9.7 Conclusion

Biodiesel is a renewable, sustainable, safe, transportable, noncombustible and biodegradable biofuel that has the potential to reduce the utilization of petroleum and reinforce energy security and economic development while lowering the environmental effect. Although biodiesel is an adequate alternative over petroleum in many aspects, it is imperiled by the excessive cost of raw materials and the lack of economically and industrially feasible techniques for its adequate production from any type of raw material. Microbes perform a crucial role in biodiesel production. Biodiesel production by indigenous strains is expensive, making it essential to transform them through a metabolic engineering strategy. Recent research focuses on employing metabolic engineering to microbial strain modification to optimize enhanced productivity and energy density at a minimum cost of production. Numerous legal actions and measures have been embraced at both regional and global levels for enhancing the use of biodiesels in order to reduce the use of petroleum. Global validation projects have been introduced and executed to evaluate the technologies developed in laboratories by government support and financial grant. Different countries governments have granted huge funds and plan projects to encourage the development of technologies and minimize production expenses. Numerous biodiesel companies have produced and commercialized biodiesel, but the process was affected by the unavailability and expensiveness of raw materials. Encouraging government policies are necessary to improve biodiesel research and make it cost-competitive with other traditional energy sources.

References

Al-Zuhair S (2007) Production of biodiesel: possibilities and challenges. Biofuels Bioprod Biorefin 1(1):57–66. https://doi.org/10.1002/bbb.2. https://www.academia.edu/21385075/Production_of_biodiesel_possibilities_and_challenges

Bhan C, Singh J (2020) Role of microbial lipases in transesterification process for biodiesel production. Environmental Sustainability 3(3):257–266. https://doi.org/10.1007/S42398-020-00119-9

Bohlouli A, Mahdavian L (2021) Catalysts used in biodiesel production: a review. Biofuels 12(8): 885–898. https://doi.org/10.1080/17597269.2018.1558836

Bušić A, Morzak G, Belskaya H, Ivančić M (2018) Recent trends in biodiesel and biogas production. Food Technol Biotechnol 56(2):152–173. https://doi.org/10.17113/ftb.56.02.18.5547

Chisti Y (2007) Biodiesel from microalgae. Biotechnol Adv 25(3):294–306. https://doi.org/10.1016/J.BIOTECHADV.2007.02.001

Dewangan A, Yadav AK, Mallick A (2018) Current scenario of biodiesel development in India: prospects and challenges. Energy Sources, Part A: Recovery, Utilization and Environmental Effects 40(20):2494–2501. https://doi.org/10.1080/15567036.2018.1502849

Fazal MA, Haseeb ASMA, Masjuki HH (2011) Biodiesel feasibility study: an evaluation of material compatibility; performance; emission and engine durability. Renew Sust Energ Rev 15(2):1314–1324. https://doi.org/10.1016/J.RSER.2010.10.004

Gebremariam SN, Marchetti JM (2018) Economics of biodiesel production: review. Energy Convers. Manag 168:74–84. https://doi.org/10.1016/j.enconman.2018.05.002

Gong Y, Jiang M (2011) Biodiesel production with microalgae as feedstock: from strains to biodiesel. Biotechnol Lett 33(7):1269–1284. https://doi.org/10.1007/s10529-011-0574-z

Gupta P, Phulara SC (2015) Metabolic engineering for isoprenoid-based biofuel production. J Appl Microbiol 119(3):605–619. https://doi.org/10.1111/jam.12871

Hasan MM, Rahman MM (2017) Performance and emission characteristics of biodiesel–diesel blend and environmental and economic impacts of biodiesel production: a review. Renew Sust Energ Rev 74:938–948. https://doi.org/10.1016/j.rser.2017.03.045

Knothe G (2008) Designer biodiesel: Optimizing fatty ester composition to improve fuel properties. Energy Fuels 22(2):1358–1364. https://doi.org/10.1021/EF700639E

Kongruang S, Roytrakul S, Sriariyanun M (2020) Renewable biodiesel production from oleaginous yeast biomass using industrial wastes. E3S Web of Conferences 141. https://doi.org/10.1051/E3SCONF/202014103010

Koppolu V, Vasigala VK (2016) Role of Escherichia coli in biofuel production. Microbiology Insights 9:29–35. https://doi.org/10.4137/MBI.S10878

Kumar M, Rathour R, Gupta J, Pandey A, Gnansounou E, Thakur IS (2020) Bacterial production of fatty acid and biodiesel: opportunity and challenges. In: Refining biomass residues for sustainable energy and bioproducts: technology, advances, life cycle assessment, and economics. Academic, New York, pp 21–49. https://doi.org/10.1016/B978-0-12-818996-2.00002-8

Li J, Li L, Tong J, Wang Y, Chen S (2012) Energy Procedia 00 (2011) 000-000 2012 International conference on future energy, environment, and materials research development on lipase-catalyzed biodiesel. Energy Procedia 16:1014–1021. https://doi.org/10.1016/j.egypro.2012.01.162

Liang MH, Jiang JG (2013) Advancing oleaginous microorganisms to produce lipid via metabolic engineering technology. Prog Lipid Res 52(4):395–408. https://doi.org/10.1016/J.PLIPRES.2013.05.002

Liu X, Sheng J, Curtiss R (2011) Fatty acid production in genetically modified cyanobacteria. Proc Natl Acad Sci U S A 108(17):6899–6904. https://doi.org/10.1073/PNAS.1103014108/-/DCSUPPLEMENTAL

Lunin VV, Sergeeva YE, Galanina LA, Mysyakina IS, Ivashechkin AA, Bogdan VI, Feofilova EP (2013) Biodiesel fuel production from lipids of filamentous fungi. Appl Biochem Microbiol 49(1):46–52. https://doi.org/10.1134/S0003683813010122

Madani M, Enshaeieh M, Abdoli A (2017) Single cell oil and its application for biodiesel production. Process Saf Environ Prot 111:747–756. https://doi.org/10.1016/J.PSEP.2017.08.027

Mallick N, Bagchi SK, Koley S, Singh AK (2016) Progress and challenges in microalgal biodiesel production. Front Microbiol 7:1019. https://doi.org/10.3389/FMICB.2016.01019/BIBTEX

Metting B, Pyne JW (1986) Biologically active compounds from microalgae. Enzym Microb Technol 8(7):386–394. https://doi.org/10.1016/0141-0229(86)90144-4

Ochsenreither K, Glück C, Stressler T, Fischer L, Syldatk C (2016) Production strategies and applications of microbial single cell oils. Front Microbiol 7:1539. https://doi.org/10.3389/FMICB.2016.01539/BIBTEX

Patel A, Arora N, Sartaj K, Pruthi V, Pruthi PA (2016) Sustainable biodiesel production from oleaginous yeasts utilizing hydrolysates of various non-edible lignocellulosic biomasses. Renew Sust Energ Rev 62:836–855. https://doi.org/10.1016/J.RSER.2016.05.014

Patel A, Karageorgou D, Rova E, Katapodis P, Rova U, Christakopoulos P, Matsakas L (2020) An overview of potential oleaginous microorganisms and their role in biodiesel and omega-3 fatty acid-based industries. Microorganisms 8(3):434. https://doi.org/10.3390/microorganisms8030434
Phukan MM, Bora P, Gogoi K, Konwar BK (2019) Biodiesel from Saccharomyces cerevisiae: fuel property analysis and comparative economics. SN Applied Sciences 1(2):1–10. https://doi.org/10.1007/S42452-019-0159-3/FIGURES/4
Ram SK, Tyagi RD (2019) Single Cell Oil Biodiesel. In: Biodiesel Production. The National Academies of Sciences, Engineering, and Medicine, Washington, pp 201–227. https://doi.org/10.1061/9780784415344.ch09
Ratledge C (1993) Single cell oils–have they a biotechnological future? Trends Biotechnol 11(7): 278–284. https://doi.org/10.1016/0167-7799(93)90015-2
Ratledge C (2002) Regulation of lipid accumulation in oleaginous microorganisms. Biochem Soc Trans 30(5):A101–A101. https://doi.org/10.1042/bst030a101
Ratledge C, Cohen Z (2008) Microbial and algal oils: do they have a future for biodiesel or as commodity oils? Lipid Tech 20(7):155–160. https://doi.org/10.1002/lite.200800044
Ratledge C, Wynn JP (2002) The biochemistry and molecular biology of lipid accumulation in oleaginous microorganisms. Adv Appl Microbiol 51:1–52. https://doi.org/10.1016/S0065-2164(02)51000-5
Rojo F (2008) Biofuels from microbes: a comprehensive view. Microb Biotechnol 1(3):208–210. https://doi.org/10.1111/j.1751-7915.2008.00024.x
Song D, Fu J, Shi D (2008) Exploitation of oil-bearing microalgae for biodiesel. Chin J Biotechnol 24(3):341–348. https://doi.org/10.1016/S1872-2075(08)60016-3
Tae Hwang H, Zhao X, Ramkrishna D, Liu D (2014) Review: lipase-catalyzed process for biodiesel production: protein engineering and lipase production. Biotechnol Bioeng 111:639–653. https://doi.org/10.1002/bit.25162
Toldrá-Reig F, Mora L, Toldrá F (2020) Developments in the use of lipase transesterification for biodiesel production from animal fat waste. Applied Sciences 10(15):5085. https://doi.org/10.3390/app10155085
Vicente G, Bautista LF, Rodríguez R, Gutiérrez FJ, Sádaba I, Ruiz-Vázquez RM, Torres-Martínez S, Garre V (2009) Biodiesel production from biomass of an oleaginous fungus. Biochem Eng J 48(1):22–27. https://doi.org/10.1016/J.BEJ.2009.07.014
Wahlen BD, Morgan MR, McCurdy AT, Willis RM, Morgan MD, Dye DJ, Bugbee B, Wood BD, Seefeldt LC (2012) Biodiesel from microalgae, yeast, and bacteria: engine performance and exhaust emissions. Energy Fuel 27(1):220–228. https://doi.org/10.1021/ef3012382
Yuzbasheva EY, Yuzbashev TV, Mostova EB, Perkovskaya NI, Sineokii SP (2014) Microbial synthesis of biodiesel and its prospects. Appl Biochem Microbiol 50(9):789–801. https://doi.org/10.1134/S0003683814090075
Abbaszaadeh A, Ghobadian B, Omidkhah MR, Najafi G (2012) Current biodiesel production technologies: a comparative review. Energy Convers Manag 63:138–148. https://doi.org/10.1016/j.enconman.2012.02.027
Adegboye MF, Ojuederie OB, Talia PM, Babalola OO (2021) Bioprospecting of microbial strains for biofuel production: metabolic engineering, applications, and challenges. Biotechnol Biofuels 14(1):1–21. https://doi.org/10.1186/s13068-020-01853-2
Atabani AE, Silitonga AS, Badruddin IA, Mahlia TMI, Masjuki HH, Mekhilef S (2012) A comprehensive review on biodiesel as an alternative energy resource and its characteristics. Renew Sust Energ Rev 16(4):2070–2093. https://doi.org/10.1016/j.rser.2012.01.003
Athar M, Zaidi S (2021) Biodiesel production technologies. In: Inamuddin MIA, Boddula R, Rezakazemi M (eds) Biodiesel technology and applications. Scrivener, Beverly, MA, pp 241–265
Dalai AK, MG Kulkarni, and LC Meher (2006) Biodiesel productions from vegetable oils using heterogeneous catalysts and their applications as lubricity additives. 2006 IEEE EIC Climate

Change Technology Conference, EICCCC 1–8. doi: https://doi.org/10.1109/EICCCC.2006.277228

Das M, Patra P, Ghosh A (2020) Metabolic engineering for enhancing microbial biosynthesis of advanced biofuels. Renew Sust Energ Rev 119:109562. https://doi.org/10.1016/j.rser.2019.109562

Ekas H, Deaner M, Alper HS (2019) Recent advancements in fungal-derived fuel and chemical production and commercialization. Curr Opin Biotechnol 57(512):1–9. https://doi.org/10.1016/j.copbio.2018.08.014

Gendy TS, El-Temtamy SA (2013) Commercialization potential aspects of microalgae for biofuel production: an overview. Egypt J Pet 22(1):43–51. https://doi.org/10.1016/j.ejpe.2012.07.001

Guerrero-Lemus R, Cañadillas-Ramallo D, Reindl T, Valle-Feijóo JM (2019) A simple big data methodology and analysis of the specific yield of all PV power plants in a power system over a long time period. Renew Sust Energ Rev 107:123–132. https://doi.org/10.1016/j.rser.2019.02.033

Ho SH, Zhang C, Tao F, Zhang C, Chen WH (2020) Microalgal torrefaction for solid biofuel production. Trends Biotechnol 38(9):1023–1033. https://doi.org/10.1016/j.tibtech.2020.02.009

Hsu A, Höhne N, Kuramochi T, Vilariño V, Sovacool BK (2020) Beyond states: harnessing sub-National Actors for the deep decarbonisation of cities, regions, and businesses. Energy Res Soc Sci 70:101738. https://doi.org/10.1016/j.erss.2020.101738

Kamm J (2004) A new class of plants for a biofuel feedstock energy crop. Applied Biochemistry and Biotechnology 113(1–3):55–70. https://doi.org/10.1007/978-1-59259-837-3_6

Khoo KS, Chew KW, Yew GY, Leong WH, Chai YH, Show PL, Chen WH (2020) Recent advances in downstream processing of microalgae lipid recovery for biofuel production. Bioresour Technol 304:122996. https://doi.org/10.1016/j.biortech.2020.122996

Ko JK, Lee JH, Jung JH, Lee SM (2020) Recent advances and future directions in plant and yeast engineering to improve lignocellulosic biofuel production. Renew Sust Energ Rev 134:110390. https://doi.org/10.1016/j.rser.2020.110390

Kumar S, Abdul Salam P, Shrestha P, Ackom EK (2013) An assessment of Thailand's biofuel development. Sustainability 5(4):1577–1597. https://doi.org/10.3390/su5041577

Maher KD, Bressler DC (2007) Pyrolysis of triglyceride materials for the production of renewable fuels and chemicals. Bioresour Technol 98(12):2351–2368. https://doi.org/10.1016/j.biortech.2006.10.025

Majidian P, Tabatabaei M, Zeinolabedini M, Naghshbandi MP, Chisti Y (2018) Metabolic engineering of microorganisms for biofuel production. Renew Sust Energ Rev 82:3863–3885. https://doi.org/10.1016/j.rser.2017.10.085

Maza DD, Viñarta SC, García-Ríos E, Guillamón JM, Aybar MJ (2021) Rhodotorula Glutinis T13 as a potential source of microbial lipids for biodiesel generation. J Biotechnol 331:14–18. https://doi.org/10.1016/j.jbiotec.2021.03.002

Mhlongo SI, Ezeokoli OT, Roopnarain A, Ndaba B, Sekoai PT, Habimana O, Pohl CH (2021) The potential of single-cell oils derived from filamentous fungi as alternative feedstock sources for biodiesel production. Front Microbiol 12:637381. https://doi.org/10.3389/fmicb.2021.637381

Montefrio MJF, Sonnenfeld DA (2011) Forests, fuel, or food? Competing coalitions and biofuels policy making in The Philippines. J Environ Dev 20(1):27–49. https://doi.org/10.1177/1070496510394321

Naghshbandi MP, Tabatabaei M, Aghbashlo M, Aftab MN, Iqbal I (2020) Metabolic engineering of microalgae for biofuel production. Methods Mol Biol 1980:153–172. https://doi.org/10.1007/7651_2018_205

Noguera J (2013) Oil prices: breaks and trends. Energy Econ 37:60–67. https://doi.org/10.1016/j.eneco.2012.12.008

Rakchai N, Kittikun AH, Zimmermann W (2016) The production of immobilized whole-cell lipase from aspergillus Nomius ST57 and the enhancement of the synthesis of fatty acid methyl esters using a two-step reaction. J Mol Catal B Enzym 133:S128–S136. https://doi.org/10.1016/j.molcatb.2016.12.006

Rezania S, Oryani B, Park J, Hashemi B, Yadav KK, Kwon EE, Hur J, Cho J (2019) Review on transesterification of non-edible sources for biodiesel production with a focus on economic aspects, fuel properties and by-product applications. Energy Convers Manag 201:112155. https://doi.org/10.1016/j.enconman.2019.112155

Rossi M, Amaretti A, Raimondi S, Leonardi A (2011) Getting lipids for biodiesel production from oleaginous fungi. Biodiesel-feedstocks and processing technologies 1:72–74

Saidur R, Abdelaziz EA, Demirbas A, Hossain MS, Mekhilef S (2011) A review on biomass as a fuel for boilers. Renew Sust Energ Rev 15(5):2262–2289. https://doi.org/10.1016/j.rser.2011.02.015

Saini R, Hegde K, Brar SK, Vezina P (2020) Advanced biofuel production and road to commercialization: an insight into bioconversion potential of Rhodosporidium Sp. Biomass Bioenergy 132:105439. https://doi.org/10.1016/j.biombioe.2019.105439

Shahid A, Rehman AU, Usman M, Ashraf MUF, Javed MR, Khan AZ, Gill SS, Mehmood MA (2020) Engineering the metabolic pathways of lipid biosynthesis to develop robust microalgal strains for biodiesel production. Biotechnol Appl Biochem 67(1):41–51. https://doi.org/10.1002/bab.1812

Sorda G, Banse M, Kemfert C (2010) An overview of biofuel policies across the world. Energy Policy 38(11):6977–6988. https://doi.org/10.1016/j.enpol.2010.06.066

Su Y, Zhang P, Yuqing S (2015) An overview of biofuels policies and industrialization in the major biofuel producing countries. Renew Sust Energ Rev 50:991–1003. https://doi.org/10.1016/j.rser.2015.04.032

Talha NS, Sulaiman S (2016) Overview of catalysts in biodiesel production. ARPN J Eng Applied Sci 11(1):439–442

Usmani Z, Sharma M, Awasthi AK, Lukk T, Tuohy MG, Gong L, Nguyen-Tri P, Goddard AD, Bill RM, Chandra Nayak S, Gupta VK (2021) Lignocellulosic biorefineries: the current state of challenges and strategies for efficient commercialization. Renew Sust Energ Rev 148:111258. https://doi.org/10.1016/j.rser.2021.111258

Wang L, Yang J (2007) Transesterification of soybean oil with Nano-MgO or not in supercritical and subcritical methanol. Fuel 86(3):328–333. https://doi.org/10.1016/j.fuel.2006.07.022

Yatim P, Mamat MN, Mohamad-Zailani SH, Ramlee S (2016) Energy policy shifts towards sustainable energy future for Malaysia. Clean Techn Environ Policy 18(6):1685–1695. https://doi.org/10.1007/s10098-016-1151-x

Yusoff MN, Mohd A, Zulkifli NWM, Sukiman NL, Chyuan OH, Hassan MH, Hasnul MH, Zulkifli MSA, Abbas MM, Zakaria MZ (2021) Sustainability of palm biodiesel in transportation: a review on biofuel standard, policy and international collaboration between Malaysia and Colombia. Bioenergy Res 14(1):43–60. https://doi.org/10.1007/s12155-020-10165-0

Chapter 10
Microbial Production of Bioactive Compounds

Luis A. Cabanillas-Bojórquez, Octavio Valdez-Baro, Erick Paul Gutiérrez-Grijalva, and J. Basilio Heredia

Abstract Bioactive compounds are natural products with high demand in the industry, which could be synthesized by various organisms such as fungi, bacteria, and plants. These metabolites comprise a wide group with different chemical structures and bio-functional properties like anti-inflammatory, antiproliferative, antimicrobial, antibacterial, antifungal, and anticancer. Recently, it has been proved that bioactive compounds such as alkaloids, terpenoids, phenolic compounds, and steroids could be used for therapeutic applications in human health. Those compounds are commonly produced by chemical synthesis or extracted from natural tissues. However, these compounds have recently been obtained by emerging biotechnological techniques such as novel cell factories and the fermentation of genetically engineered microorganisms. Biotechnological methods use fungi, bacteria, and yeast easily cultured with these techniques. Also, microorganisms can be genetically modified to enhance the synthesis of the desired metabolite. In addition, to obtain bioactive compounds, a fast production, a more efficient and promising process can be obtained with such biotechnological methods as compared to the traditional processes. Therefore, this chapter aims to describe the role of microorganisms in producing bioactive compounds with the potential to enhance human health, which may evolve in further studies to discover and develop novel nutraceutical supplements or therapeutic drugs.

Keywords Fungi · Bacteria · Alkaloids · Terpenoids · Phenolic compounds

L. A. Cabanillas-Bojórquez · O. Valdez-Baro · J. B. Heredia (✉)
Centro de Investigación en Alimentación y Desarrollo A.C, Culiacán, Sinaloa, Mexico
e-mail: jbheredia@ciad.mx

E. P. Gutiérrez-Grijalva
Cátedras CONACYT-Centro de Investigación en Alimentación, Culiacán, Sinaloa, Mexico

A. Sarkar, I. A. Ahmed (eds.), *Microbial products for future industrialization*, Interdisciplinary Biotechnological Advances,
https://doi.org/10.1007/978-981-99-1737-2_10

10.1 Introduction

Bioactive compounds are natural chemicals present in different tissues of fruits, vegetables, cereals, among others, which have been related to health benefits for humans such as anti-aging, anti-inflammatory, and antioxidant activity, as well as protection against diabetes mellitus, cancer, and neurodegenerative diseases (Chouhan et al. 2017; Wu et al. 2020). In this sense, bioactive compounds are important agents used principally in the food and pharmaceutical industries, which represent more than 50% of products used as molecules for clinical use; likewise, about 40% of bioactive compounds and derivatives have been approved by the US Food and Drug Administration for human consumption (Chen et al. 2020a). Bioactive compounds can be obtained by chemical synthesis; however, the extraction of these compounds can also be achieved by methods such as solvents extraction, ultrahigh-pressure extraction, and supercritical CO_2 extraction from natural sources (Chen et al. 2020a; Dudnik et al. 2018). On the other hand, biotechnological methods also have been studied for bioactive compounds production; in this sense, cell culture technology, metabolic engineering, systems biology, and synthetic biology have been reported (Wu et al. 2020). Furthermore, microorganisms such as bacteria, yeast, and fungi have been studied for synthesized bioactive compounds; in this sense, the bacteria *Escherichia coli*, *Streptomyces venezuelae*, and *Yarrowia lipolytica*, as well as the yeast *Saccharomyces cerevisiae* and the fungi *Tremella fuciformis* are the most used in biotechnological methods to produce bioactive compounds such as alkaloids, amino acids, terpenoids, polyphenols, and vitamins, among others (Fig. 10.1) (Dudnik et al. 2018; Pham et al. 2019; Wu et al. 2020).

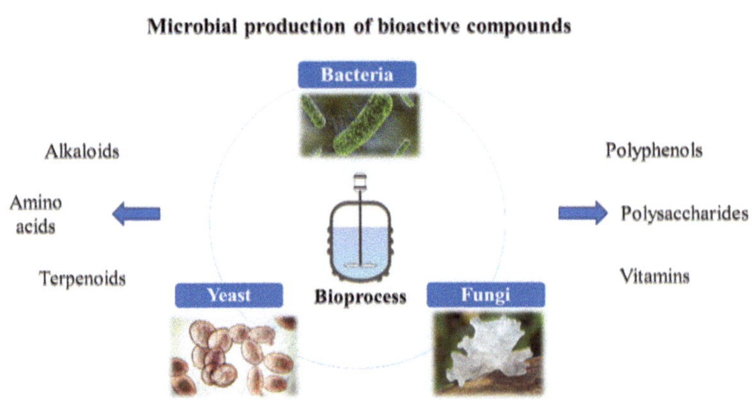

Fig. 10.1 Microbial production of bioactive compounds

10.1.1 Biotechnological Methods of Bioactive Compounds Production

10.1.1.1 Cell Culture Technology

Cell culture is a promised technology to produce bioactive compounds through a suitable host strain, which can be genetically modified to express proteins, which improve the production of the compounds of interest, likewise, both the culture conditions; thus, the strain is the most important aspect in the design of the bioprocess (Pham et al. 2019). However, the pathways for bioactive compounds production from microorganisms are not yet fully clarified; likewise, the enzymes that catalyze these compounds still need to be elucidated. In this sense, for the biosynthesis of the bioactive compounds, enzymes present in the microorganisms that carry out the pathways of the compounds of interest must be identified or designed (Li et al. 2018a). Cell culture has been used for polyphenols, terpenoids, and alkaloids, among other compounds produced from yeast, fungi, and bacteria (Cao et al. 2020; Engels et al. 2021; Li et al. 2021; Suzuki et al. 2014; Trenchard and Smolke 2015).

10.1.1.2 Metabolic Engineering

Metabolic pathway engineering is based mainly on the manipulation of the genes, either by improving the expression or elimination of genes, as well as the introduction of new genes that enhance the overproduction of bioactive compounds (Pham et al. 2019; Sun et al. 2021; Wu et al. 2020). However, the metabolic pathway engineering has some limitations, such as the complex identification of the pathways for bioactive compounds production, the lack of best conditions for the heterogeneous expression of enzymes for higher production of bioactive compounds, as well as a complex biosynthetic pathway due to cellular processes, and low yields in the production of the bioactive metabolites by the synthesis of intermediate compounds (Chen et al. 2020a). Metabolic engineering has been studied to enhance the production of polyphenols, terpenoids, and alkaloids, among others, from yeast, fungi, and bacteria (Zhang et al. 2021).

10.1.1.3 Synthetic Biology

Natural products can also be produced using synthetic biology, a tool for pathway design that uses a combination of DNA sequencing and computationally and experimentally determined genetic elements using computational tools to build vast databases of natural product biosynthetic genes (Galanie et al. 2015). Synthetic biology provides information on the sequences and genes encode enzymes to predict functions and design biosynthetic pathways for target compounds (Ceranic et al.

2021; Chen et al. 2017a). Also, the predicted biosynthetic pathways design and synthesis have been used to test the candidate enzymes' activity in the desired cell host in studies at the laboratory level. In this sense, the assembly of DNA for biosynthetic pathways construction has been carried out in yeast as *S. cerevisiae* by combining in vitro enzyme assembly and assembly methods based on recombination associated with transformation (Galanie and Smolke 2015; Grewal et al. 2018). However, this technique at the laboratory scale has found some problems due to discordance or problems in DNA assembly, which could be solved with sequences optimization and bioinformatics (Chen et al. 2020a). Furthermore, synthetic biology has been used for constructing yeast for secondary metabolites production, such as polyphenols, terpenoids, and alkaloids (Ceranic et al. 2021; Chen et al. 2017a; Galanie et al. 2015).

10.1.2 Bioactive Compounds Classification

Bioactive compounds include polyphenols, polysaccharides, amino acids, vitamins, and others that have shown beneficial effects such as cardiovascular diseases prevention and protection against noncommunicable diseases such as diabetes and cancer (Wu et al. 2020).

Polyphenols are classified as secondary metabolites that accumulate in different plant parts (Dudnik et al. 2018). Polyphenols comprise phenolic compounds, phenylpropanoids, flavonoids, stilbenes, coumarins, and their derivatives (Sun et al. 2021). In addition, this group has an aromatic ring linked to one or more hydroxyl groups, which possess beneficial activities to human health, such as antiinflammation, antioxidation, antitumor, and antibacterial activities (Wu et al. 2020).

Terpenoids, also known as terpenes or isoprenoids, are a class of bioactive compounds produced naturally by plants and microorganisms, which comprise thousands of compounds such as sterols, carotenoids, and quinines (Chen et al. 2020a; Li et al. 2021; Wu et al. 2017). These compounds have the most chemically and structurally diverse family used extensively in foods, pharmaceuticals, cosmetics, biofuels and fuel additives, and agricultural products (Chen et al. 2020a). Many terpenoids are still extracted from plants; however, terpenoid extraction has low yield and purity by abiotic or biotic factors and high consumption of resources that are technically challenging and expensive (Zhang et al. 2021). In this sense, terpenoids production from microbial using renewable resources have been an alternative strategy due to the diverse advantage such as ease and speed of microbial cultivation, the genetic manipulation of these microorganisms, and bioprocess with potential industrial scaling (Chen et al. 2020a; Zhang et al. 2021).

Alkaloids are natural products that contain nitrogen atoms and have been related to enhancing human health. In this sense, alkaloids have properties like analgesic, sedative, and anticancer agents (Chen et al. 2020a). Alkaloid production has been obtained principally by plants; however, these compounds from plants have been limited by complex biosynthesis and biotic and abiotic factors (Trenchard and Smolke 2015). In this sense, microbial biosynthesis is an alternative platform to produce alkaloids due to genetic tools for gene manipulation and pathway expression, easy scaling methods, and fast growth rates (Chen et al. 2020b; Trenchard and Smolke 2015).

Polysaccharides are sugar polymers that have been recognized as nutraceuticals because they are related to health beneficial properties such as antimicrobial, antitumoral, anticoagulant, and immunologic capacity (Wu et al. 2020). Polysaccharides are obtained from plant, animal tissues, and microorganisms such as bacteria, yeast, and fungi (Ko et al. 2019). In this sense, cell culture technology and metabolic engineering have been used for polysaccharides production from microorganisms (Ko et al. 2019; Shehata et al. 2018).

Likewise, other bioactive compounds such as peptides and vitamins are related to human health; beneficial effects have been obtained by cell culture technology and metabolic engineering from microorganisms such as bacteria and yeast (Dreyer et al. 2019; Jach et al. 2021; Kathayat et al. 2021; You et al. 2021).

10.2 Microbial Production of Alkaloids

Alkaloids are bioactive compounds that contain nitrogen heterocycle and possess various bioactivities such as antimicrobial, anticancer agents, cough suppressants, and analgesic effects (Galanie and Smolke 2015). Alkaloids are produced by plants that can be affected by abiotic and biotic factors, such as climate, cultivation and harvest methods, geography, plant material storage, among others. Moreover, plant alkaloids need subsequent purification stages, which causes low yields due to the derivatization of functional groups, obtaining a mixture of functional compounds (Galanie and Smolke 2015; Galanie et al. 2015). In this sense, biotechnological methods for alkaloids production by microbial have been widely studied due to DNA modifications on microorganisms to enhance the target compounds, as well as fast microbial growth conditions (Courdavault et al. 2021; Zhou et al. 2014). Biotechnological methods for alkaloids production have various advantages such as the use of metabolic engineering of microorganisms for alkaloids production, cell culture of modified microorganisms for bioactive compounds production, high yields, short time, and high purity (Galanie and Smolke 2015; Zhou et al. 2014). Alkaloids such as berberine, betanin, canadine, noscapine, among others, have been produced from bacteria and yeast by cell culture, metabolic engineering, and synthetic biology (Table 10.1) (Galanie and Smolke 2015; Li and Smolke 2016).

Table 10.1 Alkaloids production by microorganisms

Alkaloid compound	Microorganisms	Biotechnological strategy	Identification technique	References
Berberine	*S. cerevisiae*	Metabolic engineering and synthetic biology	LC-MS/MS	Galanie and Smolke (2015)
Betanin	*S. cerevisiae*	Metabolic engineering	LC-MS	Grewal et al. (2018)
Canadine	*S. cerevisiae*	Metabolic engineering and synthetic biology	LC-MS/MS	Galanie and Smolke (2015)
Corytuberine	*E. coli* and *S. cerevisiae*	Cell culture technology and metabolic engineering	LC-MS	Suzuki et al. (2014)
Hydrocodone	*P. putida* and yeast	Metabolic engineering and synthetic biology	LC-MS and HPLC	Ozber et al. (2020)
Hyoscyamine	Yeast	Metabolic engineering and synthetic biology	Not mentioned	Courdavault et al. (2021)
Magnoflorine	*E. coli* and *S. cerevisiae*	Cell culture technology and metabolic engineering	LC-MS	Cao et al. (2020)
Morphine	*S. cerevisiae*	Metabolic engineering	LC-MS	Thodey et al. (2014)
Noscapine	*S. cerevisiae*	Cell culture technology and metabolic engineering	LC-MS	Li et al. (2018c)
Reticuline	*S. cerevisiae*	Metabolic engineering and synthetic biology	LC-MS/MS	Galanie and Smolke (2015)
Sanguinarine	*S. cerevisiae*	Cell culture technology	LC-MS/MS	Trenchard and Smolke (2015)
Scopolamine	Yeast	Metabolic engineering and synthetic biology	Not mentioned	Courdavault et al. (2021)
Scoulerine	*E. coli*	Metabolic engineering and synthetic biology	LC-MS	Suzuki et al. (2014)
Strictosidine	*S. cerevisiae*	Cell culture technology and metabolic engineering	LC-MS	Brown et al. (2015)
Thebaine	Yeast	Metabolic engineering and synthetic biology	LC-MS/MS	Galanie et al. (2015)
Tropine	*S. cerevisiae*	Metabolic engineering	MS/MS	Srinivasan and Smolke (2019)

HPLC high-pressure liquid chromatography, *HRMS* high resolution mass spectrometry, *LC* liquid chromatography, *MS* mass spectrometry

10.2.1 Bacteria

E. coli and *Pseudomonas putida* have been studied for alkaloids production; in this sense, Suzuki et al. (2014) reported the Corytuberine production from *E. coli* by cell culture technology and metabolic engineering. They identified the alkaloid corytuberine using liquid chromatography coupled to mass spectrometry. Also, Ozber et al. (2020) reported the hydrocodone production from *P. putida*. They studied the use of metabolic engineering and synthetic biology for alkaloid production and identified hydrocodone by high-pressure liquid chromatography.

10.2.2 Yeast

E. coli is the principal yeast used for alkaloid production; in this sense, berberine, betanin, canadine, hydrocodone, morphine, pseudotropine, reticuline, scopolamine, scoulerine, and tropine have been produced from this yeast by metabolic engineering (Courdavault et al. 2021; Galanie and Smolke 2015; Grewal et al. 2018; Srinivasan and Smolke 2019; Suzuki et al. 2014; Thodey et al. 2014). Also, synthetic biology and cell culture techniques have been used for the production of corytuberine, magnoflorine, noscapine, sanguinarine, and strictosidine from *E. coli* (Brown et al. 2015; Cao et al. 2020; Li et al. 2018c; Suzuki et al. 2014; Trenchard and Smolke 2015).

10.3 Microbial Production of Polyphenols

Polyphenols are a group of bioactive compounds that have been extracted mainly by plants. Polyphenols are composed of at least one phenolic hydroxyl group, and these compounds in plants are derived from the aromatic amino acids phenylalanine and tyrosine (Dudnik et al. 2018; Wu et al. 2020). This group includes curcuminoids, flavonoids, stilbenes, phenolic acids, and styrylpyrones, which are related to beneficial effects on human Health (Dudnik et al. 2018). Polyphenols are produced by plants or chemical synthesis; however, these methods have some disadvantages such as low yield, a variation on the interest compounds by biotic and abiotic factors, and expensive synthesis and use of hazardous chemicals agents (Sun et al. 2021). Polyphenols are biosynthesized through the pathway, which consists of enzymes that catalyze the substrates and produce the different polyphenols by some mechanism. These mechanisms have been studied to produce polyphenols by metabolic strategies using artificial pathways by genes from different species and microorganisms. In this sense, the bioproduction of polyphenols has been extensively studied by metabolic engineering and synthetic biology processes for microbial cell factories to produce these compounds (Sun et al. 2021; Wu et al. 2020). The bacteria *E. coli, S. venezuelae* and *Y. lipolytica*, as well as the fungi *T. fuciformis* and the yeast *S. cerevisiae,* are the most studied microorganisms to produce polyphenols due to their adaptability, fast growth, cost-effectiveness, and well-developed genetic manipulation (Table 10.2) (Pham et al. 2019; Sun et al. 2021; Wu et al. 2020). Likewise, polyphenols have been obtained from microorganisms by metabolic engineering and synthetic biology (Dudnik et al. 2018; Wu et al. 2020; Zhou et al. 2014).

Table 10.2 Polyphenol production by microorganisms

Polyphenol compound	Microorganisms	Biotechnological strategy	Identification technique	References
Apigenin	*E. coli* and *S. cerevisiae*	Metabolic engineering and synthetic biology	HPLC	Dudnik et al. (2018)
Caffeoyl alcohol	*E. coli*	Metabolic engineering and synthetic biology	HPLC	Chen et al. (2017b)
p-Coumaric acid	*E. coli* and *S. cerevisiae*	Metabolic engineering	HPLC	Dudnik et al. (2018)
Cyanidin-3-*O*-glucoside	*E. coli* and *S. cerevisiae*	Metabolic engineering	HPLC, UPLC/MS	Dudnik et al. (2018)
Daidzein	*E. coli*	Metabolic engineering	CE and UV	Dudnik et al. (2018)
Eriodictyol	*E. coli*	Metabolic engineering	HPLC	Chouhan et al. (2017)
Gallic acid	*E. coli*	Metabolic engineering and synthetic biology	HPLC	Sun et al. (2021)
Genistein	*E. coli* and *S. cerevisiae*	Metabolic engineering	CE and UV	Dudnik et al. (2018)
Hydroxytyrosol	*E. coli*	Metabolic engineering and synthetic biology	HPLC	Li et al. (2018a, b)
Kaempferol	*E. coli*, *S. cerevisiae* and *S. venezuelae*	Metabolic engineering and synthetic biology	HPLC, CE and UV	Dudnik et al. (2018)
Luteolin	*S. cerevisiae*	Metabolic engineering	HPLC	Dudnik et al. (2018)
Naringenin	*Corynebacterium glutamicum*, *E. coli*, *S. cerevisiae* and *Y. lipolytica*	Metabolic engineering	HPLC, LC-MS	Chouhan et al. (2017)
Pelargonidin 3-*O*-glucoside	*E. coli*	Metabolic engineering	HPLC	Chouhan et al. (2017)
Pinosylvin	*E. coli*, *S. venezuelae*	Metabolic engineering	LC-MS	Dudnik et al. (2018)
Pterostilbene	*C. glutamicum*, *E. coli*	Metabolic engineering and synthetic biology	HPLC	Dudnik et al. (2018), Wu et al. 2020)
Pyrogallol	*E. coli*	Metabolic engineering and synthetic biology	UV-VIS and HPLC	Sun et al. (2021)
Quercetin	*C. glutamicum* and *S. cerevisiae*	Metabolic engineering	GC/MS	Dudnik et al. (2018)
Resveratrol	*E. coli*, *S. cerevisiae*, *S. venezuelae*, *T.*	Metabolic engineering	HPLC	Dudnik et al. (2018)

(continued)

Table 10.2 (continued)

Polyphenol compound	Microorganisms	Biotechnological strategy	Identification technique	References
	fuciformis, and *Y. lipolytica*			
Vanillin	*E. coli* and *S. cerevisiae*	Metabolic engineering	HPLC, LC-MS	Cao et al. (2020), Chen et al. (2020b)

CE capillary electrophoresis, *HPLC* high-pressure liquid chromatography, *LC* liquid chromatography, *MS* mass spectrometry, *UPLC* ultra high-performance liquid chromatography

10.3.1 Bacteria

E. coli, S. venezuelae, and *Y. lipolytica* are the most studied bacteria for polyphenol production. In this sense, Dudnik et al. (2018) reported the pinosylvin production from *S. venezuelae* by metabolic engineering, also the and resveratrol production by metabolic engineering from *S. venezuelae* and *Y. lipolytica*. On the other hand, apigenin, caffeoyl alcohol, daidzein, eriodictyol, gallic acid, genistein, hydroxytyrosol, pyrogallol, and vanillin have been obtained from *E. coli* by metabolic engineering and synthetic biology. The polyphenol identification was carried out by liquid chromatography coupled to mass spectrometry (LC-MS), ultra-high-performance liquid chromatography (UPLC), and high-pressure liquid chromatography (HPLC) (Chen et al. 2017b).

10.3.2 Others

T. fuciformis is the most studied fungus for polyphenol production. Dudnik et al. (2018) reported resveratrol production from *T. fuciformis* by metabolic engineering. The resveratrol identification was carried out by high-pressure liquid chromatography (HPLC).

On the other hand, apigenin, p-coumaric acid, cyanidin-3-*O*-glucoside, genistein, kaempferol, luteolin, naringenin, quercetin, resveratrol, and vanillin have been obtained from *S. cerevisiae* by metabolic engineering and synthetic biology. The polyphenol identification was obtained by liquid chromatography coupled to mass spectrometry (LC-MS) and high-pressure liquid chromatography (HPLC) (Dudnik et al. 2018).

10.4 Microbial Production of Terpenes

The production of terpenes in plants and microorganisms can be carried out by two metabolic routes: the route of mevalonic acid and the route of non-mevalonate or 1-deoxy-D-xylulose-5-phosphate (DXP) pathway (Chen et al. 2020a). Microorganisms can produce terpenes naturally as part of their secondary metabolism or by applying metabolic engineering and cell culture technology techniques. Metabolic engineering is used in the production of terpenes in order to maximize the production of a specific terpene, which guarantees greater purity and performance; this is carried out by mixing and modifying various genes that modify metabolic pathways to prioritize the specific path of biosynthesis of a particular terpene (Zhang et al. 2021). On the other hand, cell culture technology has been studied to produce terpenes from genetically modified microorganisms in reactors under controlled conditions for the highest concentration and yield (Engels et al. 2021; Jin et al. 2019; Wu et al. 2017).

Monoterpenes, sesquiterpenes, diterpenes, triterpenes, and tetraterpenes, among others, have been synthesized from microorganisms such as bacteria, fungi, and yeast (Table 10.3) (Godara and Kao 2021; Huang et al. 2021). In this sense, *Aspergillus calidoustus*, *Bacillus methanolicus*, *C. glutamicum*, *S. cerevisiae*, and *Y. lipolytica* have been studied for terpenes production (Engels et al. 2021; Godara and Kao 2021; Huang et al. 2021; Jin et al. 2019; Liu et al. 2021).

10.4.1 Bacteria

B. methanolicus, *B. subtilis*, *C. glutamicum*, *E. coli*, and *Synechocystis* sp. PCC 6803 are bacteria that have been reported for amorphadiene, 4,4′-diaponeurosporene, 4,4′-diapolycopene, α-carotene, β-carotene, lycopene, β-phellandrene, and taxadiene production by cell culture technology and metabolic engineering. Likewise, the terpenes identification was carried out by gas chromatography coupled to mass spectrometry (GC-MS), high-pressure liquid chromatography (HPLC), ultraviolet-visible spectroscopy (UV-vis), and ultra-high-performance liquid chromatography (UPLC) techniques (Abdallah et al. 2019; Huang et al. 2021; Li et al. 2021; Wu et al. 2017).

10.4.2 Fungi

Fungi have been used for terpenes production; in this sense, *A. calidoustus*, *Fusarium proliferatum*, and *Trichoderma viride* have been reported for the production of bovistol A, B, and C, also brasilane-type sesquiterpenoid, drimane-type sesquiterpene esters, cyathane diterpenoids, fusaproliferin, and pasteurestin C by cell culture technology and metabolic engineering. The terpenes identification was carried out

Table 10.3 Microbial production of terpenes

Terpene compound	Microorganism	Biotechnological strategy	Identification technique	References
Armillyl orsellinate	*S. cerevisiae*	Cell culture technology and metabolic engineering	GC-MS, NMR	Engels et al. (2021)
Betulinic acid Betulinic alde-hyde Betulin	*Y. lipolytica*	Cell culture technology and metabolic engineering	GC-MS	Jin et al. (2019)
Brasilane-type sesquiterpenoid	*T. viride*	Metabolic engineering	GC-MS, NMR	Sun et al. (2019)
α-Carotene	*C. glutamicum*	Cell culture technology and metabolic engineering	HPLC	Li et al. (2021)
β-Carotene	*E. coli*	Cell culture technology	HPLC	Wu et al. (2017)
	Y. lipolytica	Cell culture technology and metabolic engineering	HPLC-	Liu et al. (2021)
β-Caryophyllene	*S. cerevisiae*	Metabolic engineering and synthetic biology	GC	Godara and Kao (2021)
Drimane-type sesquiterpene esters	*A. calidoustus*	Metabolic engineering	HRESIMS, NMR	Huang et al. (2021)
Fusaproliferin	*F. proliferatum*	Metabolic engineering and synthetic biology	TLC, HRESIMS, NMR	Ceranic et al. (2021)
β-Ionone	*Y. lipolytica*	Metabolic engineering	GC-MS	Czajka et al. (2018)
Lycopene	*E. coli*	Cell culture technology	UV-Vis	Hussain et al. (2021)
Patchoulol	*S. cerevisiae*	Cell culture technology and metabolic engineering	GC-MS	Ma et al. (2019)
β-Phellandrene	*Synechocystis* sp. PCC 6803	Cell culture technology and metabolic engineering	GC-MS	Valsami et al. (2021)
Taxadiene	*B. subtilis*	Cell culture technology	GC-MS	Abdallah et al. (2019)
Nootkatone β-Nootkatol Valencene	*S. cerevisiae*	Cell culture technology and metabolic engineering	GC-MS	Ouyang et al. (2019)

GC gas chromatography, *HPLC* high pressure liquid chromatography, *HRESIMS* high resolution electrospray ionization mass spectrometry, *MS* mass spectrometry, *NMR* nuclear magnetic resonance, *TLC* thin layer chromatography, *UPLC* ultra high-performance liquid chromatography, *UV-vis* ultraviolet visible spectroscopy

by gas spectrometry coupled to mass spectrometry (GC-MS), nuclear magnetic resonance (NMR), high-pressure liquid chromatography (HPLC), high-resolution electrospray ionization mass spectrometry (HRESIMS), and thin-layer chromatography (TLC) techniques (Ceranic et al. 2021; Huang et al. 2021; Sun et al. 2019).

10.4.3 Yeast

Yeasts are one of the most widely used microorganisms in the production of metabolites such as terpenes at an industrial level, having a great advance in research and technological application, being the yeasts *S. cerevisiae* and *Y. lipolytica* the most studied (Engels et al. 2021; Jin et al. 2019). In this sense, it has been reported the production of armillyl orsellinate, betulinic acid, betulinic aldehyde, betulin, β-carotene, β-caryophyllene, β-ionone, lycopene, β-nootkatol, nootkatone, patchoulol, and valencene by cell culture technology and metabolic engineering from *S. cerevisiae* and *Y. lipolytica*. The terpenes identification was carried out by gas spectrometry coupled to mass spectrometry (GC-MS), nuclear magnetic resonance (NMR), and high-pressure liquid chromatography (HPLC) techniques (Engels et al. 2021; Godara and Kao 2021; Jin et al. 2019; Ma et al. 2019; Ouyang et al. 2019).

10.5 Microbial Production of Other Bioactive Compounds

Microorganisms have been studied to synthesize secondary metabolites such as phenolic compounds, terpenes, and alkaloids. Likewise, microorganisms also synthesize other bioactive compounds of great interest and potential as vitamins, peptides, and carbohydrates (Chen et al. 2020a; Pham et al. 2019). These bioactive compounds are mostly synthesized by bacteria, fungi, and yeast (Table 10.4).

10.5.1 Vitamins

Vitamins are an organic group that are essential for normal growth and nutrition (Wu et al. 2020). These compounds have been widely studied and obtained by several sources. In this sense, microorganisms such as bacteria and yeast have been used for vitamins production (Cao et al. 2020; Wu et al. 2020). Vitamins B1, B2, B6, B7, B9, and B12, as well as vitamin K2, has been obtained by Cell culture technology and metabolic engineering from bacteria such as *B. subtilis*, *E. coli*, *Lactobacillus* sp., and *Propionibacterium freudenreichii* (Hati et al. 2019; You et al. 2021; Zhang et al. 2020), as well as yeast (*Rhodotorula glutinis* and *Y. lipolytica*) (Hassan et al. 2020; Jach et al. 2021).

Table 10.4 Microbial production of vitamins, peptides, and carbohydrates

Bioactive compounds	Microorganisms	Biotechnological strategy	Identification technique	References
Asperphenins A and B	*Aspergillus* sp.	Metabolic engineering	LC – MS, NMR	Liao et al. (2017)
Bioactive peptides	*L. rhamnosus* and *B. lactis*	Cell culture technology	LC-MS/MS	Kathayat et al. (2021)
Chitosan-oligosaccharides	*A. griseoaurantiacus*	Cell culture technology	TLC	Shehata et al. (2018)
Exopolysaccharide	*B. mycoides*	Cell culture technology	GPC, FTIR GC-MS	Farag et al. (2020)
β-1,3-glucan water-soluble	*Agrobacterium* sp. and *T. harzianum*	Cell culture technology	GPC, MALDI-TOF MS	Liang et al. (2018)
Hemolytic lipopeptide	*B. halotolerans*	Cell culture technology	FTIR, GC-MS, HPLC	Etemadzadeh and Emtiazi (2021)
Human neutrophil peptide	*P. pastoris*	Cell culture technology and metabolic engineering	Protein assay kit	Zhang et al. (2018)
Iturin-like and bogorol-like lipopeptides	*Brevibacillus spp*	Cell culture technology	HPLC-MS/MS, FTIR, MALDI-TOF-MS	Singh et al. (2021b)
Levan	*S. cerevisiae*	Cell culture technology and metabolic engineering	HPLC	Ko et al. (2019)
PAX lipopeptides	*X. khoisanae*	Cell culture technology	UPLC-MS/MS	Dreyer et al. (2019)
Rhabdopeptides/xenortides	*E. coli*	Cell culture technology	UPLC-MS	Oestreich et al. (2021)
Serine-glycine-betaine	*M. phaseolin*	Cell culture technology	FTIR, UV-vis, NMR, LC-MS	Singh et al. (2021a)
Vitamin B1, B2, B6, B7, B9, B12	*Y. lipolytica*	Cell culture technology	AOAC official methods	Jach et al. (2021)
Vitamin B2	*B. subtilis*	Metabolic engineering	UV-Vis	You et al. (2021)
	R. glutinis	Cell culture technology	HPLC, UV-vis	Hassan et al. (2020)
Vitamin B2, B9 and B12	*Lactobacillus* sp.	Cell culture technology	HPLC, UV-vis	Hati et al. (2019)
Vitamin B12	*P. freudenreichii*	Cell culture technology	HPLC, UV-vis	Hedayati et al. (2020)
Vitamin K2	*B. subtilis*	Cell culture technology	HPLC, UV-vis	Zhang et al. (2020)

FTIR fourier transform infrared spectroscopy, *GC* gas chromatography, *GPC* gel permeation chromatography, *HPLC* high pressure liquid chromatography, *MALDI-TOF-MS* matrix-assisted laser desorption/ionization mass spectrometry, *MS* mass spectrometry, *NMR* nuclear magnetic resonance, *TLC* thin-layer chromatography, *UPLC* ultra high-performance liquid chromatography, *UV-vis* ultraviolet visible spectroscopy

10.5.2 Peptides

Peptides are short chains of amino acids (lower than 15 amino acids) linked by covalent chemical bonds that are beneficial to human health (Kathayat et al. 2021). Peptides have been obtained by bacteria and fungi (Kathayat et al. 2021; Liao et al. 2017). In this sense, *B. cereus*, *B. halotolerans*, *Bifidobacterium lactis*, *Lacticaseibacillus rhamnosus*, *Brevibacillus spp*, *E. coli*, and *Xenorhabdus khoisanae* have been studied for peptides production by cell culture technology (Dreyer et al. 2019; Etemadzadeh and Emtiazi 2021; Kathayat et al. 2021; Singh et al. 2021b). Also, *A. allahabadii*, *A. ochraceopetaliformis*, *Macrophomina phaseolina* and *Pichia pastoris* have been reported as fungi with potential for the production of different peptides by cell culture technology and metabolic engineering (Liao et al. 2017; Singh et al. 2021a; Zhang et al. 2018).

10.5.3 Carbohydrates

Carbohydrates are biomolecules with carbon, hydrogen, and oxygen atoms that have been obtained by bacteria, fungi, and yeast through cell culture technology and metabolic engineering (Cao et al. 2020; Ko et al. 2019; Liang et al. 2018; Shehata et al. 2018). In this sense, Shehata et al. (2018) reported the chitosan-oligosaccharides production by *A. griseoaurantiacus,* and thin-layer chromatography (TLC) was used for the carbohydrate identification. Likewise, exopolysaccharide and β-1,3-glucan water-soluble were obtained by *B. mycoides*, *Agrobacterium* sp., and *T. harzianum*. The identification techniques were gel permeation chromatography (GPC), Fourier transform infrared spectroscopy (FTIR), gas spectrometry coupled to mass spectrometry (GC-MS), and matrix-assisted laser desorption/ionization mass spectrometry (MALDI-TOF-MS) (Farag et al. 2020; Liang et al. 2018). On the other hand, Ko et al. (2019) reported the levan production from the yeast *S. cerevisiae* by cell culture technology and metabolic engineering; they used high-pressure liquid chromatography (HPLC) as an identification technique.

10.6 Perspectives

Bioactive compounds from plants have proven beneficial for human health, and the recent advances in microbial production of these bioactive compounds have been increasing (Chen et al. 2020a; Wu et al. 2020; Sun et al. 2021). The research in critical molecular components and enzymes that catalyze compounds from different microbial strains, as well as the novel pathways, have allowed the different bioactive compounds production by microbial fermentation, cell culture, and semisynthetic process on a small and medium scale (Ozber et al. 2020; Pham et al. 2019). Biotechnological methods improve the production of target compounds in shorter times and minimize the use of chemical agents. However, some compounds obtained by these methods are still under development due to the poorly enzymes expressed in yeast or bacteria for the bioactive compounds production as well as the native pathways are still not fully elucidated; therefore, the use in in vivo models is still under study (Ozber et al. 2020; Sun et al. 2019; Zhou et al. 2014). Likewise, tools have recently been developed to counteract the challenges of the production of bioactive compounds by microorganisms, such as the use of computational protein design to accelerate the engineering of the enzymes that limit the production rate of target compounds, as well as *in silico* genome-scale metabolic analysis for the identification of inactivated or overexpressed genes and modifying microorganisms for higher production of bioactive compounds (Suzuki et al. 2014; Wu et al. 2020; Sun et al. 2021). Also, the industrial scaling of biotechnological methods has not been achieved, and bioactive compounds are strictly regulated for their use in humans (Ozber et al. 2020; Zhou et al. 2014).

10.7 Conclusion

In recent years, the advances in bioactive compounds production from microbial have increased due to pharmaceutical demands. Biotechnological techniques have allowed an industrial scaling of microbial the production of the compounds due to the elucidation of biosynthetic pathways and the development of synthetic biology strategies through genetic manipulation to obtain bioactive compounds with higher yields in less time, in comparison with the compounds extracted from plants. Despite continuous advances in biotechnological techniques, many pathways for the synthesis of compounds with potential health effects are still unknown, and the development of effective techniques to obtain bioactive compounds from microbial.

References

Abdallah II, Pramastya H, van Merkerk R, Sukrasno, Quax WJ (2019) Metabolic engineering of *Bacillus subtilis* toward Taxadiene biosynthesis as the first committed step for Taxol production. Front Microbiol 10:218

Brown S, Clastre M, Courdavault V, O'connor SE (2015) De novo production of the plant-derived alkaloid strictosidine in yeast. PNAS 112:3205–3210

Cao M, Gao M, Suástegui M, Mei Y, Shao Z (2020) Building microbial factories for the production of aromatic amino acid pathway derivatives: from commodity chemicals to plant-sourced natural products. Metab Eng 58:94–132

Ceranic A, Svoboda T, Berthiller F, Sulyok M, Samson JM, Guldener U, Schuhmacher R, Adam G (2021) Identification and functional characterization of the gene cluster responsible for fusaproliferin biosynthesis in *Fusarium proliferatum*. Toxins 13:468

Chen Z, Shen X, Wang J, Wang J, Yuan Q, Yan Y (2017a) Rational engineering of p-hydroxybenzoate hydroxylase to enable efficient gallic acid synthesis via a novel artificial biosynthetic pathway. Biotechnol Bioeng 114:2571–2580

Chen Z, Sun X, Li Y, Yan Y, Yuan Q (2017b) Metabolic engineering of *Escherichia coli* for microbial synthesis of monolignols. Metab Eng 39:102–109

Chen R, Yang S, Zhang L, Zhou YJ (2020a) Advanced strategies for production of natural products in yeast. iScience 23:100879

Chen Y, Boggess EE, Ocasio ER, Warner A, Kerns L, Drapal V, Gossling C, Ross W, Gourse RL, Shao Z, Dickerson J, Mansell TJ, Jarboe LR (2020b) Reverse engineering of fatty acid-tolerant *Escherichia coli* identifies design strategies for robust microbial cell factories. Metab Eng 61: 120–130

Chouhan S, Sharma K, Zha J, Guleria S, Koffas MAG (2017) Recent advances in the recombinant biosynthesis of polyphenols. Front Microbiol 8:2259

Courdavault V, Cassereau J, Papon N (2021) Engineered microbes for producing anticholinergics. Chembiochem 22:1368–1370

Czajka JJ, Nathenson JA, Benites VT, Baidoo EEK, Cheng QS, Wang YC, Tang YJJ (2018) Engineering the oleaginous yeast *Yarrowia lipolytica* to produce the aroma compound beta-ionone. Microb Cell Factories 17:136

Dreyer J, Rautenbach M, Booysen E, van Staden AD, Deane SM, Dicks LMT (2019) *Xenorhabdus khoisanae* SB10 produces Lys-rich PAX lipopeptides and a Xenocoumacin in its antimicrobial complex. BMC Microbiol 19:132

Dudnik A, Gaspar P, Neves AR, Forster J (2018) Engineering of microbial cell factories for the production of plant polyphenols with health-beneficial properties. Curr Pharm Des 24:2208–2225

Engels B, Heinig U, Mcelroy C, Meusinger R, Grothe T, Stadler M, Jennewein S (2021) Isolation of a gene cluster from *Armillaria gallica* for the synthesis of armillyl orsellinate-type sesquiterpenoids. Appl Microbiol Biotechnol 105:211–224

Etemadzadeh SS, Emtiazi G (2021) In vitro identification of antimicrobial hemolytic lipopeptide from halotolerant *Bacillus* by Zymogram, FTIR, and GC mass analysis. Iran J Basic Med Sci 24: 666–674

Farag MMS, Moghannem SAM, Shehabeldine AM, Azab MS (2020) Antitumor effect of exopolysaccharide produced by *bacillus mycoides*. Microb Pathog 140:103947

Galanie S, Smolke CD (2015) Optimization of yeast-based production of medicinal protoberberine alkaloids. Microb Cell Factories 14:144

Galanie S, Thodey K, Trenchard ISIS J, Filsinger Interrante M, Smolke Christina D (2015) Complete biosynthesis of opioids in yeast. Science 349:1095–1100

Godara A, Kao KC (2021) Adaptive laboratory evolution of β-caryophyllene producing *Saccharomyces cerevisiae*. Microb Cell Factories 20:106

Grewal PS, Modavi C, Russ ZN, Harris NC, Dueber JE (2018) Bioproduction of a betalain color palette in *Saccharomyces cerevisiae*. Metab Eng 45:180–188

Hassan S, Shetaia Y, Shafei MS, Elrefai HA, Elaasser M (2020) Optimization of riboflavin production by *Rhodotorula glutinis* using statistical design. Egypt Pharm J 19:312–320

Hati S, Patel M, Mishra BK, Das S (2019) Short-chain fatty acid and vitamin production potentials of *Lactobacillus* isolated from fermented foods of Khasi tribes, Meghalaya, India. Ann Microbiol 69:1191–1199

Hedayati R, Hosseini M, Najafpour GD (2020) Optimization of semi-anaerobic vitamin B12 (*cyanocobalamin*) production from rice bran oil using *Propionibacterium freudenreichii* PTCC1674. Biocatal Agric Biotechnol 23:101444

Huang Y, Hoefgen S, Valiante V (2021) Biosynthesis of fungal Drimane-type Sesquiterpene esters. Angew Chem Int Ed 60:23763–23770

Hussain MH, Hong Q, Zaman WQ, Mohsin A, Wei YL, Zhang N, Fang HQ, Wang ZJ, Hang HF, Zhuang YP, Guo MJ (2021) Rationally optimized generation of integrated *Escherichia coli* with stable and high yield lycopene biosynthesis from heterologous mevalonate (MVA) and lycopene expression pathways. Synth Syst Biotechnol 6:85–94

Jach ME, Sajnaga E, Janeczko M, Juda M, Kochanowicz E, Baj T, Malm A (2021) Production of enriched in B vitamins biomass of *Yarrowia lipolytica* grown in biofuel waste. Saudi J Biol Sci 28:2925–2932

Jin CC, Zhang JL, Song H, Cao YX (2019) Boosting the biosynthesis of betulinic acid and related triterpenoids in *Yarrowia lipolytica* via multimodular metabolic engineering. Microb Cell Factories 18:18

Kathayat D, Closs G, Helmy YA, Deblais L, Srivastava V, Rajashekara G (2021) In Vitro and in vivo evaluation of *Lacticaseibacillus rhamnosus* GG and *Bifidobacterium lactis* Bb12 against avian pathogenic *Escherichia coli* and identification of novel probiotic-derived bioactive peptides. Probiotics Antimicrob proteins 14:1012–1028

Ko H, Bae JH, Sung BH, Kim MJ, Kim CH, Oh BR, Sohn JH (2019) Efficient production of Levan using a recombinant yeast *Saccharomyces cerevisiae* hypersecreting a bacterial levansucrase. J Ind Microbiol Biotechnol 46:1611–1620

Li YR, Smolke CD (2016) Engineering biosynthesis of the anticancer alkaloid noscapine in yeast. Nat Commun 7:12137

Li S, Li Y, Smolke CD (2018a) Strategies for microbial synthesis of high-value phytochemicals. Nat Chem 10:395–404

Li X, Chen Z, Wu Y, Yan Y, Sun X, Yuan Q (2018b) Establishing an artificial pathway for efficient biosynthesis of hydroxytyrosol. ACS Synth Biol 7:647–654

Li Y, Li S, Thodey K, Trenchard I, Cravens A, Smolke CD (2018c) Complete biosynthesis of noscapine and halogenated alkaloids in yeast. PNAS 115:E3922

Li C, Swofford CA, Ruckert C, Chatzivasileiou AO, Ou RW, Opdensteinen P, Luttermann T, Zhou K, Stephanopoulos G, Prather KLJ, Zhong Johnson EZL, Liang SL, Zheng SP, Lin Y, Sinskey AJ (2021) Heterologous production of alpha-carotene in *Corynebacterium glutamicum* using a multi-copy chromosomal integration method. Bioresour Technol 341:125782

Liang Y, Zhu L, Gao MJ, Wu JR, Zhan XB (2018) Effective production of biologically active water-soluble-1,3-glucan by a coupled system of *Agrobacterium* sp and *Trichoderma harzianum*. Prep Biochem Biotechnol 48:446–456

Liao L, Bae S, Won TH, You M, Kim SH, Oh D-C, Lee S, Oh K-B, Shin J (2017) Asperphenins A and B, Lipopeptidyl benzophenones from a marine-derived *Aspergillus* sp. Fungus Org Lett 19: 2066–2069

Liu L, Qu YL, Dong GR, Wang J, Hu CY, Meng YH (2021) Elevated beta-carotene production using codon-adapted CarRA&B and metabolic balance in engineered *Yarrowia lipolytica*. Front Microbiol 12:627150

Ma B, Liu M, Li ZH, Tao XY, Wei DZ, Wang FQ (2019) Significantly enhanced production of Patchoulol in metabolically engineered *Saccharomyces cerevisiae*. J Agric Food Chem 67: 8590–8598

Oestreich AM, Suli MI, Gerlach D, Fan R, Czermak P (2021) Media development and process parameter optimization using statistical experimental designs for the production of nonribosomal peptides in *Escherichia coli*. Electron J Biotechnol 52:85–92

Ouyang X, Cha Y, Li W, Zhu C, Zhu M, Li S, Zhuo M, Huang S, Li J (2019) Stepwise engineering of *Saccharomyces cerevisiae* to produce (+)-valencene and its related sesquiterpenes. RSC Adv 9:30171–30181

Ozber N, Watkins JL, Facchini PJ (2020) Back to the plant: overcoming roadblocks to the microbial production of pharmaceutically important plant natural products. J Ind Microbiol Biotechnol 47: 815–828

Pham JV, Yilma MA, Feliz A, Majid MT, Maffetone N, Walker JR, Kim E, Cho HJ, Reynolds JM, Song MC, Park SR, Yoon YJ (2019) A review of the microbial production of bioactive natural products and biologics. Front Microbiol 10:1404

Shehata AN, Abd El Aty AA, Darwish DA, Wahab WAA, Mostafa FA (2018) Purification, physicochemical and thermodynamic studies of antifungal chitinase with production of

bioactive chitosan-oligosaccharide from newly isolated *Aspergillus griseoaurantiacus* KX010988. Int J Biol Macromol 107:990–999
Singh G, Singh J, Singamaneni V, Singh S, Gupta P, Katoch M (2021a) Serine-glycine-betaine, a novel dipeptide from an endophyte *Macrophomina phaseolina*: isolation, bioactivity and biosynthesis. J Appl Microbiol 131:756–767
Singh SS, Sharma D, Baindara P, Choksket S, Harshvardhan M, S. M., Grover, V. & Korpole, S. (2021b) Characterization antimicrobial studies of Iturin-like and Bogorola-like lipopeptides from *Brevibacillus* spp. strains GI9 and SKDU10. Front Microbiol 12:729026
Srinivasan P, Smolke CD (2019) Engineering a microbial biosynthesis platform for de novo production of tropane alkaloids. Nat Commun 10:3634
Sun X, Cai YS, Yuan YJ, Bian GK, Ye ZL, Deng ZX, Liu TG (2019) Genome mining in *trichoderma viride* j1-030: discovery and identification of novel sesquiterpene synthase and its products. Beilstein J Org Chem 15:2052–2058
Sun XX, Li XL, Shen XL, Wang J, Yuan QP (2021) Recent advances in microbial production of phenolic compounds. Chin J Chem Eng 30:54–61
Suzuki S, Koeduka T, Sugiyama A, Yazaki K, Umezawa T (2014) Microbial production of plant specialized metabolites. Plant Biotechnol 31:465–482
Thodey K, Galanie S, Smolke CD (2014) A microbial biomanufacturing platform for natural and semisynthetic opioids. Nat Chem Biol 10:837–844
Trenchard IJ, Smolke CD (2015) Engineering strategies for the fermentative production of plant alkaloids in yeast. Metab Eng 30:96–104
Valsami EA, Pateraki A, Melis A, Ghanotakis DF (2021) Heterologous beta-phellandrene production by alginate immobilized *Synechocystis* sp. PCC 6803. J Appl Phycol 33:2157–2168
Wu T, Ye LJ, Zhao DD, Li SW, Li QY, Zhang BL, Bi CH, Zhang XL (2017) Membrane engineering–A novel strategy to enhance the production and accumulation of beta-carotene in *Escherichia coli*. Metab Eng 43:85–91
Wu X, Zha J, Koffas MAG (2020) Microbial production of bioactive chemicals for human health. Curr Opin Food Sci 32:9–16
You JJ, Yang C, Pan XW, Hu MK, Du YX, Osire T, Yang TW, Rao ZM (2021) Metabolic engineering of *Bacillus subtilis* for enhancing riboflavin production by alleviating dissolved oxygen limitation. Bioresour Technol 333:125228
Zhang XL, Jiang AM, Qi BH, Yu H, Xiong YY, Zhou GL, Qin MS, Dou JF, Wang JF (2018) Secretion expression of human neutrophil peptide 1 (HNP1) in *Pichia pastoris* and its functional analysis against antibiotic-resistant *Helicobacter pylori*. Appl Microbiol Biotechnol 102:4817–4827
Zhang C, Ren HX, Zhong CQ (2020) Economical production of vitamin K-2 using wheat starch wastewater. J Clean Prod 270:122486
Zhang G, Wang H, Zhang Z, Verstrepen KJ, Wang QH, Dai ZJ (2021) Metabolic engineering of *Yarrowia lipolytica* for terpenoids production: advances and perspectives. Crit Rev Biotechnol 42(4):618–633
Zhou J, Du G, Chen J (2014) Novel fermentation processes for manufacturing plant natural products. Curr Opin Biotechnol 25:17–23

Chapter 11
Future Marine Microbial Products for the Pharmaceuticals Industry

Puja Dokania, Rushikesh Fopase, G. Swagathnath, Vivekanand, Kriti Gupta, Pooja Pabari, Krishna Kalyani Sahoo, and Angana Sarkar

Abstract The marine environment addresses endless and different assets of the pharmaceutical industries for new drugs in the terms of battling significant illnesses like cancer, malaria, etc. It additionally offers a natural asset involving an assortment of oceanic creatures such as microorganisms. These marine microorganisms are screened for anticancer, antifungal, antimicrobial, antibacterial, immunomodulatory, neuroprotective, analgesic, anti-inflammatory, and antimalarial properties. They are utilized in the pharmaceutical industry for the development of new drugs widely across the world. The enormous development of the world's total population has overburdened the current assets for drugs. Furthermore, subsequently, the manufacturers of drugs are generally looking out for new resources to build effective as well as safe drugs to fulfill the rising requests of the total population. Marine-derived products are viewed as a novel method for satisfying the worldwide need for pharma, food, as well as energy. According to this viewpoint, this chapter explores the future marine microbial-derived products, their resources, as well as their applications in the pharmaceutical industries. Apart from these it also describes the economic contribution of the marine microbial-derived products to global needs.

Keywords Marine-derived microbial products · Pharmaceutical industry · Drugs · Marine microbes

11.1 Introduction

The oceans cover about 70% area of the planet Earth. It contains very diverse life forms that can be utilized in the study of marine biotechnology. As the field of biotechnology is advancing, the interest in enzymes having novel properties is also

P. Dokania · R. Fopase · G. Swagathnath · Vivekanand · K. Gupta · P. Pabari · K. K. Sahoo · A. Sarkar (✉)
Department of Biotechnology and Medical Engineering, National Institute of Technology, Rourkela, India
e-mail: sarkara@nitrkl.ac.in

A. Sarkar, I. A. Ahmed (eds.), *Microbial products for future industrialization*, Interdisciplinary Biotechnological Advances,
https://doi.org/10.1007/978-981-99-1737-2_11

increasing. Animals, plants, and microorganisms form the basis for the isolation of enzymes; out of these three, microbes due to their wide diversity, high mass production, and easiness in genetic manipulation act as the most common enzyme source. The enzymes derived from marine microbes are drawing more thoughtfulness concerning new enzymes as they are comparatively more active and stable than plant- or animals-derived enzymes (Robinson 2015). As compared to the earthly atmosphere, the nautical atmosphere provides marine microbes that live in exceptional habitats and inhibit exclusive genetic assemblies. High saline, low temperatures, high pressure, and low light at low depth, such conditions are endured by the microorganisms living in such extreme areas (Trincone 2013). Due to such extreme conditions, the microorganisms that adapt to such areas anchor enzymes with a unique and appealing feature, which has resulted in a growing interest in marine microbes for enzymes (Arnosti et al. 2014). The complexity of the marine environment, from nutrient-rich areas to nutrient-scarce areas, explains the dissimilarities between the enzymes produced by marine microbes and homologous enzymes from terrestrial microbes.

Marine microorganisms like bacteria, fungi, and actinomycetes have provided a wide range of enzymes with extraordinary activities and some of them have been utilized in industrial applications as well. The enzyme technology from the marine microbes has been boosted in the past few years resulting in valuable products that are utilized in food additives, fine chemicals, and pharmaceuticals (Beygmoradi and Homaei 2017). The marine empire is a rich source of potential products that are of importance to mankind. Due to the gap that lies between the finding and creating a commercial product, limited marine products have extended up to commercialization. The research areas of biology, chemistry, and engineering can be assimilated for the development of bioprocess methodologies for the production of marine products.

Bioactive compounds, present in nutraceuticals (nutrition + pharmaceuticals), provide health benefits beyond basic nutrition. They are naturally occurring compounds present in food like peptides, polysaccharides, fatty acids, vitamins, and phytochemicals (Waters et al. 2010). An extensive study has been conducted on the bioactive compounds derived from marine bacteria and fungi. Physical, chemical, and biological properties have been studied, and thus concluded that these compounds have functionality as a secondary metabolite. This makes the bioactive compounds a potential source of organic products having applications in pharmaceuticals, food industries, and cosmetics (Zhang and Kim 2012).

This review presents the application of marine microbes in the field of pharmaceutical and food industries. The main focus is on enzymes, antioxidants, bioactive compounds, and fatty acids derived from marine organisms.

11.2 Protease

Protease accounts for greater than 60% of global industrial enzyme sales. They have a wide range of applications in the food, detergent, leather industry, and pharmaceutical industries for the production of anti-inflammatory and digestive drugs. Proteases derived from microorganisms are eco-friendly and are commercially important. An alkaline protease was first isolated from *Bacillus licheniformis*, by Dane in 1960. Nobou Kato, in 1972, extracted a new kind of alkaline protease from *Psychrobacter* thriving in the marine environment, and furthermore, proteases gradually began to be isolated from microorganisms thriving in the marine environment. An alkaline protease obtained from a bacterial strain living in a symbiotic relationship with a marine shipworm in the Gland of Deshayes was found to have cleansing activity (Zhang and Kim 2012). Strain N1-35, isolated after screening 30 types of marine bacterial strains from fish, mud, sea-water, and various other samples followed by UV mutagenesis, synthesized a protease that was significantly advantageous over the ones produced from terrestrial microorganisms. *Aureobasidium pullulans,* a yeast strain obtained from the China Yellow Sea saltern, had a maximum alkaline protease yield of 623.1 U/mg protein (7.2 U/mL) (Anithajothi et al. 2014).

An enzyme that carries out proteolysis, that is, catabolizes the protein by hydrolyzing the peptide bonds, which connect the amino acids across the polypeptide chain to form the protein, is called a protease. Proteases are necessary for several physiological processes such as apoptosis and in the life cycle of disease-causing organisms including the replication of retrovirus, processing of polypeptide hormones, signal transduction, blood clotting cascade, cell division, protein turnover, digestion of food proteins, and so on (Devlin et al. 2007).

11.2.1 Sources of Proteases

Proteases are naturally produced by a wide range of living species. Some major proteases synthesized by plants include keratinases, bromelain, and papain (Mahendran et al. 2010); rennin, pepsin, chymotrypsin and pancreatic trypsin are some common proteases of animal origin. But the quantity of proteases being produced from plant and animal sources is not enough to satisfy the global protease demand. Because of their broad biochemical diversity and ease of genetic manipulation, microorganisms are drawing greater attention from protease producers (Asha and Palaniswamy 2018). Proteases of microbial origin include carboxypeptidases Y, endoproteinase, thermolysin, collagenase, dipase, Rhino virus3C, TEV protease, TVMV protease, WNV protease, proteinase K, subtilisin, endoproteinase (Feijoo-Siota et al. 2014) (Fig. 11.1).

Proteases are involved in the various processes taking place to maintain the life cycle of organisms, which cause various diseases. Hence these proteases are

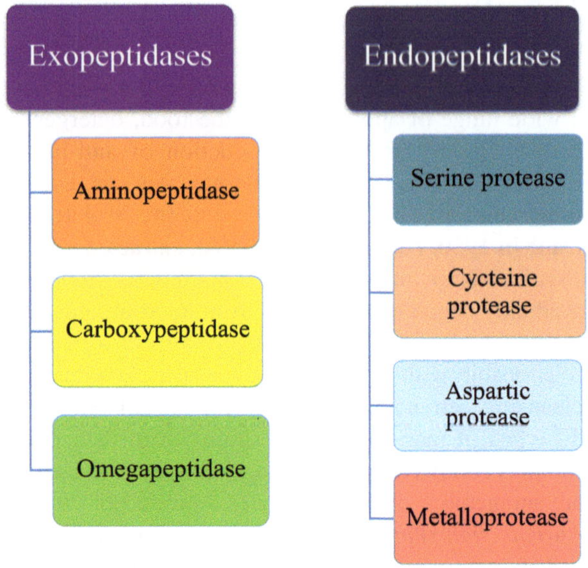

Fig. 11.1 Classification of proteases on the basis of their mode of action (exopeptidases) and their amino acid group (endopeptidases) application in pharmaceutical

potential constituents for the production of therapeutic agents targeting deadly diseases like AIDS and cancer. Proteases derived from microorganisms are being widely used for the treatment of different health-associated disorders such as cardiovascular disorders, inflammation, necrotic wounds, cancer, and so on. Proteases are being utilized as agents to stimulate the immune system. When protease was used concomitantly with an antibiotic, it led to an increase in the concentration of the antibiotic at the target site. Pharmaceutical industries also use proteases for producing ointments seeking to remove the damaged tissue or foreign objects from wounds. Proteases are also utilized in contact lenses cleansing solutions and denture cleansers (Singh et al. 2016).

Proteases are produced by microorganisms with a wide range of specificities and diversity. This is beneficial in producing therapeutic agents, which are effective against a broad range of diseases. Proteases derived from *Aspergillus oryzae* when administered orally act as a digestive aid to treat certain syndromes arising from the deficiency of lytic enzymes. Collagenase or subtilisin produced from *Clostridia* sp. is utilized in conjugation with a broad-spectrum of antibiotics for treating wounds and burns. Lymphocytic leukemia can be treated by removing asparagine from the bloodstream by using *Escherichia Coli*-derived asparaginase (Ghasemian et al. 2019). *Conidiobolus coronatus* produces an alkaline protease that can be utilized in place of trypsin for animal cell culture (Sutar et al. 1992).

Slow-release dosage forms used in therapeutic applications are prepared using alkaline proteases like collagenase. *Aspergillus niger LCF9* produces a semi-alkaline protease possessing high collagenolytic properties, which can be used in hydrolyzing different types of collagen without releasing any amino acid and thus produces peptides of low molecular weight having therapeutic applications. *Bacillus subtilis*

316M produced a protease, elastoterase, possessing high elastolytic properties. It was used for treating purulent wounds and burns, furuncles, carbuncles, and deep abscesses, by immobilizing the protease on a bandage. *Bacillus* spp. *CK11-4* has been found to be safe for therapeutic applications in humans (Silva et al. 2006). Kim et al. (1996), reported an alkaline protease possessing fibrinolytic property, which was then subsequently found to be a potent thrombolytic agent (Table 11.1).

Inulin is used for formulating ultra-high-fructose syrup. Fructose syrup is effective in diabetic patients, increases the absorption of iron in children, is used for the diet of an obese person due to its sweetening capacity, functions as dietary fibers because of its fat-like design, and is effective in preventing cancer of the colon. It provides stimulation of Bifidobacterium growth in the large and small intestine and absorption of calcium in postmenopausal women.

Instead of sucrose, fructose is used in foods, beverages, and pharmaceuticals. But there are some drawbacks related to the chemical approach of fructose production (Naveed et al. 2021).

11.3 Chitinases

Chitin, the second richest polymer on Earth, is a polysaccharide that contains 1, 4-linked N-acetyl-D-glucosamine moieties and a perpetual root of renewable raw elements. These have their antitumor and antimicrobial activities but are not water-soluble. The Chitinase enzyme, isolated from many microorganisms, hydrolyzes the β-1,4-linkages in the chitin and results in products with moderate molecular weight (free N-acetyl glucosamine). The crab shell powder and shrimp shell powder have the marine carbon source of the chitinase enzyme. *Alcaligenes faecalis AU02*, isolated from seafood industrial effluent, was cultured in the powdered media for the production of the chitinase in the media. Later, the enzyme has been extracted and the activity measured. From the marine *Streptomyces* sp. *DA11*chitinase has been extracted through the genetic engineering method where a gene has been isolated and expressed. It has been shown that chitinase produced by this species shows antifungal activity against *Candida albicans* and *Aspergillus niger* (Jahromi and Barzkar 2018).

Chitinase has been produced in different batches and continuous fermentation process with respect to the organisms used for the production process. The highest enzyme activity was achieved in *Penicillium janthineflum* about 686 mU/mL and the enzyme production process is done through a batch process with a reactor volume of 2/3 L. The temperature and the pH of the production conditions are 28 °C and 4.0, respectively. For the increased production, along with the corn steep liquor, the carbon source colloidal chitin is also added. The other two methods of producing chitinase do not yield the enzyme as high as the previous method of the batch culture where the yield is 11.8 mU/mL and 18–19 mU/mL from the organisms *Paenibacillus* sp. And *Verticillium lecanii*, respectively (Nampoothiri et al. 2004).

Table 11.1 Protease production from different marine microorganisms

Enzyme	Marine microorganisms	Source	Purification method	Parameters for maximum activity	References
Alkaline protease	*Bacillus cereus*	Marsa-Matrouh beaches	Ammonium sulfate precipitation and Sephadex G-200	pH 10, 50 °C and Cu^{2+} ions	Abou-Elela et al. (2011)
	Aureobasidium pullulans	Sediment of saltern in Qingdao	–	pH 9.0, 45 °C	Chi et al. (2007)
	Teredinobacter turnirae	Shipworm	–	–	Elibol and Moreira (2005)
	Pseudoalteromonas sp. 129-1	Seawater of South China Sea	Ammonium sulfate precipitation, anion exchange chromatography, and gel filtration	pH 8.0, 50 °C	Wu et al. (2015)
	Aureobasidium pullulans N13d	–	–	–	Ni et al. (2009)
	Aureobasidium pullulans 10	–	–	–	
	Aureobasidium pullulans HN2-3	–	–	–	
	Leucosporidium antarcticum 171	–	–	–	
Alkaline serine protease	*Engyodontium album BTMFS10*	Marine sediment of Cochin coast	Ammonium sulfate precipitation and ion-exchange chromatography	pH 11 and 60 °C, Co^{2+}	Chellappan et al. (2006)
	Gamma-Proteobacteria	Sediment of the Lothian Island	Acetone precipitation, anion exchange chromatography, sephadex gel filtration	pH 9.0, 30 °–70 °C, Ba^{2+}, Ca^{2+}	Sana et al. (2006)
	Psychrophilic antarctic yeast Leucosporidium antarcticum 171	–	–	–	Turkiewicz et al. (2003)
	Bacillus flexus APCMST-RS2P	Sediment samples of Rajakkamangalam coast	Ammonium sulfate precipitation and sephadex G-75 gel filtration chromatography	pH 8.0, 40 °C, BaCl2 and CaCl2	Maruthiah et al. (2014)

Thermo-tolerant alkaline protease	*Roseobacter sp. MMD040*	Marine sponge Fasciospongia cavernosa	Ammonium sulfate precipitation and dialysis	pH 6–9, 50 °C	Shanmughapriya et al. (2008)
	Roseobacter MMD040 (sponge-associated bacteria)	–	–	–	Shanmughapriya et al. (2008)
Cold-active alkaline serine protease	*Penicillium chrysogenum FS010*	Huanghai Sea	Ammonium sulfate precipitation, DEAE-IEC and sephadex G-100 gel chromatography	pH 9.0, 15 °–35 °C, Mg^{2+}, Ca^{2+}	Zhu et al. (2009)
Acid protease	*Metschnikowia reukaufii W6b*	–	–	–	Li et al. (2010a, 2010b)
Fibrinolytic enzyme (enzyme group-serine protease)	*Bacillus clausii ICTF-1*	–	–	–	Mahajan et al. (2012)
Protease (enzyme group-hydrolases)	*Codium divaricatum* (green alga)	–	–	–	Matsubara et al. (2000)
Caseinase, gelatinase	*Streptomyces fungicidicus*	Bay of Bengal	Ammonium sulfate precipitation and dialysis	pH 9 and 40 °C	Ramesh et al. (2009)

Table 11.2 Chitinase production

Organism	Process	Enzyme activity	References
Penicillium janthineflum	Batch	686 U/L	Fenice et al. (1998)
Paenibacillus sp.	Batch	11.8 mU/mL	Kao et al. (2007)
Verticillium lecanii	Batch	18–19 mU/mL	Liu et al. (2003)
Beauveria bassiana	Solid state	246.6 U/g	Chellappan et al. (2006)
Beauveria bassiana	Petri plates/culture dishes	248 U/g	Chysirichote et al. (2014)

In the batch solid-state process, *Beauveria bassiana* is used in the lab-scale production of the chitinase enzyme in the 250 mL Erlenmeyer flask. This also yielded the highest production yield up to 246.6 U/g of IDS but didn't take to the large-scale production process. Similarly, in another case of small-scale culturing, Beauveria *bassiana* was grown in the culture plates with the yield of 248 U/g of IDS. In the continuous culturing of *Paenibacillus* sp., the chitinase enzyme production yield is up to 42,800 mU/mL (Du et al. 2021).

More studies have been done to optimize the media composition and the different media on the chitinase production on different microorganisms such as *Vibrio* sp. and *Pantoea dispersa* (Muffler et al. 2015) (Table 11.2).

11.3.1 Pharmaceutical Uses of Chitinases

Chitinases are known to have different roles in the field of medicines such as chemotherapeutic targets, asthma treatment, allergies, and proteins having chitinase activity in humans, which play a role as biomarkers in terms of human disease. Apart from that, chitinase is used in treating fungal infections as it is an antifungal agent. Due to the antitumor characteristics of chitooligosaccharides, it also has huge potential in cancer treatments. In the airways, asthma shows close relativity with the increased activity of Th2 lymphocytes as an immune response. In Th2-related airway inflammation, chitinases, for example, AMCase and TKL-40, show grateful evidence as a biomarker. During inflammation, an increased level of expression of such chitinases makes them a potential biomarker in the treatment of asthma. Chemotherapeutic drugs mostly have several side effects, so developing modern cancer research plays a key role in overcoming this problem (Beygmoradi et al. 2018).

11.4 Fatty Acids Isolated from Marine Microorganisms

11.4.1 Polyunsaturated Fatty Acids (PUFAs) from Marine Microbes

Polyunsaturated fatty acids (PUFAs) hold great importance because they are essential for human health. Various important regulatory molecules inside the body use PUFAs as precursors for their synthesis. Cells of the brain of the retina also comprise PUFA components in their membranes. PUFAs also help in the prevention of cardiovascular diseases disease. PUFAs are also produced by marine Gram-negative bacteria growing in the gut flora of fishes that survive in deep, low-temperature waters (Kawamoto et al. 2009). These marine bacteria can also be grown on a large scale using fermentation methods and thus represent a contamination-free, sustainable, and renewable source of PUFAs for application in animal and human dietary supplements. Eicosapentaenoic acid (EPA) and Docosahexaenoic acid (DHA) are two most important types of PUFAs. EPA-producing bacteria can be easily isolated from marine organisms, sea sediments, sea ice, sea water, and certain fishes such as deep-sea fish or blue-backed fish. The method for isolation of EPA-producing bacteria involves a culture of source sample on marine agar plates, isolation of single colonies, and regeneration of biomass for analyzing the content of EPA. EPA-producing bacteria can be screened using various methods such as gas chromatography-mass spectroscopy (GC-MS), gas chromatography (GC), and thin-layer chromatography (Ryan et al. 2010).

Long-chain polyunsaturated fatty acids consist of a long chain of hydrocarbon composed of 18 or greater atoms of carbon and a carboxylate group at the terminus consisting of two or more double bonds. The classification of fatty acids is done as per the location of the initial-most double bond when counted starting from the methyl termini. For example, in ω-3 PUFA the first double bond is present after three carbon atoms when counted along the methyl end. The same rule applies to ω-6 and ω-9. The symbol n can be used in place of ω for the classification of PUFAs. The symbol Δ is used to classify fatty acids when the position of the first double bond is counted from the carboxylate terminus (Sijtsma and de Swaaf 2004). The most studied PUFAs within the ω-3 group are α-linolenic acid, docosahexaenoic acid (DHA), and eicosapentaenoic acid (EPA). DHA and EPA undergo esterification to produce lipid molecules inside living cells.

For the commercial production of DHA, marine microorganisms like *Crypthecodinium cohnii* (De Swaaf et al. 2003) and *Schizochytrium* spp. are being used widely. *Schizochytrium* sp. contains lipids that constitute over 70% of its weight and DHA comprises 35% of the total fatty acids. In the marine dinoflagellate, *C. cohnii* DHA constitutes 25–60%, while the derived oil contains less than 1% of other PUFAs (Moi et al. 2018) (Fig. 11.2).

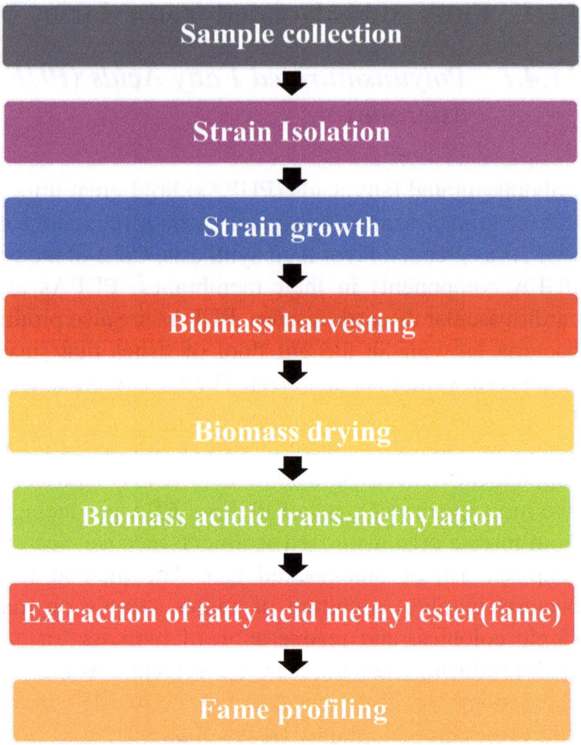

Fig. 11.2 Generalized bioprocess flow diagram for fatty acids

11.4.2 Application in the Pharmaceutical Industry

Owing to the fact that marine n-3 fatty acids target the ARA metabolism, which has been recognized to be associated with rheumatoid arthritis, extensive research is going on to discover the therapeutic potential of marine n-3 fatty acids in curing the disease. Researchers have shown that the application of marine n-3 fatty acids reduced the levels of eicosanoids and inflammatory cytokines. Consequently, cartilage destruction and pain will be reduced. This may help rheumatoid arthritis patients to minimize the use of pain-killing drugs like NSAIDs. PUFAs such as DHA helped in decreasing the infiltration of eosinophils into the lungs and enhanced the functioning of lungs in an ovalbumin-sensitized mice model, in which asthma was induced by exposing it to ovalbumin, which shows the significant effect of DHA in curing asthma. The diets of the patients suffering from inflammatory bowel disease (IBD) are supplemented with EPA and DHA, which helps in reducing inflammation. This is also associated with improved gut mucosal histology, improved clinical score, improved sigmoidoscopic score, decreased use of corticosteroids, and a lower rate of relapse in IBD patients.

EPA, DHA, and GLA have been observed to possess anti-inflammatory and anticholesterolaemic activities. Administration of ω-3 PUFAs has been shown to improve inflammatory conditions. However, the intake of ω-3 and/ω-6 fatty acids in excessive amounts may lead to interventions in the homeostasis of eicosanoids. A wide range of PUFA lipids have also been found to possess antiaging, anticancer, antithrombotic, immunosuppressant, and immunostimulant properties and thus have a considerable medical application (Jovanovic et al. 2021).

11.5 Antioxidants

Antioxidants have a major role in protection against excess reactive oxygen species. Reactive oxygen species may form by the normal metabolism of oxygen due to light energy transfer or lipid peroxidation and may leak the mitochondria with disturbing cell structure and DNA same as done by radiation. Antioxidants neutralize oxidative species in different ways based on which they are categorized. Oxygen scavenging antioxidants donate their extra electron to reactive oxygen and make them stable. Also, in transitional metal ions, chelator antioxidants chelate the metal ions and make them stable. The quencher type absorbs the energy from high-energy oxidants. Radical antioxidants are known as chain-breaking antioxidants. The property of neutralizing reactive species makes them play an important role in pharmaceutical products. Antioxidants such as carotenes, some photosynthetic pigments, and some vitamins are widely used in pharmaceuticals and as food additives. A list of some of these compounds is given in the following table.

Also, they act against oxidative rancidity as well as peroxidation products such as hydroxyl radicals and hydrogen peroxide (H_2O_2), superoxide anions that cause the deterioration of some foods. Antioxidants such as carotenes, some photosynthetic pigments, and some vitamins are widely used in pharmaceuticals and as food additives. *Dunaliella salina* is popularly used for the production of both *cis* and *trans* forms of β carotene with a productivity of up to 14% of its dry weight and high bioavailability and bio-efficacy. Along with it, vitamin C can be obtained from *Monostroma undulatum* and *Undaria pinnatifida. Porphyridium cruentum* produces α and γ-tocopherol, which have antioxidant properties. Microorganism species such as *Dunaliella salina, Haematococcus pluvialis, Chlorella* sp., etc., are applied for the production of multiple metabolites such as carotenoids, vitamin C, vitamin E, chlorophylls, violaxanthin, zeaxanthin, and many more. Different environments are provided for the production of these products. It may reduce the cost and viability of the strain (Olatunde and Benjakul 2022) (Table 11.3).

11.5.1 Applications in the Pharmaceutical Industry

Due to the feature of reducing the oxidative species, antioxidants are one of the key ingredients used for various kinds of nutraceuticals. Carotenes play a major role in protection against free oxygen radicals. β carotene has the ability to neutralize free

Table 11.3 List of antioxidants and their source as well as usage in pharmaceutical and as food additives

Antioxidants	Sources	Applications	References
Cartenoids, α-carotene, β-carotene	*Chlorella vulgaris, Sarcina maxima, Dunaliella salina, Haematococcus pluvialis, Laminaria digitate, Chlorella pyrenoidosa, Ascophyllum nodosum, Chlorococcum, Laminaria saccharina, Gracilaria changgi*	Applied to a range of food and beverage products to improve their appearance. Antioxidant, anti-inflammatory, anticancer, immunomodulatory, prevent neurodegenerative and cardiovascular diseases	Cha et al. (2008)
Vitamin C	*Ulva* sp., *Monostroma undulatum, Undaria pinnatifida, Ascophyllum nodosum, Porphyra umbilicalis, Chaetoceros muelleri, Thalassiosira pseudonana*	Free radical scavenging activity, activation of protein synthesis of the connective tissues	Aklakur (2018)
Vitamin E	*Porphyridium cruentum, Laminaria ochroleuca, Saccorhiza polychides, Himanthalia elongate, Tetraselmis suecica, Porphyridium cruentum*	Films, coatings, bacon and meat, dairy products, oils	
Chlorophyll	*Dunaliella salina, Himanthalia elongate, Porphyridium cruentum, Chlorella vulgaris, Porphyridium cruentum*	Natural food and beverage colorants, reduces blood glucose level	Zhang et al. (2021)
Fucoxanthin	*Tetraselmis suecica, Hijikia fusiformis, Undaria pinnatifida, Fucus serratus, Padina tetrastromatic*	Apoptosis-inducing effect on human leukocytes cells, antioxidant effect	Xia et al. (2013)
Fucoidan	*Ascophyllum nodosum, Cladosiphon kamuranus, Saccharina japonica, Undaria pinnatifida*	Skin protection, antiaging, anti-inflammatory, anticoagulant, food additives	Vo and Kim (2013)

radicals that may lead to many life-threatening diseases such as cancer, heart diseases, and arthritis. Also, it helps in reducing the premature aging caused by UV exposure. Vitamin A having β-carotene as a precursor functions as an antioxidant and is used in cosmetic products. Vitamin C has antioxidant properties, which are largely exploited in the industrial sector. Vitamin C extracted from Crithmum maritimum along with other by-products helps in tone and elasticity improvement of the skin. Many other photosynthetic pigments from red and blue-green algae, aquatic plants, microalgae, and seaweed have also proven effective against oxidative stress, inflammatory reaction, and some cancers. *Dunaliella salina* can produce a significant amount of xanthophyll, especially zeaxanthin, which has potential against many diseases. *Haematococcus pluvialis* produces loroxanthin and *Ascophyllum nodosum* produces violaxanthin, which can act as an antioxidant. Other pigments such as Lutein are used as colorants for drugs and cosmetics.

Fucoxanthin, the major carotenoid obtained from seaweeds has been found to help in reducing blood glucose levels. Also, its metabolites fucoxanthin and halocynthiaxanthin proved to inhibit various cancer cells in humans such as leukemia, breast cancer, and colon cancer. Chlorophylls are green-colored pigments found in higher plants, algae, as well as in cyanobacteria. Chlorophyll has shown strong antioxidant properties due to its porphyrin ring structure essentially in dark. It reacts with peroxyl radicals and the charge transfer complex reaction inactivates the radicals. Derivative of chlorophyll *a*, pheophorbide has proven to show more antioxidant properties compared to chlorophyll *a*. Fucoidan, a polysaccharide mainly found in brown algae, has shown antioxidant activity helping in skin protection and antiaging. Other than this some other properties such as anti-inflammatory, anticoagulant, and anti-tumor are shown by fucoidan, which is used in various drug preparations (Kumar et al. 2021).

11.6 Lipase

Lipase (E.C.3) acts on an ester bond of the carboxylic esters and hydrolyzes triacylglycerols to fatty acids, diacylglycerol, monoacylglycerol, and glycerol. The primary reaction of the lipase is the hydrolysis of fat. Apart from hydrolysis of triglycerides, lipases catalyze esterification, inter-esterification, acidolysis, alcoholysis, and aminolysis. Many useful properties like broad substrate specificity and optimum activity at a wide range of temperatures make lipase a most important group of the biocatalyst. The commercial use of lipase is a multi-dollar business that comprises a variety of various applications including biopolymer and biodiesel synthesis, and the production of agrochemicals, cosmetics, pharmaceuticals, and flavors (Navvabi et al. 2018; Deng et al. 2016).

11.6.1 Application in the Pharmaceutical Industry

Cold active lipases have very specific region-selective reactions and broad substrate specificity and recognition, which is an important trait for biomedical applications. The single isomer chiral drug can be best prepared using the cold-active lipase in the organic solvents. Smidt (1996) prepared amines that are optically active by using cold-active lipases from *C. antarctica* or *Pseudomonas* sp. on stereospecific N-acetyl amines. These amines are important intermediates in the preparation of insecticides and pharmaceuticals. The major advantages of using these cold-active lipases were their economic feasibility, usage simplicity, and easy availability. A wide range of nitrogenated compounds could be prepared using this lipase for the pharmaceutical (Choudhury and Bhunia 2015).

11.7 Marine Actinobacteria Metabolites

About 9% of bacteria dwelling on marine patches are comprised of actinomycetes. Polyketide Synthase (PKS) pathways and Non-Ribosomal Polyketide Synthase (NRPS) are present in the actinomycetes, isolated from the ocean depths. These pathways symbolize the production of secondary metabolites (Habbu et al. 2016). Bioactive compounds, having antibacterial, antifungal, antitumor, anticancer, etc. properties, are produced by the marine actinobacteria. On the basis of chemical activity, these compounds are segregated as (Fig. 11.3, Table 11.4).

11.8 Marine-Derived Natural Products: From Ocean to Pharmaceutical Industry

On the basis of pharmaceuticals production, Marine microbes have played a very essential role. Organisms do not need the secondary metabolites as their basic or primary metabolic processes but these metabolites are believed to confer some evolutionary advantages. Many marines-derived natural products show their identity by not only inhibiting cell division but also being involved in different kinds of human disease processes through interacting or preventing encroachment with similar or same receptors or enzymes.

At the beginning of the twentieth century, microbes have been discovered as the manufacturer of therapeutical agents. Consequently, toward a significant

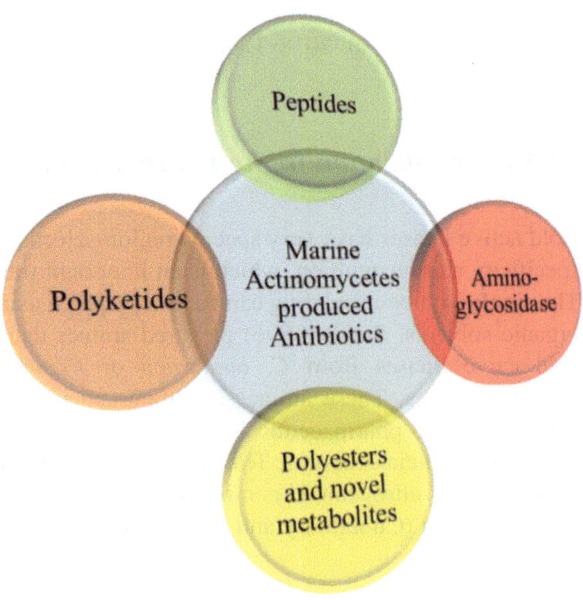

Fig. 11.3 Segregation of marine actinomycetes producing antibiotics based on chemical activity

Table 11.4 List of marine actinobacteria-derived bioactive compounds and their role in antibacterial, antifungal, antitumor, and anticancer properties

	Antibiotics	Species	Activity/treatment	References
Aminoglycosidase	Streptomycin	*Streptomyces griseus*	Bactericidal antibiotic	Reilly (1947)
	Tobramycin	*S. tenebrarius*	Infections caused by bacteria	Lancini and Lorenzetti (2013)
	Kanamycin	*S. kanamyceticus*		Nickel et al. (1985)
	Paromomycin	*S. krestomuceticus*		Manikprabhu and Li (2015)
	Spectinomycin	*S. spectabilis*		Davidson and Rosenson (2009)
	Neomycin	*Streptomyces fradiae*	Creams, ointments and eye drops	Lu et al. (2009)
	Chloramphenicol	*Streptomyces Venezuelae*	Typhoid	Vora and Raikwar (2013)
	Tetracycline	*Streptomyces* sp.	Bacterial infections	González and Moran (1997)
	Anthracyclines	*Streptomyces peucetius*	Cancer chemotherapy	Fujiwara et al. (1985)
	Vancomycin	*Amycolatopsis orientalis*	Bacterial (gram +) infections	D'Costa et al. (2007)
	Cephamycins	*Streptomyces*	Infections by bacteria	Manivasagan et al. (2013)
	Nystatin (Fungicidin)	Some species of *Streptomyces*	Polyene antifungal medication	Holz and Finkelstein (1970)
	Actinomycin	Soil bacteria of the genus *Streptomyces*	Older chemotherapy drugs	Hollstein (1974)
Peptides	Piperazimycin A	*Streptomyces* sp.	In-vitro cytotoxicity	Miller et al. (2007)
Polyketides	SBR-22	*Streptomyces psammoticus*	Antibacterial activity	Sujatha et al. (2005)
	Actinofuranones A and B	*Streptomyces genus*	In-vitro cytotoxicity against mouse macrophages and splenocyte T-cells	Dharmaraj (2010)
	Daryamides A, B and C	*Streptomyces* CNQ-085	Cytotoxicity against the human colon carcinoma cell line HCT-116	Manivasagan et al. (2014)
	Phaeochromycins F	*Streptomyces* sp. DSS-18	Cytotoxic effect	Li et al. (2008)

(continued)

Table 11.4 (continued)

	Antibiotics	Species	Activity/treatment	References
Polyesters	Bonactin	*Streptomyces* sp. BD21–2	Antimicrobial and antifungal activity	Manivasagan et al. (2014)
Novel metabolites	ZHD-0501	*Actinomadura* sp.	Anticancer	Han et al. (2005)
	Chinikomycin	*Streptomyces* sp.		Li et al. (2005)
	Mechercharmycins	*Thermoactinomyces* sp.		Lam (2006)
	Salinosporamide A	*Salinispora tropica*		Lam et al. (2007)
	Lodopyridone	*Saccharomonospora* sp.		Maloney et al. (2009)
	Streptokordin	*Streptomyces* sp.	Antitumor	Jeong et al. (2006)
	Piericidins	*Streptomyces* sp.		Hayakawa et al. (2007)
	Butenolides	*Streptoverticillium luteoverticillatum*		Li et al. (2006)
	Arenicolides A–C	*Salinispora arenicola*	Mild cytotoxicity	Williams et al. (2007)
	Salinosporamides B & C	*Salinispora tropica*	Cytotoxicity	Reed et al. (2007)
	Proximicins	*Verrucosispora* sp.	Antibacterial, anticancer	Fiedler et al. (2008)
	Caboxamycin	*Streptomyces* sp. *NTK 937*		Hohmann et al. (2009)
	Daryamides	*Streptomyces* sp.	Antifungal; anticancer	Asolkar et al. (2006)
	Staurosporine	*Streptomyces* sp.	Antitumor; phytotoxicity	Wu et al. (2006)
	Resistoflavin methyl ether	*Streptomyces* sp.	Antibacterial; anti-oxidative	Kock et al. (2005)
	Salinosporamide A	*Salinispora tropica*	Anticancer; antimalarial	Prudhomme et al. (2008)
	2-Allyloxyphenol	*Streptomyces* sp.	Antimicrobial; food preservative; oral disinfectant	Arumugam et al. (2018)
	Saliniketal	*Salinispora arenicola*	Cancer chemoprevention	Jensen et al. (2007)
	Sporolide A	*Salinispora tropica*	Unknown	
	Cyanosporaside A	*Salinispora pacifica*		
	Salinispyrone	*Salinispora pacifica*		
	Sesquiterpene	*Streptomyces* sp.		Wu et al. (2006)

contribution to drug development, novel compounds have been produced through the development of marine biotechnology. Over the past decades, as a tapped and significant source for novel bioactive compounds, marine microbes have got acknowledged. Agricultural fungicides, bio-fertilizers, shrimp feed supplements, and cholesterol-reducing drugs have been identified as well as isolated successfully from marine microbes and likewise, the industrial potential to develop anticancer agents has been validated by pharmaceuticals. Apart from these marine-derived microbes such as bacteria plays a promising role in the discovery of antimicrobial natural product. For example, bacicyclin, a novel cyclic peptide, and Anthracimycin B, a polyketide obtained from Bacillus sp. Strain and marine-derived Streptomyces cyaneofuscatus, respectively, show antibacterial activity. Other than this, there is some research application of marine-derived microorganism, for example, bioluminescent jellyfish *A. Victoria*-derived product green fluorescent protein (GFP) play a tremendous role as a reporter gene in numerous studies by regulating gene expression. Besides this, it also helps in differentiation in both embryos and tissue culture cells.

11.9 Challenges of Marine-Derived Pharmaceuticals

The Marine biosphere is one of the largest on the Earth. Living conditions are fundamentally different from those of the terrestrial environment. Specific secondary metabolites production is one of the most valuable mechanisms adapted by marine organisms for their survival in the sea. Interestingly, these metabolites act as a possible drug for humans as they possess biological activities. Among the marine-derived products, some of them are used to lower triglyceride levels in the blood and chronic pain and also for the treatment of viral diseases. Few other marine-derived products are interestingly involved as experimental tools as well as in diagnostics. There are some challenges that have to be faced by our society due to the development of a drug from marine sources, and especially because of the problem related to supply (Lindequist 2016).

11.10 Conclusion and Future Prospective

The enormous potential of the use of marine natural products has been demonstrated through the advancement in the development, approval, as well as therapeutic use of marine drugs. Nowadays, novel pharmaceutical responses are evidently in need of the response of pharmaceutical responses. The resistivity of fungi and bacteria to antibiotics, the availability of treatments for some diseases like parasites diseases such as malaria, unavailability of the medicines for the treatment of new drugs and viruses for the treatment of cancer, are required additional developmental focus and research in discovering the bioactive molecules in a novel aspect. Oceans, the home to a substantial portion of the world's biodiversity, are a very important source of drugs as well as drug leads and are yet to be explored. In the future, the discovery of new leads will be derived through marine-derived products and they will act as a supplement to the conventional methods including high-throughput analysis, such as genome analysis and metagenomics, as well as the identification and dereplication tools. Hereby it has been shown that in the path of developing new pharmaceuticals, not only marine-derived microbial products but also a combination of this with the rational approach as well as high throughput approach plays a key role (Wang et al. 2022)

Acknowledgments The authors would like to acknowledge the help and the platform provided by National Institute of Technology (NIT) Rourkela and also express their gratitude to Dr. Angana Sarkar, corresponding author, for the guidance throughout writing this chapter. Authors would also like to thank their friends and family who supported and offered deep insight into the study.

References

Abou-Elela GM, Ibrahim HA, Hassan SW, Abd-Elnaby H, El-Toukhy NM (2011) Alkaline protease production by alkaliphilic marine bacteria isolated from Marsa-Matrouh (Egypt) with special emphasis on Bacillus cereus purified protease. Afr J Biotechnol 10(22):4631–4642

Aklakur M (2018) Natural antioxidants from sea: a potential industrial perspective in aquafeed formulation. Rev Aquac 10(2):385–399

Anithajothi R, Nagarani N, Umagowsalya G, Duraikannu K, Ramakritinan CM (2014) Screening, isolation and characterization of protease producing moderately halophilic microorganism Halomonas meridiana associated with coral mucus. Toxicol Environ Chem 96(2):296–306

Arnosti C, Bell C, Moorhead DL, Sinsabaugh RL, Steen AD, Stromberger M, Wallenstein M, Weintraub MN (2014) Extracellular enzymes in terrestrial, freshwater, and marine environments: perspectives on system variability and common research needs. Biogeochemistry 117(1): 5–21

Arumugam V, Venkatesan M, Ramachandran S, Sundaresan U (2018) Bioactive peptides from marine ascidians and future drug development–a review. Int J Pept Res Ther 24(1):13–18

Asha B, Palaniswamy M (2018) Optimization of alkaline protease production by Bacillus cereus FT 1 isolated from soil. J Appl Pharm Sci 8(02):119–127

Asolkar RN, Jensen PR, Kauffman CA, Fenical W (2006) Daryamides A−C, weakly cytotoxic polyketides from a marine-derived actinomycete of the genus Streptomyces strain CNQ-085. J Nat Prod 69(12):1756–1759

Beygmoradi A, Homaei A (2017) Marine microbes as a valuable resource for brand new industrial biocatalysts. Biocatal Agric Biotechnol 11:131–152

Beygmoradi A, Homaei A, Hemmati R, Santos-Moriano P, Hormigo D, Fernández-Lucas J (2018) Marine chitinolytic enzymes, a biotechnological treasure hidden in the ocean? Appl Microbiol Biotechnol 102(23):9937–9948

Bijina B, Chellappan S, Krishna JG, Basheer SM, Elyas KK, Bahkali AH, Chandrasekaran M (2011) Protease inhibitor from Moringa oleifera with potential for use as therapeutic drug and as seafood preservative. Saudi J Biol Sci 18(3):273–281

Cha KH, Koo SY, Lee DU (2008) Antiproliferative effects of carotenoids extracted from Chlorella ellipsoidea and Chlorella vulgaris on human colon cancer cells. J Agric Food Chem 56(22): 10521–10526

Chellappan S, Jasmin C, Basheer SM, Elyas KK, Bhat SG, Chandrasekaran M (2006) Production, purification and partial characterization of a novel protease from marine Engyodontium album BTMFS10 under solid state fermentation. Process Biochem 41(4):956–961

Chi Z, Ma C, Wang P, Li HF (2007) Optimization of medium and cultivation conditions for alkaline protease production by the marine yeast Aureobasidium pullulans. Bioresour Technol 98 (3):534–538

Choudhury P, Bhunia B (2015) Industrial application of lipase: a review. Biopharm J 1(2):41–47

Chysirichote T, Reiji T, Asami K, Ohtaguchi K (2014) Quantification of the glucosamine content in the filamentous fungus Monascus ruber cultured on solid surfaces. J Basic Microbiol 54(5): 350–357

D'Costa VM, Griffiths E, Wright GD (2007) Expanding the soil antibiotic resistome: exploring environmental diversity. Curr Opin Microbiol 10(5):481–489

Davidson MH, Rosenson RS (2009) Novel targets that affect high-density lipoprotein metabolism: the next frontier. Am J Cardiol 104(10):52E–57E

Deng D, Zhang Y, Sun A, Liang J, Hu Y (2016) Functional characterization of a novel marine microbial GDSL lipase and its utilization in the resolution of (±)-1-phenylethanol. Appl Biochem Biotechnol 179(1):75–93

De Swaaf ME, Sijtsma L, Pronk JT (2003) High-cell-density fed-batch cultivation of the docosahexaenoic acid producing marine alga Crypthecodinium cohnii. Biotechnol Bioeng 81 (6):666–672

Devlin MG, Gasser RB, Cocks TM (2007) Initial support for the hypothesis that PAR2 is involved in the immune response to Nippostrongylus brasiliensis in mice. Parasitol Res 101(1):105–109

Dharmaraj S (2010) Marine Streptomyces as a novel source of bioactive substances. World J Microbiol Biotechnol 26:2123–2139. https://doi.org/10.1007/s11274-010-0415-6

Du J, Duan S, Miao J, Zhai M, Cao Y (2021) Purification and characterization of chitinase from Paenibacillus sp. Biotechnol Appl Biochem 68(1):30–40

Elibol M, Moreira AR (2005) Optimizing some factors affecting alkaline protease production by a marine bacterium Teredinobacter turnirae under solid substrate fermentation. Process Biochem 40(5):1951–1956

Feijoo-Siota L, Blasco L, Luis Rodriguez-Rama J, Barros-Velázquez J, de Miguel T, Sánchez-Pérez A, Villa G, T. (2014) Recent patents on microbial proteases for the dairy industry. Recent Adv DNA Gene Seq 8(1):44–55

Fenice M, Leuba JL, Federici F (1998) Chitinolytic enzyme activity of Penicillium janthinellum P9 in bench-top bioreactor. J Ferment Bioeng 86(6):620–623

Fiedler HP, Bruntner C, Riedlinger J, Bull AT, Knutsen G, Goodfellow M, Jones A, Maldonado L, Pathom-Aree W, Beil W, Schneider K (2008) Proximicin A, B and C, novel aminofuran antibiotic and anticancer compounds isolated from marine strains of the actinomycete Verrucosispora. J Antibiot 61(3):158–163

Fujiwara A, Hoshino T, Westley JW (1985) Anthracycline antibiotics. Crit Rev Biotechnol 3(2): 133–157

Ghasemian A, Al-marzoqi AH, Al-abodi HR, Alghanimi YK, Kadhum SA, Shokouhi Mostafavi SK, Fattahi A (2019) Bacterial L-asparaginases for cancer therapy: current knowledge and future perspectives. J Cell Physiol 234(11):19271–19279

González JM, Moran MA (1997) Numerical dominance of a group of marine bacteria in the alpha-subclass of the class Proteobacteria in coastal seawater. Appl Environ Microbiol 63(11): 4237–4242

Habbu P, Warad V, Shastri R, Madagundi S, Kulkarni VH (2016) Antimicrobial metabolites from marine microorganisms. Chin J Nat Med 14(2):101–116

Han XX, Cui CB, Gu QQ, Zhu WM, Liu HB, Gu JY, Osada H (2005) ZHD-0501, a novel naturally occurring staurosporine analog from Actinomadura sp. 007. Tetrahedron Lett 46(36): 6137–6140

Hayakawa Y, Shirasaki S, Shiba S, Kawasaki T, Matsuo Y, Adachi K, Shizuri Y (2007) Piericidins C7 and C8, new cytotoxic antibiotics produced by a marine Streptomyces sp. J Antibiot 60(3): 196–200
Hohmann C, Schneider K, Bruntner C, Brown R, Jones AL, Goodfellow M, Krämer M, Imhoff JF, Nicholson G, Fiedler HP, Süssmuth RD (2009) Albidopyrone, a new α-pyrone-containing metabolite from marine-derived Streptomyces sp. NTK 227. J Antibiot 62(2):75–79
Hollstein U (1974) Actinomycin. Chemistry and mechanism of action. Chem Rev 74(6):625–652
Holz R, Finkelstein A (1970) The water and nonelectrolyte permeability induced in thin lipid membranes by the polyene antibiotics nystatin and amphotericin B. J Gen Physiol 56(1): 125–145
Jahromi ST, Barzkar N (2018) Marine bacterial chitinase as sources of energy, eco-friendly agent, and industrial biocatalyst. Int J Biol Macromol 120:2147–2154
Jensen PR, Williams PG, Oh DC, Zeigler L, Fenical W (2007) Species-specific secondary metabolite production in marine actinomycetes of the genus Salinispora. Appl Environ Microbiol 73(4):1146–1152
Jeong SY, Shin HJ, Kim TS, Lee HS, Park SK, Kim HM (2006) Streptokordin, a new cytotoxic compound of the methylpyridine class from a marine-derived Streptomyces sp. KORDI-3238. J Antibiot 59(4):234–240
Jovanovic S, Dietrich D, Becker J, Kohlstedt M, Wittmann C (2021) Microbial production of polyunsaturated fatty acids—high-value ingredients for aquafeed, superfoods, and pharmaceuticals. Curr Opin Biotechnol 69:199–211
Kao PM, Huang SC, Chang YC, Liu YC (2007) Development of continuous chitinase production process in a membrane bioreactor by Paenibacillus sp. CHE-N1. Process Biochem 42(4): 606–611
Kawamoto J, Kurihara T, Yamamoto K, Nagayasu M, Tani Y, Mihara H et al (2009) Eicosapentaenoic acid plays a beneficial role in membrane organization and cell division of a cold-adapted bacterium, Shewanella livingstonensis Ac10. J Bacteriol 191(2):632–640
Kim W, Choi K, Kim Y, Park H, Choi J, Lee Y, Oh H, Kwon I, Lee S (1996) Purification and characterization of a fibrinolytic enzyme produced from Bacillus sp. strain CK 11-4 screened from Chungkook-Jang. Appl Environ Microbiol 62(7):2482–2488
Kock I, Maskey RP, Biabani MA, Helmke E, Laatsch H (2005) 1-Hydroxy-1-norresistomycin and resistoflavin methyl ether: new antibiotics from marine-derived streptomycetes. J Antibiot *58*(8):530–534
Kumar A, Singh SK, Kant C, Verma H, Kumar D, Singh PP, Modi A, Droby S, Kesawat MS, Alavilli H, Bhatia SK (2021) Microbial biosurfactant: a new frontier for sustainable agriculture and pharmaceutical industries. Antioxidants 10(9):1472
Lam KS (2006) Discovery of novel metabolites from marine actinomycetes. Curr Opin Microbiol 9(3):245–251
Lam KS, Tsueng G, McArthur KA, Mitchell SS, Potts BC, Xu J (2007) Effects of halogens on the production of salinosporamides by the obligate marine actinomycete Salinispora tropica. J Antibiot 60(1):13–19
Lancini G, Lorenzetti R (2013) Biotechnology of antibiotics and other bioactive microbial metabolites. Springer, Boston, MA. https://doi.org/10.1007/978-1-4757-9522-6
Li F, Maskey RP, Qin S, Sattler I, Fiebig HH, Maier A, Zeeck A, Laatsch H (2005) Chinikomycins A and B: isolation, structure elucidation, and biological activity of novel antibiotics from a marine Streptomyces sp. isolate M045, 1. J Nat Prod 68(3):349–353
Li DH, Zhu TJ, Liu HB, Fang YC, Gu QQ, Zhu WM (2006) Four butenolides are novel cytotoxic compounds isolated from the marine-derived bacterium, Streptoverticillium luteoverticillatum 11014. Arch Pharm Res 29(8):624–626
Li J, Lu CH, Zhao BB, Zheng ZH, Shen YM (2008) Phaeochromycins F–H, three new polyketide metabolites from Streptomyces sp. DSS-18. Beilstein J Org Chem 4(1):46

Li M, Liu GL, Chi Z, Chi ZM (2010a) Single cell oil production from hydrolysate of cassava starch by marine-derived yeast Rhodotorula mucilaginosa TJY15a. Biomass Bioenergy 34(1): 101–107

Li J, Chi Z, Wang X (2010b) Cloning of the SAP6 gene of Metschnikowia reukaufii and its heterologous expression and characterization in Escherichia coli. Microbiol Res 165(3): 173–182

Lindequist U (2016) Marine-derived pharmaceuticals–challenges and opportunities. Biomol Ther 24(6):561–571

Liu BL, Kao PM, Tzeng YM, Feng KC (2003) Production of chitinase from Verticillium lecanii F091 using submerged fermentation. Enzym Microb Technol 33(4):410–415

Lu Y, Dong X, Liu S, Bie X (2009) Characterization and identification of a novel marine Streptomyces sp. produced antibacterial substance. Mar Biotechnol 11(6):717–724

Mahajan PM, Nayak S, Lele SS (2012) Fibrinolytic enzyme from newly isolated marine bacterium Bacillus subtilis ICTF-1: media optimization, purification and characterization. J Biosci Bioeng 113(3):307–314

Mahendran S, Sankaralingam S, Shankar T, Vijayabaskar P (2010) Alkalophilic protease enzyme production from estuarine Bacillus aquimaris. World J Fish Mar Sci 2:436–443

Maloney KN, MacMillan JB, Kauffman CA, Jensen PR, DiPasquale AG, Rheingold AL, Fenical W (2009) Lodopyridone, a structurally unprecedented alkaloid from a marine actinomycete. Org Lett 11(23):5422–5424

Manikprabhu D, Li W (2015) Antimicrobial agents from actinomycetes. In: Antimicrobials Synthetic and Natural Compounds, pp 99–116

Manivasagan P, Venkatesan J, Sivakumar K, Kim SK (2013) Retracted: marine actinobacterial metabolites: current status and future perspectives. Microbiol Res 168(6):311–332

Manivasagan P, Kang KH, Sivakumar K, Li-Chan EC, Oh HM, Kim SK (2014) Marine actinobacteria: an important source of bioactive natural products. Environ Toxicol Pharmacol 38(1):172–188

Maruthiah T, Esakkiraj P, Immanuel G, Palavesam A (2014) Alkaline serine protease from marine Bacillus flexus APCMST-RS2P: purification and characterization. Curr Biotechnol 3(3): 238–243

Matsubara K, Hori K, Matsuura Y, Miyazawa K (2000) Purification and characterization of a fibrinolytic enzyme and identification of fibrinogen clotting enzyme in a marine green alga, Codiumdivaricatum. Comp Biochem Physiol B: Biochem Mol Biol 125(1):137–143

Miller ED, Kauffman CA, Jensen PR, Fenical W (2007) Piperazimycins: cytotoxic hexadepsipeptides from a marine-derived bacterium of the genus Streptomyces. J Organomet Chem 72(2):323–330

Moi IM, Leow ATC, Ali MSM, Rahman RNZRA, Salleh AB, Sabri S (2018) Polyunsaturated fatty acids in marine bacteria and strategies to enhance their production. Appl Microbiol Biotechnol 102(14):5811–5826

Muffler K, Sana B, Mukherjee J, Ulber R (2015) Marine enzymes–production & applications. In: Springer handbook of marine biotechnology. Springer, Berlin, Heidelberg, pp 413–429

Nampoothiri KM, Baiju TV, Sandhya C, Sabu A, Szakacs G, Pandey A (2004) Process optimization for antifungal chitinase production by Trichoderma harzianum. Process Biochem 39(11): 1583–1590

Naveed M, Nadeem F, Mehmood T, Bilal M, Anwar Z, Amjad F (2021) Protease—a versatile and ecofriendly biocatalyst with multi-industrial applications: an updated review. Catal Lett 151(2): 307–323

Navvabi A, Razzaghi M, Fernandes P, Karami L, Homaei A (2018) Novel lipases discovery specifically from marine organisms for industrial production and practical applications. Process Biochem 70:61–70

Ni X, Yue L, Chi Z, Li J, Wang X, Madzak C (2009) Alkaline protease gene cloning from the marine yeast Aureobasidium pullulans HN2-3 and the protease surface display on Yarrowia lipolytica for bioactive peptide production. Mar Biotechnol 11(1):81–89

Nickel JC, Ruseska I, Wright JB, Costerton JW (1985) Tobramycin resistance of Pseudomonas aeruginosa cells growing as a biofilm on urinary catheter material. Antimicrob Agents Chemother 27(4):619–624

Olatunde OO, Benjakul S (2022) Antioxidants from crustaceans: a panacea for lipid oxidation in marine-based foods. Food Rev Int 38(1):1–31

Prudhomme J, McDaniel E, Ponts N, Bertani S, Fenical W, Jensen P, Le Roch K (2008) Marine actinomycetes: a new source of compounds against the human malaria parasite. PLoS One 3(6): e2335

Ramesh S, Rajesh M, Mathivanan N (2009) Characterization of a thermostable alkaline protease produced by marine Streptomyces fungicidicus MML1614. Bioprocess Biosyst Eng 32(6): 791–800

Reed KA, Manam RR, Mitchell SS, Xu J, Teisan S, Chao TH, Deyanat-Yazdi G, Neuteboom ST, Lam KS, Potts BC (2007) Salinosporamides D–J from the marine actinomycete Salinispora tropica, bromosalinosporamide, and thioester derivatives are potent inhibitors of the 20S proteasome. J Nat Prod 70(2):269–276

Reilly HC (1947) Isolation of streptomycin-producing strains of Streptomyces griseus. J Bacteriol 54:27

Robinson PK (2015) Enzymes: principles and biotechnological applications. Essays Biochem 59: 1–41

Ryan J, Farr H, Visnovsky S, Vyssotski M, Visnovsky G (2010) A rapid method for the isolation of eicosapentaenoic acid-producing marine bacteria. J Microbiol Methods 82(1):49–53

Sana B, Ghosh D, Saha M, Mukherjee J (2006) Purification and characterization of a salt, solvent, detergent and bleach tolerant protease from a new gamma-Proteobacterium isolated from the marine environment of the Sundarbans. Process Biochem 41(1):208–215

Shanmughapriya S, Krishnaveni J, Selvin J, Gandhimathi R, Arunkumar M, Thangavelu T, Thangavelu GS, Kiran & Natarajaseenivasan, K. (2008) Optimization of extracellular thermotolerant alkaline protease produced by marine Roseobacter sp. (MMD040). Bioprocess Biosyst Eng 31(5):427–433

Sijtsma L, De Swaaf ME (2004) Biotechnological production and applications of the ω-3 polyunsaturated fatty acid docosahexaenoic acid. Appl Microbiol Biotechnol 64(2):146–153

Silva CJ, Zhang Q, Shen J, Cavaco-Paulo A (2006) Immobilization of proteases with a water soluble–insoluble reversible polymer for treatment of wool. Enzym Microb Technol 39(4): 634–640

Singh R, Mittal A, Kumar M, Mehta PK (2016) Microbial proteases in commercial applications. J Pharm Chem Biol Sci 4(3):365–374

Smidt H, Fischer A, Fischer P, Schmid RD (1996) Preparation of optically pure chiral amines by lipase-catalyzed enantioselective hydrolysis of N-acyl-amines. Biotechnol Tech 10:335–338

Sujatha P, Raju KB, Ramana T (2005) Studies on a new marine streptomycete BT-408 producing polyketide antibiotic SBR-22 effective against methicillin resistant Staphylococcus aureus. Microbiol Res 160(2):119–126

Sutar II, Srinivasan MC, Vartak HG (1992) Production of an alkaline proteinase fromConidiobolus coronatus and its use to resolvedl-phenylalanine anddl-phenylglycine. World J Microbiol Biotechnol 8(3):254–258

Trincone A (2013) Marine enzymes for biocatalysis: sources, biocatalytic characteristics and bioprocesses of marine enzymes. Elsevier

Turkiewicz M, Pazgier M, Kalinowska H, Bielecki S (2003) A cold-adapted extracellular serine proteinase of the yeast Leucosporidium antarcticum. Extremophiles 7(6):435–442

Vo TS, Kim SK (2013) Fucoidans as a natural bioactive ingredient for functional foods. J Funct Foods 5(1):16–27

Vora VR, Raikwar MK (2013) Determination of chloramphenicol and thiamphenicol residues in fish, shrimp and milk by ESI-LCMSMS. Int J Agric Food Sci Technol 4:2249–3050

Wang J, Zhu L, Li T, Li X, Huang K, Xu W (2022) Multiple functionalities of functional nucleic acids for developing high-performance lateral flow assays. TrAC Trends Anal Chem 148: 116529

Waters AL, Hill RT, Place AR, Hamann MT (2010) The expanding role of marine microbes in pharmaceutical development. Curr Opin Biotechnol 21(6):780–786

Williams PG, Asolkar RN, Kondratyuk T, Pezzuto JM, Jensen PR, Fenical W (2007) Saliniketals A and B, bicyclic polyketides from the marine actinomycete Salinispora arenicola. J Nat Prod 70 (1):83–88

Wu SJ, Fotso S, Li F, Qin S, Kelter G, Fiebig HH, Laatsch H (2006) N-carboxamido-staurosporine and selina-4 (14), 7 (11)-diene-8, 9-diol, new metabolites from a marine Streptomyces sp. The Journal of Antibiotics 59(6):331–337

Wu S, Liu G, Zhang D, Li C, Sun C (2015) Purification and biochemical characterization of an alkaline protease from marine bacteria Pseudoalteromonas sp. 129-1. J Basic Microbiol 55(12): 1427–1434

Xia S, Wang K, Wan L, Li A, Hu Q, Zhang C (2013) Production, characterization, and antioxidant activity of fucoxanthin from the marine diatom Odontella aurita. Mar Drugs 11(7):2667–2681

Zhang C, Kim SK (2012) Application of marine microbial enzymes in the food and pharmaceutical industries. Adv Food Nutr Res 65:423–435

Zhang H, Zhou Y, Kong Q, Dong W, Lin Z (2021) Toxicity of naphthenic acids on the chlorophyll fluorescence parameters and antioxidant enzyme activity of Heterosigma akashiwo. Antioxidants 10(10):1582

Zhu HY, Tian Y, Hou YH, Wang TH (2009) Purification and characterization of the cold-active alkaline protease from marine cold-adaptive Penicillium chrysogenum FS010. Mol Biol Rep 36(8):2169–2174

Chapter 12
Microbial Pigments and Paints for Clean Environment

Soma Ghosh and Suchetana Banerjee

Abstract Microbial pigments, due to their several beneficial properties, are increasingly gaining attention in various industries such as nutraceutical, pharmaceutical, food and beverages, textile, wood dyeing, and paint industries etc. The eco-friendly nature, involvement of lesser synthetic chemicals, less harmful waste generation during the production process, and use of organic waste materials as substrate for microbial growth has revolutionized the increased production of microbial pigments. These pigments have been often used as antimicrobial, strong antioxidants, anticancer, antiartherosclerotic, antidiabetic, and immunomodulatory agents. Several biotechnological developments have been made through genetic engineering, heterologous expression of the biochemical pathways, omics-based technology development, and optimization of the fermentation techniques and extraction procedures have been done to improve strain selection and increase product purification. However, industrial-scale production of the pigments faces several challenges such as instability to environmental condition, reactivity, regulatory hurdles, co-production of toxins, choice of fermenters, optimization of conditions, etc. Successful development of technologies for large-scale production of microbial pigments and plausible interventions in its transition from laboratory to industry shall overcome the rampant usage of harmful synthetic pigments by mankind.

Keywords Microbial pigments · Pigments in food and beverages · Pharmaceutical use · Genetic engineering · Statistical optimization techniques · Challenges in pigment production

---

S. Ghosh (✉)
CSIR-National Environmental Engineering Research Institute, Kolkata Zonal Centre, Kolkata, India

Presented at: Department of Biotechnology, School of Life Science and Biotechnology, Adamas University, Barasat, West Bengal, India

S. Banerjee
Department of Biotechnology, School of Life Science and Biotechnology, Adamas University, Barasat, West Bengal, India

Department of Polymer Science and Technology, University of Calcutta, Kolkata, India

A. Sarkar, I. A. Ahmed (eds.), *Microbial products for future industrialization*, Interdisciplinary Biotechnological Advances,
https://doi.org/10.1007/978-981-99-1737-2_12

12.1 Introduction

Pigments are chemical compounds used inevitably in various spheres of human life including food, cosmetics, pharmaceuticals, etc. Since 1850, synthetic pigments were extensively utilized because of their cost effectiveness and ease of production and minimal amounts needed for color development (Sen et al. 2019). However, with increasing concern regarding unsuitability of these synthetic compounds toward human health due to their carcinogenicity and teratogenicity, the use of synthetic pigments is being banned (Babitha 2009). Moreover, the production of synthetic pigments involves use of several hazardous chemicals, which are being expelled in the waste water generated from the production unit, hence causing environmental contamination and pollution (Liu et al. 2021). Hence, the need of natural pigments, i.e., coloring agents of biological origin emerged, which were found to be eco-friendly, biodegradable, and with several health benefits. Natural pigments are majorly produced by plants, fungi, algae, and bacterial spp., which apart from parting colors to the substrate have also shown several other roles. Biological pigments have shown various therapeutic roles such as antimicrobial activity, nutritional roles, anticancer agents, antioxidants, antiviral, antimalarial, antiproliferative roles, immunosuppressant, etc. (Kumar et al. 2015; Manivasagan et al. 2018). The use of biotech-based colors in food is desirable as biological pigments have shown several nutritively beneficial roles toward human health such as pro-vitamins, antioxidants, and tumor-inhibiting agent (Rodriguez-Amaya 2019). Natural or bio-based colors are gaining industrial importance in food, pharmaceutical, cosmetic, and paint or dying applications. However, the color instability and production-leakage of pigments obtained from plant or animal sources are the limiting factors to promote their application (Méndez et al. 2011). Hence, pigments of microbial origin are favorable because of the fast growth of microorganisms, with high production rate of nontoxic eco-friendly compounds (Salama et al. 2021). Taxonomically, large groups of pigment-producing microorganisms prevail in various ecological environments like river, sea, ocean, soil, sediment, etc. Large-scale production of these pigments has been possible using bioreactors and several optimization techniques. In this chapter, the novel biological pigments, their sources, vivid applications, and technologies involved in large-scale pigment production has been discussed. The chapter also sheds light on the challenges faced during microbial pigment production and plausible green interventions have been suggested.

12.2 Microbial Pigments

Historically, a large variety of plant pigments were explored and was applied as natural food colorant, dye, and in medicinal uses. However, the major drawbacks of using plant pigments are nonavailability of the plant species throughout the year, long growing times, and pigment stability and solubility. Moreover, long-term large-scale use of plant species for pigment production might lead to loss of the plant species (Narsing Rao et al. 2017).

1. Microbial pigments have gained immense popularity and are an emerging field of research with wide applications in various industries. Besides having therapeutic and nutritional values, these have also proved to be promising coloring agents used as dyes, inks, paints, and as beneficial food additives. Use of microbial pigments have been preferred over plant pigments due to their ease of production on low-cost medium, easy propagation, and weather independent growth, easier genetic manipulation techniques, availability of simple and fast culturing and long-term storage techniques, and inherent fermentation capability of microorganisms (Kumar et al. 2015; Venil et al. 2013). Several fermentation techniques such as solid substrate fermentation or submerged fermentation are used to cultivate several pigment-producing microbes (Tuli et al. 2014). Microbial pigments are classified under six different categories based on their structural characteristics (Delgado-Vargas et al. 2010). They are: (1) chlorophylls and heme colors as tetrapyrrole derivatives; (2) carotenoids and iridoids as isoprenoid derivatives; (3) pterins, purines, flavins, phenoxazines, phenazines, and betalains as N-heterocyclic compounds apart from tetrapyrroles; (4) anthocyanins and alternative flavonoid pigments as benzopyran derivatives; (5) benzoquinone, naphthoquinone, anthraquinone as quinines; and (6) eumelanins, phaeomelanins, and allomelanins are few indolic polymers classified as melanins (Liu et al. 2018).

   Most important factors responsible for microbial pigment production are temperature, pH, carbon source, type of fermentation, minerals, and nitrogen source available in the growth medium, moisture content, and rate of aeration (Kumar et al. 2015). Microbial pigments are produced by a variety of microorganisms like fungal sp., i.e., yeast and molds, algae, bacteria, etc. (Indra Arulselvi et al. 2014) (Table 12.1). These pigments are basically fermentation-derived products and their production can be successfully optimized for better quality and quantity for industrial use. However, a suitable microbial agent for pigment production must be nonpathogenic and non-toxicogenic.

   (a) *Bacterial pigments*: Bacteria are known to make varieties of pigments like prodigiosin, carotenoids, flexirubin, violacein, riboflavin, quinones, monascins, melanins, phycocyanin, and pyocyanin. Carotenoids are the major class of pigments comprising astaxanthin, zeaxanthin, canthaxanthin, lycopenes, and β-carotene produced by different genera such as *Flavobacterium, Sphingobacterium, Actinobacteria, Agrobacterium aurantiacum, Serratia, Chromobacterium, Bacillus, Pseudomonas, Cyanobacteria*, etc. Canthaxanthin have also been found to be produced by few archaea such as *Haloferax alexandrinus*. These are the secondary metabolites produced by heterotrophic bacteria playing fundamental roles in cell adaptability under harsh environmental conditions such as UV irradiation or oxidative damage. Bacterial pigments like rubrolone, staphyloxanthin, xanthomonadins, and riboflavin are produced by bacterial sp. Like *Streptomyces echinoruber, Staphylococcus aureus, Xanthomonas oryzae*, and

Table 12.1 List of pigments produced by various classes of organisms and their structures

Pigments	Bacteria	Fungi	Algae	Structure
β-carotene	*Erwinia uredovora, Pantoea agglomerans, Rhodococcus maris*	*Blakeslea trispora, Mucor circinelloides, Phycomyces blakesleeanus, Xanthophyllomyces dendrorhous*	*Dunaliela salina*	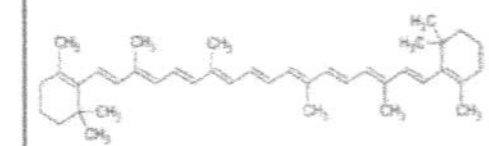
Zeaxanthin	*Flavobacterium, Paracoccus zeaxanthinifaciens*		*Synechococcus* sp., *Synechocystis* sp.	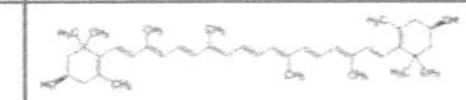
Canthaxanthin	*Brevibacterium, Corynebacterium, Rhodococcus, Bradyrhizobium, Halobacterium, Haloferax alexandrinus, Micrococcus roseus*	*Monascus roseus*	*Hematococcus, Chlorella, Chlamydomonas, Scenedesmus, Ankistrodesmus, Dictyococcus cinnanarinus*	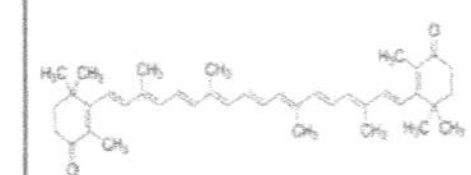
Astaxanthin	*Agrobacterium aurantiacum, Halobacterium, Brevundimonas, Paracoccus, Micrococcus, Altererythrobacter ishigakiensis, Sphingomicrobium*	*Haematococcus Pluvialis*	*Hemacoccus pluvialis*	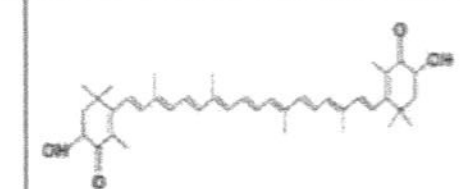
Rubrolone	*Streptomyces echinoruber, Dactylosporangium vinaceum*			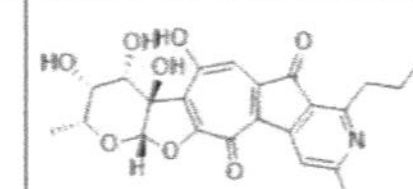
Staphyloxanthine	*Staphylococcus aureus*			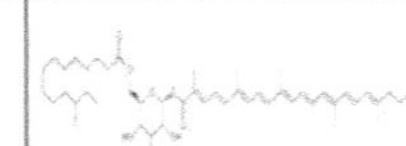
Xanthomonadins	*Xanthomonas oryzae, Xylella fastidiosa, Pseudoxanthomonas* spp.			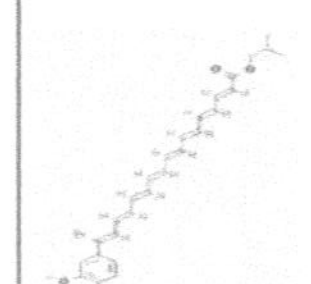

Flexirubin	*Chitinophaga pinensis, C. filiformes, C. pinensis, Chryseobacterium artocarpi* CECT 8497			
Riboflavin	*Bacillus subtilis*	*Ashbya gossypii*		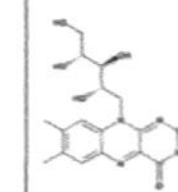
Prodigiosin	***Rugamonas rubra, Streptoverticillium rubrireticuli,*** *Vibrio gaogenes, Alteromonas rubra, Serratia marcescens, Serratia rubidaea*			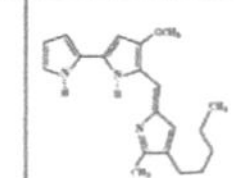
Phycocyanin				
Melanin	*Vibrio cholerae, Shewanella colwelliana, Aeromonas salmonicida, Alteromonas nigrifaciens, Marinomonas mediterranea,*			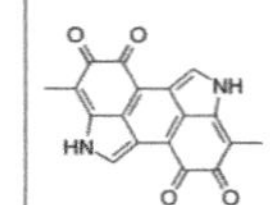
Violacein	*Chromobacterium violaceum, Janthinobacterium lividum, Pseudoalteromonas, Alteromonas luteviolacea, Duganella*			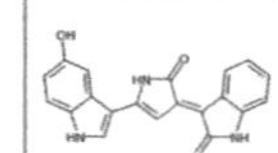

Bacillus subtilis, respectively. Another important pigment prodigiosin acts as the chemotaxonomic marker for phyla *Bacteroides* and *Cytophaga – Flavobacteria* group with several pharmacological importance. Phycocyanin is the characteristic marker of chlorophyll A containing cyanobacterial spp. like *Aphanizomenon flos-aquae* and *Spirulina*. Melanins are another set of pigments comprising of eumelanins (blackish brown), phaeomelanins (reddish or yellowish), and allomelanins (dark brownish to black) produced by various bacterial genera such as *Vibrio, Shewanella, Aeromonas, Alteromonas, Macinomonas, Actinobacteria*, etc. A purple pigment known as violacein has shown dyeing properties and have been used in bacterial dyes and inks. These are produced by members of *Chromobacterium violaceum, Janthinobacterium lividum, Pseudoalteromonas, Alteromonas luteviolaceae, Duganella*, etc. (Durán et al. 2012). Bacterial pigments are widely preferred in industries due to various advantages such as nontoxic in nature, easy to separate from cell biomass, and comparatively easier genetic manipulation techniques to increase yield (Prasanna et al. 2007). Bacterial pigments have found a wide array of applications in laboratory, pharmaceuticals, and various industrial aspects. Bacterial pigments are of great taxonomic significance, which acts as characteristic markers helpful in identification and classification of bacterial taxa. These bioactive compounds also find immense application in pharmaceutical industry as they have several therapeutic properties such as resistance to phagocytosis, protection against UV rays, antimicrobial activity against other microbes, antioxidant properties, heavy metal resistance, nutrient and energy acquisition, ,roles in photosynthesis etc. Industrially, these pigments have also proven useful as alternatives to color additives of plant origin, in paint formulations, textile dyeing, as food colorants, in therapeutics, in cosmetics, as index of any oil spillage, as well as acting as biosensors for environmental (water, air, soil) pollution.

(b) *Fungal pigments*: Fungal pigments are mostly carotenoids, melanins, flavins, phenazines, quinines, monascins, violacein, and indigo (Lagashetti et al. 2019; Venil et al. 2014). More than 200 fungal strains belonging to a wide array of families like *Monascaceae, Cordycipitaceae, Xylariaceae, Chaetomiaceae, Sordoriaceae,* and *Chlorociboriaceae,* have been described as potent pigment producers. Besides, yeasts belonging to genera *Rhodotorula, Sporidiobolus, Sporobolomyces, Xanthophyllomyces*, and *Pichia* have also been identified as prolific pigment producers (Paillié-Jiménez et al. 2020). In fact, yeasts are more advantageous for pigment production as compared to filamentous fungi w.r.t toxicity, faster growth rates, and unicellularity of the cells. Yeasts have been found to produce pigments which have no harmful health concerns. However, since the pigments are produced intracellularly, proper extraction methodologies including cell disruption are required (Sánchez-Muñoz et al. 2020).

The most widely used pigment β-carotene has been found to be majorly produced by *Blakeslea trispora, Mucor circinelloides,* and *Xanthophyllomyces dendrorhous.* Members of zygomycetes and ascomycetes have been found to be major producers of carotenoids (Narsing Rao et al. 2017). Canthaxanthin is majorly produced by an orange and dark pink pigmented bacteriochlorophyll containing photosynthetic strain *Bradyrhizobium* and *Halobacterium* spp. Other pigments such as anthraquinones, naphthaquinones, naphthalene melanin, flavin, chrysophenol, helminthosporin, tritisporin, erythroglaucin, and cyndontin are learnt to be produced by genera *Fusarium, Eurotium, Curvularia, Drechslera. Monascus purpureus* have been known to produce a mixture of six major related pigmented polyketides, which are characteristically insoluble in acid and known as azaphilone mix. The major pigments comprising the mix are a red colored rubropunctatin and monoscorubin, yellow colored monoscoflavin and ankaflavin (Joshi et al. 2003). A study has shown use of microalgae *S. japonica* as substrate for pigment production by *Talaromyces amestolkiae GT1* using solid state fermentation technique without the use of added salt or N_2 source to the medium (General et al. 2014). Fungal pigments have been also used in wood dyeing industry through inoculation methods. Few well-known fungal species used in wood dyeing are *Scytalidium cuboideum, S. ganodermophthorum, Arthrographis cuboidea, Xylaria polymorpha, Penicillium griseofulvum, Monascus ruber, Ceratocystis* sp., *Ophiostoma* sp., *Lasiodiplodia theobromae, Chlorociboria aeruginascens, Phialocephala* sp., and *Trametes versicolor* (Liu et al. 2021).

(c) *Algal pigments*: Algae form another major class of microorganism producing several pigments. *Dunalia salina* of class Chlorophyta produce β-carotene. Another class of algae Rhodophyta are associated with the production of phycobiliproteins, i.e., phycocyanins and phycoerythrins (Gogate and Joshi 2020). Phycocyanins have been reported to be derived from various kind of *Arthrospira* and several other microalgae. *Haematococcus pluvialis,* a green algae, is known to produce astaxanthin. Canthaxanthin is produced by several other algal taxa such as *H. lacustris, Chlorella, Chlamydomonas, Scenedesmus, Ankistrodesmus* and, *Dictyococcus cinnanarinus* (Rebelo et al. 2020).

12.3 Technologies Involved in Microbial Pigment Production and Extraction

Several biotechnological and genetic manipulation techniques are employed to enhance microbial pigment production. Media composition, growth conditions, and genetics of the microbial species used for pigment production are the major factors influencing quality and quantity of pigment produced. Hence, these parameters can be statistically optimized and monitored to bring about significant increase in pigment production. Various statistical experimental designs such as response surface methodology (RSM), Plackett Burman design, Central Composite design, Artificial Neural Networks (ANN), etc., have been used to optimize the media composition through regression analysis (Nigam and Luke 2016). Several advanced techniques such as high hydrostatic pressure (HHP), pulse electric field (PEF), sonication-assisted extraction, gamma irradiation, enzymatic extraction, and membrane technology have been adopted for color extraction. Other green technologies include ultrasound-assisted extraction (UAE), microwave-assisted extraction (MAE), pressurized liquid extraction (PLE), supercritical fluid extraction (SFE), and ionic liquid-assisted extraction (ILAE).

Immobilization culture by adding silica gel, aerosols, or alginate beads have been proved to be effective in enhancing cell proliferation and thus increased production of pigment prodigiosin (Lin et al. 2019; Yamashita et al. 2001). Exposure to stressed conditions like elevated temperatures, osmotic pressures, metabolic inhibition, and presence of heavy metals have also been found to increase secondary metabolite and pigment production via fermentation techniques.

Manipulations through genetic and metabolic engineering have also resulted in increase in microbial pigment production. In fungal strain *Blakeslea trispora*, β-carotene, and lycopene production are optimized through lycopene cyclase inhibitor and controlled oxygen transfer rate, respectively (Choudhari et al. 2008). Lycopene production in yeast *S. cerevisiae* is enhanced by its genome modification, enzymatic modulation, or combined modification of lycopene biosynthesis pathways. Production of astaxanthin by *S. cerevisiae* have been enhanced by increasing the acetyl coA pool. Multiplex automated genomic engineering have been used in several bacterial strains to improve pigment production with an objective to maximize carbon substrate-based flux targeting production of the compound of interest besides diminishing flux causing yield of unnecessary by-products. Genomic manipulation of different carbon assimilation pathways such as Embden Meyerhoff pathway, NADPH and ATP production pathways, blocking pentose phosphate pathway, augmentation of MEP pathway, integration/replacement of *crt* genes, etc., have been performed to elevate the production of bacterial carotenoids and lycopene production in *E. coli, Corynebacterium glutamicum, Rhodobacter sphaeroides,* and *Yarrowia lypolitica* (Usmani et al. 2020). Common mutagenic agents utilized for the production of biopigments are UV radiation, gamma radiation, microwaves, atmospheric and room temperature plasma, ethyl methanesulfonate, and N-methyl-N-nitro-N-nitrosoguanidine (Sánchez-Muñoz et al. 2020).

Newer screening and quick detection methods have been developed to differentiate between useful and undesired compounds or metabolites generated. For example, the condensed handheld Raman spectrometer, which uses a 532 nm excitation laser to detect common and uncommon carotenoids, bacterioruberin, and other known pigments (Jehlička and Oren 2013; Kumar et al. 2015). Intelligent screening based on earlier information on the toxic metabolite pathway of the pigment producing microorganisms to rule out toxic and pathogenic pigment producers is another technique for selecting useful strains. Mass spectrometry with electrospray ionization HPLC, LCMS, nuclear magnetic resonance (NMR), and UV-VIS spectra are also used for faster identification of pigments and pigment producers. Even known compounds within relatively complex mixtures can be identified without the need for purification (Elyashberg et al. 2002; Smedsgaard and Frisvad 1996). Cost-effective downstream natural purification processes such as one involving raisins with higher absorption capacity has been used to separate prodigiosin from the culture media yielding concentrated semi-purified compound (Wang et al. 2004).

Strain manipulation through metabolic engineering to overproduce pigments or produce variants of pigment by changing the pigments' molecular structure and color by regulating the biosynthetic pathways have been extensively performed (Sen et al. 2019). Genetic manipulation has been done to produce kalafungin, a bright yellow polyketide from a blue pigment Actinorhodin, by *Streptomyces coelicolor*. Kalafungin is in turn been used to produce antraquinone, a reddish-yellow colored compound (Bartel et al. 1990; McDaniel et al. 1993). Heterologous expression by varying the substrates for cellular growth has been used to express or manipulate biosynthetic pathways in pigment producing cell factories (Nielsen and Nielsen 2017). Various low-cost technologies involving use of waste organic substrates such as sugarcane bagasse (SCB) hydrolysate, jackfruit seed, marine algal biomass *Saccharina* (Laminaria) *japonica* and kinnow (*Citrus reticulata*) peel has been used in solid state fermentation and submerged fermentation to obtain thermally stable microbial pigments (General et al. 2014; Panesar 2014; Subhasree et al. 2011; Terán Hilares et al. 2018).

12.4 Challenges in Microbial Pigment Production

Natural pigments have been gaining increasing popularity due to their environment-friendly properties such as safety, nontoxicity, noncarcinogenicity, biodegradability, and nutritional values. These natural pigments are now industrialized, as they have wide varieties and distribution, faster rate of reproduction, easier cultivation, and space restriction (Liu et al. 2021). However, various challenges exist in microbial pigment production such as regulatory hurdles, cost intensiveness of the production process, availability of a safe solvent for extraction, higher dose requirements, deodorization issues, reactiveness and lesser stability on certain food matrices, and sensitivity to environmental stress (Sen et al. 2019).

Development of any new microbial pigments and their transition from petridish to market face a lot of regulatory hurdles due to their physical and chemical attributes, toxicity, stability issues, low shelf life, etc. (Malik et al. 2012). Few pigments developed from *Monascus* are prohibited in developed countries as they have mycotoxin but are used as food colorant in several places like, Asia (Dufossé et al. 2005). Cost of pigment production, extraction, preservation, and requirement of higher dosage to develop the desired hue also sums up to a considerable amount, which is sometimes 5–20 times higher than that of synthetic colorants, which also becomes a limiting criteria for regular use of these colors in industries (Sen et al. 2019; Sigurdson et al. 2017). Development of undesirable flavors and odors, weak tinctorial strength, and loss of coloration upon reaction with some matrices are few other challenges that often restrict the use of microbial pigments in food industries. Moreover, sensitivity of natural pigments toward various environmental conditions such as light, pH, UV, temperature, oxygen, and heat or presence of some reactive species such as metal ions, proteins, or organic compounds also limit their long-term use (Lakshmi 2014; Rodriguez-Amaya 2019). For example, vitamin C enhances the stability of beverages in the presence of carotenoids; however, it causes degradation of anthocyanins (Kirti et al. 2014). Other microbial pigments like carotenoids, chlorophyll, and anthocyanins also encounter similar limitations. Carotenoids and chlorophylls become unstable and are prone to undergo enzymatic degradation or degrade due to factors like light, oxygen, heat, or acid (Gutiérrez et al. 2013; Mao et al. 2009). Encapsulation of the active compound in nanoemulsions of wall material with properties like low viscosity, emulsifying properties, biodegradability, low hygroscopicity, as well as low cost is a well-known alternative technique to increase the shelf life, stability, and controlled release of the pigment and suppressing the aroma (Venil et al. 2020). Microencapsulation of microbial pigments as core material with maltodextrins, modified starch, inulin, furcellaran, etc., as core material have been found to increase solubility, stability, and extended shelf life (Özkan and Ersus Bilek 2014). Another important challenge faced by microbial pigment producing industries is the availability of an appropriate organic solvent for extraction. Apart from water and ethanol, all other solvents are mostly synthetic, which primarily limit their use in natural pigment production (Sen et al. 2019). Dichloromethane is the most popular solvent used in pigment industry, which however has serious health hazards and highly toxic and pungent smell. It has been classified as class 2A carcinogen by World Health Organization (WHO) and International Agency for Research on Cancer (IARC) (WHO 2016). Use of organic solvents remains a cost-intensive procedure as a lot of solvent is consumed to purify a small amount of pigment besides posing environmental pollution.

Other production factors responsible for desired productivity of microbial pigments are types of bioreactors selected (e.g., conventional bioreactors, air lift and stirred-tank reactor, and trickle-bed reactor), mode of fermentation (batch, feed-batch, or continuous), and physicochemical and biological conditions in fermentation process (Venil et al. 2014). Hence, future prospects including development of technologies of microbial pigment production with respect to strain selection, pigment characteristics, and extraction procedures must be determined to overcome the challenges faced by pigment industries till date.

12.5 Applications of Microbial Pigments

12.5.1 Food and Beverage Industries

Several fermentation-derived pigments, are used in food industry, for example, a fungal β-carotene obtained from *Blakeslea trispora* used in Europe or an Asian origin of pigment known as *Monascus*, etc. (Wissgott and Bortlik 1996) (Downham and Collins 2000). Such kind of pigments not only offer a good appearance to the food items but also sometimes are known to have neutraceutical values, *viz.*, antibiotic, antiviral, antioxidants, antiparasitic, anticancerous, etc. A good example of such natural pigment is the Monascus red pigments, produced as MFR (Monascus Fermented Rice), which contains monocolins, which is known to be able to reduce the LDL-cholesterol while increasing the HDL-cholesterol, and thus improving the organoleptic characteristics of any food product (Kumar et al. 2015). Other food grade pigments obtained from bacteria are zeaxanthin from *Flavobacterium* sp., canthaxanthin from the photosynthetic bacterium, *Bradyrhizobium* sp., or the extremely halophilic bacterium, *Halobacterium* sp., astaxanthin from *Agrobacterium aurantiacum*, etc. (Dufossé 2006). The primary concern for a microbial pigment to be employed as food or beverage colorant is to evaluate the co-production of toxins by the species used.

Various microbial pigments are currently under use in food industry and are subject to research for their better applicability and shelf life. β-carotenes are lipid soluble, yellow-orange-red isoprenoid polyenes, with a highly conjugated ring structure, which contributes to its higher sensitivity to light and oxygen and hence, lesser stability. However, due to its neutraceutical values such as antioxidant and anticancerous properties and retinol precursor with prominent role in vitamin A synthesis, these are preferred in food industries for supplementation and fortification and are generally used in encapsulated forms (Zakynthinos and Varzakas 2016). *Blakeslea trispora*, *Mucorcircinelloides*, and *Phycomyces blakesleeanus* are few well-known fungal species producing β-carotenes (Kumar et al. 2015). Another soil strain *Penicillium oxalicum,* produces a red pigment, containing chromophore of anthraquinone type known as Arpink Red, which is used in food products as per recommendation by the Codex Alimentarius Commmision (Rotterdam meeting, March 11–15, 2002). Due to various toxicological data found on arpink Red such as acute oral toxicity and micronucleus in mice, acute dermal and eye irritation/corrosion, antitumor effectiveness, antibiotic activity, reverse mutation assay through AMES test on *Salmonella typhimurium*, and estimation of 5 mycotoxins, dosage recommendations have been made by the Codex Alimentarius Commission. Concentrations of 100 mg/kg in meat, meat products, and nonalcoholic beverages; 150 mg/kg in milk products and ice creams; 200 mg/kg in alcoholic drinks; and 300 mg/kg in confectionaries has been allowed. Another pigment Riboflavin commonly known as vitamin B 12 is a popular yellow food colorant having important roles in metabolic processes, e.g., the electron transport chain, DNA damage repair, lipids metabolism, and is responsible for maintaining health of eyes (Sánchez-Muñoz et al. 2020). It is used in food products like an additive to drinks, desserts items, ice cream, and tablets (Aberoumand 2011). Cheese, pastas, sauces, and cereals also have riboflavin (Nigam and Luke 2016). Several microorganisms are known to produce riboflavin via the process of fermentation, which can be categorized into three kinds, viz., weak overproducer, moderate overproducer, and strong overproducer (Stahmann et al. 2000; Santos et al. 2005). Likewise, the strain *Ashbya gossypii* is known to produce high level of riboflavin, via fermentation, with greater genetic stability. Riboflavin production has been improved four times by heterologous expression of a purinogenic enzyme and a transcriptional factor ortholog genes from *Debaryomyces hansenii* in *C. famata* and overexpressing these two genes (Dmytruk et al. 2011).

Monascus spp. a member of class *Ascomycetes* and family *Monascaceae,* is a xerophillic fungus known to produce industrially important red, orange, and yellow colorants. *M. pilosus, M.purpureus, M. ruber*, and *M. froridanus* often isolated from oriental food are the four important species used in production of *Monascus* pigments. The members of this genus are characteristically able to undergo esterification to synthesize pigments from polyketide chromophores and β-keto acids (Silbir and Goksungur 2019). The pigments produced by *Monascus purpureus* are classified into six types: (Aberoumand 2011) red pigment (rubropunctamin, $C_{21}H_{26}NO_4$, and monascorubramine, $C_{23}H_{27}NO_4$); (Afra et al. 2017) orange pigment (rubropunctatin, $C_{21}H_{22}O_5$ and monascorubrin, $C_{23}H_{26}O_5$); and (Ahmad et al. 2012) yellow pigment (monascin, $C_{21}H_{26}O_5$ and ankaflavin, $C_{23}H_{30}O_5$). Another important characteristic of such fungal sp. is the production of angkak, which has the ability to convert starchy substrates into several metabolites, for example, vitamins, enzymes, alcohols, antibiotics, antihypertensive, flavor compounds, fatty acids, pigments, flocculants, etc. A fermented rice product known as Red Yeast Rice (RYR) or angkak, produced traditionally by fermenting cooked rice kernels with yeast Monascus spp., has anticholesterol activity and many other active constituents such as unsaturated fatty acid, sterols, B-complex vitamins, and statin-like structures (Kumar et al. 2015). Lycopene, also known as psi-carotene, is a red colored open-chain unsaturated carotenoid and is an acyclic isomer of beta-carotene. It is found to be longer than most of the other carotenoids and insoluble in water as well as found to be very sensitive to heat and oxidation. A study showed that lycopene cis-isomers were more stable and have higher antioxidant potential in comparison to all-trans lycopene. Another study used cheap corn fiber as their substrate and extracted lycopene from a genetically modified fungus *Fusarium sporotrichioides* that acted as an effective colorant with antioxidant properties (Kumar et al. 2015).

12.5.2 Therapeutic Applications

12.5.2.1 As Antimicrobial Agents

Microbial pigments have shown various antimicrobial properties, which have sometimes gained preference over various next-generation antibiotics as currently high amount of cases of multidrug resistance can be seen among several pathogenic microorganisms and the lack of antibiotics to combat such pathogens (Dhanasekaran et al. 2015; Prestinaci et al. 2015). Pigments such as carotenoids, prodigiosin, pyocyanin, melanins, flavins, quinones, monascins, violacein, and indigo have been reported as good antimicrobial agents (Malik et al. 2012). Prodiginine compounds like prodigiosin, undecylprodigiosin, cycloprodigiosin, heptylprodigiosin, nonylprodigiosin, cyclononylprodigiosin, and cyclomethyl-decylprodigiosins exhibit antibacterial activities against different gram-negative and gram-positive bacterial members as well as antifungal activities against *Coccidioides, Candida, Didymella, Aspergillus, Penicillium, Saccharomyces, Cryptococcus, Histoplasma, Trichophyton, and Verticillium.* Violacein and flexirubin pigments isolated from *Janthinobacterium* sp. Ant5-2 and *Flavobacterium* sp. Ant342 have also demonstrated antituberculosis activity by inhibiting *Mycobacterium tuberculosis* growth. Violacein also shows antiparasital and antiprotozoal activities (Ramesh et al. 2019). Antiviral activities by various pigments such as phenazine compounds synthesized by *Pseudomonas* and *Streptomyces,* violacein against herpes simplex virus, poliovirus, and simian rotavirus SA II and quinones such as benzoquinones, naphthoquinones, and anthraquinones have been demonstrated to exhibit antiviral properties (Ramesh et al. 2019). Anti-HIV properties have been detected in pigments extracted from pigmented *Phoma* sp., which have been shown to inhibit HIV virus integrase. Antimalarial activity has been reported in violacein and prodigiosin compounds against *Plasmodium falciparum* strains in vitro (Kim et al. 1999; Lazaro et al. 2002). Anti-trypanosomal activity has been reported in prodigiosin against *Trypanosoma cruzi,* which causes cell death via inhibiting mitochondrial function that interferes with the processes of oxidative phosphorylation. Phenazine compounds from *Pseudomonas fluorescens* have shown antinematodal activity by suppressing the hatching of egg and also increasing rate of juvenile mortality of root the knot nematode, *Meloidogyne incognita,* in vitro (Rao et al. 2021) (Table 12.2).

12.5.2.2 As Anticancer Agents

Cancer, a noncommunicable disease has been the leading cause of cases of human morbidity and mortality. Although several anticancer drugs have been developed till date, a majority of them suffers from the drawbacks of their huge side effects as well as great resistance toward treatment procedures and therapies (Foo and Michor 2014). Therefore, modern day research is focusing on developing anticancer drugs that are nontoxic in nature. Pigments like violacein with low toxicity and photodynamic properties may have possible applications as a photochemotherapeutic drug (Durán 1980). Several pigmented compounds belonging to classes of polyketide, pyrroloiminoquinone, indolocarbazole, butenolide, phenoxazinone, alkaloid, phycobiliprotein, terpenoid dihydroquinones, phenazine, peptides, indole, and pyrrole alkaloid produced by marine microbes such as *Pseudoalteromonas*, *Actinomycetes*, cyanobacterial species, and other bacterial species exert potential antitumor activities (Soliev and Enomoto 2013). In such scenario, several studies on microbial pigments as anticancer drugs against different kinds of cancer are found to show promising anticancerous properties. Table 12.3 enlists the source microorganisms and their pigments along with their target cells.

Table 12.2 Major antimicrobial roles of microbial pigments, source organisms, and their target organisms

Pigments	Source microorganism	Effective against	References
Antibacterial			
Carotenoids	*Holomonas* sp.	*Klebsiella* sp., *S. aureus,* and *P. aeruginosa*	Ravikumar et al. (2016))
	Kocuria roseus, Bacillus sp., *Corynebacterium* sp., *Staphylococcus* sp., *Brevibacterium* sp.	*S. aureus* ATCC 25923 (MSSA), *S. xylosus, Paenibacillus macerans,* along with *Citrobacter divesus*	Azman et al. (2018)
Prodigiosin	*S. marcescens* UFPEDA 398	Oxacillin-resistant *S. aureus*	Lapenda et al. (2015)
	Vibrio ruber DSM 14379	*Bacillus subtilis* and *E. coli*	Danevcic et al. (2016)
	Marine *S. marcescens* IBRL USM 84	Gram-positive bacteria	Ibrahim (2008)
	–	*Candida albicans, E. coli,* and *S. aureus*	Suryawanshi et al. (2017)
	–	*S. pyogenes, Klebsiella pneumoniae, S. aureus,* and *P. aeruginosa*	Nwankwo et al. (2017)
	Streptomyces sp. JS520	*Bacillus, Micrococcus* sp., and *C. albicans*	Stankovic et al. (2012)
Methanolic extract of prodigiosin		*C. albicans* and *Crypto-coccus neoformans*	Danevcic et al. (2016)
Melanin	*Pseudomonas balearica* strain U7	*Erwinia chrysanthemi* and *Erwinia carotovora, S. aureus, E. coli,* and *C. albicans*	
	Streptomyces strain	*E. coli*	Vasanthabharathi et al. (2011)
Violacein	*C. violaceum* ATCC 12472	*S. aureus*	Batista and da Silva Neto (2017)
	–	Biofilm-forming and non-biofilm-forming *Staphylo-coccus epidermidis* strains ATCC 35984 and 12228	Dodou et al. (2017)
	–	Avirulent *Mycobacterium tuberculosis* H37Ra strain	de Souza et al. (1999)
	Janthinobacterium sp. Ant5-2	Avirulent *M. tuberculosis*	Mojib et al. (2010)
Pyocyanin		Urinary tract pathogens; *S. aureus, S. saprophyticus, S. epidermidis, E. coli,* and *C. freundii*	Ahmed et al. (2013)
		Group of Gram-positive bacteria and *C. albicans*	El-Shouny et al. (2011)

(continued)

Table 12.2 (continued)

Pigments	Source microorganism	Effective against	References
Antiviral			
Cycloprodigiosin hydrochloride (cPrG-HCl)	*Pseudomonas denitrificans*	Potent antimalarial activity, better than antimalarial agent chloroquine	Kim et al. (1999)
Violacein		Can inhibit the growth of mouse and human-derived *Plasmodium* parasites, effective against young and mature forms of the human parasite, effective against chloroquine-sensitive and resistant strains of *Plasmodium falciparum*	Lopes et al. (2009)
Antimalarial			
Violacein	*Pseudomonas denitrificans*	*Plasmodium falciparum*	Kim et al. (1999)
Prodigiosin			Lazaro et al. (2002)

12.5.2.3 As Antioxidants

Oxidative stress is caused because of the generation of free radicals increasing the possibilities of chronic diseases, *viz.*, cardiovascular, cancer, neurological disorders, diabetes, as well as autoimmune diseases, etc. (Phaniendra et al. 2015), and plays important role in endothelial dysfunction (Schramm et al. 2012), lung disease (Paola Rosanna and Salvatore 2012), gastrointestinal dysfunction (Kim et al. 2012), and atherosclerosis (Hulsmans et al. 2012). Potential antioxidation properties have been found in various microbial pigments like violacein, carotenoids, anthocyanins, and naphthoquinone (Sen et al. 2019). Violacein produced by *Pseudoalteromonas* and *Chromobacter violaceum* and staphyloxanthin produced by *Staphylococcus aureus* are powerful antioxidants, which provides protection against oxidative damage (Durán et al. 2012; Kurjogi et al. 2010). Several other pigments also known to act as antioxidants are astaxanthin, granadaene, canthaxanthin, lycopene, riboflavin, β-carotene, torularhodin, etc. (Table 12.4).

12.5.2.4 As Anti-Inflammatory and Anti-Allergic Agents

Mechanical injuries induce degrees of immunological defense mechanism known as, inflammation. This happens because of the expression of variety of cytokines, chemokines, and their receptors (Menichini et al. 2009). Several microbial pigments have exhibited anti-inflammatory responses through inhibition of various

Table 12.3 Potential microbial pigments with their anticancerous properties

Bacterial pigments	Source microorganism	Effective against	References
Carotenoid	*Arthrobacter* sp. G20	Esophageal cancer cell line (KYSE30)	Afra et al. (2017)
	Haloferax volcanii	Human liver carcinoma cell lines HepG2	Sikkandar et al. (2013)
Methanolic extract of carotenoid	*Kocuria* sp. QWT-12	Breast cancer cell lines MCF-7, lung cancer cells and breast cancer cells (A549 and MDA-MB-468), breast cancer cells (MDA-MB-231)	Rezaeeyan et al. (2017)
Yellow carotenoid pigment from	*Streptomyces griseoaurantiacus* JUACT 01	Cervical cancer cells (HeLa) and HepG2 cells	Prashanthi et al. (2015)
Prodigiosin	*S. marcescens*	Human cervical cancer cell (HeLa) and human laryngeal cancer cells (Hep2)	Maheswarappa et al. (2013)
	Pseudoalteromonas sp. 1020R	U937 leukemia cells	Wang et al. (2012)
	–	MCF7 breast cancer cell lines	Bhagwat and Padalia (2020)
	Vibrio sp. C1-TDSG02-1	Human oral squamous carcinoma cells (OSCC)	Cheng et al. (2017)
	–	GLC4 small cell lung cancer cell line	Llagostera et al. (2003)
Cycloprodigiosin hydrochloride (cPrG-HCl)	*Pseudoalteromonas denitrificans*	Liver cancer cells; Huh-7, HCC-M, HCC-T, dRLh-84, and H-35, hepatocellular carcinoma; HepG2, hepatoblastoma	Yamamoto et al. (1999)
Melanin	*Streptomyces glaucescens* NEAE-H	HFB4 skin cancer cell line	El-Naggar and El-Ewasy (2017)
Dihydroxyphenylalanine (DOPA) melanin	*Streptomyces* sp. MVCS6	Cervical cancer cell line	Sivaperumal et al. (2015)
Violacein	*C. violaceum*	Chinese hamster lung fibroblast V79 cells, leukemia cell lines	da Silva Melo et al. (2000), Melo et al. (2003), Ferreira et al. (2004)
	–	Ehrlich ascites tumor (EAT)	Bromberg et al. (2010)
Pyocyanin	*P. aeruginosa*	HepG2 human hepatoma cells	Zhao et al. (2014)

(continued)

Table 12.3 (continued)

Bacterial pigments	Source microorganism	Effective against	References
	Mutant strain *P. aeruginosa* S300-8	Pelvic rhabdomyosarcoma (RD) cells	Hassani et al. (2012)
	–	Glioblastoma cells (U87MG)	Vipin et al. (2017)

inflammatory signal pathways. Scytonemin pigment extracted from cyanobacterial sp. have shown anti-inflammatory as well as antiproliferative responses (Stevenson et al. 2002). *Micrococcus* sp., a marine pigmented bacteria, is known to have strong anti-inflammatory along with wound healing properties accompanied with antibacterial property of the pigment (Srilekha et al. 2018). The *Monascus* pigment, Monascin, exhibited anti-inflammatory activity by inhibiting inflammatory signal pathways of PKC (protein kinase C) and JNK (c-Jun N-terminal kinase) phosphorylation (Ramesh et al. 2019). Other pigments belonging to carotenoid and fucoxanthin from cyanobacteria are known to possess anticancer, anti-inflammatory, and anti-obesity properties (El-Naggar and El-Ewasy 2017).

12.5.2.5 As Metabolic Helpers

Microbial pigments also show various metabolism-enhancing properties. Supplementation with red mold dioscorea in rats with diabetes mellitus exhibiting high-density lipoprotein and decreased levels of triglycerides (TG) and glycosylated hemoglobin (HbA1c) have shown antihypertriglyceridemia activity (Pan et al. 2011). Antiatherosclerotic activity of ankaflavin, astaxanthin, and monascin has been found to prevent chances of fatty liver, lipid plaque, and also increased high-density lipoprotein cholesterol by enhancement of oxidative stress, inflammation, lipid metabolism, and glucose metabolism (Kishimoto et al. 2016; Lee et al. 2013). Astaxanthin, tambjamin, beta-carotene, and undecylprodigiosin have also displayed antiproliferative activity against various human cell lines such as T and B lymphocytes, MCF-7, and MDA-MB231cell lines (Lee et al. 2013; Teo et al. 2005). Antiaging properties have been established in astaxanthin and water-soluble phycobiliproteins (Ananya and Ahmad 2014; Capelli et al. 2013). Monascin and ankaflavin from various sources have demonstrated anti-obesity properties via reduction of the preadipocyte proliferation, by inhibiting cholesterol and triglyceride content, reduction of triglyceride accumulation; and suppression of the expression of adipocyte specific transcription factors (Kim et al. 2010). Besides, monascins have also shown antidiabetic activity by showing antioxidant effects, protection of pancreatic β-cells, and control of hyperglycemia by decreasing blood glucose and serum-free fatty acid levels in streptozotocin-induced diabetic rats (Lee et al. 2013; Pan et al. 2011).

Table 12.4 Microbial pigments with antioxidant properties with their mode of action

Bacterial pigments	Source organism	Mode of action	References
Carotenoids	*Kocuira marina DAGII, Meiothermus,* and *Thermus* sp.	Potent antioxidant activities	Kumar Samanta et al. (2016)
	Pedobacter	Strong antioxidant capacity and protects against oxidative damage	Correa-Llantén et al. (2012)
Staphyloxanthin (carotenoids)	*Staphylococcus aureus*	Can protect S. aureus itself against oxidative stress	Clauditz et al. (2006)
Xanthomonadin (carotenoids)	–	Inhibits photodynamic lipid peroxidation in liposome and offers protection against photodamage	Rajagopal et al. (1997)
Flexirubin (carotenoids)	*Fontibacter flavus* YUAB-SR-25	Effective against 2, 2-diphenyl-1-picryl-hydrazyl-hydrate (DPPH), hydroxyl radical, nitric oxide, and inhibition of lipid peroxidation	Prabhu et al. (2013)
(3R)-saproxanthin (rare monocyclic C40 carotenoids)	*Rubritalea squalenifaciens, Planococcus maritimus*	Potent antioxidant activities	Shindo and Misawa (2014)
(3R, 2′S)-myxol (rare monocyclic C40 carotenoids)			
Diapolyconedioc acid xylosylesters (new C30 carotenoids)			
Methyl 5-glucosyl-5, 6-dihydro-apo-4, 4′- lycopenoate (new C30 carotenoids)			
Melanin	*Streptomyces glaucescens NEAE-H, Pseudomonas* sp. *and Bifidobacterium infantis*	Antioxidant	Tarangini and Mishra (2014), Llagostera et al. (2003), Huang et al. (2011)
Pyomelanin	*Burkholderia cenocepacia C5424*	Antioxidant	Keith et al. (2007)
Violacein	*C. violaceum*	Gives protection against oxidative damage in gastric ulceration	Antonisamy and Ignacimuthu (2010)

12.5.2.6 As Bio-Indicators

Other than the applications of being antimicrobial agents, pigments, antioxidants, and anticancer agents, etc., microbial pigments can also be used as bio-indicators. These bio-indicators help to assess environmental health and also are capable of detecting environmental changes. Certain environmental factors control the presence of bio-indicators, viz., transmission of light, water, temperature, suspended solids, etc. Bio-indicators are known to detect the natural state of certain region or the level/ degree of certain environmental contaminations (Khatri and Tyagi 2015). The advantages of using bio-indicators are, (1) determination of biological effects is possible, (2) helps determine synergetic and antagonistic effects of pollutants on a particular creature, (3) helps early-stage diagnosis as well as to monitor harmful effects of toxicants in plants, humans, etc., (4) easy to produce, and (5) cost effective viable alternative, compared to specialized measuring systems. Microbial indicators are often used as health indicators of aquatic and terrestrial ecosystems. They are gaining popularity due to their abundance in nature. For example, few microorganisms are capable of detecting toxic materials like, cadmium, benzene in their surroundings and developing stress proteins that can be used as early warning indicators (Khatri and Tyagi 2015). Microbes are also capable of producing fluorescent pigments that can be used to check progress of specific reactions. For example, phycoerythrin is used to predict the rate of peroxy radical scavenging in human plasma (DeLange and Glazer 1989). Another example of heavy metal detecting microorganism is the *Vogesella indigofera* species, which produces a blue pigment under normal environmental growth condition, but stops their pigment production when they get exposed to heavy metals, such as hexavalent chromium (Gu and Cheung 2001). Microbial pigments can also be used to monitor temperature variations. For example, *Pantoea agglomerans* microbe, produce a blue pigment when their surrounding temperatures is ≥ 10 °C and so they are used as temperature indicators for the maintenance of foods and clinical materials at low temperature storage conditions (Fujikawa and Akimoto 2011).

12.6 Microbial Paints

Bioproduction and application of biopaints in various industries such as wood dyeing and textile industries are growing day by day to avoid harmful effects of chemical dyes and paints including environmental pollution, huge generation of chemically hazardous wastewater, and human health hazards. Microbial pigments from *Chryseobacterium* (FL-flexirubin type) and *Deinococcus* (DX-deinoxanthin type) species are mixed in casein-based paints to use them functionally as painting materials for dyeing and painting (Park et al. 2020). The traditional wood processing industry uses effective ways of wood dyeing to improve their surface quality, offers decorative effect, as well as improves wood value. Currently, this has become an important area of research to investigate upon techniques to process fast-growing wood along with contributing to financial revenue of forestry processing system

(Yang et al. 2019). Traditional wood dyeing involves processes like physical and chemical dyeing, using synthetic chemical dyes (Jamili et al. 2019). The major disadvantage of such methods was the huge discharge of processing water as well as wastewater from the production process, and also the use of complex raw materials that required increased costs for wastewater treatment (Li et al. 2019). Along with these, the traditional wood dyeing methods also caused environmental pollution as well as caused harm to human health via wastewater pollution, release of nondegradable toxic volatile organic compounds (VOCs), etc., that requires more manpower and huge financial resources to deal with it (Jegatheesan et al. 2016; Zhao et al. 2019). Due to increasing awareness regarding harmful impacts of chemical dyes and environmental pollution, use of biopaints is gaining popularity. However, establishment of microbial dying technologies still is in developmental stage and has to overcome several challenges.

Several fungi and bacteria are known to secrete pigments that can be produced via cultivation and fermentation under specific conditions and be used in microbial dyeing procedures. Few of the fungal species that are commonly used in wood dyeing industry are, pink, red, blue, or yellow pigment producing microorganism, *Scytalidium cuboideum*, used on sugar maple or pine woods, red color producing *Penicillium griseofulvum*, blue and green color producing *Chlorociboria aeruginascens*, black color producing *Trametes versicolor*, etc. (Robinson et al. 2009, 2014a; Zhu et al. 2018). The wood dying method works via two different paths, one where pigment producing microorganism are inoculated directly on the material offering certain pattern randomness and also different color depth, and then the material is sterilized for ecological and environmental protection (Liu et al. 2020). The other method involves laboratory cultivation of the microorganism to fully secrete their pigments, which is then extracted via physical, mechanical, or chemical methods. Organic solvents are then further used to purify or modify the pigments and then used to treat wood materials, textiles, stone materials, etc. Such microbial pigments are gaining popularity due to their biocompatibility, biodegradability, as well as their potential use in medicine or pharmaceuticals (Robinson et al. 2014b).

A fine artwork in the Bilbao Fine Arts Museum dated between 1560 and 1570, which originated from Augsburg, displays seven wood species and a blue-green colored piece that was dyed by xylindein produced by *Chlorociboria* sp. (Gutierrez and Robinson 2017). Another artwork dated 1600, located at the National Museum of Decorative Arts, showcases blue-green artwork with different vegetations and architectural elements made via the *Chlorociboria* sp. (Gutierrez and Robinson 2017).

Textile industries are also using natural colors to dye variety of fabrics to avoid production of undesirable toxic or hazardous chemicals, which are also known to cause various skin irritations and harmful allergies to our body. For example, *Vibrio* spp. is capable of producing bright red color pigment, known as prodigiosin that is used to dye many fibers, viz. wool, nylon, acrylics, silk, etc. (Sánchez-Muñoz et al. 2020). Another textile dyeing colorant, produced from *Janthinobacterium lividum*, is used for coloring natural fibers like silk, cotton, wool, as well as synthetic fibers like nylon, vinylon, etc. (Vinotha et al. 2019). Applications of prodigiosin and violacein pigments for making of a specific kind of motif named batik on cotton fabric popularly used by the Southeast Asian women can be observed (Venil et al. 2013). The pigment prodigiosin is also used in coloring candles, paper, soap, or pencil case pouch as well as can also be used as ink in ball point and highlighter pens (Ahmad et al. 2012).

12.7 Concluding Remarks and Green Future Prospective

Microbial pigments with enormous application potential in various fields including pharmaceutical and nutritional industries, food and beverages, candidate bioindicative roles, and paint and dyeing industries, etc., is a promising green method to reduce the harmful effects of synthetic pigmented compounds, rigorously used in various sectors, on environment and human health. Extensive future research is, however, required to develop and standardize the cultivation protocols including strain selection, choice of fermentation methodology, use of low cost substrates, designing protocols for enhancement of the shelf life, and minimizing the use of toxic synthetic organic solvents to maximally replace the low-cost synthetic pigments. Various biotechnological advances need to be made using omics-based information, genetic engineering for strain improvement, development of heterologous engineered expression systems, and adequate toxicity evaluation of the product for successful marketing of the product. The transition of the pigment from lab to market has also to undergo various regulatory protocols, which must be taken care of while developing production strategies. Successful replacement of the synthetic dyes with natural ones derived from microbial origin shall be an ecofriendly endeavor reducing the generation of various toxic by-products while in their production and several health effects when consumed through food or neutraceuticals. Besides, full-fledged introduction of natural colorants in foods, processed foods, and pharmaceuticals shall slowly enhance the health status of the population with reduced side effects caused due to synthetic pigments.

Acknowledgments DBT BioCaRe fellowship to Dr. Soma Ghosh by Department of Biotechnology, Govt. of India (BT/PR31826BIC/101/1211/2019) is highly acknowledged. The Knowledge Resource Centre, CSIR NEERI is duly acknowledged.

References

Aberoumand A (2011) A review article on edible pigments properties and sources as natural biocolorants in foodstuff and food industry. World J Dairy Food Sci 6(1):71–78

Afra S, Makhdoumi A, Matin MM, Feizy J (2017) A novel red pigment from marine Arthrobacter sp. G20 with specific anticancer activity. J Appl Microbiol 123(5):1228–1236. https://doi.org/10.1111/JAM.13576

Ahmad WA, Ahmad WY, Zakaria ZA, Yusof NZ (2012) Application of bacterial pigments as colorant. Ch. 2. In: Briefs in Molecular Science, New York

Ahmed N, Singh J, Kour H, Gupta P (2013) Naturally occurring preservatives in food and their role in food preservation. Int J Pharm Biol Arch 4(1):22–30

Ananya AK, Ahmad IZ (2014) Cyanobacteria the blue green algae and its novel applications: a brief review. Int J Innov Appl Stud 7(1):251

Antonisamy, P., & Ignacimuthu, S. (2010). Immunomodulatory, analgesic and antipyretic effects of violacein isolated from Chromobacterium violaceum. Phytomedicine, 17(3-4), 300–304

Azman AS, Mawang CI, Abubakar S (2018) Bacterial pigments: the bioactivities and as an alternative for therapeutic applications. Nat Prod Commun 13(12):1934578X1801301240

Babitha S (2009) Microbial pigments. In: Biotechnology for agro-industrial residues utilisation. Springer, Dordrecht, pp 147–162

Bartel PL, Zhu CB, Lampel JS, Dosch DC, Connors NC, Strohl WR et al (1990) Biosynthesis of anthraquinones by interspecies cloning of actinorhodin biosynthesis genes in Streptomycetes: clarification of actinorhodin gene functions. J Bacteriol 172(9):4816–4826. https://doi.org/10.1128/JB.172.9.4816-4826.1990

Batista JH, da Silva Neto JF (2017) *Chromobacterium violaceum* pathogenicity: updates and insights from genome sequencing of novel *Chromobacterium* species. Front Microbiol 8:2213

Bhagwat A, Padalia U (2020) Optimization of prodigiosin biosynthesis by Serratia marcescens using unconventional bioresources. J Genet Eng Biotechnol 18(1):1–9

Bromberg N, Dreyfuss JL, Regatieri CV, Palladino MV, Durán N, Nader HB, Haun M, Justo GZ (2010) Growth inhibition and pro-apoptotic activity of violacein in Ehrlich ascites tumor. Chem Biol Interact 186(1):43–52. https://doi.org/10.1016/j.cbi.2010.04.016

Capelli B, Bagchi D, Cysewski GR (2013) Synthetic astaxanthin is significantly inferior to algal-based astaxanthin as an antioxidant and may not be suitable as a human nutraceutical supplement. Forum Nutr 12:145–152

Cheng MF, Lin CS, Chen YH, Sung PJ, Lin SR, Tong YW, Weng CF (2017) Inhibitory growth of oral squamous cell carcinoma cancer via bacterial prodigiosin. Mar Drugs 15(7):224. https://doi.org/10.3390/md15070224

Choudhari SM, Ananthanarayan L, Singhal RS (2008) Use of metabolic stimulators and inhibitors for enhanced production of β-carotene and lycopene by *Blakeslea trispora* NRRL 2895 and 2896. Bioresour Technol 99(8):3166–3173. https://doi.org/10.1016/J.BIORTECH.2007.05.051

Clauditz A, Resch A, Wieland KP, Peschel A, Götz F (2006) Staphyloxanthin plays a role in the fitness of *Staphylococcus aureus* and its ability to cope with oxidative stress. Infect Immun 74(8):4950–4953. https://doi.org/10.1128/IAI.00204-06

Correa-Llantén DN, Amenábar MJ, Blamey JM (2012) Antioxidant capacity of novel pigments from an Antarctic bacterium. J Microbiol 50(3):374–379. https://doi.org/10.1007/S12275-012-2029-1

da Silva Melo P, Maria SS, de Vidal BC, Haun M, Durán N (2000) Violacein cytotoxicity and induction of apoptosis in V79 cells. In Vitro Cell Dev Biol Anim 36(8):539–543

Danevcic T, Vezjak MB, Tabor M, Zorec M, Stopar D (2016) Prodigiosin induces autolysins in actively grown *Bacillus subtilis* cells. Front Microbiol 7:27. https://doi.org/10.3389/FMICB.2016.00027

de Souza AO, Aily DCG, Sato DN, Durán N (1999) Atividade da violaceina in vitro sobre o Mycobacterium tuberculosis H37RA. Rev Inst Adolfo Lutz 58(1):59–62

DeLange RJ, Glazer AN (1989) Phycoerythrin fluorescence-based assay for peroxy radicals: a screen for biologically relevant protective agents. Anal Biochem 177(2):300–306. https://doi.org/10.1016/0003-2697(89)90056-0

Delgado-Vargas F, Jiménez AR, Paredes-López O, Francis FJ (2010) Natural pigments: carotenoids, anthocyanins, and Betalains — characteristics, biosynthesis, processsing, and stability. Crit Rev Food Sci Nutr 40(3):173–289. https://doi.org/10.1080/10408690091189257

Dhanasekaran D, Thajuddin N, Panneerselvam A (2015) Antimicrobials: synthetic and natural compounds. CRC Press

Dmytruk K, Sybirna NO, Fedorovych D, Dmytruk KV, Yatsyshyn VY, Sybirna NO, Fedorovych DV, Sibirny AA (2011) Metabolic engineering and classic selection of the yeast Candida famata (Candida flareri) for construction of strains with enhanced riboflavin production. Metab Eng 13: 82–88. https://doi.org/10.1016/j.ymben.2010.10.005

Dodou HV, De Morais Batista AH, Sales GWP, De Medeiros SC, Rodrigues ML, Nogueira PCN, Silveira ER, Nogueira NAP (2017) Violacein antimicrobial activity on *Staphylococcus epidermidis* and synergistic effect on commercially available antibiotics. J Appl Microbiol 123(4):853–860. https://doi.org/10.1111/jam.13547

Downham A, Collins P (2000) Colouring our foods in the last and next millennium. Int J Food Sci Technol 35(1):5–22. https://doi.org/10.1046/J.1365-2621.2000.00373.X

Dufossé L (2006) Microbial production of food grade pigments. Food Technol Biotechnol 44(3): 313–323

Dufossé L, Galaup P, Yaron A, Arad SM, Blanc P, Murthy KNC, Ravishankar GA (2005) Microorganisms and microalgae as sources of pigments for food use: a scientific oddity or an industrial reality? Trends Food Sci Technol 16(9):389–406. https://doi.org/10.1016/J.TIFS.2005.02.006

Durán, N. (1980). Bacterial chemistry. I: Studies of a potential phototherapeutic substance from *Chromobacterium violaceum*

Durán M, Ponezi AN, Faljoni-Alario A, Teixeira MFS, Justo GZ, Durán N (2012) Potential applications of violacein: a microbial pigment. Med Chem Res 21(7):1524–1532. https://doi.org/10.1007/S00044-011-9654-9

El-Naggar NEA, El-Ewasy SM (2017) Bioproduction, characterization, anticancer and antioxidant activities of extracellular melanin pigment produced by newly isolated microbial cell factories *Streptomyces glaucescens* NEAE-H. Sci Rep 7:42129. https://doi.org/10.1038/srep42129

El-Shouny WA, Al-Baidani ARH, Hamza WT (2011) Antimicrobial activity of Pyocyanin produced by Pseudomonas aeruginosa isolated from surgical wound-infections. Int J Pharm Med Sci 1(1):1–07

Elyashberg ME, Blinov KA, Williams AJ, Martirosian ER, Molodtsov SG (2002) Application of a new expert system for the structure elucidation of natural products from their 1D and 2D NMR data. J Nat Prod 65(5):693–703

Ferreira CV, Bos CL, Versteeg HH, Justo GZ, Durán N, Peppelenbosch MP (2004) Molecular mechanism of violacein-mediated human leukemia cell death. Blood 104(5):1459–1464. https://doi.org/10.1182/blood-2004-02-0594

Foo J, Michor F (2014) Evolution of acquired resistance to anti-cancer therapy. J Theor Biol 355: 10–20. https://doi.org/10.1016/j.jtbi.2014.02.025

Fujikawa H, Akimoto R (2011) New blue pigment produced by Pantoea agglomerans and its production characteristics at various temperatures. Appl Environ Microbiol 77(1):172–178

General T, Prasad B, Kim HJ, Vadakedath N, Cho MG (2014) Saccharina japonica, a potential feedstock for pigment production using submerged fermentation. Biotechnol Bioprocess Eng 19(4):711–719. https://doi.org/10.1007/S12257-013-0709-2

Gogate PR, Joshi SM (2020) Process intensification aspects of extraction of pigments from microalgae. In: Pigments from Microalgae Handbook, pp 309–324. https://doi.org/10.1007/978-3-030-50971-2_13

Gu JD, Cheung KH (2001) Phenotypic expression of *Vogesella indigofera* upon exposure to hexavalent chromium, Cr6+. World J Microbiol Biotechnol 17:475–480

Gutierrez PTV, Robinson SC (2017) Determining the presence of spaltedwood in Spanish marquetry woodworks of the 1500s through the 1800s. CoatingsTech 7(11):188. https://doi.org/10.3390/coatings7110188

Gutiérrez FJ, Albillos SM, Casas-Sanz E, Cruz Z, García-Estrada C, García-Guerra A et al (2013) Methods for the nanoencapsulation of β-carotene in the food sector. Trends Food Sci Technol 32(2):73–83

Hassani HH, Hasan HM, Al-Saadi A, Ali AM, Muhammad MH (2012) A comparative study on cytotoxicity and apoptotic activity of pyocyanin produced by wild type and mutant strains of *Pseudomonas aeruginosa*. Eur J Exp Biol 2012(5):1389–1394

Huang HC, Chiu SH, Ke HJ, Chiu SW, Wu SY, Chang TM (2011) Antimelanogenic and antioxidant activities of *Bifidobacterium infantis*. Afr J Microbiol Res 5(20):3150–3156

Hulsmans M, Van Dooren E, Holvoet P (2012) Mitochondrial reactive oxygen species and risk of atherosclerosis. Curr Atheroscler Rep 14(3):264–276. https://doi.org/10.1007/S11883-012-0237-0

Ibrahim ASS (2008) Production of carotenoids by a newly isolated marine *Micrococcus* sp. Food and Agricultural Organization of United Nations

Indra Arulselvi P, Umamaheswari S, Ranandkumar SG, Karthik C, Jayakrishna C (2014) Screening of yellow pigment producing bacterial isolates from various eco-climatic areas and analysis of the carotenoid produced by the isolate. J Food Process Technol 5(292):2

Jamili F, Mirjalili M, Zamani HA (2019) Antibacterial wood-plastic composite produced from treated and natural dyed wood fibers. Polym Polym Compos 27(6):347–355. https://doi.org/10.1177/0967391119847537

Jegatheesan V, Pramanik BK, Chen J, Navaratna D, Chang CY, Shu L (2016) Treatment of textile wastewater with membrane bioreactor: a critical review. Bioresour Technol 204:202–212. https://doi.org/10.1016/j.biortech.2016.01.006

Jehlička J, Oren A (2013) Raman spectroscopy in halophile research. Front Microbiol 4:380. https://doi.org/10.3389/FMICB.2013.00380

Joshi VK, Attri D, Bala A, Bhushan S (2003) Microbial pigments. Indian J Biotechnol 2(2003): 362–369

Keith KE, Killip L, He P, Moran GR, Valvano MA (2007) Burkholderia cenocepacia C5424 produces a pigment with antioxidant properties using a homogentisate intermediate. J Bacteriol 189(24):9057–9065. https://doi.org/10.1128/JB.00436-07

Khatri N, Tyagi S (2015) Influences of natural and anthropogenic factors on surface and groundwater quality in rural and urban areas. Front Life Sci 8(1):23–39

Kim HS, Hayashi M, Shibata Y, Wataya Y, Mitamura T, Horii T, Kawauchi K, Hirata H, Tsuboi S, Moriyama Y (1999) Cycloprodigiosin hydrochloride obtained from Pseudoalteromonas denitrificans is a potent antimalarial agent. Biol Pharm Bull 22(5):532–534. https://doi.org/10.1248/bpb.22.532

Kim JH, Kim YO, Jeun J, Choi DY, Shin CS (2010) L-Trp and L-Leu-OEt derivatives of the Monascus pigment exert high anti-obesity effects on mice. Biosci Biotechnol Biochem 74(2): 304–308

Kim YJ, Kim EH, Hahm KB (2012) Oxidative stress in inflammation-based gastrointestinal tract diseases: challenges and opportunities. J Gastroenterol Hepatol 27(6):1004–1010. https://doi.org/10.1111/j.1440-1746.2012.07108.x

Kirti K, Amita S, Priti S, Mukesh Kumar A, Jyoti S (2014) Colorful world of microbes: carotenoids and their applications. Adv Biol 2014:1–13. https://doi.org/10.1155/2014/837891

Kishimoto Y, Yoshida H, Kondo K (2016) Potential anti-atherosclerotic properties of astaxanthin. Mar Drugs 14(2):35

Kumar A, Shankar Vishwakarma H, Singh J, Kumar M (2015) Microbial pigments: production and their applications in various industries. Int J Pharm Biol 5(1):203–212

Kurjogi MM, Sanakal RD, Kaliwal BB (2010) Antibiotic susceptibility and antioxidant activity of Staphylococcus aureus pigment staphyloxanthin on carbon tetrachloride (ccl4) induced stress in swiss albino mice. Int J Biotechnol Appl 2(2):33–40

Lagashetti AC, Dufossé L, Singh SK, Singh PN (2019) Fungal pigments and their prospects in different industries. Microorganisms 7(12):604

Lakshmi C (2014) Food coloring: the natural way. Res J Chem Sci 4(2):87–96

Lapenda JC, Silva PA, Vicalvi MC, Sena KXFR, Nascimento SC (2015) Antimicrobial activity of prodigiosin isolated from *Serratia marcescens* UFPEDA 398. World J Microbiol Biotechnol 31(2):399–406. https://doi.org/10.1007/S11274-014-1793-Y

Lazaro JEH, Nitcheu J, Predicala RZ, Mangalindan GC, Nesslany F, Marzin D, Concepcion GP, Diquet B (2002) Heptyl prodigiosin, a bacterial metabolite, is antimalarial in vivo and non-mutagenic in vitro. J Nat Toxins 11(4):367–377. https://europepmc.org/article/med/12503881

Lee CL, Hung YP, Hsu YW, Pan TM (2013) Monascin and ankaflavin have more anti-atherosclerosis effect and less side effect involving increasing creatinine phosphokinase activity than monacolin K under the same dosages. J Agric Food Chem 61(1):143–150

Li K, Liu Q, Fang F, Luo R, Lu Q, Zhou W, Huo S, Cheng P, Liu J, Addy M, Chen P, Chen D, Ruan R (2019) Microalgae-based wastewater treatment for nutrients recovery: a review. Bioresour Technol 291:121934. https://doi.org/10.1016/j.biortech.2019.121934

Lin C, Jia X, Fang Y, Chen L, Zhang H, Lin R, Chen J (2019) Enhanced production of prodigiosin by *Serratia marcescens* FZSF02 in the form of pigment pellets. Electron J Biotechnol 40:58–64. https://doi.org/10.1016/J.EJBT.2019.04.007

Liu Q, Xiao J, Liu B, Zhuang Y, Sun L (2018) Study on the preparation and chemical structure characterization of melanin from boletus griseus. Int J Mol Sci 19(12):3736. https://doi.org/10.3390/IJMS19123736

Liu Y, Yu Z, Zhang Y, Qi C, Tang R, Zhao B, Wang H, Han Y (2020) Microbial dyeing—infection behavior and influence of *Lasiodiplodia theobromae* in poplar veneer. Dyes Pigments 173:107988. https://doi.org/10.1016/j.dyepig.2019.107988

Liu Y, Yu Z, Zhang Y, Wang H (2021) Microbial dyeing for inoculation and pigment used in wood processing: opportunities and challenges. Dyes Pigments 186:109021. https://doi.org/10.1016/J.DYEPIG.2020.109021

Llagostera E, Soto-Cerrato V, Montaner B, Pérez-Tomás R (2003) Prodigiosin induces apoptosis by acting on mitochondria in human lung cancer cells. Ann N Y Acad Sci 1010:178–181. https://doi.org/10.1196/ANNALS.1299.030

Lopes SCP, Blanco YC, Justo GZ, Nogueira PA, Rodrigues FLS, Goelnitz U, Wunderlich G, Facchini G, Brocchi M, Duran N, Costa FTM (2009) Violacein extracted from *Chromobacterium violaceum* inhibits *Plasmodium* growth in vitro and in vivo. Antimicrob Agents Chemother 53(5):2149–2152. https://doi.org/10.1128/AAC.00693-08

Maheswarappa G, Kavitha D, Vijayarani K, Kumanan K (2013) Prodigiosin as anticancer drug produced from bacteria of termite gut. Indian J Basic Appl Med Res 1:257–266

Malik K, Tokkas J, Goyal S (2012) Microbial pigments: a review. Int J Microbial Res Technol 1(4):361–365

Manivasagan P, Bharathiraja S, Santha Moorthy M, Mondal S, Seo H, Dae Lee K, Oh J (2018) Marine natural pigments as potential sources for therapeutic applications. Crit Rev Biotechnol 38(5):745–761

Mao L, Xu D, Yang J, Yuan F, Gao Y, Zhao J (2009) Effects of small and large molecule emulsifiers on the characteristics of β-carotene nanoemulsions prepared by high pressure homogenization. Food Technol Biotechnol 47(3):336–342

McDaniel R, Ebert-Khosla S, Hopwood DA, Khosla C (1993) Engineered biosynthesis of novel polyketides. Science 262(5139):1546–1550. https://doi.org/10.1126/SCIENCE.8248802

Melo PS, Justo GZ, De Azevedo MBM, Durán N, Haun M (2003) Violacein and its β-cyclodextrin complexes induce apoptosis and differentiation in HL60 cells. Toxicology 186(3):217–225. https://doi.org/10.1016/S0300-483X(02)00751-5

Méndez A, Pérez C, Montañéz JC, Martínez G, Aguilar CN (2011) Red pigment production by *Penicillium purpurogenum* GH2 is influenced by pH and temperature. J Zhejiang Univ Sci B 12(12):961–968. https://doi.org/10.1631/JZUS.B1100039

Menichini F, Conforti F, Rigano D, Formisano C, Piozzi F, Senatore F (2009) Phytochemical composition, anti-inflammatory and antitumour activities of four Teucrium essential oils from Greece. Food Chem 115(2):679–686

Mojib N, Philpott R, Huang JP, Niederweis M, Bej AK (2010) Antimycobacterial activity in vitro of pigments isolated from Antarctic bacteria. Anton Leeuw Int J Gen Mol Microbiol 98(4):531–540. https://doi.org/10.1007/S10482-010-9470-0

Narsing Rao MP, Xiao M, Li WJ (2017) Fungal and bacterial pigments: secondary metabolites with wide applications. Front Microbiol 8:1113. https://doi.org/10.3389/FMICB.2017.01113/BIBTEX

Nielsen JC, Nielsen J (2017) Development of fungal cell factories for the production of secondary metabolites: linking genomics and metabolism. Synth Syst Biotechnol 2(1):5–12. https://doi.org/10.1016/J.SYNBIO.2017.02.002

Nigam PS, Luke JS (2016) Food additives: production of microbial pigments and their antioxidant properties. Curr Opin Food Sci 7:93–100. https://doi.org/10.1016/J.COFS.2016.02.004

Nwankwo IU, Itaman VO, Chidiebere OL, Nwachukwu MP (2017) Evaluation of antimicrobial activity of prodigiosin produced from Serratia marcescens against some pathogenic bacteria. Futo J Series 3:93–102

Özkan G, Ersus Bilek S (2014) Microencapsulation of natural food colourants. Int J Nutr Food Sci 3(3):145

Pailliè-Jiménez ME, Stincone P, Brandelli A (2020) Natural pigments of microbial origin. Front Sustain Food Syst 4:160

Pan TM, Shi YC, Liao JW (2011) Antihypertriglyceridemia and anti-inflammatory activities of Monascus-fermented dioscorea in streptozotocin-induced diabetic rats. Exp Diabetes Res 2011: 710635. https://doi.org/10.1155/2011/710635

Panesar R (2014) Bioutilization of kinnow waste for the production of biopigments using submerged fermentation. Int J Food Nutr Sci 3(1):9

Paola Rosanna D, Salvatore C (2012) Reactive oxygen species, inflammation, and lung diseases. Curr Pharm Des 18(26):3889–3900. https://doi.org/10.2174/138161212802083716

Park SA, Ahn SY, Choi KY (2020) Functional microbial pigments isolated from *Chryseobacterium* and *Deinococcus* species for bio-paint application. Biotechnol Bioprocess Eng 25(3):394–402. https://doi.org/10.1007/S12257-019-0372-3

Phaniendra A, Jestadi DB, Periyasamy L (2015) Free radicals: properties, sources, targets, and their implication in various diseases. Indian J Clin Biochem 30(1):11–26. https://doi.org/10.1007/S12291-014-0446-0

Prabhu S, PD R, Young CC, Hameed A, Lin SY, AB A (2013) Zeaxanthin production by novel marine isolates from coastal sand of India and its antioxidant properties. Appl Biochem Biotechnol 171(4):817–831

Prasanna R, Sood A, Suresh A, Nayak S, Kaushik BD (2007) Potentials and applications of algal pigments in biology and industry. Acta Bot Hungar 49(1–2):131–156. https://doi.org/10.1556/ABOT.49.2007.1-2.14

Prashanthi K, Suryan S, Varalakshmi KN (2015) In vitro anticancer property of yellow pigment from Streptomyces griseoaurantiacus JUACT 01. Braz Arch Biol Technol 58(6):869–876. https://doi.org/10.1590/S1516-89132015060271

Prestinaci F, Pezzotti P, Pantosti A (2015) Antimicrobial resistance: a global multifaceted phenomenon. Pathog Glob Health 109(7):309–318. https://doi.org/10.1179/2047773215Y.0000000030

Rajagopal L, Sundari CS, Balasubramanian D, Sonti RV (1997) The bacterial pigment xanthomonadin offers protection against photodamage. FEBS Lett 415(2):125–128. https://doi.org/10.1016/S0014-5793(97)01109-5

Ramesh C, Vinithkumar NV, Kirubagaran R, Venil CK, Dufossé L (2019) Multifaceted applications of microbial pigments: current knowledge, challenges and future directions for public health implications. Microorganisms 7(7):186. https://doi.org/10.3390/MICROORGANISMS7070186

Rao AS, Deka SP, More SS, Nair A, More VS, Ananthjaraju KS (2021) A comprehensive review on different microbial-derived pigments and their multipurpose activities. In: Microbial Polymers, pp 479–519. https://doi.org/10.1007/978-981-16-0045-6_20

Ravikumar S, Uma G, Gokulakrishnan R (2016) Antibacterial property of halobacterial carotenoids against human bacterial pathogens. J Sci Ind Res 75:253–257. http://nopr.niscair.res.in/handle/123456789/34058

Rebelo BA, Farrona S, Rita Ventura M, Abranches R (2020) Canthaxanthin, a red-hot carotenoid: applications, synthesis, and biosynthetic evolution. Plan Theory 9(8):1039. https://doi.org/10.3390/PLANTS9081039

Rezaeeyan Z, Safarpour A, Amoozegar MA, Babavalian H, Tebyanian H, Shakeri F (2017) High carotenoid production by a halotolerant bacterium, Kocuria sp. strain QWT-12 and anticancer activity of its carotenoid. EXCLI J 16:840–851. https://doi.org/10.17179/excli2017-218

Robinson SC, Laks PE, Turnquist EJ (2009) A method for digital color analysis of spalted wood using Scion image software. Materials 2(1):62–75. https://doi.org/10.3390/ma2010062

Robinson SC, Hinsch E, Weber G, Leipus K, Cerney D (2014a) Wood colorization through pressure treating: the potential of extracted colorants from spalting fungi as a replacement for woodworkers' aniline dyes. Materials 7(8):5427–5437

Robinson SC, Tudor D, Zhang WR, Ng S, Cooper PA (2014b) Ability of three yellow pigment producing fungi to colour wood under controlled conditions. Int Wood Prod J 5(2):103–107. https://doi.org/10.1179/2042645313Y.0000000060

Rodriguez-Amaya DB (2019) Update on natural food pigments - A mini-review on carotenoids, anthocyanins, and betalains. Food Res Int 124:200–205. https://doi.org/10.1016/J.FOODRES.2018.05.028

Salama HA, Badr AN, Elkhadragy MF, Hussein AMS, Shaban IAS, Yehia HM (2021) New antifungal microbial pigment applied to improve safety and quality of processed meat-products. Microorganisms 9(5):989

Samanta AK, Chaudhuri S, Dutta D (2016) Antioxidant efficacy of carotenoid extract from bacterial strain Kocuria marina DAGII. Mater Today: Proc 3(10):3427–3433

Sánchez-Muñoz S, Mariano-Silva G, Leite MO, Mura FB, Verma ML, Da Silva SS, Chandel AK (2020) Production of fungal and bacterial pigments and their applications. In: Biotechnological Production of Bioactive Compounds, pp 327–361. https://doi.org/10.1016/B978-0-444-64323-0.00011-4

Santos MA, Mateos L, Stahmann K-P, Revuelta J-L (2005) Insertional mutagenesis in the vitamin B_2 producer fungus Ashbya gossypii. In: Microbial Processes and Products, pp 283–300. https://doi.org/10.1385/1-59259-847-1:283

Schramm A, Matusik P, Osmenda G, Guzik TJ (2012) Targeting NADPH oxidases in vascular pharmacology. Vasc Pharmacol 56(5–6):216–231. https://doi.org/10.1016/J.VPH.2012.02.012

Sen T, Barrow CJ, Deshmukh SK (2019) Microbial pigments in the food industry—challenges and the way forward. Front Nutr 6:7. https://doi.org/10.3389/FNUT.2019.00007/BIBTEX

Shindo K, Misawa N (2014) New and rare carotenoids isolated from marine bacteria and their antioxidant activities. Mar Drugs 12(3):1690–1698. https://doi.org/10.3390/md12031690

Sigurdson GT, Tang P, Giusti MM (2017) Natural colorants: food colorants from natural sources. Annu Rev Food Sci Technol 8:261–280. https://doi.org/10.1146/ANNUREV-FOOD-030216-025923

Sikkandar S, Murugan K, Al-Sohaibani S, Rayappan F, Nair°, A. (2013) Halophilic bacteria-A potent source of carotenoids. J Pure Appl Microbiol 7(4):2825–2830

Silbir S, Goksungur Y (2019) Natural red pigment production by *Monascus purpureus* in submerged fermentation systems using a food industry waste: Brewer's spent grain. Foods 8(5):161. https://doi.org/10.3390/foods8050161

Sivaperumal P, Kamala K, Rajaram R (2015) Bioactive DOPA melanin isolated and characterised from a marine actinobacterium *Streptomyces* sp. MVCS6 from Versova coast. Nat Prod Res 29(22):2117–2121. https://doi.org/10.1080/14786419.2014.988712

Smedsgaard J, Frisvad JC (1996) Using direct electrospray mass spectrometry in taxonomy and secondary metabolite profiling of crude fungal extracts. J Microbiol Methods 25(1):5–17. https://doi.org/10.1016/0167-7012(95)00073-9

Soliev AB, Enomoto K (2013) Antitumor pigments from marine bacteria. In: Kim S (ed) Marine biomaterials: characterization, isolation and applications, pp 149–171

Srilekha V, Krishna G, Seshasrinivas V, Singaracharya MA (2018) Evaluation of wound healing and anti-inflammatory activity of a marine yellow pigmented bacterium, *Micrococcus* sp. Indian J Geo-Mar Sci 47(12):2454–2464. http://nopr.niscair.res.in/handle/123456789/45437

Stahmann KP, Revuelta JL, Seulberger H (2000) Three biotechnical processes using Ashbya gossypii, Candida famata, or Bacillus subtilis compete with chemical riboflavin production. Appl Microbiol Biotechnol 53(5):509–516. https://doi.org/10.1007/S002530051649

Stankovic N, Radulovic V, Petkovic M, Vuckovic I, Jadranin M, Vasiljevic B, Nikodinovic-Runic J (2012) *Streptomyces* sp. JS520 produces exceptionally high quantities of undecylprodigiosin with antibacterial, antioxidative, and UV-protective properties. Appl Microbiol Biotechnol 96(5):1217–1231. https://doi.org/10.1007/S00253-012-4237-3

Stevenson CS, Capper EA, Roshak AK (2002) Scytonemin— a marine natural product inhibitor of kinases key in hyperproliferative inflammatory diseases. Inflamm Res 51(2):112–114

Subhasree RS, Dinesh Babu P, Vidyalakshmi R, Mohan VC (2011) Effect of carbon and nitrogen sources on stimulation of pigment production by *Monascus purpureus* on jackfruit seeds. Int J Microbiol Res 2(2):184–187. https://www.cabdirect.org/cabdirect/abstract/20123370325

Suryawanshi RK, Patil CD, Koli SH, Hallsworth JE, Patil SV (2017) Antimicrobial activity of prodigiosin is attributable to plasma-membrane damage. Nat Prod Res 31(5):572–577. https://doi.org/10.1080/14786419.2016.1195380

Tarangini K, Mishra S (2014) Production of melanin by soil microbial isolate on fruit waste extract: two step optimization of key parameters. Biotechnol Rep 4(1):139–146. https://doi.org/10.1016/j.btre.2014.10.001

Teo ITN, Chui CH, Tang JCO, Lau FY, Cheng GYM, Wong RSM, Kok SHL, Cheng CH, Chan ASC, Ho KP (2005) Antiproliferation and induction of cell death of Phaffia rhodozyma (*Xanthophyllomyces dendrorhous*) extract fermented by brewer malt waste on breast cancer cells. Int J Mol Med 16(5):931–936. https://doi.org/10.3892/ijmm.16.5.931

Terán Hilares R, de Souza RA, Marcelino PF, da Silva SS, Dragone G, Mussatto SI, Santos JC (2018) Sugarcane bagasse hydrolysate as a potential feedstock for red pigment production by *Monascus ruber*. Food Chem 245:786–791. https://doi.org/10.1016/j.foodchem.2017.11.111

Tuli HS, Chaudhary P, Beniwal V, Sharma AK (2014) Microbial pigments as natural color sources: current trends and future perspectives. J Food Sci Technol 52(8):4669–4678. https://doi.org/10.1007/S13197-014-1601-6

Usmani Z, Sharma M, Sudheer S, Gupta VK, Bhat R (2020) Engineered microbes for pigment production using waste biomass. Curr Genomics 21(2):80–95. https://doi.org/10.2174/1389202921999200330152007

Vasanthabharathi V, Lakshminarayanan R, Jayalakshmi S (2011) Melanin production from marine *Streptomyces*. Afr J Biotechnol 10(54):11224–11234. https://doi.org/10.5897/AJB11.296

Venil CK, Zakaria ZA, Ahmad WA (2013) Bacterial pigments and their applications. Process Biochem 48(7):1065–1079. https://doi.org/10.1016/J.PROCBIO.2013.06.006

Venil CK, Aruldass CA, Dufossé L, Zakaria ZA, Ahmad WA (2014) Current perspective on bacterial pigments: emerging sustainable compounds with coloring and biological properties for the industry – an incisive evaluation. RSC Adv 4(74):39523–39529. https://doi.org/10.1039/C4RA06162D

Venil CK, Dufossé L, Renuka Devi P (2020) Bacterial pigments: sustainable compounds with market potential for pharma and food industry. Front Sustain Food Syst 4:100. https://doi.org/10.3389/FSUFS.2020.00100/FULL

Vinotha M, Prabhavathi P, Vijayaraghavan R, Kumar D (2019) Bacterial pigments and their application in textile industries using mordants. J Adv Sci Res 10(03 Suppl. 1):139–145

Vipin C, Ashwini P, Kavya A, Rekha P (2017) Overproduction of Pyocyanin in *Pseudomonas aeruginosa* by supplementation of pathway precursor shikimic acid and evaluation of its activity. Res J Pharm Technol 10(2):533. https://doi.org/10.5958/0974-360x.2017.00106.8

Wang X, Tao J, Wei D, Shen Y, Tong W (2004) Development of an adsorption procedure for the direct separation and purification of prodigiosin from culture broth. Biotechnol Appl Biochem 40(3):277–280. https://doi.org/10.1042/BA20030210

Wang Y, Nakajima A, Hosokawa K, Soliev AB, Osaka I, Arakawa R, Enomoto K (2012) Cytotoxic prodigiosin family pigments from *Pseudoalteromonas* sp. 1020R isolated from the Pacific coast of Japan. Biosci Biotechnol Biochem 76(6):1229–1232. https://doi.org/10.1271/bbb.110984

Wissgott U, Bortlik K (1996) Prospects for new natural food colorants. Trends Food Sci Technol 7(9):298–302. https://doi.org/10.1016/0924-2244(96)20007-X

World Health Organization. (2016). WHO expert consultation: available evidence for the future update of the WHO Global Air Quality Guidelines (AQGs): meeting report Bonn, Germany 29 September-1 October 2015 (No. WHO/EURO: 2016-4105-43864-61762). World Health Organization. Regional Office for Europe

Yamamoto C, Takemoto H, Kuno K, Yamamoto D, Tsubura A, Kamata K, Hirata H, Yamamoto A, Kano H, Seki T, Inoue K (1999) Cycloprodigiosin hydrochloride, a new H+/cl- symporter, induces apoptosis in human and rat hepatocellular cancer cell lines in vitro and inhibits the growth of hepatocellular carcinoma xenografts in nude mice. Hepatology 30(4):894–902. https://doi.org/10.1002/HEP.510300417

Yamashita M, Nakagawa Y, Li H, Matsuyama T (2001) Silica gel-dependent production of Prodigiosin and Serrawettins by *Serratia marcescens* in a liquid culture. Microbes Environ 16(4):250–254. https://doi.org/10.1264/JSME2.2001.250

Yang Y, Di Zeng G, Zhang Y, Xue R, Hu YJ (2019) Molecular and biochemical characterization of carbendazim-resistant *Botryodiplodia theobromae* field isolates. Plant Dis 103(8):2076–2082. https://doi.org/10.1094/PDIS-01-19-0148-RE

Zakynthinos G, Varzakas T (2016) Carotenoids: from plants to food industry. Curr Res Nutr Food Sci 4(1):38–51

Zhao J, Wu Y, Alfred AT, Wei P, Yang S (2014) Anticancer effects of pyocyanin on H ep G 2 human hepatoma cells. Lett Appl Microbiol 58(6):541–548

Zhao Y, Liu D, Huang W, Yang Y, Ji M, Nghiem LD et al (2019) Insights into biofilm carriers for biological wastewater treatment processes: current state-of-the-art, challenges, and opportunities. Bioresour Technol 288:121619. https://doi.org/10.1016/j.biortech.2019.121619

Zhu T, Liu S, Ren K, Chen J, Lin J, Li J (2018) Colorability of dyed wood veneer using natural dye extracted from *Dalbergia cochinchinensis* with different organic solvents. Bioresources 13(4): 7197–7211

Chapter 13
Microbial Production of Polyhydroxyalkanoate (PHA)

Pooja Pawar, Anupama Shrivastav, and Vijay Jagdish Upadhye

Abstract Over the last few decades, tremendous progress has been made in the field of environmentally friendly polymer research. One of the most well-known families of such biopolymers is bacterially generated polyhydroxyalkanoates (PHAs), which have been known since the 1920s. However, it is only during the 1990s that significant research has been conducted in this area. Since then, various areas of investigation into these fascinating polymers have been discovered. However, no thorough review of the attempts undertaken thus far has been carried out. As a result, we conducted an unbiased search of current literature to uncover a trend. Microbial biopolyesters known as polyhydroxyalkanoates (PHA) are used as "green plastics". In terms of production, implementing and using an integrated system to separate their by-products (intracellular and extracellular) could be cost-effective. We focused in this review on a variety of microorganisms that exist in different habitats and utilise carbon as polyhydroxyalkanoate granule, as well as the variables that govern their production and composition.

Keywords Polyhydroxyalkanoate granules · Microbial biopolyesters · Biodegradable products · PHA producers

---

P. Pawar
Parul Institute of Applied Sciences, Parul University, Vadodara, India

A. Shrivastav (✉)
Parul Institute of Applied Sciences, Parul University, Vadodara, India

Faculty of Life Health and Allied Sciences, ITM Vocational University, Vadodara, India
e-mail: anupamam@itmvu.in

V. J. Upadhye
Center of Research for Development (CR4D), Parul Institute of Applied Sciences (PIAS), Parul University (DSIR-SIRO Recognized), Vadodara, Gujarat, India

A. Sarkar, I. A. Ahmed (eds.), *Microbial products for future industrialization*, Interdisciplinary Biotechnological Advances,
https://doi.org/10.1007/978-981-99-1737-2_13

13.1 Introduction

Since their discovery in 1920, polyhydroxyalkanoates (PHAs) have been studied. In the early 1990s, there was an exponential increase in scientific publications (Guzik et al. 2020). Since then, more and more scientific papers have appeared each year, expanding our understanding of the field of biopolymers. A comprehensive literature review is a crucial tool for learning about a topic (Guzik et al. 2020). Bioplastics have received a lot of attention in recent years as consumer views towards biodegradable products have altered around the world (Surendran et al. 2020).

Bioplastics have substantial advantages over regular, petrochemical plastics due to their inherent biodegradability, sustainability and ecologically beneficial qualities (petro-plastics) (Heng et al. 2016; Sudesh et al. 2011). These fascinating biopolymers could be a more environmentally friendly alternative to petro-plastics, helping to conserve valuable fossil fuel resources while lowering greenhouse gas emissions (Surendran et al. 2020). As a result, they are a key long-term development invention (Hassan et al. 2013). Sugarcane, polyhydroxyalkanoates (PHAs), polylactides, polyglycolic acids, aliphatic polyesters and polysaccharides, among other biomaterials, are used to make bioplastics (Lee 1996; Divya et al. 2013). Due to their biodegradable, thermoplastic and mechanical qualities, PHAs are one of the most researched classes of bioplastics that are projected to replace some of today's petroplastics (e.g. flexibility, adaptability, elasticity, etc.) (Nielsen et al. 2017; Steinbüchel 2001; Lee et al. 2008). The market opportunity for PHA is expected to reach over USD 98 million by 2024, according to the PHA Market Research Report (Market 2019). PHAs are a form of biodegradable carbon storage polymer that looks like polyesters and has mechanical properties similar to those of petrochemical-based polymers (Mumtaz et al. 2010). In reaction to a variety of stress factors (e.g. plentiful carbon or inadequate phosphate, sulphur, nitrogen or oxygen), diverse bacteria may produce these renewable polyesters, which protect from nutritional deprivation and hostile environments (Prieto et al. 2016). PHAs are made from more than 55 genera of Gram-negative and Gram-positive bacteria (Doi 1990; Ng and Sudesh 2016). Because of its capacity to thrive on many carbon substrates and be cultured to high cell density, *Ralstonia eutropha*, *Alcaligenes eutrophus* and *Hydrogenomonas eutrophus* have attracted the most attention among the various types of PHA-producing bacteria (Bhubalan et al. 2010; Devi-Nair et al. 2013). The properties and activities of PHA polymers, as well as their biodegradability, are influenced by over 155 verified distinct PHA monomer subunits (Nielsen et al. 2017; Surendran et al. 2020). PHA biosynthesis PHAs are polyesters with a linear chain made up of hydroxy acid (HA) monomers joined by an ester bond (Surendran et al. 2020).

This bond is formed when the carboxyl group of one monomer is joined to the hydroxyl group of another monomer nearby (Philip et al. 2007). A variety of microbiological organisms produce PHA in the environment. Bacteria, cyanobacteria and archaebacteria are all members of this category. PHAs are classified as either short-chain length (scl-PHAs) or medium-chain length (mcl-PHAs) based on the number of carbons in the monomers (Surendran et al. 2020). Scl-PHAs

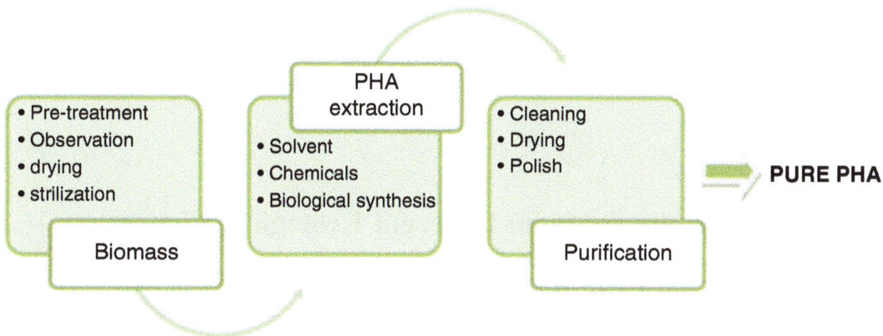

Fig. 13.1 Production of PHA

have 3–5 carbon atoms and are produced by a variety of bacteria (Surendran et al. 2020). *C. necator*, the model bacterium utilised in many PHA research, produces scl-PHAs. The most common type of scl-PHA generated is a homopolymer of poly (3-hydroxybutyrate) [P(3HB)], which is formed completely of 3-hydroxybutyrate (3HB) monomers. *Pseudomonas* spp. produce mcl-PHAs, which are monomers with 6–14 carbon atoms (Mozejko-Ciesielska and Kiewisz 2016).

Microorganisms may biosynthesise copolymers from a variety of substrates. PHA copolymers are made up of several different monomers (Surendran et al. 2020). Microorganisms can convert carbon sources into scl copolymers, such as poly (3-hydroxybutyrate-co-3-hydroxyvalerate) [P(3HB-co-3 HV)] or poly (3-hydroxybutyrate-co-4-hydroxybutyrate) [P(3HB-co-4HB)], and mcl copolymers, like poly(3-hydroxyhexanoate-co-3-hydroxyoctanoate) (Mozejko-Ciesielska and Kiewisz 2016) or poly(3-hydroxybutyrateco-3- hydroxyhexanoate) [P(3HB-co-3HHx)] which is containing both scl and mcl monomers, such as Mozejko-Ciesielska and Kiewisz (2016) and Surendran et al. (2020).

Scl-mcl-PHAs are particularly intriguing due to their biophysical features, which are similar to those of various common polymers. The enzyme PHA synthase is in charge of polymerising PHA from (R)-3-hydroxyacyl-CoA (Surendran et al. 2020). This synthase can be divided into classes I–IV based on the amino acid sequence, subunits and substrate preference (Surendran et al. 2020). Class I, III and IV PHA synthases polymerise scl monomers, but class II polymerises mcl monomers. The sole subunit (PhaC) of PHA synthases in classes I and II has a molecular weight of 61–73 kDa (Qi and Rehm 2001). The bacteria *C. necator* has class I PHA synthase, while *Pseudomonas putida* has class II PhaC (Surendran et al. 2020). To function properly, class III and IV synthases require two types of subunits. Class III PHA synthases have subunits PhaC and PhaE, while class IV PHA synthases have subunits PhaC and PhaR. PhaC and PhaE are class III synthases with molecular weights of 40.3 and 20–40 kDa, respectively (Surendran et al. 2020).

Class IV synthases PhaC and PhaR, on the other hand, have molecular weights of 41.5 and 22 kDa, respectively. *Bacillus megaterium* and *Allochromatium vinosum* are bacteria that make class III and IV synthases, respectively (Surendran et al.

2020). Only the crystal structures of scl-PHA synthases from *Chromobacterium* sp. USM2 and *Chromobacterium necator* H16 have been determined thus far (Kim et al. 2017; Wittenborn et al. 2016; Chek et al. 2017; Surendran et al. 2020) (Fig. 13.1).

13.2 PHA Producers from Different Ecological Niches

13.2.1 Photosynthetic Bacteria (PHB)

Cyanobacteria are photosynthetic prokaryotes that synthesise PHA by oxygenic photosynthesis and have a short generation time. According to studies, some cyanobacteria have the potential to store PHAs in their native state. PHA, a species-specific enzyme induced by phosphate deficiency and an excess of reducing equivalents, was studied in numerous *Cyanobacteria*. In phosphate-restricted conditions, *Synechococcus* sp. MA19 which accumulated up to 55% of CDW (cell dry weight), *Nostoc muscorum* and *Spirulina platensis* all formed PHB (Nishioka et al. 2001; Panda et al. 2005). PHA, a species-specific enzyme induced by phosphate deficiency and an excess of reducing equivalents, was studied in numerous Cyanobacteria. In phosphate-restricted conditions, *Synechococcus* sp. MA19 (which accumulated up to 55% of CDW), *Nostoc muscorum* and *Spirulina platensis* all formed PHB (Nishioka et al. 2001; Panda et al. 2005). Phosphate nitrogen and gas-exchange limiting conditions were introduced to *Synechocystis* sp. that had been pregrown in glucose PCC 6803 (Panda and Mallick 2007). Pregrown in glucose, PCC 6803 cells accumulated PHA up to 29% of CDW under phosphorus shortage conditions and supplemented with 0.4% acetate under phosphorus-deficient conditions (Melnicki et al. 2009; Panda et al. 2006). *Synechocystis* sp. UNIWG and *Synechocystis* sp. PCC 6803 accumulated PHB up to 14% and 15% of CDW, respectively, under nitrogen-limiting circumstances (Wu et al. 2002; Yew et al. 2005). When nitrogen, phosphorus and gas-exchange limiting conditions were applied to *Synechocystis* sp. PCC 6803, PHB synthesis increased (Panda and Mallick 2007). In sulphur-deficient conditions, PHB concentrations increased 3.5-fold (Melnicki et al. 2009). Under mixotrophy, chemoheterotrophy and nitrogen-limiting conditions, *Nostoc muscorum* may produce five times more PHB than under photo-autotrophic conditions, according to a study (Sharma and Mallick 2005). PHB accumulation in *Nostoc muscorum* was found to be influenced by external pH, carbon sources, light-dark cycles, nitrogen equilibrium, nitrogen and phosphorus levels (Sharma and Mallick 2005). Cyanobacteria's ability to manufacture PHB from sunlight-derived energy can save money while also reducing CO_2, a "greenhouse gas" (Saharan et al. 2014).

13.2.2 Plant Growth-Promoting Rhizobia (PGPR)

The rhizosphere is the soil near plant roots that may host bacteria that secrete extracellular chemicals that promote root and plant growth (Saharan et al. 2011). The rhizosphere, which acts as a hub for microbial interaction and nutrient recycling, receives nutrients through the roots' exudates (Rana et al. 2011). In addition to PGPR and antagonistic action, the rhizosphere has been discovered to harbour hidden or untapped reserves of PHA accumulators. PHA producers have been identified in *Burkholderia terricola*, *Pseudomonas brassicacearum*, *Lysobacter gummosus*, *Pseudomonas extremaustralis* and *Pseudomonas orientalis* utilising PCR methods with phaC as the target gene (Gasser et al. 2009).

A well-known, widely employed in commercial agriculture, plant growth-promoting rhizobacterium (PGPR) *Azospirillum brasilense* that contributes to its ability to adapt to the rhizosphere habitat and drive plant growth has been identified in a recent study (Saharan et al. 2014). Mutants of the phaC and phaZ genes were found to be less tolerant to stress conditions such as osmotic pressure, H_2O_2, UV radiation, heat, osmotic shock and desiccation when compared to wild-type *Azospirillum brasilense* (Kadouri et al. 2005, 2003). Increased root colonisation, plant growth promotion, survival, chemotaxis, motility and cell multiplication are all advantages of PHA synthesis. PHA production can also aid in the development of more dependable, efficient and long-lasting *A. brasilense* inoculants (Kadouri et al. 2005). It was previously discovered that rhizosphere soil produces less PHA than non-rhizosphere soil using cultivation-dependent approaches. Sugar beet, oilseed rape and wheat rhizospheres produce higher PHA, according to cultivation-independent and molecular approaches. The presence of considerable carbon and inorganic nutrients in root exudates despite their restricted status shows that roots of oily and carbohydrate-producing plants have more PHA producers (Gasser et al. 2009). PHB production was linked to famine survival in *Sinorhizobium meliloti*, which produced up to 0.25 pg cell of PHB to support growth and survival during the starving stage in alfalfa. Later, it was shown that PHB accumulation in rhizobia is a quantitative trait that is controlled by a variety of genes (Ratcliff et al. 2008). In *Azotobacter beijerinckii*, *Azotobacter insignia* and Rhizobium ORS571, a low oxygen environment and a lowered redox potential promote PHA formation in the nodule (Saharan and Badoni 2007). Rhizobia create nitrogen-fixing nodules in legumes, such as *Sinorhizobium*, *Bradyrhizobium*, *Rhizobium*, *Azorhizobium* and *Mesorhizobium*. PHA that has been stored helps with colonisation, nodulation and nitrogen fixation (Saharan et al. 2014). PHB is produced by both *Rhizobia* linked to roots that are actively participating in nodulation (rhizobia-legume symbiosis) and free-living nitrogen-fixing organisms. The type of nodule produced influences the ability to accumulate PHB, with determinate and indeterminate nodules differing in the presence of persistent meristems (Trainer and Charles 2006). Firmicutes and Proteobacteria were found. PHB had been reported to accumulate in the root nodules of eight leguminous plants (Kumbhakar et al. 2012; Saharan et al. 2014).

13.2.3 Hydrocarbon Degraders

Environmental stress, such as the presence of xenobiotic chemicals, causes resident organisms' physiological responses to be diverted to create more PHA. Oil-contaminated locations have a higher carbon concentration (84%) and a lower nitrogen content (>1%), posing a serious challenge for cells to generate PHA (Saharan et al. 2014). *Pseudomonas*, *Acinetobacter*, *Brochothrix*, *Sphingobacterium*, *Burkholderia*, *Caulobacter*, *Ralstonia* and *Yokenella* species capable of generating PHA while decomposing oil have been isolated from oil-contaminated areas (Dalal et al. 2010). When glucose is utilised as the carbon source, *Bacillus cereus* FA11, isolated from trinitrotoluene-contaminated soil (providing the appropriate stress conditions), has been found at pH 7 and 30^0 C; it can synthesise copolymer (3HB-co-3 HV) up to 48.43% (Masood et al. 2012). PHB was deposited extracellularly by a mutant of the marine bacterium *Alcanivorax borkumensis* SK2-fed aliphatic hydrocarbons in one study (Sabirova et al. 2006). Construction of a bioprocess that allows *Rhodococcus aetherivorans* IAR1 to manufacture copolymer P(3HB-co-HV) and triacylglycerols (TAGs) from waste containing toluene as the primary carbon source results in cost savings and effective waste treatment (Hori et al. 2009; Saharan et al. 2014).

13.2.4 Halophiles

Archaea are considered extremophiles because they have been found in hot springs, marshlands, oceans, salt lakes and other extreme environments, as well as producing PHA. Salts are required for the survival of these species. They grow best with 5% salt NaCl (w/v) and can tolerate up to 10% salt NaCl (w/v) (Oren 2008). *Halobacterium marismortui*, which was investigated using the freeze-fracture technique, was the first reported PHB buildup by archaea in the Dead Sea in 1970 (Saharan et al. 2014). When nutrients are scarce and carbon sources are plentiful, extremely halophilic archaebacteria (Halobacteriaceae) synthesise PHB. *Haloferax mediterranei* produces 60–65% PHA of its cell dry weight when grown in phosphate-restricted circumstances with glucose or sucrose as the optimal carbon source and a salt concentration of 25% (w/v) (CDW). *Halomonas boliviensis* LC1, a moderate halophile growing at 3–15% (w/v) salt concentration, produces greater PHB, up to 56% of CDW, when grown on starch hydrolysate as substrate. Pretreatment converts more complex forms of organic molecules into more easily digestible forms (simpler forms). Maltose from starch hydrolysates is preferred by *H. boliviensis* LC1. It has been discovered that limiting oxygen boosts PHA synthesis (Quillaguaman et al. 2005). In the presence of ample butyric acid and sodium acetate as carbon sources, *H. boliviensis* LC1 could synthesise considerable amounts of PHB, up to 88% of CDW, under nutrient-limited conditions where 0.1% w/v yeast was extracted during the stationary phase (Quillaguaman et al. 2006). When grown in nutrient-restricted environments and with ample carbon, *Halopiger aswanensis* (strain 56T) cells produced significant amounts of polyhydroxybutyrate (Hezayen et al. 2001). Very importantly, the Halophilic archaeon *Haloarcula marismortui* has been shown to produce PHA under shaking flask conditions utilising vinasses (a by-product of the ethanol industry) as a substrate. PHB production was up to 24% and 30% of CDW when 10% raw vinasses and 100% pretreated vinasses were used, respectively (Pramanik et al. 2012; Saharan et al. 2014).

13.2.5 Producers of Antibiotics

Streptomyces is a filamentous Gram-positive bacteria that thrives in an aerobic environment and produces useful compounds. PHA is a granular form of PHA that is synthesised intracellularly in *Streptomyces*, and carbon units are needed for antibiotic production and sporulation (Saharan et al. 2014). PHB synthesis in *Streptomyces griseorubiginosus* was enhanced by nitrogen supplementation and a greater carbon source, with up to 9.5% of mycelial dry mass produced during the early stationary phase (Saharan et al. 2014). PHB production in a variety of *Streptomyces* strains has been investigated for its potential significance in antibiotic synthesis. PHB was discovered to be produced during the exponential phase of *Streptomyces coelicolor* A3 (2) M145, and acetyl-CoA was used as a precursor for the manufacture of three antibiotics during the stationary phase: actinorhodin, Y-actinorhodin and undecylprodigiosin (Verma et al. 2003). For the manufacture of actinomycin A, *Streptomyces antibioticus* utilised stored PHB. In *Streptomyces venezuelae* and *Streptomyces hygroscopicus*, PHB was discovered to be a source of acetoacetyl-Co-A and butyryl-Co-A, which are used in the production of chloramphenicol and macrolide ascomycin FK520, respectively (Ranada and Vining 1993; Wu et al. 2000). PHB is a main metabolite that acts as a building block for the creation of other metabolites (Valappil et al. 2007). *Streptomyces aureofaciens* 84/25, *S. fradiae*, *S. griseus*, *S. olivaceous*, *S. parvus*, *S. albus* and other bacteria that use glucose as a carbon source have been found to make PHB (Verma et al. 2003; Saharan et al. 2014).

13.2.6 Activated Sludge

From various industries, a constant influx of industrial wastes is a source of concern for the environment in terms of dangerous pollutant elements. Organic content is generally high in waste, while nutritional content is low (Saharan et al. 2014). This unbalanced nutritional state promotes PHA production but has a low rate of organic matter (substrate) consumption and storage (Beccari et al. 2002; Dircks et al. 2001). After being subjected to various treatments such as enzymatic treatment, heat treatment, anaerobic treatment, mechanical treatment and so on, pretreated trash of various types was employed to solve these concerns. Simpler wastes are more easily absorbed and retained by cells, such as volatile fatty acids (VFAs) and sugar moieties (Albuquerque et al. 2007; Beccari et al. 2009; Bengtsson et al. 2008). Some investigations used a three-stage process, starting with waste treatment, enriching activated sludges and finally batch manufacturing of PHA with treated waste as substrate and enriched activated sludge as inocula (Saharan et al. 2014). AN/AE and ADF can enrich activated sludge in SBR (sequencing batch reactor) (Serafim et al. 2008). The composition of VFAs has an impact on the polymer's composition and microstructure. As a result, it is critical to think about it first. In an aerobic microenvironment, the consortium accumulated more PHA near about 39.6% when fermented food waste was used as a substrate than when unfermented food waste was used; PHA accumulation was 35.6%. Unlike unfermented food waste, which includes complex organic substrates that must be degraded before being used, fermented food waste contains VFA that is ready to absorb (Reddy and Mohan 2012; Saharan et al. 2014). Anaerobic and aerobic systems collaborate. Concurrent PHA synthesis can be achieved by using enriched activated sludge and waste treatment that alternates aerobic and anaerobic conditions, such as improved biological phosphate removal (EBPR) (Saharan et al. 2014). Wallen and Rohwedder were the first to report PAOs (phosphorus accumulating organisms) synthesising PHA (Wallen and Rohwedder 1974). PAOs which are also known as phosphorus-accumulating organisms use energy from polyphosphate degradation to accumulate PHA in anaerobic conditions (phosphorus release phase) and then replenish polyphosphate and use the stored form to maintain normal metabolic activity (PHA) (van Loosdrecht et al. 1997; Santos et al. 1999). Phosphate thrown into the stream might be used to fertilise farm crops (le Corre et al. 2009) and then to restore polyphosphate and maintain normal metabolic activity and use the stored form (PHA) (van Loosdrecht et al. 1997, Santos et al. 1999). Phosphate that has been discharged into a stream could be used to fertilise crops in the field (le Corre et al. 2009). *Candidatus*, *Competibacter phosphatis* and *Defluviicoccus vanus* are examples of glycogen-accumulating organisms (GAOs) found in SBR under AN/AE conditions (Wong et al. 2004; Bengtsson et al. 2010; Meyer et al. 2006; Bengtsson 2009). GAOs absorb volatile fatty acids using glycogen as energy using propionyl-CoA and acetyl-CoA, while synthesising PHA. Glycogen is transformed to PHA as well. PHA accumulated in the cell is required to sustain proper cell activity and repair glycogen under aerobic conditions. Stored glycogen is utilised in both anaerobic and aerobic circumstances, depending on the stoichiometry. When anaerobic and aerobic circumstances were alternated, 3HB and 3 HV produced up to 49% of CDW, but only 3HB produced up to 60% of CDW in aerobic conditions, indicating that GAOs followed different metabolic routes (Bengtsson 2009). GAOs may make PHA from fermented cane molasses as well as synthetic carbon sources like acetate and propionate (Bengtsson et al. 2010; Dai et al. 2007a, 2008, 2007b; Pisco et al. 2009; Saharan et al. 2014) Fig. 13.2.

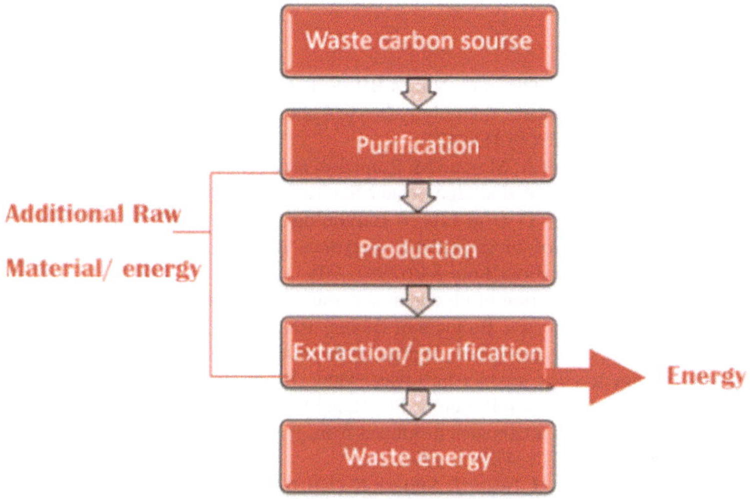

Fig. 13.2 PHA production through waste carbon source

13.2.7 *Conclusion*

To solve the environmental difficulties generated by the manufacturing and building of traditional plastics, bioplastics must be generated at a low cost and with minimal environmental impact. The search for potential PHA-producing strains is still on, and the creation of efficient polyhydroxyalkanoate-generating bacteria is urgently needed. PHA is a biodegradable microbe-stored material that can be used instead of petroleum-based polymers. A technique must be carefully studied while constructing an integrated system minimising production costs for the separation of high-valued microbial-produced products. When there is an abundance of carbon in the environment, some microorganisms are said to create PHA intracellularly while also producing other metabolites. These vital biologically produced by-products necessitate the rapid construction of industrial processes that will contribute to high-cost production. The emergence of study into the simultaneous creation of polymeric compounds (both intracellularly and extracellularly) offers up new avenues for understanding metabolic connections and ecological connotations like diversity, evolution and role. By having these high-valued endopolymers and exopolymers manufactured simultaneously using the same organisms under optimal conditions using the household, agricultural, industrial or industrial effluent waste, we can address environmental pollution, manufacturing costs and market commercialisation. As a result, we can argue if garbage is truly "waste" in terms of the development of high-value items from the bacteria that live in it.

References

Albuquerque MGE, Eiroa M, Torres C, Nunes BR, Reis MAM (2007) Strategies for the development of a side stream process for polyhydroxyalkanoate (PHA) production from sugar cane molasses. J Biotechnol 130(4):411–421

Beccari M, Dionisi D, Giuliani A, Majone M, Ramadori R (2002) Effect of different carbon sources on aerobic storage by activated sludge. Water Sci Technol 45(6):157–168

Beccari M, Bertin L, Dionisi D et al (2009) Exploiting olive oil mill effluents as a renewable resource for production of biodegradable polymers through a combined anaerobic-aerobic process. J Chem Technol Biotechnol 84(6):901–908

Bengtsson S (2009) The utilization of glycogen accumulating organisms for mixed culture production of polyhydroxyalkanoates. Biotechnol Bioeng 104(4):698–708

Bengtsson S, Werker A, Christensson M, Welander T (2008) Production of polyhydroxyalkanoates by activated sludge treating a paper mill wastewater. Bioresour Technol 99(3):509–516

Bengtsson S, Pisco AR, Johansson P, Lemos PC, Reis MAM (2010) Molecular weight and thermal properties of polyhydroxyalkanoates produced from fermented sugar molasses by open mixed cultures. J Biotechnol 147(3–4):172–179

Bhubalan K, Rathi D-N, Abe H, Iwata T, Sudesh K (2010) Improved synthesis of P(3HB-co-3HV-co-3HHx) terpolymers by mutant Cupriavidus necator using the PHA synthase gene of Chromobacterium sp. USM2 with high affinity towards 3HV. Polym Degrad Stab 95:1436–1442. https://doi.org/10.1016/j.polymdegradstab.2009.12.018

Chek MF, Kim S-Y, Mori T, Arsad H, Samian MR, Sudesh K et al (2017) Structure of polyhydroxyalkanoate (PHA) synthase PhaC from Chromobacterium sp. USM2, producing biodegradable plastics. Sci Rep 7:5312. https://doi.org/10.1038/s41598-017-05509-4

Dai Y, Yuan Z, Wang X, Oehmen A, Keller J (2007a) Anaerobic metabolism of Defluviicoccus vanus related glycogen accumulating organisms (GAOs) with acetate and propionate as carbon sources. Water Res 41(9):1885–1896

Dai Y, Yuan Z, Jack K, Keller J (2007b) Production of targeted poly(3-hydroxyalkanoates) copolymers by glycogen accumulating organisms using acetate as sole carbon source. J Biotechnol 129(3):489–497

Dai Y, Lambert L, Yuan Z, Keller J (2008) Microstructure of copolymers of polyhydroxyalkanoates produced by glycogen accumulating organisms with acetate as the sole carbon source. Process Biochem 43(9):968–977

Dalal J, Sarma PM, Lavania M, Mandal AK, Lal B (2010) Evaluation of bacterial strains isolated from oil-contaminated soil for production of polyhydroxyalkanoic acids (PHA). Pedobiologia 54(1):25–30

Devi-Nair R, Jutemar EP, Maurer FHJ, Sudesh K (2013) Biosynthesis of P(3HB-co-3HV-co-3HHp) terpolymer by *Cupriavidus necator* PHB__4 transformant harboring the highly active PHA synthase gene of Chromobacterium sp. USM2. Malays J Microbiol 9:140–146

Dircks K, Henze M, van Loosdrecht MCM, Mosbæk H, Aspegren H (2001) Storage and degradation of poly-β-hydroxybutyrate in activated sludge under aerobic conditions. Water Res 35(9): 2277–2285

Divya G, Archana T, Manzano R (2013) Polyhydroxyalkanoates—a sustainable alternative to petro-based plastics. J Pet Environ Biotechnol 4:2. https://doi.org/10.4172/2157-7463.1000143

Doi Y (1990) Microbial polyesters. Wiley VCH, New York, NY

Gasser I, Muller H, Berg G (2009) Ecology and characterization ¨ of polyhydroxyalkanoate-producing microorganisms on and in plants. FEMS Microbiol Ecol 70(1):142–150

Guzik M, Witko T, Steinbüchel A, Wojnarowska M, Sołtysik M, Wawak S (2020) What has been trending in the research of Polyhydroxyalkanoates? A Systematic Review. Front Bioeng Biotechnol 8(959):1–15

Hassan MA, Yee L-N, Yee PL, Ariffin H, Raha AR, Shirai Y et al (2013) Sustainable production of polyhydroxyalkanoates from renewable oil-palm biomass. Biomass Bioenergy 50:1–9. https://doi.org/10.1016/j.biombioe.2012.10.014

Heng K-S, Ong SY, Sudesh K (2016) Efficient biosynthesis and recovery of polyhydroxyalkanoate. Malays J Microbiol 12:383–398

Hori K, Abe M, Unno H (2009) Production of triacylglycerol and poly(3-hydroxybutyrate-co-3-hydroxyvalerate) by the toluenedegrading bacterium *Rhodococcus aetherivorans* IAR1. J Biosci Bioeng 108(4):319–324

F. F. Hezayen, M. C. Gutierrez, A. Steinbuchel, B. J. Tindall, ¨ and B. H. A. Rehm, "*Halopiger aswanensis* sp. nov., a polymer producing and extremely halophilic archaeon isolated from hypersaline soil," Int J Syst Evol Microbiol, vol. 60, no. 3, pp. 633–637, 2001

Kadouri D, Jurkevitch E, Okon Y (2003) Involvement of the reserve material poly-β-hydroxybutyrate in Azospirillum brasilense stress endurance and root colonization. Appl Environ Microbiol 69(6):3244–3250

Kadouri D, Castro-Sowinski S, Jurkevitch E, Okon Y (2005) Ecological and agricultural significance of bacterial polyhydroxyalkanoates. Crit Rev Microbiol 31(2):55–67

Kim J, Kim Y-J, Choi SY, Lee SY, Kim K-J (2017) Crystal structure of Ralstonia eutropha polyhydroxyalkanoate synthase C-terminal domain and reaction mechanisms. Biotechnol J 12:1600648. https://doi.org/10.1002/biot.201600648

Kumbhakar S, Singh PK, Vidyarthi AS (2012) Screening of root nodule bacteria for the production of polyhydroxyalkanoate (PHA) and the study of parameters influencing the PHA accumulation. Afr J Biotechnol 11(31):7934–7946

le Corre KS, Valsami-Jones E, Hobbs P, Parsons SA (2009) Phosphorus recovery from wastewater by struvite crystallization: a review. Crit Rev Environ Sci Technol 39(6):433–477

Lee SY (1996) Bacterial polyhydroxyalkanoates. Biotechnol Bioeng 49:1–14

Lee W-H, Loo C-Y, Nomura CT, Sudesh K (2008) Biosynthesis of polyhydroxyalkanoate copolymers from mixtures of plant oils and 3-hydroxyvalerate precursors. Bioresour Technol 99: 6844–6851. https://doi.org/10.1016/j.biortech.2008.01.051

Market MA (2019) Polyhydroxyalkanoate (PHA) Market by type (short chain length, medium chain length). In: Production method (sugar fermentation, vegetable oil fermentation, methane fermentation), application, and region—global forecast to 2024 [online]. Markets and Markets Research Private Ltd.

Masood F, Hasan F, Ahmed S, Hameed A (2012) Biosynthesis and characterization of poly (3-hydroxybutyrate-co-3- hydroxyvalerate) from Bacillus cereus FA11 isolated from TNT contaminated soil. Ann Microbiol 62(4):1377–1384

Melnicki MR, Eroglu E, Melis A (2009) Changes in hydrogen production and polymer accumulation upon sulfur-deprivation in purple photosynthetic bacteria. Int J Hydrog Energy 34(15): 6157–6170

Meyer RL, Saunders AM, Blackall LL (2006) Putative glycogen-accumulating organisms belonging to the Alphaproteobacteria identified through rRNA-based stable isotope probing. Microbiology 152(2):419–429

Mozejko-Ciesielska J, Kiewisz R (2016) Bacterial polyhydroxyalkanoates: still fabulous? Microbiol Res 192:271–282. https://doi.org/10.1016/j.micres.2016.07.010

Mumtaz T, Yahaya NA, Abd-Aziz S, Yee PL, Shirai Y, Hassan MA (2010) Turning waste to wealth-biodegradable plastics polyhydroxyalkanoates from palm oil mill effluent–a Malaysian perspective. J Clean Prod 18:1393–1402. https://doi.org/10.1016/j.jclepro.2010.05.016

Ng L-M, Sudesh K (2016) Identification of a new polyhydroxyalkanoate (PHA) producer Aquitalea sp. USM4 (JCM 19919) and characterization of its PHA synthase. J Biosci Bioeng 122:550–557. https://doi.org/10.1016/j.jbiosc.2016.03.024

Nielsen C, Rahman A, Rehman AU, Walsh MK, Miller CD (2017) Food waste conversion to microbial polyhydroxyalkanoates. Microb Biotechnol 10:1338–1352. https://doi.org/10.1111/1751-7915.12776

Nishioka M, Nakai K, Miyake M, Asada Y, Taya M (2001) Production of poly-β-hydroxybutyrate by thermophilic *cyanobacterium, Synechococcus* sp. MA19, under phosphate-limited conditions. Biotechnol Lett 23(14):1095–1099

Oren A (2008) Microbial life at high salt concentrations: phylogenetic and metabolic diversity. Saline Systems 4:2
Panda B, Mallick N (2007) Enhanced poly-β-hydroxybutyrate accumulation in a unicellular cyanobacterium, Synechocystis sp. PCC 6803. Lett Appl Microbiol 44(2):194–198
Panda B, Sharma L, Mallick N (2005) Poly-β-hydroxybutyrate accumulation in Nostoc muscorum and *Spirulina platensis* under phosphate limitation. J Plant Physiol 162(12):1376–1379
Panda B, Jain P, Sharma L, Mallick N (2006) Optimization of cultural and nutritional conditions for accumulation of poly-β-hydroxybutyrate in Synechocystis sp. PCC 6803. Bioresour Technol 97(11):1296–1301
Philip S, Keshavarz T, Roy I (2007) Polyhydroxyalkanoates: biodegradable polymers with a range of applications. J Chem Technol Biotechnol 82:233–247. https://doi.org/10.1002/jctb.1667
Pisco AR, Bengtsson S, Werker A, Reis MAM, Lemos PC (2009) Community structure evolution and enrichment of glycogen-accumulating organisms producing polyhydroxyalkanoates from fermented molasses. Appl Environ Microbiol 75(14):4676–4686
Pramanik A, Mitra A, Arumugam M et al (2012) Utilization of vinasse for the production of polyhydroxybutyrate by *Haloarcula marismortui*. Folia Microbiol 57(1):71–79
Prieto A, Escapa IF, Martínez V, Dinjaski N, Herencias C, de la Peña F et al (2016) A holistic view of polyhydroxyalkanoate metabolism in pseudomonas putida. Environ Microbiol 18:341–357. https://doi.org/10.1111/1462-2920.12760
Qi Q, Rehm BHA (2001) Polyhydroxybutyrate biosynthesis in Caulobacter crescentus: molecular characterization of the polyhydroxybutyrate synthase. Microbiology 147:3353–3358. https://doi.org/10.1099/00221287-147-12-3353
Quillaguaman J, Hashim S, Bento F, Mattiasson B, Hatti-Kaul R (2005) Poly(β-hydroxybutyrate) production by a moderate halophile, *Halomonas boliviensis* LC1 using starch hydrolysate as substrate. J Appl Microbiol 99(1):151–157
Quillaguaman J, Delgado O, Mattiasson B, Hatti-Kaul R (2006) Poly(β-hydroxybutyrate) production by a moderate halophile, *Halomonas boliviensis* LC1. Enzym Microb Technol 38(1–2): 148–154
Rana A, Saharan B, Joshi M, Prasanna R, Kumar K, Nain L (2011) Identification of multi-trait PGPR isolates and evaluating their potential as inoculants for wheat. Ann Microbiol 61(4): 893–900
Ranada N, Vining LC (1993) Accumulation of intracellular carbon reserves in relation to chloramphenicol biosynthesis by Streptomyces venezuelae. Can J Microbiol 39(4):377–383
Ratcliff WC, Kadam SV, Denison RF (2008) Poly-3- hydroxybutyrate (PHB) supports survival and reproduction in starving rhizobia. FEMS Microbiol Ecol 65(3):391–399
Reddy MV, Mohan SV (2012) Influence of aerobic and anoxic microenvironments on polyhydroxyalkanoates (PHA) production from food waste and acidogenic effluents using aerobic consortia. Bioresour Technol 103(1):313–321
Sabirova JS, Ferrer M, Lunsdorf H et al (2006) Mutation in a ¨ "tesB-like" hydroxyacyl-coenzyme A-specific thioesterase gene causes hyperproduction of extracellular polyhydroxyalkanoates by Alcanivorax borkumensis SK2. J Bacteriol 188(24):8452–8459
Saharan BS, Badoni P (2007) Poly-β-hydroxy production using Azotobacter species from contaminated sites. Environ Ecol 25(3):737–740
Saharan BS, Sahu RK, Sharma D (2011) A review on biosurfactants: fermentation, current developments and perspectives. Genet Eng Biotechnol J 2011:1–4
Saharan BS, Grewal A, Kumar P (2014) Biotechnological production of polyhydroxyalkanoates: a review on trends and latest developments. Chinese J Biol 2014:1–19
Santos MM, Lemos PC, Reis MAM, Santos H (1999) Glucose metabolism and kinetics of phosphorus removal by the fermentative bacterium Microlunatus phosphovorus. Appl Environ Microbiol 65(9):3920–3928
Serafim LS, Lemos PC, Albuquerque MGE, Reis MAM (2008) Strategies for PHA production by mixed cultures and renewable waste materials. Appl Microbiol Biotechnol 81(4):615–628

Sharma L, Mallick N (2005) Enhancement of poly-β-hydroxybutyrate accumulation in Nostoc muscorum under mixotrophy, chemoheterotrophy and limitations of gas exchange. Biotechnol Lett 27(1):59–62

Steinbüchel A (2001) Perspectives for biotechnological production and utilization of biopolymers: metabolic engineering of polyhydroxyalkanoate biosynthesis pathways as a successful example. Macromol Biosci 1:1–24. https://doi.org/10.1002/1616-5195(200101)1:13.0.CO;2-B

Sudesh K, Bhubalan K, Chuah J-A, Kek Y-K, Kamilah H, Sridewi N et al (2011) Synthesis of polyhydroxyalkanoate from palm oil and some new applications. Appl Microbiol Biotechnol 89:1373–1386. https://doi.org/10.1007/s00253-011-3098-5

Surendran A, Lakshmanan M, Chee JY, Sulaiman AM, Thuoc DV, Sudesh K (2020) Can polyhydroxyalkanoates be produced efficiently from waste plant and animal oils? Front Bioeng Biotechnol 8(169):1–15

Trainer MA, Charles TC (2006) The role of PHB metabolism in the symbiosis of rhizobia with legumes. Appl Microbiol Biotechnol 71(4):377–386

Valappil SP, Boccaccini AR, Bucke C, Roy I (2007) Polyhydroxyalkanoates in Gram-positive bacteria: insights from the genera Bacillus and Streptomyces. Antonie Van Leeuwenhoek 91(1): 1–17

van Loosdrecht MCM, Hooijmans CM, Brdjanovic D, Heijnen JJ (1997) Biological phosphate removal processes. Appl Microbiol Biotechnol 48(3):289–296

Verma S, Bhatia Y, Valappil SP, Roy I (2003) A possible role of poly-3-hydroxybutyric acid in antibiotic production in Streptomyces. Arch Microbiol 179(1):66–69

Wallen LL, Rohwedder WK (1974) Poly-β-hydroxyalkanoate from activated sludge. Environ Sci Technol 8(6):576–579

Wittenborn EC, Jost M, Wei Y, Stubbe J, Drennan CL (2016) Structure of the catalytic domain of the class I polyhydroxybutyrate synthase from *Cupriavidus necator*. J Biol Chem 291:25264–25277. https://doi.org/10.1074/jbc.M116.756833

Wong M-T, Tan FM, Ng WJ, Liu W-T (2004) Identification and occurrence of tetrad-forming Alphaproteobacteria in anaerobic-aerobic activated sludge processes. Microbiology 150(11): 3741–3748

Wu K, Chung L, Revill WP, Katz L, Reeves CD (2000) The FK520 gene cluster of Streptomyces hygroscopicus var. ascomyceticus (ATCC 14891) contains genes for biosynthesis of unusual polyketide extender units. Gene 251(1):81–90

Wu G, Boa T, Shen Z, Wu Q (2002) Sodium acetate stimulates PHB biosynthesis in Synechocystis sp. PCC, 6803. Tsinghua Sci Technol 7(4):435–438

Yew SP, Sau MH, Yong KH, Abed RMM, Sudesh K (2005) Morphological studies of Synechocystis sp. UNIWG under polyhydroxyalkanoate accumulating conditions. Malays J Microbiol 1(1):48–52

Chapter 14
Organic Acid and Solvent Production from Microbial Fermentation

Yasmeen Shaikh and Mayuri R. Jagtap

Abstract Organic acids are broadly distributed in nature, and people have used them in their natural forms since ancient times. Organic acids are the intermediates or products of many metabolic pathways and serve as basic compounds in the chemical industry for a wide range of polymer and solvent production processes. They account for a large segment of the global fermentation market, and for many of them, microbiological production is a sustainable economic alternative to chemical synthesis. Fermentation is defined as the chemical changes that occur in an organic substrate as a result of the action of microbial enzymes. This feature of microorganisms is being used in the fermentation of various organic acids. Growing environmental concerns and depletion of limited resources have led to the development of various microbial approaches that have transformed many fermentation processes by optimizing fermentation processes and improving recovery and purification methods. Microbial patented acids include citric acid (CA), itaconic acid, lactic acid, malic acid, tartaric acid, gluconic acid, mevalonic acid, salicylic acid, gibberellic acid and propionic acid. This chapter focuses on the production and application of citric acid, acetic acid and lactic acid, which are currently manufactured by fermentation in large quantities and offer potential for future developments.

Keywords Organic acids · Fermentation · Microorganisms · Biochemical changes · Citric acid · Lactic acid · Acetic acid

14.1 Introduction

Citric acid (C6H8O7, 2-hydroxy-1,2,3-propane tricarboxylic acid), the most significant organic acid produced in nature and widely employed in the food and pharmaceutical industries, is a natural component and common metabolite of plants and

Y. Shaikh · M. R. Jagtap (✉)
Department of Zoology, Dr. Rafiq Zakaria College for Women, Aurangabad, India

A. Sarkar, I. A. Ahmed (eds.), *Microbial products for future industrialization*, Interdisciplinary Biotechnological Advances,
https://doi.org/10.1007/978-981-99-1737-2_14

animals. They are obtained as the end-products or as the intermediate components of a specific biochemical cycle. They are one of the versatile ingredients in food and beverage industries. During the first half of the twentieth century, advances in chemical synthesis provided new production tactics that have become economically competitive and changed many fermentation processes by optimizing fermentation processes and improving the recovery and purification processes. Further trends inside the biotechnology field, environmental pressures and the growing demand for organic acids, numerous efforts and research have been made to increase the yield and productivity. The organic acids mentioned right here consist of the ones that are presently manufactured via way of means of fermentation in massive quantities and are gaining significance due to their demand in the modern biotechnology market.

14.2 Citric Acid

14.2.1 Introduction to Citric Acid

Citric acid ($C_6H_8O_7$, 2-hydroxy-1,2,3-propane tricarboxylic acid), a natural constituent and common metabolite of plants and animals, is the most important organic acid produced in tonnage and is extensively used in food and pharmaceutical industries. The name is derived from Latin word *citrus*, which refers to trees of the genus *Citrus*, including lemon trees. Citric acid in its pure form is readily soluble in water and colourless. It is solid at room temperature. Citric acid has a melting point of 153 °C and it decomposes at higher temperatures. It has a molecular weight of 210.14 g/mol. It possesses three different pK_a values, at pH 3.1, 4.7 and 6.4, owing to the presence of three functional groups of carboxylic acid in its structure. Anhydrous CA is highly soluble in water, freely soluble in ethanol and sparingly soluble in ether, whereas monohydrate CA is soluble in water and sparingly soluble in ether. Once dissolved in water, it shows weak acidity but a strong acid taste that affects sweetness and provides a fruity tartness. Citric acid is one of the most common products which have a never-ending demand in the global market. It is the most versatile and widely used organic acid. It plays an important role in pharmaceutical (10%), food (60%) and chemical cosmetic industries and other industries for applications such as acidulation, antioxidant, flavour, enhancement, preservation and plasticization and as a synergistic agent.

Citric acid can be derived from natural sources such as lemon, lime and orange or synthetic sources such as chemical reaction and microbial fermentation. It is commonly produced by submerged fermentation by using *Aspergillus niger* or *Candida* sp. from different sources of carbohydrates, such as molasses and starch-based media. Citric acid is formed as an intermediate in the Krebs cycle, but it is accumulated in greater quantities in the fungus *Aspergillus niger*; this is because of the fact that this organism has capacity to utilize varieties of substrates due to its well-developed enzymatic system. This aspect of the fungus is being exploited for the fermentative production of citric acid. It is also found as a natural product in many fruits especially citrus fruits. Citric acid can also be produced from glycerol, but it is very expensive. Though *Aspergillus wentii*, *A. clavatus*, *Penicillium divaricatum*, *P. citrinum*, *P. luteum*, *Mucor piriformis*, *Citromyces pleffencinus*, *Candida guilliermondii*, *Saccharomyces lipolytica*, *Trichoderma viride*, *Arthrobacter paraffinicius and Corynebacterium spp.* can produce citric acid, *A. niger* is employed extensively.

14.2.2 Historical Developments

Karl Wilhelm Scheele, a Swedish chemist, pioneered the method of extracting citric acid from lemon juice in 1784. This method was adopted in the commercial production of citric acid in England around 1826, utilising lemons imported from Italy. For nearly 100 years, Italian producers held a monopoly on its production, and it was sold at a high price. This led significant efforts around the world to find alternative ways of production, including chemical and microbiological processes. A German botanist, Wehmer, first observed the feasibility of obtaining citric acid through the fermentation of a sugar medium containing inorganic salts with *Penicillium glaucum*. Two years after this discovery, Wehmer successfully isolated two strains which were able to produce citric acid. These strains were later named *Citromyces* spp. (*Penicillium*). However, industrial trials did not succeed due to contamination problems and long duration of fermentation. Although methods were well developed to synthesize citric acid using chemical means also, better successes were achieved using microbial fermentations, and over the period of time, this technique has become the method of ultimate choice for its commercial production, mainly due to economic advantage of biological production over chemical synthesis. Much attention has been paid on research to improve the microbial strains and to maintain their production capacity.

14.2.3 Microorganisms Used for Citric Acid Production

A large number of microorganisms including bacteria, fungi and yeasts have been employed to produce citric acid. Most of them, however, are not able to produce commercially acceptable yields. This fact could be explained by the fact that citric acid is a metabolite of energy metabolism, and its accumulation rises in appreciable amounts only under conditions of drastic imbalances. The fungus *Aspergillus niger* has remained the organism of choice for commercial production. The main advantages of using this microorganism are (a) its ease of handling, (b) its ability to ferment a variety of cheap raw materials and (c) high yields.

Species of *Aspergillus* such as *A. wenti*, *A. foetidus*, *A. aculeatus*, *A. awamori*, *A. fonsecaeus*, *A. phoenicis* and *A. carbonari*, as well as *Trichoderma viride* and *Mucor piriformis*, have been found to produce significant amounts of citric acid.

Table 14.1 Citric acid (CA)-producing microorganism

Microbes	CA-producing species	References
Bacteria	*Bacillus licheniformis*, *Arthrobacter paraffinens*, *Corynebacterium* sp., *Bacillus subtilis*, *Brevibacterium flavum*, *Corynebacterium* sp. and *Penicillium janthinellum*	–
Fungi	*Aspergillus niger*, *A. aculeatus*, *A. awamori*, *A. carbonarius*, *A. wentii*, *A. foetidus* and *Penicillium janthinellum*	–
Yeast	*Saccharomycopsis lipolytica*, *Candida tropicalis*, *C. oleophila*, *C. guilliermondii*, *C. parapsilosis*, *C. citroformans*, *Hansenula anomala* and *Yarrowia lipolytica*	–

Besides fungi and bacteria, yeast species such as *Candida*, *Candida oleophila*, *Candida guilliermondii*, *Yarrowia lipolytica*, *Torulopsis*, *Hansenula*, *Debaryomyces*, *Torula*, *Pichia*, *Kloeckera*, *Saccharomyces* and *Zygosaccharomyces* are capable of producing citric acid from *n*-alkanes and carbohydrates. Mutagenesis has been used in recent years to improve the citric acid-producing strains so that they can be used in industrial applications. The most common methods include the use of mutagens to induce mutations on the parental strains. As such, strains of this microorganism can be improved to create industrial strains for use in commercial production. Different mutagens, including radiation, such as ultraviolet, X-rays and gamma rays and chemicals, such as ethyl methane sulphonate and diethyl sulphonate, have been used to induce the mutation of *A. niger* (Table 14.1).

14.2.4 Biochemical Aspect of CA Production

During glycolysis, pyruvate produced is oxidized and combined with coenzyme A to form CO_2, acetyl coenzyme A (acetyl-CoA) and nicotinamide adenine dinucleotide (NAD) + hydrogen (H) (NADH). Normally, oxaloacetate would largely be supplied through the completion of tricarboxylic acid (TCA) cycle, allowing recommencement of the cycle by condensing with acetyl – CoA to form citrate, catalysed by citrate synthase. Pyruvic acid produced from glucose is not only decarboxylated to acetyl-CoA by the pyruvate dehydrogenase complex, but it is also partially carboxylated to oxaloacetic acid during the idiophase by the action of pyruvate carboxylase. CA produced by the combination of acetyl-coA and oxaloacetate is then transformed through a reaction sequence that yields two molecules of CO_2 and regenerates the four-carbon oxaloacetate again, at each turn of the cycle, one molecule of acetic acid enters, two molecules of adenosine triphosphate (ATP) and CO_2 are formed, and a molecule of oxaloacetate is utilized to form (Fig. 14.1).

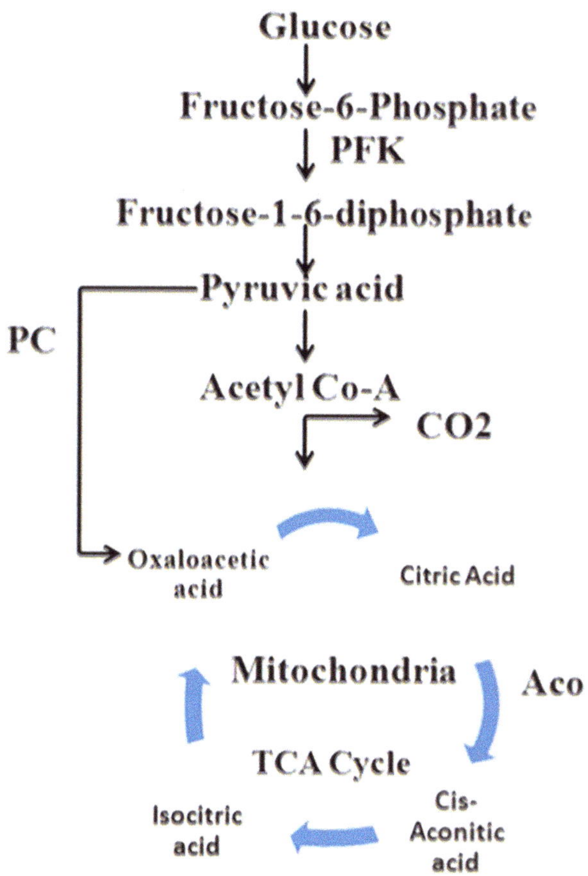

Fig. 14.1 Production of citric acid by *Aspergillus niger (PFK* phosphofructokinase, *PC* pyruvate carboxylase, *ACO* aconitase)

14.2.5 Fermentation Processes

The industrial citric acid production can be carried out in three different ways:

1. Submerged fermentation
2. Surface fermentation
3. Solid-state fermentation

Submerged Fermentation

The submerged fermentation (SmF) process is the commonly employed technique for citric acid production. Eighty per cent of the world's production is estimated to be from the submerged method. Several advantages such as higher yields and productivity and lower labour costs and lower contamination risks are the main reasons for this submerged fermentation which was developed after surface fermentation. Submerged fermentation is mostly operated as a batch system; however, continuous systems are possible and are used in practice. Submerged fermentation also includes the shake flask technique, which is usually used for the optimization of fermentation conditions. This is basically an Erlenmeyer flask which is placed on a shaker and stirred continuously throughout the fermentation process.

In this method, *A. niger* is made to grow uniformly dispersed throughout the liquid production medium. Fermentation is generally carried out in large fermenters having 4 klt a capacity of thousands of gallons which are provided with mechanical agitator and sparges. Fermenters are made of high-grade steel and require provision of aeration system, which can maintain a high dissolved oxygen level. It involves low labour cost, longer incubation period, more energy consumption and sophisticated techniques. Two types of fermenters, conventional stirred fermenters and tower fermenters, are employed, although the latter is preferred due to the advantages it offers on price, size and operation; three factors are important for production in submerged culture process. They are quality of the metal used for the construction of fermenter, mycelium structure and oxygen supply. Candida lipolytica, an alkane utilizing fungus, can also be employed in citric acid production under continuous fermentation.

Surface Fermentation

Surface fermentation, also known as liquid surface culture, was the original citric acid industrial production technique. The first individual process for citric acid production was the liquid surface culture (LSC), which was introduced in 1919 by the Société des Produits Organiques in Belgium and in 1923 by the Chas Pfizer & Co. in the USA. After that, other methods of fermentation, such as submerged fermentation, were developed. Even though in recent years submerged fermentation has gained popularity, there are still small- and medium-scale industries that make use of this method. Surface fermentation offers advantages such as lower installation and energy costs (as it does not require energy for aeration and agitation) and is also foam free.

This method consists of two phases, both of which are characterized by a rapid uptake of carbohydrates. In the classical process for citric acid manufacture, the culture solution is held in shallow trays (capacity of 50–100 L), and the fungus develops as a mycelial mat on the surface of the medium. The second phase utilizes carbohydrates by converting them to citric acid. The trays are made of high-purity aluminium or special grade steel and are mounted one over another in stable racks. Since the chamber needs to be effectively ventilated, sterile air is aseptically passed continuously over the medium surface through a bacteriological filter. The air supplies the required oxygen demand of the microorganisms due to the highly aerobic nature of the process; it also controls the humidity and temperature (via evaporative cooling). After completion of the second phase, the citric acid is recovered by washing the mycelial mats, and the impregnated citric acid is extracted.

The process is conventionally performed in fermentation chambers, using trays made from materials such as special-grade steel, high-purity aluminium or polyethylene. However, stainless steel trays are preferred, as they are resistant to deformation with prolonged use. The fermentation chambers are provided with an effective air circulation in order to control temperature and humidity. Fermentation chambers are always in aseptic conditions, which might be conserved principally during the first two days when spores germinate. It is estimated that about 80% of the world's citric acid production is obtained by submerged fermentation in stirred tanks of 40–200 m^3 or larger airlift fermenters of 200–900 m^3 capacity.

Submerged fermentation can be carried out in batch, fed batch or continuous systems, although the batch mode is more frequently used. Contamination is also an important issue in surface cultures. Yeasts, penicillia, lactic bacteria and species of *Aspergillus* are the most common sources of contamination. Though it is an old process, it is still employed. It is a kind of stationary fermentation process.

Solid-State Fermentation

The solid-state process, or 'Koji' fermentation, originates from Japan, which has an abundance of agro-industrial residues/wastes.

Solid-state fermentation (SSF) has been termed as an alternative method to produce citric acid from agro-industrial residues. This process involves the cultivation of microorganisms in the absence of free liquid on moist solid materials (Käppeli et al. 1978). The solid materials act as a physical support and source of nutrients for the microorganism. SSF can be carried out using several raw materials. Generally, the substrate is moistened to about 70% moisture depending on the substrate absorption capacity. The initial pH is normally adjusted to 4.5–6.0, and the temperature of incubation can vary from 28 to 30 °C. Under optimal conditions, the process should be completed in four days. The most common organism is *A. niger*.

However, there also have been reports with yeasts. The main advantage of solid-state fermentation is its superior yield and the ability to utilize inexpensive and widely available agro-industrial residues as substrates for bio-production, making it more environmentally friendly than submerged fermentation. One of the important advantages of SSF process is that the presence of trace elements may not affect citric acid production so harmfully as it does in SmF. There is no need for pretreatment as the system is less sensitive to the presence of trace elements compared to submerged fermentation.

Consequently, substrate pre-treatment is not required. Different types of fermenters such as conical flasks, glass incubators and trays, etc. have been used for citric acid fermentation in SSF.

One limitation of this method is that it does not completely utilize available nutrients owing to poor heat and oxygen transfer in the substrate. There is also a limited pool of viable microorganisms, and strains with large nitrogen and phosphorus requirements cannot be used.

14.2.5.1 Production with *A. niger*

1. Inoculum Production

Mycelial mats called pellets are used as inoculum for fermentation in this process. Suitable and high-yielding strains of *A. niger* are selected from a stock culture. The spores are induced to germinate in a seed fermenter. A nutrient solution containing 15% sugar from molasses is used in this seed fermenter. To induce the formation of mycelial pellets, cyanide ions are added to the medium.

Pellet formation largely depends upon the concentration of cyanide ions in the medium. Lower yield of citric acid occurs if the cyanide ions are in less concentration. This is because lower concentrations of cyanide ions induce formation of normal mycelium instead of pellets. The spores germinate at 32 °C and form pellets of 0.2–0.5 mm diameter within 24 h. During this period, the pH falls to 4.3. These pellets are then used as inoculums for production fermenters.

2. Preparation of Medium

   Beet molasses substrate (12–15%, reducing sugar content) nutritive salts, such as ammonium nitrate or potassium dihydrogen phosphate, are added; pH of the substrate is maintained at 5.5–5.9.

3. Fermentation Process

   Mostly fermenters used for citric acid production are constructed in the range of 10–220 klt.

   The mycelial pellets developed in the seed tank are transferred aseptically to the fermenters and incubated at a constant temperature, 30 °C. The structure of the mycelium that forms in the fermenter is vital to a successful production process. Little citric acid is produced if the mycelium is loose and filamentous with limited branches and no chlamydospores.

   Optimal citric acid is formed if the mycelium is in the form of pellets. The ratio of iron to copper in the medium determines the nature of mycelium. Although *A. niger* requires relatively little oxygen, it is sensitive to oxygen deficiency. There must be minimum oxygen concentration of 20–25% of the saturation value throughout the fermentation process. A short interruption in the oxygen supply ceases the production irreversibly. The aeration rate should be 0.2–1.0 volume per min during the acid production phase. Due to low viscosity, stirring is not necessary. Thus, although some plants use stirred fermenters, airlift reactors can also be used.

   Foaming is a problem in the submerged culture process. However, it can be controlled by adding antifoam agents such as lard oil at frequent intervals. A foam chamber, one-third of the size of the fermenter volume is needed in both airlift and stirred bioreactors. Mechanical antifoam devices can also be used. Progress of the fermentation process is monitored regularly by calculating the content of sugar and citric acid in the fermentation.

14.2.5.1.1 Product Recovery

The biomass is separated by filtration. Then the liquid broth is transferred to the recovery process. The recovery of citric acid from liquid fermentation is generally accomplished by three basic procedures, precipitation, extraction and adsorption (mainly using ion-exchange resins).

14.2.5.1.2 Purification

Purification is a simple way of getting pure citric acid followed by two simple techniques:

1. Precipitation
2. Filtration:

   - Precipitation is the most commonly used technique.
   - Precipitation is the classical method, and it is performed by the addition of calcium oxide hydrate (milk of lime) to form the slightly soluble tri-calcium citrate tetra hydrate. The precipitated tri-calcium citrate is removed by filtration and washed several times with water. It is then treated with sulphuric acid forming calcium sulphate, which is filtered off. Mother liquor containing citric acid is treated with active carbon and passed through cation and anion exchangers. Several anion-exchange resins are commercially available.

14.2.5.1.3 Further Purification

After purification, it can be produced in two forms:

1. Monohydrate
2. Anhydrous

Monohydrate

- It contains one water molecule for every citric acid molecule.
- It requires repeated crystallization until the water content is approximately 7.5–8.
- The liquor is concentrated in vacuum crystallizers at 20–25 °C, forming citric acid monohydrate.

Anhydrous

- Processed to remove all water from the end product
- Prepared by dehydrating the monohydrate citric acid product at a temperature above 36.6 °C (Fig. 14.2)

14.2.5.2 Production with Yeasts

The yeast species which were reported to produce citric acid are *Candida* (*Yarrowia*) *lipolytica*, *Candida guilliermondii*, *Candida oleophila*, *Candida intermedia*,

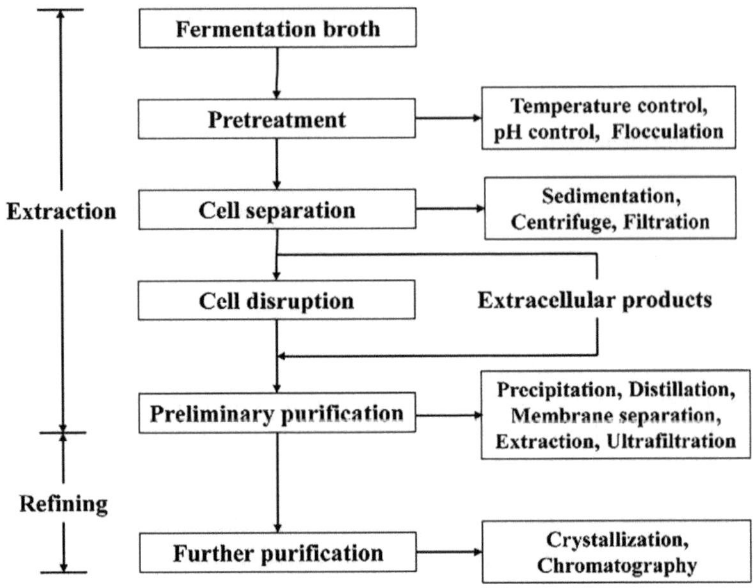

Fig. 14.2 Recovery process of organic acid from fermentation broth

Candida paratropicalis, *Candida zeylanoides*, *Candida catenulata*, *Candida parapsilosis*, *Pichia anomala* and some *Rhodotorula* species.

Among the yeast species, *Yarrowia lipolytica* is known as a potential producer of citric acid.

The main advantages of using yeasts are mentioned as follows:

1. Yeasts are characterized by greater resistance to high substrate concentrations than fungi.
2. With comparable conversion rates, yeasts have greater tolerance to metal ions that allow the use of less refined substrates.
3. Using yeasts also gives a better process control due to their unicellular nature.

And the major disadvantage of using yeasts is the simultaneous production of citric and isocitric acids. It is reported that the ratio of citric/isocitric acid can vary between 1:1 and 20:1 according to the yeast strain, carbon source and micronutrient concentration. Selection of a yeast strain with high citric acid production and giving high citric acid/isocitric acid ratios have been reported as the principal step of a citric acid production process. Citric acid production by yeasts is exclusively carried out by submerged cultivation. The submerged fermentation process is desirable because of its higher efficacy due to higher susceptibility to automatization.

14.2.6 *Factors Affecting Citric Acid Production*

14.2.6.1 Medium and Its Components

Carbon source: Sucrose was the most favourable carbon source followed by glucose, fructose and galactose. Galactose contributed to a very low growth of fungi and did not favour citric acid accumulation. Other sources of carbon such as sorbose, ethanol, cellulose, mannitol, lactic, malic and a-ketoglutaric acid allow a limited growth and low production. Starch, pentoses (xyloses and arabinoses), sorbitol and pyruvic acid slow down growth, though the production is minimal. There are some critical factors (costs, need of pretreatment), which should be considered for substrate determination. One another aspect is the presence of trace elements, which can act as inhibitors or stimulants. Consequently, sometimes it is necessary to conduce a pretreatment, for example, precipitation of trace metals of molasses by potassium ferrocyanide.

Nitrogen source: Citric acid production is directly influenced by the nitrogen source. Nitrogen source: Usually ammonium sulphate or ammonium nitrate has been used as a nitrogen source, for example, urea, ammonium sulphate, ammonium chloride, peptone, malt extract, etc. Their consumption lowers the pH of the medium to below 2 which is an additional prerequisite of citric acid fermentation. The concentration of nitrogen source required for citric acid fermentation is 0.1–0.4 N/L. A high nitrogen concentration increases fungal growth and the consumption of sugars but decreases the amount of citric acid produced.

Phosphorous source: Potassium dihydrogen phosphate has been reported to be the most suitable phosphorous source. Phosphorous at concentration of 0.5–5.0 g/L was required by the fungus in a chemically defined medium for maximum production of citric acid. Phosphate is known to be essential for the growth and metabolism of *A. niger*. Phosphate plays a key role in secondary metabolite production. Low levels of phosphate favour citric acid production.

Trace elements: Trace element nutrition is probably the main factor influencing the yield of citric acid. A number of divalent metals such as zinc, manganese, iron, copper and magnesium have been found to affect citric acid production by *A. niger*. Zinc favoured the production of citric acid if added with KH_2PO_4. Copper was found to complement the ability of iron at optimum level, to enhance the biosynthesis of citric acid. Manganese deficiency resulted in the repression of the anaerobic and TCA cycle enzymes with the exception of citrate synthetase. Magnesium is required both for growth and for citric acid production. Optimal concentration of magnesium sulphate was found in the range of 0.02–0.025%.

Miscellaneous: Some compounds which are inhibitors of metabolism such as calcium fluoride, sodium fluoride and potassium fluoride have been found to accelerate the citric acid production, while potassium ferrocyanide has been found to decrease the yield. There are many compounds, which act in many ways to favour citric acid accumulation. Some of them are capable to impair the action of metal ions, and other toxic compounds influence growth during the initial phase. Some of these are 4-methyl-umbelliferone, 3-hydroxi-2-naphtoic, benzoic acid, 2-naphtoic acid, iron cyanide, quaternary ammonium compounds, amine oximes, starch, EDTA, vermiculite, etc.

14.2.6.2 Process Parameters

pH of culture medium: The pH of the medium is important in two stages of the process. All fermentations start from spores and their germination requires pH > 5. The absorption of ammonia by germinating spores causes release of protons, thus lowering the pH and improving the production of citric acid. The low pH value during the production phase (pH ≤ 2) reduces the risk of contamination by other microorganisms and inhibits the production of unwanted organic acids (gluconic and oxalic acids), which makes the product recovery easier.

Aeration: Aeration has been shown to have a determinant effect on citric acid fermentation. Increased aeration rates led to enhanced yields and reduced fermentation time. The high demand of oxygen is fulfilled by constructing appropriate aeration devices, which is also dependent on the viscosity of the fermentation broth. This is an additional reason why small compact pellets are the preferred mycelial forms of *A. niger* during fermentation. Aeration is performed during the whole fermentation with the same intensity through the medium at a rate of 0.5–1.5 vvm. High aeration rates lead to high amounts of foam, especially during the growth phase. Therefore, the addition of antifoaming agents and the construction of mechanical "defoamers" are required to tackle this problem.

14.2.7 Uses of Citric Acid in Industries

1. It is used in the production of carbonated beverages.
2. It is used as a chelating and sequestering agent in the tanning and textile industry.
3. Citrate esters are used as plasticizer.
4. It is abundantly used in food industry as an acidulent in the preparation of food items like jams, preserved fruits and fruit juices, etc.
5. It is used in frozen foods to prevent its change in colour and flavour.
6. It is used in metal painting industry.
7. It is used in pharmaceutical industry.
8. It is used in the manufacture of astringent, hair rinsers and hair setting fluids.
9. It is used in beverage industry as a preservative to prevent oxidation of alcohol and emulsifier of dairy products like cheese and ice creams.
10. It is used as preservative and to prevent change in colour and flavour and in the oxidation of alcohol.

14.3 Acetic Acid: Biosynthesis and Fermentation Process

14.3.1 Introduction to Acetic Acid

Acetic acid is a commodity chemical with the global demand of approximately 15 million tons per year with several applications in the chemical and food industry. Acetic acid (CH_3COOH) also called ethanoic acid is an organic compound. Acetic acid is a clear liquid with a pungent odour, sharp taste and melting point of 16.73 °C and boils at 117.9 °C. Pure acetic acid is often called glacial acetic acid. Glacial acetic acid is an excellent polar protic solvent that is frequently used as a solvent for recrystallization to purify organic compounds. Vinegar is traditionally the product of acetous fermentation of natural alcoholic substrates. Vinegar is a product obtained biologically by oxidative conversion of ethanol-containing solutions by AAB (acetic acid bacteria).

Vinegar fermentation is one of the oldest fermentations known to man. It is formed naturally due to spoilage of wine. Therefore, literally vinegar means "sour wine." Vinegar was produced only for local consumption until the middle ages. Depending on the substrate, vinegars can be classified as fruit, starch or spirit substrate vinegars. Industrial vinegar manufacturing processes fall into three main categories: slow processes, quick processes and submerged processes. The ancient uses of vinegar which can be seen from various records include a wide variety of uses including use as a food condiment and treatment of wounds and as a cosmetic aid.

Industrially, acetic acid is used in the preparation of metal acetates, used in some printing processes; vinyl acetate, employed in the production of plastics; cellulose acetate, used in making photographic films and textiles; and volatile organic esters (such as ethyl and butyl acetates), widely used as solvents for resins, paints and lacquers. Biologically, acetic acid is an important metabolic intermediate, and it occurs naturally in body fluids and in plant juices. Acetic acid has several applications in food industry and is traditionally known as vinegar. Its pathway is conversion of glucose to ethanol and ethanol to acetic acid. In first step, *Saccharomyces cerevisiae* (yeast) converts fermentable sugar of molasses into ethanol and carbon dioxide. In the second step, *Acetobacter aceti* (acetic acid bacteria) converts ethanol into acetic acid and water. After completing process, the separation of product is carried out via centrifugation. Mixture of acetic acid and water is separated by distillation.

Acetic acid is produced via fermentation. Commercially acetic acid is produced by two methods, surface fermentation process and submerged fermentation process. From 1949, submerged process was employed in the fermentative production of vinegar. Presently, both these processes, that is, surface fermentation and submerged fermentation, are employed worldwide. The surface fermentation is used still today because of the better flavour of the product.

14.3.2 Acetic Acid Bacteria (AAB)

They are a group of Gram-negative, rod-shaped and obligate aerobes which oxidize sugars or ethanol and produce acetic acid during fermentation. AAB are catalase-positive and oxidase-negative, ellipsoidal- to rod-shaped cells that can occur singly, in pairs or chains. They are also mesophilic microorganisms, and their optimum growth temperature is between 25 and 30 °C. The optimum pH for their growth is 5.0–6.5, but they can also grow at lower pH values. Several species of acetic acid bacteria are used in industry for production of certain foods and chemicals. The principal bacteria are *Acetobacter aceti*. The main yeasts are *Saccharomyces cerevisiae*. It is important to maintain an acidic environment to suppress the growth of undesirable organisms and to encourage the presence of desirable acetic acid-producing bacteria. The microbial oxidation of ethanol to acetic acid is an aerobic fermentation that has high oxygen requirement. Acetobacter bacteria are employed for industrial production of vinegar. The genus name *Acetobacter* was put forward by Beijerinck in 1900. Acetobacter bacteria can be divided into two groups—Gluconobacter and Acetobacter. Gluconobacter oxidizes ethanol to acetic acid, while Acetobacter oxidizes ethanol first to acetic acid and then to CO_2 and H_2O. The principal bacteria are *Acetobacter aceti*.

Although a variety of bacteria can produce acetic acid, mostly members of Acetobacter, Gluconacetobacter and Gluconobacter are used commercially.

14.3.3 Acetic Acid Fermentation

Fermentative route is mostly adapted for the generation of food-grade acetic acid that is vinegar. This process mainly involves the use of renewable carbon resources such as apple, grape, pears, honey, cane, coconut, date, syrup, cereals, hydrolysed starch, beer and wine.

Production of acetic acid can be carried out by aerobic and anaerobic fermentation. Anaerobic process is a one-stage process carried out by *Clostridium*. Aerobic fermentation has two stage processes. Glucose is converted into ethanol by *S. cerevisiae*. And in the second stage, ethanol is converted into acetic acid by *Acetobacter aceti*. The fermentation is usually initiated by yeasts which break down glucose into ethyl alcohol with the liberation of carbon dioxide gas. Following on from the yeasts, *Acetobacter aceti* oxidize the alcohol to acetic acid and water.

$$\text{Yeast reaction}: \underset{\text{Glucose}}{C_6H_{12}O_6} \rightarrow 2CO_2 + \underset{\text{ethanol}}{2CH_3CH_2OH}$$

$$\text{Acetobacter aceti reaction}: \underset{\text{Ethanol}}{CH_3CH_2OH} + O_2 \rightarrow \underset{\text{Acetic acid}}{CH_3COOH} + H_2O$$

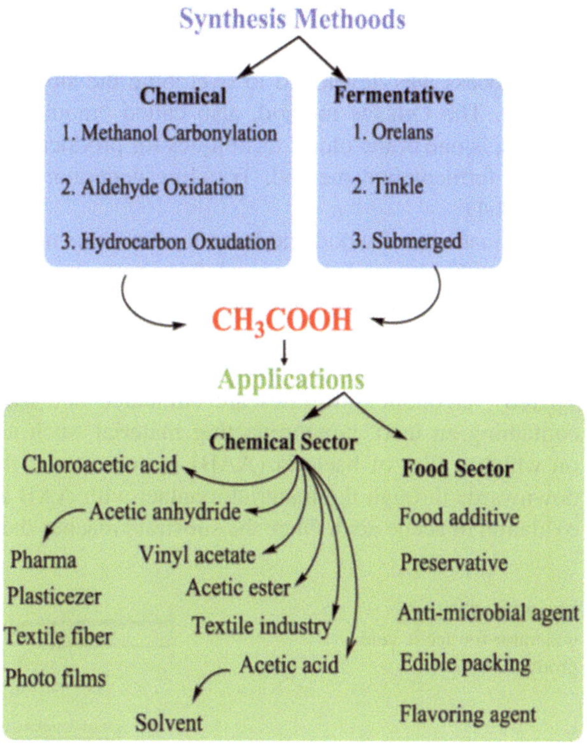

Fig. 14.3 Methods for production of acetic acid

The *Acetobacter aceti* are dependent upon the yeasts to produce an easily oxidisable substance (ethyl alcohol). It is not possible to produce vinegar by the action of one type of microorganism alone. For ethanol fermentation, it is essential to have sugar concentration around 20%, and also pH and temperature should be maintained around 5 and 32 °C, respectively. For a good fermentation, it is required to have an alcohol concentration of 10–13%. If the alcohol content is much higher, the alcohol is incompletely oxidized to acetic acid. If it is lower than 13%, there is a loss of vinegar because the esters and acetic acid are oxidized. In addition to acetic acid, other organic acids are formed during the fermentation which become esterified and contribute to the characteristic odour, flavour and colour of the vinegar. Also pH and temperature should be maintained around 4.5 and 30 °C, respectively. In general, the yield of acetic acid from glucose is approximately 60%. That is, three parts of glucose yield two parts acetic acid. For analysis of ethanol and acetic acid, gas chromatography and high-performance liquid chromatography methods are used (Fig. 14.3).

14.3.3.1 Surface Fermentation Process

This process was developed to overcome the slow rate of acetification in Orleans process. The Orléans method, also called "mother of vinegar system" or French method, is one of the oldest techniques for producing vinegar and is an example of a surface fermentation method. Trickling generator is generally used in this process (Fig. 14.4).

It is made up of wood and has a total volume up to 60 m^3 and its inner surface is lined with birch wood shavings. A false bottom supports the coils of wood shavings and separates them from the collection chamber which occupies about one-fifth of the total capacity of the generator. The rationale of the trickling generator processes is to increase the surface area for oxygen contact with the fermenting vinegar. Thereby, alcoholic substrates are circulated and trickled through vessels or vats containing an inert, noncompacting material, such as wood shavings or charcoal, on which a film of bacteria (AAB) is present. As the alcoholic substrate trickles downwards through the material, contact with AAB and oxygen results in efficient oxidation to acetic acid. Once the substrate reaches the bottom, it is recirculated over

Fig. 14.4 Trickling generator for acetic acid production

the bed to promote increased transformation to acetic acid. Thus, the process is repeated again and again until 88–90% of alcohol is changed to acetic acid. The starting material should contain both acetic acid and ethanol for optimal growth of *Acetobacter*. Air is forced through the false bottom-up through the set-up. The cooling water in the heat exchanger is used to regulate the temperature in the generator so that it is between 29 °C and 35 °C; this is determined with thermometers placed at different levels of the generator. Presently, higher-yielding strains are employed in vinegar fermentation which are able to yield 13–14% of acetic acid.

The process has the drawback of accumulating gelatinous material on the surface the membrane, which reduces the rate of reaction over the period.

Must measure three parameters:

(a) The circulation of the mash.
(b) The flow of cooling water through the heat exchange.
(c) The amount of air delivered through the system. If the air flow rate is too high alcohol and vinegar are lost in effluent air.

14.3.3.2 Submerged Fermentation Process

This is the most widely used method and has a high yield along with a fast rate of oxidation as compared to the previous method. This method is 30 times faster than the Orleans method with higher efficiency for production of acetic acid. Materials with low alcohol concentration such as fruits, wines and special mashes were first used in the initial stages of submerged fermentation. Fermenters constructed with stainless steel are employed and they are stirred from the bottom. Aeration is provided with a suction rotor, with the incoming air coming down through a pipe from the top of the vessel. Heat exchanger is provided to control the temperature along with foam eliminators. Domestic vinegar is produced through semicontinuous fully automatic process under continuous stirring and aeration (Fig. 14.5).

The starting materials 100 mL of 5% ethanol are used in the process to get 7–10 g acetic acid/100 mL. The oxidative process occurs at the liquid–air interface of the bubbles, where the AAB convert ethanol into acetic acid, with limited production of other metabolites. The fermentation process is carried up to 35 h at 40 °C temperature. The ethanol concentration is continuously measured. The yield of acetic acid is about 98%. The pure substrates are required to achieve the high quality of acetic acid. This fermentation process is much economical, of simple design with easy process control. A much smaller space is occupied (about one-sixth) in comparison with the trickling generator. It is easy and cheap to change from one type of vinegar to another. And continuous production and automation can take place more easily than with trickling. A disadvantage of this rapid method is that the high airflow leads to significant stripping of the volatile components from the original substrate, producing a more organoleptically limited product. Despite this, the rapidity of the process (vinegar can be produced in 24-h cycles) and the high acidity achievable (acetic acid levels of up to 23–25%, compared to 6–13% with traditional systems)

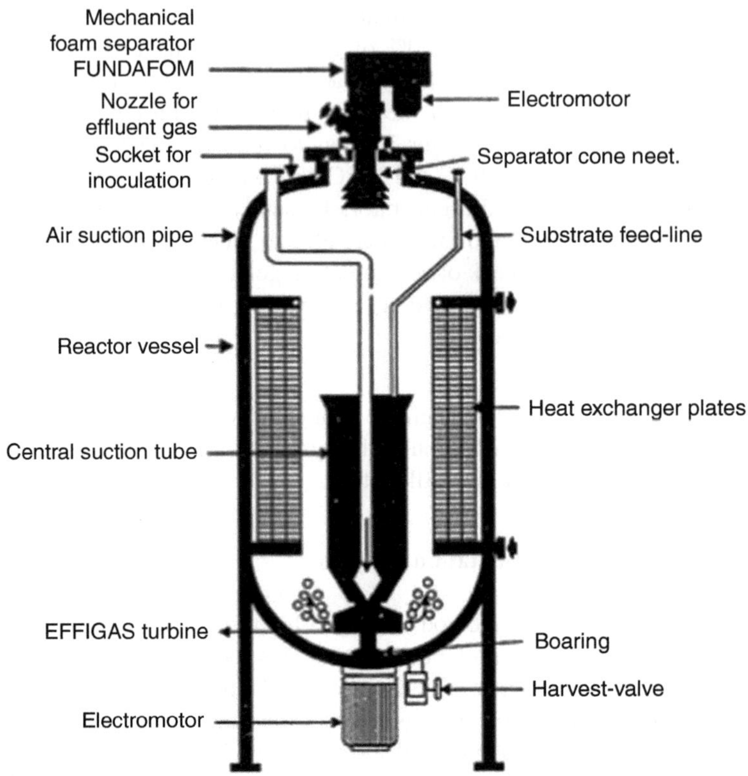

Fig. 14.5 Submerged fermenter for acetic acid production

are key advantages. Higher acidity makes transportation more cost-effective by reducing water transport.

14.3.3.3 Application of Acetic Acid in Food Industry

Direct applications of acetic acid are reported from ancient times. Over the period, applications of acetic have diversified as per the demands of modern life:

1. It was used as a medicine and food preservative.
2. Using different concentrations, it is utilized in food additives, food preservation, antimicrobial agent, acidulant, flavour and taste enhancer.
3. It was used as edible packaging material.
4. It was used as artificial food ripening agent.
5. Some of the applications such as acidulant and as acetification agents are also of great importance.

14.4 Lactic Acid

14.4.1 Introduction

Lactic acid (LA) or 2-hydroxypropionic acid is the simplest hydroxycarboxylic acid and perhaps the most widely occurring in nature. It is a natural organic acid with a long history of applications in the food, pharmaceutical, textile and chemical industries.

LA is a chiral molecule with two optical active isomeric forms (known as "enantiomers"): the dextrorotatory form, L (+)-lactic acid or (S)-lactic acid, and the levorotatory form, D(−)-lactic acid or (R)-lactic acid. Due to the presence of two reactive functional groups—one hydroxyl (–OH) and one carboxyl (–COOH)—lactic acid can undergo a variety of chemical conversions to produce, for example, acrylic acid (via dehydration), 1, 2-propanediol (via hydrogenation), acetaldehyde (via decarboxylation), 2, 3-pentanedione (via condensation) and poly lactic acid (PLA) which is a biodegradable and biocompatible polymer that is used in a wide variety of applications. Its uses range from packaging and fibres to foams (Abdel-Rahman et al. 2013) and applications in biomedical devices. Lactic acid is a yellow to colourless liquid (at 15 °C and 1 atm) and is odourless. It is the simplest hydroxycarboxylic acid.

LA can be produced via chemical or microbial fermentation routes. It is naturally produced by a wide spectrum of microbes including bacteria, yeast and filamentous fungi. In general, bacteria ferment C5 and C6 sugars to lactic acid by either homo- or heterofermentative mode. Starchy materials such as corn, maize, rice, wheat, potato, barley and cassava are in general good substrates for LA production; but they need to be pretreated by physicochemical methods and amylolytic enzymes prior to fermentation. Moreover, the use of these feedstocks for LA production imposes the competition of LA industry vs. food sector as these are used for human and animal feeding as well. Lignocellulosic materials such are sawdust, poplar trees, sugarcane bagasse, brewer's spent grains, etc. are the most promising feedstocks for LA production because they are abundant, potentially inexpensive and have high carbohydrate content and they do not outcompete with the food chain supply. However, the main challenge is to break down the polymeric matrix into fermentative sugars.

L(+)-Lactic acid is found in living organisms more often than D (−)-lactic acid. In the human body, only L(+)-lactic acid is produced during muscle contraction. For applications in food and in medicine, L(+)-lactic acid is preferred because the metabolic conversion of L(+)-lactic acid in the body is faster than for D(−)-lactic acid. Different methods of production result in different amounts of isomers. In lactic acid production by chemical synthesis, only a racemic mixture is obtained, where the concentrations of the isomers are equal, whereas fermentation allows for producing one isomer in a greater amount (Fig. 14.6).

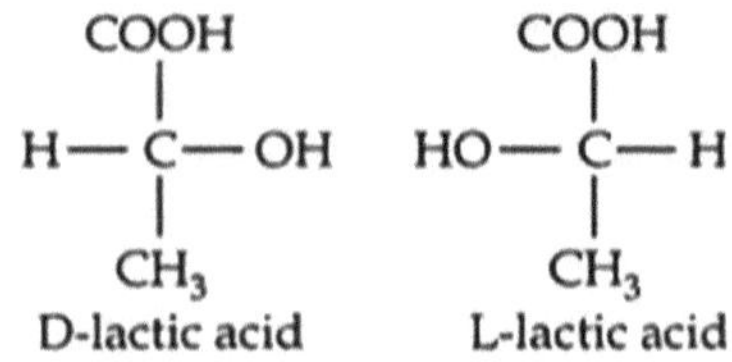

Fig. 14.6 Structure of lactic acid asymmetric carbon

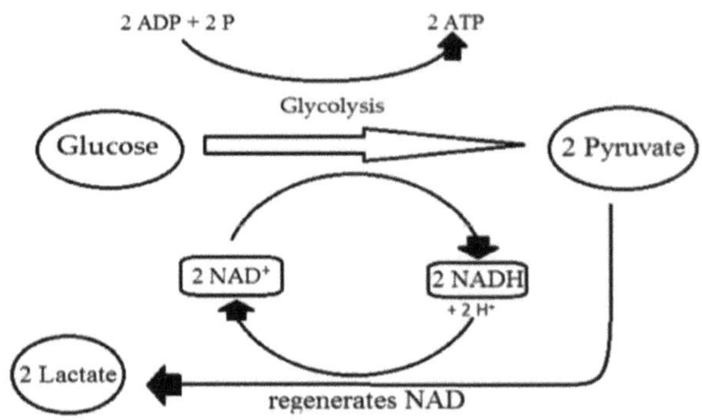

Fig. 14.7 Lactic acid fermentation

14.4.2 Historic Development

Lactic acid was first discovered in sour milk in 1780 by the Swedish chemist Scheele (Lima et al.2001). In 1839, Fremy carried out lactic acid fermentation with various carbohydrates, such as sugar, milk, starch and dextrin. In 1857, Pasteur discovered that lactic acid was not a component of milk but a metabolite that certain microorganisms produced by fermentation.

In the last few decades, the production of lactic acid has substantially increased primarily because of the development of new uses and products. Global lactic acid demand was estimated to be 714.2 kilo tons in 2013, and it is expected to grow annually by 15.5% to reach 1960.1 kilo tons by 2020. The three largest consumer markets in the world are the United States (31% of total lactic acid consumption in 2013), followed by China and Western Europe.

14.4.3 Production of Lactic Acid

Microbial fermentation offers advantages, including cheap renewable substrates, low production temperatures and low energy consumption. Because of these advantages, it is the production process that is used most often. Worldwide, lactic acid production from microbial fermentation accounts for around 90% of the total lactic acid production (Fig. 14.7).

Lactic acid-producing microorganisms are classified into bacteria, fungi and yeast. The choice of which type of microorganism to use depends primarily on the

Table 14.2 Characteristics of selected bacteria and moulds of interest in lactic acid production

Microorganism	Lactic acid isomer	Fermentation pattern	Raw material
Bacteria: *Lactobacillus amylophilus*	L(-)	Homofermentative	Starch
L. amylovorus	DL	Homofermentative	Starch
L. casei subsp. *Rhamnosus* (*L. delbrueckii* NRRL B 445)	D(-)	Facultative heterofermentative	Glucose, sucrose (molasses)
L. delbrueckii subsp. *bulgaricus*	DL	Homofermentative	Cheese whey and permeate (lactose)
L. helveticus	L(+)	Homofermentative	Cheese whey and permeate (lactose)
Moulds: *Rhizopus arrhizus and R. oryzae*	L(+)	Homofermentative	Glucose and starch

carbohydrate that is to be fermented, as a microorganism's metabolism differs with different sources of carbon. Most of the lactic acid productions are done industrially by the use of lactic acid-producing bacteria. Lactic acid-producing bacteria are classified into four main categories, which are lactic acid bacteria (LAB), *Escherichia coli*, *Corynebacterium glutamicum* and *Bacillus* strains.

Lactic acid bacteria can produce lactic acid by anaerobic glycolysis with high yield and productivity. Most LAB are facultative anaerobic, catalase negative, nonmotile and non-spore forming. They normally have a high acid tolerance and can survive at pH 5 and lower. Their high acid tolerance gives them a competitive advantage over other bacteria. The optimal temperature for growth varies between the genera from 20 to 45 °C. Based on the fermentation end product,

lactic acid bacteria are grouped into two types: homofermentative and heterofermentative (Table 14.2).

Homofermentative LAB convert glucose almost exclusively to lactic acid. Glycolysis results in lactic acid as the end product of glucose metabolism. Two molecules of lactic acid are produced per glucose molecule, which results in a yield of more than 0.90 g/g. Hence, homofermentative LAB are used in commercial production of lactic acid. Some of the homofermentative lactic acid bacteria used in the production of lactic acid are *Lactobacillus delbrueckii*, *Lactococcus lactis, Lactobacillus casei*, *Lactobacillus helveticus* and *Lactobacillus acidophilus*.

Heterofermentative LAB catabolize glucose into ethanol, CO_2 and lactic acid. Heterofermentative LAB produces fewer yields due to formation of by-products. *Lactobacillus pentosus*, *Lactobacillus bifermentans* and *Lactobacillus brevis* are some of the examples of heterofermentative bacteria.

Fungi, like *Rhizopus* spp., can metabolize different renewable carbon resources, with advantageously amylolytic properties to produce lactic acid. Fungal fermentation uses chemically defined medium, and so, the purification of products is simple. This is a major advantage in food industry.

Additionally, yeasts can tolerate environmental restrictions (e.g. acidic conditions), being the wild-type low lactic acid producers that have been improved by genetic manipulation. Microalgae and cyanobacteria, as photosynthetic microorganisms, can be an alternative lactic acid producers without carbohydrate feed costs.

14.4.4 Fermentation Methods for Lactic Acid Production

Fermentation of lactic acid, like any other fermentation processes, is dependent on factors such as raw materials used, nutrients present in media and the microorganisms used.

Three different methods of fermentation are practiced, namely, batch fermentation, fed-batch fermentation and continuous fermentation.

14.4.4.1 Batch Fermentation

Lactic acid is usually produced in batch mode. It is the most commonly practised fermentation process as it is simple to perform.

In batch fermentation, all the required materials such as carbon source, nitrogen source and other components are added prior to beginning of the fermentation process. In this process, all of the substrates gets used, whereas in a continuous process, there is a residual substrate concentration that is always present. The choice of operation mode depends on the costs of the substrate and the capital investment. If the substrate is expensive, the yield is maximized, by either batch or semicontinuous operation.

The major advantage of batch fermentation is that it prevents contamination to a good extent when compared to the other methods as it is a closed system, and so, high concentration of lactic acid is produced. Fermentation in batch mode has superior conversion and yield compared to continuous fermentation, but the volumetric productivity is lower.

The drawbacks of batch fermentation include low productivity due to substrate inhibition. Batch fermentation is mainly of two types, which are solid-state fermentation (SSF) and separate hydrolysis and fermentation (SHF).

14.4.4.2 Fed-Batch Fermentation

In fed-batch fermentation, all the required raw materials such as carbon source, nitrogen source and other required components are added during fermentation process at regular intervals of time without removal of fermentation broth. This type of fermentation is especially useful to maintain low substrate concentration by supplying nutrients to the fermentation culture which in turn reduces substrate inhibition. Ding and Tan reported that lactic acid production by *Lactobacillus casei* using fed-batch fermentation was found to be more efficient in production of lactic acid than other methods. They used 1% yeast extract and glucose as raw materials and obtained maximum lactic acid concentration of 210 g/L and L-lactic acid concentration of 180 g/L at the rate of production 2.14 g/L/h, and the yield of about 90.3% of L-lactic acid was obtained. Dong-Mei et al. (2003) used *Lactobacillus lactis* for the production of L-lactic acid and obtained about 97% yield of L-lactic acid at the rate of 2.2 g/L/h.

14.4.4.3 Continuous Fermentation

Continuous fermentation involves addition of fresh medium to the fermenter while withdrawing the already existing broth at the same rate. Due to this, the concentration of substrates and products is maintained constant. The advantages of continuous fermentation include prevention of end-product inhibition, less frequency to shutdown and less decrease in productivity during lag phase, inoculation of culture is done once only, high product yield can be obtained and it saves time and involves less labour work. Continuous fermentation suffers from a few drawbacks such as contamination and requirement of field operator with expertise, and it is expensive to perform. Shibata et al. reported the production of lactic acid at the rate of 1.56 g/L/h using *Enterococcus faecium* by continuous fermentation. Ahring et al. reported the production of lactic acid using *Bacillus coagulans* (strain AD) at the productivity of 3.69 g/L/h using continuous fermentation.

14.4.5 Medium and Manufacturing Process of Lactic Acid

To satisfy the nutritional requirements of the microorganisms, the medium should be prepared in such a way as to contain carbon source, nitrogen source, mineral source and source for growth factors. Glucose, maltose, lactose, etc. or crude substrates such as corn starch, potato and rice, whey, molasses, sulphite waste liquor, etc. are used as carbon source. The culture medium contains semi-refined sugar (molasses or whey contains semi-refined sugar), molasses or whey starch, maltose, lactose, sucrose and calcium carbonate with ammonium hydrogen phosphate. The malt sprouts are mixed and pH is kept between 5.5 and 6.5. Pretreatment of starchy material by amylase or by sulphuric acid is necessary to bring about hydrolysis and to change complex starch into simple sugars like glucose and maltose. The sugar content in the medium is maintained at 5–12%. Though various ammonium salts can be used, ammonium hydrogen phosphate is usually employed as nitrogen source in the fermentation of lactic acid. About 0.25% of ammonium salts are added to the medium. Based on the requirement of the selected microorganism, minerals are added to the medium. Calcium carbonate is also added as a buffering agent to control pH. *Lactobacillus* requires vitamins, especially vitamin B complexes for proper growth. These are added to the medium in the form of crude vegetable sources like the rootlets of germinating seeds of barley.

14.4.6 Fermentation Process

Lactic acid production is carried in 25–125 klt fermenters. Lactic acid is quite corrosive; hence metals are avoided, and consequently wooden fermenters are used. The pH of the fermentation broth is maintained at 5.5–6.5 by using calcium carbonate. Maintenance of pH is essential because it helps not only in the increase of fermentation rate but also in the increase of the yield. Maintenance of low pH values also helps in the elimination of bacterial contamination and also in the sterilization at low temperatures.

Suspension of pure bacterial culture is prepared from the suitable high-yielding strain of stock culture. The selection of species of *Lactobacillus* largely depends upon the nature of the carbon source being used in the fermentation. *Lactobacillus delbrueckii* and *Lactobacillus leichmannii* are employed when glucose is used as substrate, *Lactobacillus bulgaricus* when whey is used as substrate and *Lactobacillus pentosus* when sulphite waste liquor is used as substrate. These organisms are facultative rather than obligate anaerobes, and therefore, bioreactors need not be run with complete oxygen exclusion.

The selected *Lactobacillus* culture is transferred to large vessel and the temperature is maintained at 45–50 °C. Each stage of culture building requires 16–18 h depending upon the volume of fermentation broth. The inoculum volume is usually 5% and the fermentation is carried out for 5–10 days. The sugar be reduced to 0.11% or less during the fermentation because residual sugar makes the recovery of better quality of lactic acid difficult. The fermentation broth is gently agitated to keep the uniform distribution of nutrients and air.

As more and more lactic acid is formed, it may become toxic to the organism. Hence, neutralization of accumulated acid is done by the continuous addition of calcium hydroxide. Generally, the fermentation duration is 5–10 days. It can also be completed in 72 h if 12–13% of glucose is used in the medium and pH is maintained at 6.3–6.5 by continuous neutralization of the acid formed (Fig. 14.8).

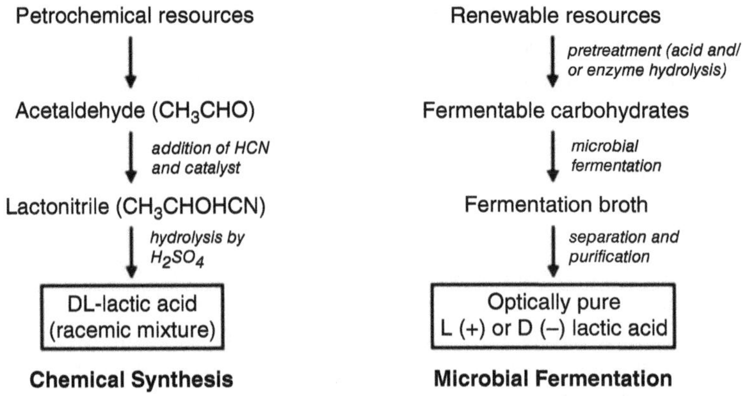

Fig 14.8 Overview of the two manufacturing methods of lactic acid: chemical synthesis (**a**) and microbial fermentation (**b**)

14.4.7 Recovery of Lactic Acid

Lactic acid process in a classical way involves a series of downstream treatments such as precipitation, conventional filtration, acidification, carbon adsorption, evaporation, crystallization and others.

A general method of recovery process, described here, has the following steps:

1. To the fermentation medium, $CaCO_3$ is added; pH is adjusted to 10, and broth is heated and filtered. Lactic acid is converted to calcium lactate.
2. The H_2SO_4 is added to remove Ca as $CaSO_4$.
3. Lactic acid is recrystallized as calcium lactate.
4. The washed filtrate is treated with activated carbon to remove organic impurities.
5. It is finally purified by passing through ion-exchange resins.

In the solvent extraction procedure, free lactic acid is extracted with isopropyl ether directly.

Lactic acid is also extracted and purified by some other methods such as the following:

1. Zinc salt of lactic acid is made which has low solubility and can be recovered.
2. Lactic acid is directly extracted with isopropyl ether by countercurrent method.
3. Methyl ester of lactic acid is prepared which is easily hydrolysed while boiling in water. Methyl alcohol is recovered.
4. High vacuum distillation of lactic acid.

14.4.8 Uses of Lactic Acid

1. It is a weak acid with good solvent properties.
2. Since it is a weak acid with good solvent properties, it polymerizes readily for the production of polymers.
3. It provides acidity in foods and beverages and served as a preservative in food stuff.
4. Calcium lactate is employed as baking powder and as a source of calcium in pharmaceutical industries.
5. Pure lactic acid is used in plastic industry.
6. Ethyl lactate is used in preparation of anti-inflammatory drugs.
7. Certain other industries such as textile and laundry use lactic acid in fabric treatment.
8. It is used as a blood coagulant and dietary calcium source.
9. It is used in the manufacture of cellophane resins, biodegradable plastics and some herbicides and pesticides.
10. It is used in food industry as acidulent as it has strong odour or flavour. Ethyl and butyl lactate are used as flavouring ingredients.

14.5 Other Organic Acids

14.5.1 Pyruvic Acid

Pyruvic acid ($CH_3COCOOH$) is the simplest of the alpha-keto acids, with a carboxylic acid and a ketone functional group. Pyruvate, the conjugate base, CH_3COCOO^-, is intermediate in several metabolic pathways throughout the cell. Pyruvic acid is an important keto-carboxylic acid and can be manufactured by both chemical synthesis and biotechnological routes.

Pyruvic acid supplies energy to living cells through the citric acid cycle (also known as the Krebs cycle) when oxygen is present (aerobic respiration); when oxygen is lacking, it ferments to produce lactic acid. It is the output of the anaerobic metabolism of glucose known as glycolysis. One molecule of glucose breaks down into two molecules of pyruvate, which are then used to provide further energy in one of two ways. Pyruvate is converted into acetyl-coenzyme A, which is the main input for a series of reactions known as the Krebs cycle. Pyruvate is also converted to oxaloacetate by an anaplerotic reaction, which replenishes Krebs cycle intermediates; also, oxaloacetate is used for gluconeogenesis. These reactions are named after Hans Adolf Krebs, the biochemist, awarded the 1953 Nobel Prize for physiology, jointly with Fritz Lipmann, for research into metabolic processes. The cycle is also known as the citric acid cycle or tri-carboxylic acid cycle, because citric acid is one of the intermediate compounds formed during the reactions.

If insufficient oxygen is available, the acid is broken down anaerobically, creating lactate in animals and ethanol in plants and microorganisms. Pyruvate from glycolysis is converted by fermentation to lactate using the enzyme lactate dehydrogenase and the coenzyme NADH in lactate fermentation. Alternatively, it is converted to acetaldehyde and then to ethanol in alcoholic fermentation. Pyruvate is a key intersection in the network of metabolic pathways. Pyruvate can be converted into carbohydrates via gluconeogenesis, to fatty acids or energy through acetyl-CoA, to the amino acid alanine and to ethanol. Therefore, it unites several key metabolic processes. Pyruvate is used in food, cosmetics and pharmaceutical and agricultural applications. Microbial production of pyruvate from either yeast or bacteria relies on restricting the natural catabolism of pyruvate, while also limiting the accumulation of the numerous potential by-products. Pyruvate can be made by classical chemical synthesis or by microbial fermentation, where the latter approach is preferred, especially for sensitive applications such as use in food. The fermentative approach for producing pyruvate was developed much later, probably due to the difficulties associated with getting microorganisms to secrete a central metabolite.

Japan submitted several patent applications for fermentative production of pyruvate using the multivitamin auxotrophic yeast *Candida glabrata* (previously named *Torulopsis glabrata*). *C. glabrata* continues to be one of the best pyruvate producers available. *It* is a good pyruvate producer as it possesses a high tolerance to low pH, which lowers product recovery costs, since it can grow on relatively simple and cheap media.

All the fermentation experiments are performed in a bioreactor equipped with standard control units for stirring speed, temperature, pH, aeration, etc. The pH and dissolved oxygen need to be monitored during the fermentation, and samples are withdrawn regularly to determine glucose, cell density and pyruvate concentration.

14.5.2 Succinic Acid

Succinic acid, which is also known as butanedioic acid, 1,2-ethanedicarboxylic acid and amber acid, occurs in nature as such or in various forms of its esters and is widely utilized in the food industry, pharmaceuticals and agriculture. Traditional applications of succinic acid include food additives, detergents, cosmetics, pigments, toners, cement additives, soldering fluxes and pharmaceutical intermediates.

It is used as a platform chemical for the synthesis of several chemical compounds including polybutylene succinate, depict acid, 1,4-butanediol, tetrahydrofuran and gamma butyrolactone that have industrial values. The commodity chemicals, derived from succinic acid, are further utilized for the manufacturing of biodegradable polyesters, neutralizing agents, pesticides, herbicides, solvents in chemical and battery industries. Before the development of fermentation processes for its production, succinic acid was manufactured by catalytic hydrogenation of maleic anhydride, which is a fossil-based chemical. The biotechnological production of succinic acid is more advantageous over chemical approaches as microbial production offers high yield, purity and cost effectiveness. As an intermediate of several biochemical pathways, succinic acid is produced by many microorganisms. The first isolated microorganism with commercial potential was *Anaerobiospirillum succiniciproducens*, which was discovered by the Michigan Biotechnology Institute. Though succinic acid is synthesized in almost all animals, plants and microbes, bacteria and but fungi are recognized for considerable commercial production of succinic acid due to technical and economical advantages.

14.5.2.1 Biosynthetic Pathway

Succinate can be produced from glucose under both anaerobic and aerobic conditions. Under both conditions, glucose is degraded into phosphoenol pyruvate (PEP) and finally to pyruvate by glycolysis. Through anaerobic fermentation, PEP is the substrate for a carboxylase-catalysed anaplerotic reaction and can be converted to oxaloacetate by PEP carboxylase or PEP carboxykinase (pck). Pyruvate, the end-product of glycolysis, can also be incorporated with CO_2 by pyruvate carboxylase or malic enzyme, forming oxaloacetate or malate. Both oxaloacetate and malate may be further converted to fumarate. Fumarate can be finally reduced to succinate by fumarate reductase.

Under aerobic conditions, succinate is an intermediate of both tricarboxylic acid (TCA) cycle and glyoxylate shunt, but no wild-type microorganism could accumulate it in large quantities. Recently, the oxidative pathway for succinate production has been artificially constructed in *E. coli*, *Saccharomyces cerevisiae* and *Corynebacterium glutamicum* by deleting the critical gene, succinate dehydrogenase, in the TCA cycle. As two carbons are lost as CO_2 in the TCA cycle, the glyoxylate shunt is employed to bypass the steps in the TCA cycle to improve the atom economy. The TCA cycle and the glyoxylate shunt separate at the isocitrate point. In order to block the metabolic flux through the TCA cycle, two key regulatory points, sdh and isocitrate dehydrogenase (icd), must be disrupted. Thus, isocitrate enters the glyoxylate shunt and undergoes cleavage into succinate and glyoxylate, which is catalysed by isocitrate lyase. Then glyoxylate condenses with acetyl-CoA, yielding malate. Finally, malate enters the residual TCA cycle to regenerate isocitrate. For both anaerobic and aerobic metabolisms, branch pathways leading to the formation of formate, acetate, ethanol and lactate also exist to compete with the succinate-producing pathway (Fig. 14.9).

Several fungal, yeast and bacterial microbes have been investigated for succinic acid production and *Actinobacillus succinogenes*, *Basfia succiniciproducens*,

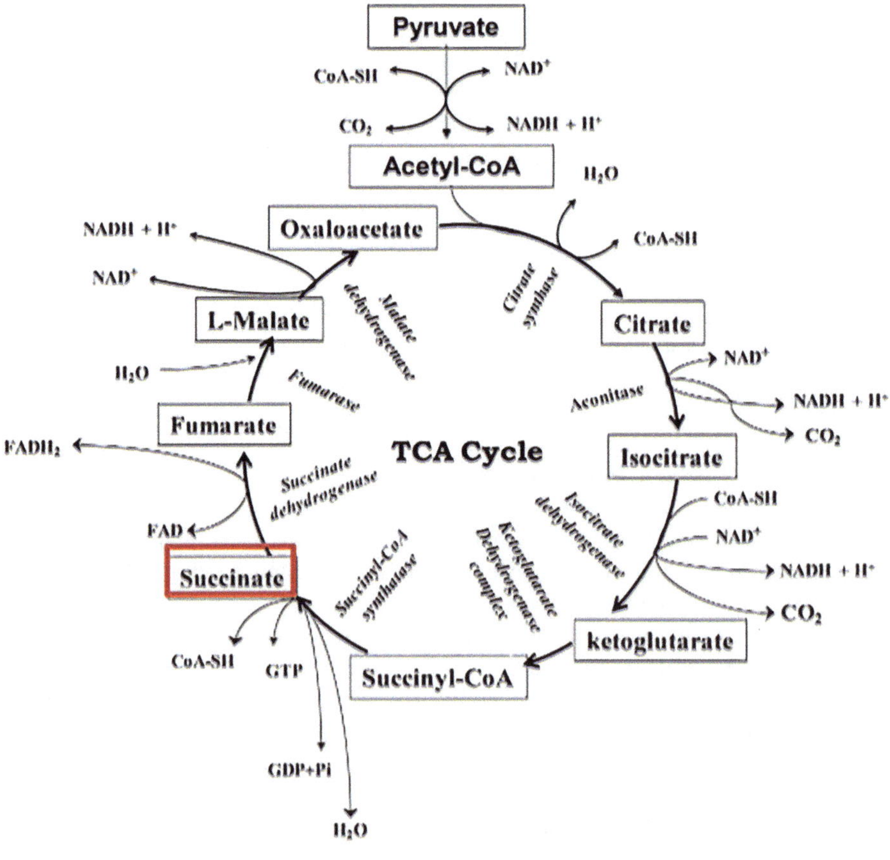

Fig. 14.9 Metabolic pathways for the biosynthesis of succinate

Table 14.3 The typical microorganisms used for fermentative succinate production

	Type	Species	Oxygen requirement
Natural producers	Bacteria	*Actinobacillus succinogenes*	Facultative anaerobe
	Bacteria	*Anaerobiospirillum succiniciproducens*	Strict anaerobe
	Bacteria	*Mannheimia succiniciproducens*	Facultative anaerobe
	Bacteria	*Bacteroides fragilis*	Strict anaerobe
	Bacteria	*Enterococcus faecalis*	Facultative anaerobe
	Bacteria	*Klebsiella pneumoniae*	Facultative anaerobe
	Bacteria	*Succinivibrio dextrinosolvens*	Strict anaerobe
	Fungi	*Aspergillus niger*	Facultative anaerobe
	Fungi	*Paecilomyces variotii*	Facultative anaerobe
	Fungi	*Penicillium simplicissimum*	Facultative anaerobe
Engineered producers	Bacteria	*Escherichia coli*	Strict anaerobe
	Bacteria	*Corynebacterium glutamicum*	Aerobe
	Yeast	*Saccharomyces cerevisiae*	Strict anaerobe

Corynebacterium glutamicum, *Pichia kudriavzevii*, *Escherichia coli*, *Saccharomyces cerevisiae* and *Mannheimia succiniciproducens* are primarily used in large commercial production. *Escherichia coli* is one of the most studied microorganisms for succinate production.

Saccharomyces cerevisiae, or other yeasts such as *Yarrowia*, can thrive under slightly acidic conditions, and methods for their metabolic engineering are well established (Anuradha et al. 1999). *S. cerevisiae* is an industrial organism capable of fermentative production and tolerance of high sugar concentrations. Similar to *E. coli*, the organism has the potential to produce succinic acid both aerobically and anaerobically (Table 14.3).

Batch or fed-batch fermentation is the preferred mode of operation for succinate production. However, continuous production of succinate is likely to outperform batch processing, especially when considering the projections of the following downstream processing. During the past few years, continuous succinate fermentation has been carried out by researchers from different institutes, but only the natural producers such as *M. succiniciproducens*, *A. succiniciproducens* and *A. succinogenes* have been tested. Cell recycle bioreactors with integrated membrane separation system are required for continuous succinate production. Glucose is generally used as the carbon source for succinate fermentation.

Applications

1. Succinic acid is a precursor to some polyesters and a component of some alkyd resins.
2. Succinic acid also serves as the bases of certain biodegradable polymers, which are of interest in tissue engineering applications.
3. It also has been used as a food additive and dietary supplement.
4. Succinic acid is used primarily as an acidity regulator in the food and beverage industry.
5. It is also available as a flavouring agent, contributing a somewhat sour and astringent component to umami taste.
6. As an excipient in pharmaceutical products, it is also used to control acidity. Drugs involving succinate include metoprolol succinate, sumatriptan succinate, doxylaminc succinatc or solifcnacin succinate.

References

Abdel-Rahman MA, Tashiro Y, Sonomoto K (2013) Recent advances in lactic acid production by microbial fermentation processes. Biotechnol Adv 31:877–902

Anuradha R, Suresh AK, Venkatesh KV (1999) Simultaneous saccharification and fermentation of starch to lactic acid. Proc Biochemist 35(3-4):367–375

Crolla A, Kennedy KJ (2001) Optimization of citric acid production from *Candida lipolytica Y-1095* using *n*-paraffin. J Biotechnol 89:27–40

Dong-Mei Z, Huai-Man C, Shen-Qiang W, Chun-Rong Z (2003) Effects of organic acids, o-phenylenediamine and pyrocatechol on cadmium adsorption and desorption in soil. Water Air Soil Pollut 145:109–121

Chapter 15
Microbial Biomaterials and Their Industrial Applications

Chitra Bhattacharya and Mousumi Das

Abstract The present era is in quest of sustainable utility and management of natural resources. Biomaterial in origin is referred as naturally synthesized from living sources like plant and animal and even from microbes and agrowaste utilization too, which is primarily designed to augment, replace, or repair body tissues or organ. Modern concept of biomaterial encompasses a wider approach of application from food to fashion, architecture to agriculture, environment to industry, biofuel to "biofaber, automotive, nanostructures, and so on and so forth. Besides plant and animal sources, a myriad types of microbial biomaterial synthesis are dependent on a unique microbial metabolic and enzyme activity, genetic diversity, and biodegradability nature of the material itself. Besides this, genetically engineered microbes are not far behind for the efficient application in sectors like liquid fuel, functional biomaterials, nanoscale structure, biocomposites, etc. A diverse group of microbial sources like bacteria, fungi, algae, actinomycetes, yeast, etc. as biomaterial agents are widely exploited nowadays to meet the targets of sustainable development goal in abovementioned practices and so on to be mentioned. The present chapter focuses on a detailed overview of basic understanding of biomaterials, categorization of it including source, mode of operation, and involvement of microbial metabolism in usage of substrates and its heterogenous dimension in areas of cement, plastic, drug-delivery carrier, carriers of few important biomolecules in medicine or edible coating in food product packaging practices, etc.

Keywords Construction · Food packaging · Bio-cement · Enzyme · Responsive · Bioplastic

C. Bhattacharya · M. Das (✉)
Department of Microbiology, Atmiya University, Rajkot, India
e-mail: chitra.bhattacharya@atmiyauni.ac.in; mousumi.das@atmiyauni.ac.in

A. Sarkar, I. A. Ahmed (eds.), *Microbial products for future industrialization*, Interdisciplinary Biotechnological Advances,
https://doi.org/10.1007/978-981-99-1737-2_15

15.1 Introduction

The fascinating disclosure of microbes and their metabolites with unflinching utilization in living world especially for human beings is always proven to be a boon. To be precise, concept of biomaterials is *natural in origin* and catabolized by varied living organisms. They have immense biotechnological applications. Although preliminary attention of biomaterials focused on medical applications, eventually in course of time, the unique properties bestowed it to be wider in utilization for areas such as industry like food cosmetics, beverages, plastic, alloy, automotive, and so many, to name a few, or a swath of environmental sustainability and protection. A huge sector of bioeconomy is now influenced by these elements (Pellicer et al. 2017). A plethora of microbial biomaterials are available in nature. The application of biomaterials is age old, but the advancement in utility due to advanced science has bolstered at myriad novel application sectors such as computing, automation, and artificial intelligence (AI) and many more. There is a paradigm shift from a restrictive utility of conventional applications in laundry detergents, biofuel, and high-performance enzymes to genetically engineered microbe exploitation in fabrics, nylon, etc. Few worth mentioning are bacterial cement, mushroom leather, bio-based plastics, rubber, composites, etc. for their explicit usage in newly oriented industries. There are mainly three biomaterial products meeting the demand of market, namely, drop-ins, bio-replacement, and bio-better. These products have a plethora of usages in apparel, automotive, electronics, fast-moving consumer goods (FMCG), and packaging, respectively. It is fascinating to know that there are certain biopolymers native in biodegradability property, explored from various bacteria, including plants and animals as helpful renewable resources in myriad application sectors. Environmental influence with chemical structure is plausibly one of a few important reasons, for which biodegradable nature persists among them and remarkable utilization takes place in myriad ways. There are certain important industries where application of biomaterials such as in food industry is one of the interesting avenues. Human-friendly, sustainable application of biomaterials in preservation and maintenance to enrich the shelf life of various types of packaged food products also nowadays is a hype in commercial market. A huge number and varieties of living organisms like plants, animal, and even many known and also untapped microbial source exploitation spurred the market of necessary commodities fabricated with different types of biomaterials. They can be terrestrial sources or marine origin or can be agricultural by-products. The present chapter focuses on basic overview on microbial biomaterials and their usages, with a highlight on areas of applications of diverse biomaterials where microbial biomass can also be a source too.

15.2 Microbial Biomaterials

The concept behind biomaterial can be illustrated as any material natural or man-made till it can assist in some clinical or surgical purposes. The broad arena covers the fundamentals of medicine, biology, chemistry, tissue engineering, and material science. The green revolution and introduction of sustainability in science and routine life had increased the demand of biocompatible, biodegradable, and bioresorbable materials' usage in recent few years. Originally, suitability of an ideal biomaterials lies in the intrinsic properties like nonimmunogenicity, biocompatibility, and biodegradability, followed by preferable functionalization with bioactive proteins and chemicals. In particular, biodegradability is one of the essential properties of the biomaterials (Cao and Wang 2009). There are three major types of biomaterials, namely, polymers, metals, and ceramics (including carbons, glass ceramics, and glasses) (Ige et al. 2012). Few important commercial employability of biomaterials are noteworthy as follows (Sanjaykumar et al. 2020).

15.2.1 Automobile

Utility of bioplastics in car seat cover, seats, and armrests and in many other automobile parts was reported (car manufacturing companies: Toyota; Daimler-Benz, Germany; Volkswagen; etc.), although the origin was agrowaste. Improved syntheses of microbial bioplastics from genetical engineered strains of *Pseudomonas oleovorans* and *P. putida* were well documented. The polyhydroxyalkanoate catabolism influenced enzyme engineering in its functional groups and had improvised the utilization and yield (Luengo et al. 2003).

15.2.2 Rubber

Replacement of petroleum source for rubber to biomass-based source to abate the pollution is a clear indication toward a sustainable green Earth approach. In good years, US manufacturer of "biotred," a biotier employing nanodroplets of starch, had reflected the less resistance in rolling. Interestingly, bacteria happened to be explored as alternative source of natural rubber (monopoly of *Hevea brasiliensis cis*, 1-4-polyisoprene) by a very sophisticated fabrication of polyhydroxyalkanoic acid in a narrow temperature range, further extended by cross-linking and modified to an elastomer (elastomers are solid polymeric materials, water insoluble, and soluble in few organic solvents). An advanced employment of isoprenoid and polyisoprene biosynthesis pathways stimulated the yield of high molecular weight isoprenoids. The use of genetically engineered strain for greater flexibility of natural rubber production has been attempted successfully (Steinbüchel 2003), yet many questions need to be explored by thorough research pertaining to challenges in metabolic pathways for synthesis of elastomer, enzyme coding gene responsible for the same, etc.

15.2.3 Biocomposites

These blended configurations have revolutionized the usage of biomaterials one step ahead. Flax, hemp, jute, kenaf, etc. are some of the common biofibers popularly used in automobile industry (Ford Motor: heat shields and noise shields are made up of hemp twine and hemp mats). Fungi is one of the most exploited organisms whose mycelium-based foams and sandwich composites are new areas of biocomposites. The lingo-cellulosic wastes as prime substrates utilized for the increased biomass and followed by final product shaping it are utilized for insulating panels, packaging materials, bricks, or new design objects. A unique feature of this wood-degrading basidiomycota fungus is its substrate matrix which got invaded by the hyphae, followed by the development of massive tight net structure. In course of time, the same substrate happens to be replaced somewhat by the fungal biomass. The resulting mycelium is now able to strongly cement the substrate itself. This results in a biocomposite material (Girometta et al. 2019). Myriad sectors, like aerospace, automobiles, packaging, electronics, and civil sectors, are more prominent to be reported for the application of biocomposites from microbial sources. The significant physicomechanical features like high specific mechanical properties, thermal insulation, CO_2 neutrality, good damping properties, high health safety, good fatigue, abrasive and corrosion resistance, availability, high acoustic insulation, low density, less production energy, and lightweightness make them more acceptable than the native forms. To reduce the carbon footprints, the European Automobiles Manufacturers Association's and the European Commission's had targeted for the support of biocomposites.

15.2.4 Biopolymers

The heterogenous applications of biopolymers such as hydrocolloid/lipid composite materials and synthetic polymer/biopolymer mixed materials are now well known. While biodegradability of synthetic polymers is achieved by incorporating microbial degradation of the same and in that starch is most popular followed by cellulose/polyurethane mixtures, gluten/synthetic resin mixtures, vegetable protein/vinyl compound mixtures, lipid/synthetic polymer mixtures, etc. in food packaging industry, these materials now inevitably form the point of economic and sustainable utilization of resources and save the "Mother Earth." Single or combination of two or more materials (proteins, lipids, polysaccharides) can be applied as wrappers in the packaging of food products (Jensen et al. 2015).

15.2.5 Industrial Application of Popular Biopolymers

A huge sector in food industries is dealing with edible films and coatings. Their main advantage is to aid in protection and retain freshness of meat, seafood, fruit, vegetables, grains, confectionery products, and food mixtures that can be canned, deep-frozen, or kept in some other forms of storage. When the questions come on the source of microbes, a diverse sector of microbes like marine organisms, bacteria, microalgae, and seaweed-derived polysaccharide based on various biomaterials is reportedly used in food industries especially in packaging where essentially packaging application of starch, galactomannans, cellulose, and carrageenan are common. Cellulose as edible coating in packaging of food products or in manufacturing and processing of edible cutlery is reported. Chitin usage to prolong the shelf life and antimicrobial efficacy to be promoted is also reported. In similar manner, application of galactomannans, carrageenan films, konjac, starch, pectin, and alginate in edible coating for less lipid oxidation promoting prolonged shelf life or in providing surface consistency of coating paper is widely used. A promising approach of nanoscale research and multifarious packaging ideas such as microemulsions and nanofibers are now stepping in the sustainable world of food packaging (Pellicer et al. 2017). The use of hydrocolloid or lipid composite films acts as an obstacle against water vapor, while the hygroscopic components (proteins or polysaccharides) are a barrier against oxygen and carbon dioxide, where the advantage lies in utility of these elements such as film's matrix (Imran et al. 2012). Translucent and moisture permeability makes the protein collagen and gelatin a preferable choice for biodegradable film formation in dairy industry products' shelf life and maintenance (Chao et al. 2015). Gluten from wheat and zein from corn after alcohol treatment achieving solubility property are used in food industries (Wang and Padua 2005). Likewise, numerous sources of protein from casein, whey, soy, etc. are also used in food packaging industries and bio-based membrane in air filter to trap the pollutants, respectively, based on their physical and mechanical properties' utilization.

15.2.6 Chitosan and Their Industrial Application

The nature of aqueous solubility of chitosan in contrast with chitin is being insoluble due to altered structural configuration (chitosan is a linear polysaccharide composed of randomly b (1!4)-linked D-glucosamine (deacetylated unit) and N-acetyl-D-glucosamine (deacetylated unit); Barikani et al. (2014) recognized it as a food additive by FDA (Food and Drug Administration). Various remarkable properties like membrane barriers, encapsulation, film-forming properties, etc. are made available in chitosan as a food packaging material (Lopez et al. 2014). Chitosan is also used as air filter to trap bacteria (Andrew and Dhakal 2022). In spite of numerous valued products and their commercial importance, a lacunae of utilization of biopolymers is mainly caused by low thermal stability, flammability, high processing cost, and low production volume. These are susceptible to microbial contamination (e.g., fungi and bacterial). So addition of antimicrobial and antifungal components before practice with it is recommended.

15.3 Enzyme-Responsive Biomaterials

There is also an another special type of biomaterials referred in research articles as "enzyme-responsive biomaterials," a means to regulate the tissue development used in regenerative medicine and tissue engineering studies. Enzymes are functional proteins and have an ability to work in ambient conditions like low temperatures, pH in a range between 5 and 8, and in an aqueous environment (Brouns and Dankers 2020). Any biological processes is influenced by enzymatic action and by nature being functional protein; these biocatalysts have the ability to modify the biomolecules to have an outcome of cell responsiveness through the modified biomaterials. A variety of natural and synthetic polymers, cross-linked hydrogels, and self-assembled peptide nanostructure systems can be introduced with enzymatic activity for enhanced biodegradation and an efficient release of therapeutic agents. It can be also used for disease diagnosis (Thomas 2021). Small molecules and proteins inhibit the protease enzymes which plays an important role in the field of biotechnological processes such as manufacture of medicine, agriculture, and molecular farming. Proteolytic enzymes such as proteases, proteinases, and peptidases all are helping to the regulation and reduction of the extreme conditions for unnecessary protein degradation. This proteolytic process permits to control both the functionalities and to rectify the temporal-spatial location of proteins (Santamaría et al. 2014). Protease inhibitors show the significant proteolytic regulatory action. Szałapata et al. (2020) reportedly found the application of immobilized form of three serine protease inhibitors on the surface of various biomaterials commonly used in regenerative medicine. Recent enzyme engineering like self-immobilization, nanostructured advanced biomaterials, and nanozyme technology was also documented as most efficient method for abatement of phenolic pollutants (Cloete et al. 2010). Interestingly to note, laccase-modified silica material implementation in wastewater phenolic removal was also well cited (Galliker et al. 2010). Starting from medicine, diagnostics, fabrication in engineering of medical devices, biosensor, or drug delivery systems, a marked revolution has been noticed in last few years pertaining to applications of biomaterials. But this is not the end, the progression is still on, and few such landmark sectors where biomaterials are doing a game-changing application are presented in Table 15.1 represented in consolidated manner.

Table 15.1 Classes of biomaterials and their properties with special areas of applications

Biomaterial type	Source	Key features of the biomaterial	Applications	References
Biopolymers				
PLA (polylactic acid) and PBS (polybutylene succinate)	*Candida antarctica* (Novozyme 435) and *Streptococcus* sp.	Bio-based "synthetic polymers"/ biodegradable and compostable polymers	Food packaging, automotive	Tawakkal et al. (2014)
Bio-PP, bio-PE, and bio-PET	Acetogenic bacterial species such as *Moorella thermoacetica* and *Clostridium ljungdahlii*	Nonbiodegradable and substituting petrochemical plastics	Food packaging	Islam et al. (2017)
Bioengineering polymers				
PHA	Bacteria species of *Pseudomonas*, *Azotobacter*, *Burkholderia*, *Halomonas*, *Aeromonas*, *Rhodobacter*, etc.	Natural polyester polymers	Packaging materials, consumer and household goods, furniture, sports, health and safety automotive and transport sector, construction and building sector, printing and dye	Elmowafy et al. (2019) and Anjum et al. (2016)
PHB	Actinobacteria—*Rhodococcus* sp.	Wastewater source microbe; PHB content of 46.8 ± 2 wt% DCW (dry cell weight) recovered	Bioremediation and bioconversion	Trakunjae et al. (2021)
PHB	Bacteria—*Alcaligenes* sp.	PHB produced as source of carbon and electron sink in bacteria; PHB content of 2.1 g/L recovered	Food packaging material	Sharma (2019)
Vegetable origin biopolymer				
EPS (exopolysaccharide)	Bacteria	Glucose, mannose, fructose, and galactose and also uronic acid and other noncarbohydrate compounds, such as acetate, pyruvate, succinate, and phosphate as components	Wastewater treatment, cosmetic, pharmaceutical, and food industry (flocculant, thickener and emulsifier, edible coating)	Dwiyantari and Bayu (2020)
"Bacterial cellulose"	*Gluconacetobacter xylinus* and *G. hansenii*		Stabilizing agent or thickener in maintaining viscosity in food	Shoda and Sugano (2005)

(continued)

Table 15.1 (continued)

Biomaterial type	Source	Key features of the biomaterial	Applications	References
		Production process optimization and genetic engineering; two parameters used for this		
"Kefiran" and "dextran"	*Leuconostoc* sp. and *Lactococcus* sp.	Microencapsulation of kefir grain by spray drying	Preservation of fermented food from microbes like this type usually carried out more flexibly	Coma et al. (2019)
Pullulan	*Aureobasidium pullulans* P 56	Cost-effective productivity achieved by the mutant strain	Food production and packaging	Goksungur et al. (2011)
Levan	*Bacillus polymyxa*	Sucrose, sugarcane juice, and sugar beet molasses (substrates used)	Levan nanostructured systems in various forms like gel, matrix, etc. used in drug delivery, coating, adhesive, etc.	Edmilson et al. (2020)
Gellan	*Sphingomonas paucimobilis*	Flow behavior patterns of high acyl gellan interesting to note for better utility	Gellan gum—food polysaccharide gelling agent	Huang et al. (2020)
Microbial biomass elements—proteins and polysaccharides				
Biomass—yeast	*Saccharomyces cerevisiae*	Easy availability of yeast protein and polysaccharide from other production (brewing) industries where yeast is used as *source organism as residual source*. Homogenization, heat, and enzyme treatment allow the film formation	Encapsulation of compounds, coatings, etc. in food packaging industry (casted films—integrated 10% wt dry yeast biomass)	Delgado et al. (2018)
Filamentous fungi biomaterials	*Ganoderma lucidum* and *Pleurotus ostreatus*	Cheap, user-friendly recovery of mycelia and controlled and tunable growth; symbiotic relation with materials used as feed (potato dextrose broth), giving rise mycelial network, lipid concentration	Application in building sector, packaging sector, biodegradable, and cost-effective	Haneef et al. (2017)

		increase in *Ganoderma* sp. and protein concentration, and decreased in *Pleurotus* sp.		
Filamentous fungi	Ascomycete fungus—*Paecilomyces variotii*	Multicomponent film	Sterilization on filmogenic dispersion	Martinez et al. (2017)
Yeast biomass	*Saccharomyces cerevisiae*	Yeast cell wall elements: beta-glucans and proteins, used in biodegradable material synthesis; Mannoproteins; technological applications (adsorption of ochratoxin A, inhibition of crystallization of tartrate, and flocculation of yeasts, etc.)	Biosorption for removal of heavy metals, in wastewater treatment plants, biocontrol agents and pharmaceutical applications	Caridi (2006)
Yeast biomass	*Pleurotus nebrodensis*	Vegetable fiber (cellulose nanofibril)—increased the physicomechanical property (hybrid composites); as filler glued by fungal mycelium; shows acoustic absorbance; substrate optimization influence the tensile strength of mycelia effecting in biomaterial functionality	Construction material	Sun et al. (2019) and Jones et al. (2020)
Yeast biomass	*Basidiomycota* and *Ascomycota* fungus	Efficient renewable and degradable property	Wound dressing material and foam substitute in bumpers or roofs, doors, engine bays, etc. of cars	Kofuji et al. (2010) and Cerimi et al. (2019)
Yeast biomass	Wood-degrading fungus—*Basidiomycota*	Efficient renewable and degradable property	Packaging material in food wrapping and substitute of polystyrene and polyurethane	Girometta et al. (2019)
Fungal biomass	*Basidiomycota*	Electrical circuit board and ecovative patent	Metal salts of copper and aluminum as coating over mycelia	Cerimi et al. (2019)

(continued)

Table 15.1 (continued)

Biomaterial type	Source	Key features of the biomaterial	Applications	References
Fungal biomass	–	Construction and building material	Insulating materials, SIPS (structural insulated panels)	Zeller and Zocher (2012)
Fungal biomass	–	Textile and paper industry	Fungal pulp used in textile	Cerimi et al. (2019)

15.4 Myco-architecture

Nowadays, there is an extensive need to sustainable building materials for the construction industry as we are facing major problems such as global warming depletion of natural resources and fossil fuels. It is tough to accomplish net-zero level energy for buildings. Macroscopic fungi (mushrooms) can play an important role in the field of architecture; basidiomycetes mycelium have the potential to be used in complex material for construction of buildings because of their strong binding capacity to break down the complex nutrient substrate such as sawdust, straw, grains, etc. by using enzymes that can convert cellulose substrate into strong compound chitin, with a focus on vegetative part of the basidiomycetes mycelium, for the exploration of the alternate uses as a living brick (Banupriya et al. 2018).

Mycelium can form the whole structure of the fungal colony through the mycelium fungus which can absorb complex nutrients such as carbohydrates, fat, and protein from its soil surface. Fungal mycelium forages the organic materials and transforms them to the cohesive matter, which act as a living glue. This occurs over a two-stage process. The processes of absorption of nutrient having two stages are as follows:

1. In the first stage, hyphae can excrete the extracellular enzymes that can hydrolyze the complex material (polymers) into the food source (monomers).
2. In the second stage, mycelium absorbs these monomers by active diffusion transport pathways.

15.4.1 Processing of the Myco-bricks

For the preparation of myco-bricks, fungal mycelium collection (spawn) of necessary substance in the wheat or rice is needed. Substrates that can be utilized include paddy straw, fine paddy powder, and sawdust. Some of the spawn would be transferred to the substrate and kept it in cool place for 5–7 days at 28 °C. These are the general steps to allow the substrate to hold the fungal tissue. Then selected mycelium are located in layers in the mold and kept aside for about seven days for the complete growth; the obtained brick is essential to pass the burning process in a hot air oven at 100 °C for 30–40 min, and when taken out, let it cool (Santhosh et al. 2018). The mycelium acts as a living three-dimensional matrix that fixes the substrate particles composed to produce a condensed and impenetrable rigid structure. Figure 15.1 elaborates the stages of myco-brick formation in detail.

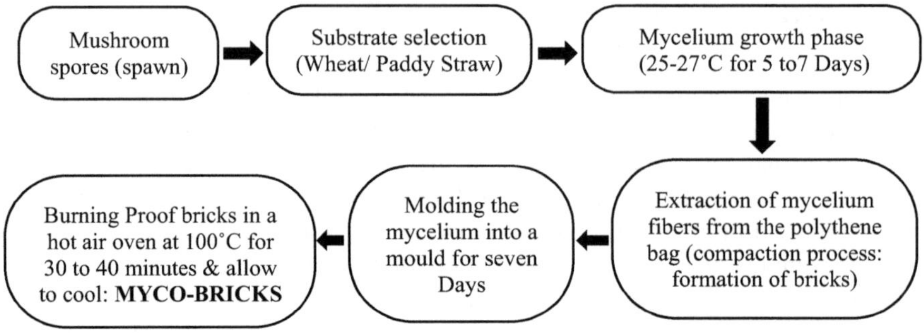

Fig. 15.1 Schematic representation of myco-brick preparation process (Santhosh et al. 2018)

15.4.2 Benefits of Myco-bricks

The development of myco-bricks can take place in the construction of biofabrication and saving of huge overwhelming of fossil fuels. For the biofabrication, lower energy is required for the massive reduction in construction waste. Myco-bricks are biodegradable and eco-friendly and can be prepared by using agrowaste organic materials. Biofabrication process is a carbon neutral building; it can eliminate artificial insulation in walls and use particle board bearing structures. Mycelium bricks contain the anti-termite ability. Myco-brick can be 200,000 times softer than steel and 10,000 times less inflexible than traditional housing brick (Hill 2016).

15.5 Bacterial Biopolymers: Functional Biomaterials

Exploration of new targets for antimicrobial drugs, utilizing synthetic biology for designing and production of innovative biomaterials from bacterial species, is mainly focused on applications in pathogenicity. In context with this, one of the popular strains of *Bacillus* sp. is discussed in this chapter pertaining to the metabolic pathways by which the functional biomaterials synthesized followed by discussion on few important biomaterials available, and its application on various manners was also elaborated as follows.

15.5.1 Bacillus subtilis

Bacillus subtilis belongs to the genus *Bacilli* having the aerobic nature, gram-positive, thick peptidoglycan cell wall layer. *Bacillus* sp. are significantly isolated from the soil and widely used for the production of heterologous proteins (Earl et al. 2008). For their growth and reproduction, they produce numerous enzymes that have the capacity to degrade various substrates in the stressed condition also. For the high secretion of protein ability, physiological characters have been employed in the production of medicinal proteins and industrially important biocatalyst as enzymes. For these reasons, it has been widely used to produce heterologous proteins (Chen et al. 2016; Aslankoohi et al. 2015).

15.5.2 Protein Excretion Systems of Bacillus subtilis

Microbes especially *Bacillus subtilis* have the property to secrete proteins which led them to be widely used in the production of industrially important enzymes. Generally, *B. subtilis* exhibit the three classical protein secretion pathways which include general protein secretion (Sec) pathway, twin-arginine translocation (Tat) pathway, and ATP binding cassette transporters (ABC) (Ling Lin et al. 2007).

15.5.2.1 Sec Pathway

Sec pathways work on the unfolded substrate. It is the main protein transport channel pathway that transports the huge number of exported proteins to the signal recognition site. For the activation of Sec pathways, essential elements are signal recognition particle (SRP), translocase enzyme, type I signal peptidase enzyme, and cytoplasmic chaperones. Sec pathways carry two routes for the translocation of synthesized precursor proteins as follows:

1. In the first step, signal peptide gets recognized by the SRP with the help of cytoplasmic chaperone than it will transfer to the membrane which is bind with the FstY, and then precursor protein gets transferred to the channel of the Sec translocase complex.
2. In the second route, intracellular chaperones can be used to completely fold mechanisms to prevent the translocation ability of precursor proteins and its transfer to the Sec translocase complex. Then N-terminal signal peptide sequence is inhibited by the signal peptidases. Finally, precursor protein is gathered in the extracellular space with the assistance of extracellular cytoplasmic chaperones (Zhang et al. 2020, Yuan et al. 2020).

15.5.2.2 Tat Pathway

In contrast, Tat pathway carriages tightly folded proteins which are having the preserved twin-arginine motif in the signal peptide sequences. Formerly in the translocation process, the precursor proteins folded in the cytoplasm with the assistance of cofactors and evacuated through the Tat translocase complex by using energy of pH gradient through cytoplasmic membrane. After translocation process, type I signal peptidase enzymes developed the signal peptide and folded mature proteins and then excrete out of the cell (Ling Lin et al. 2007).

15.5.2.3 ABC Transporters

ATP binding cassette (ABC) transporter pathways transport the various substrate molecules such as ions, amino acids, antibiotics, peptides, proteins, and polysaccharides. It contains two transmembrane domains (TMD) that can express substrate binding sites and two soluble nucleotide binding domains (NBD) which can act like a motor domain and export and import of all the molecules (Song et al. 2015).

15.5.3 Industrially Important Chemicals Produced by **Bacillus subtilis**

Microbe play an important role in various industrial aspects; it becomes the major microbial cell factory for various industrial products such as vitamins, enzymes, heterologous proteins, amino acids, and antibiotics. In this chapter, we are going to deal with the two industrially important products that are as follows:

15.5.3.1 Lichenase

Lichenase has the enzyme commission number (EC) 3.2.1.73 that contains mixed linked β-glucan (MLG) endo-hydrolase that is obtained from plant and microorganism and utilized for the production of biofuel production. Due to the poor thermal stability, Wang et al. modify the usages of SpyTag/SpyCatcher-mediated cyclization and non-chromatographic ITC purification obtained from the cyclo-lichenase with specific heat resistive quality in lichenase that can be employed for the preparation of biofuel. Lichenase plays the key role in the addition of catalytic action by inducing the collagen triple helix domain and essential oligomers, such as comp and foldon, trigger the target enzyme trimerization, and thus enhance the lichenase activity and increase the production of biofuel (Wang et al. 2016).

15.5.3.2 Hyaluronic Acid (HA)

Hyaluronic acid is commonly called as hyaluronate; it is naturally obtained from mucopolysaccharide which is widely employed in pharmaceutical, cosmetic, and food industries. It has the high-value glycosaminoglycan which contains repeating fragments of N-acetylglucosamine and glucuronic acid and is found in the many tissues of human bodies including lungs, brain, kidneys, and muscles; the maximum concentration of HA occurs in vitreous humor, umbilical cords, and synovial fluid (Necas et al. 2008). Mainly two methods are employing for the extraction of hyaluronate; previously, extraction of HA is based on animal tissues of shark cartilage fermentation. However, this method contains disadvantages as follows: product extracted from animal tissues has poor quality, low yield production contains some additional compounds of proteins which pose a threat to allergies of peoples, and the production time was very high and reduced the amount in purification, whereas in another method, production of hyaluronate is from microbial fermentation. In the studies of Li et al. (2019), *Bacillus subtilis* has the property to produce HA with different molecular weights and titers at various temperatures. Similarly, some of the other bacteria such as GRAS (generally recognized as safe), such as *Lactobacillus lactis* or *Corynebacterium glutamicum*, are also capable to produce hyaluronic acid (Cheng et al. 2016). In the current studies, employment of hyaluronic acid (HA) as a food additive for meat emulsions to prepare a novel functional food enhanced the better quality and water binding capacity (Marzena et al. 2016). In the recent research studies, hyaluronic acid plays a vital role in the food additives to provide the highly hydrophilic nature, which permits to trap approximately 1000 times of weight of water by forming a special coil structure and provide the lubricant of meat (sausages). Addition of 0.05 and 0.1 g kg^{-1} hyaluronate could deliver the stability of meat emulsion by triggering the water discharge from the product and decreased the sensory scores of the produced sausages. In addition, hyaluronate is the only economically viable option for announcing the product on the market (Sheng et al. 2014).

15.6 Conclusion

The extent of microbial biomaterial exploration from single arena of tissue engineering and clinical devices now had swath of regions of applications in automotive, nanostructures, biofabric, biofuel, construction material, "designer material," environment pollution monitoring, and management devices. The avenue is now enriched with advanced implications of source organisms (genetically engineered microbes) besides conventional plants and animal tissues or parts or components as counterparts (Ige et al. 2012). The synthetic material had got beautifully replaced by the biomaterials even in medicine, clinical sectors, automotive, laundry, power house, packaging area, etc. at a large extent. It is also noteworthy that the stringency in utilization of resources and dependency also got relaxed in terms of wider availability of biomaterials in various application sectors. Enzymes such as functional protein are well known, but the advanced science has also contributed for the surface modification (Thomas 2021) of biomaterial structures by fabricating and tailoring, to multiply its usages in bioreactors, for treatment of wastewater or, as immobilized matrices, clinical practices at large extent. Biopolymer also changed its definition of usages simply not being restricted to clothing and food but approaching toward architectural materials, advanced nanostructured edible coating in food packaging industries, medical surgery kits, devices, etc. The polysaccharides, lipids, and proteins are now fabricated to "designer materials" in various food packaging, biofuel, fabric, and many such industrial sectors. The futuristic research of microbial biomaterials is now widening toward nanoengineering, enzyme-responsive biomaterials, biosorption technology by application of microbial biosorbent, functional biomaterials availed from unconventional microbes, and wiser application of microbial pathways like *Bacillus* sp., *E. coli*, *Pseudomonas* sp., etc.

References

Andrew JJ, Dhakal HN (2022) Sustainable biobased composites for advanced applications: recent trends and future opportunities—a critical review. Composites Part C:Open Access 7:100220. https://doi.org/10.1016/j.jcomc.2021.100220

Anjum A, Zuber M, Zia KM, Noreen A, Anjum MN, Tabasum S (2016) Microbial production of polyhydroxyalkanoates (PHAs) and its copolymers: a review of recent advancements. Int J Biol Macromol 89:161–174. https://doi.org/10.1016/j.ijbiomac.2016.04.069

Aslankoohi E, Rezaei MN, Vervoort Y, Courtin CM, Verstrepen KJ (2015) Glycerol production by fermenting yeast cells is essential for optimal bread dough fermentation. PLoS One 10: e0119364

Banupriya R, Nagarajan R, Nijanthan G, Pandi Maharajan M (2018) Mycel bricks—an eco-friendly building material using macroscopic fungi. Int J Comput Eng Sci 8(4):16807–16810

Barikani M, Oliaei E, Seddiqi H, Honarkar H (2014) Preparation and application of chitin and its derivatives: a review. Iran Polym J 23(4):307–326

Brouns J, Dankers PYW (2020) Introduction of enzyme-responsivity in biomaterials to achieve dynamic reciprocity in cell-material interactions. Biomacromolecules 22(1):4–23. https://doi.org/10.1021/acs.biomac.0c00930

Cao Y, Wang B (2009) Biodegradation of silk biomaterials. Int J Mol Sci 10(4):1514–1524

Caridi A (2006) Enological functions of parietal yeast mannoproteins. Antonie Van Leeuwenhoek 89:417–422

Cerimi K, Akkaya KC, Pohl C, Schmidt B, Neubauer P (2019) Fungi as source for new bio-based materials: a patent review. Fungal Biol Biotechnol 6:17

Chao Q, Xiaojing L, Na J et al (2015) Rheological properties and microstructure characterization of normal and waxy corn starch dry heated with soy protein isolate. Food Hydrocoll 48:1–7

Chen J, Zhu Y, Fu G, Song Y, Jin Z, Sun Y, Zhang D (2016) High-level intra- and extra-cellular production of d-psicose 3-epimerase via a modified xylose-inducible expression system in *Bacillus subtilis*. J Ind Microbiol Biotechnol 43:1577–1591

Cheng F, Gong Q, Yu H, Stephanopoulos G (2016) High-titer biosynthesis of hyaluronic acid by recombinant *Corynebacterium glutamicum*. Biotechnol J: Early version available online 11: 574–584

Cloete TE, Kwaadsteniet MD, Botes M, Lpez-Romero JM (2010) Nanotechnology in water treatment applications. Caister Academic press, Norfolk, UK

Coma ME, Peltzer MA, Delgado JF, Salvay AG (2019) Water kefir grains as an innovative source of materials: study of plasticiser content on film properties. Eur Polym J 120:109234

Delgado JF, Peltzer MA, Salvay AG, de la Osa O, Wagner JR (2018) Characterization of thermal, mechanical and hydration properties of novel films based on Saccharomyces cerevisiae biomass. Innov Food Sci Emerg Technol 48:240–247

Dwiyantari W and Bayu M (2020). The application of microbial extracellular polymeric substances in food industry. IOP Conference Series: Earth and Environmental Science 426. 012181. https://doi.org/10.1088/1755-1315/426/1/012181

Earl AM, Losick R, Kolter R (2008) Ecology and genomics of *Bacillus subtilis*. Trends Microbiol 16:269–275

Edmilson C, de Siqueira J, de Souza R, Pinheiro IO, Formiga FR (2020) Levan-based nanostructured systems: an overview. Int J Pharm 580:119242., ISSN 0378-5173. https://doi.org/10.1016/j.ijpharm.2020.119242

Elmowafy EM, Tiboni M, Soliman ME (2019) Biocompatibility, biodegradation and biomedical applications of poly (lactic acid)/poly(lactic-co-glycolic acid) micro and nanoparticles. J Pharm Investig 49:347–380

Galliker P, Hommes G, Schlosser D, Corvini PF, Shahgaldian P (2010) Laccase-modified silica nanoparticles efficiently catalyze the transformation of phenolic compounds. J Colloid Interface Sci 349:98–105

Girometta C, Picco AM, Baiguera RM, Dondi D, Babbini S, Cartabia M, Savino E (2019) Physico-mechanical and thermodynamic properties of mycelium-based biocomposites: a review. Sustain For 11:281

Goksungur Y, Uzuno Gulları P, Dagbagh S (2011) Optimization of pullulan production from hydrolysed potato starch waste by response surface methodology. Carbohydr Polym 83:1330–1337

Haneef M, Ceseracciu L, Canale C, Bayer IS, Heredia-Guerrero JA, Athanassiou A (2017) Advanced materials from fungal mycelium: fabrication and tuning of physical properties. Sci Rep 7:41292

Hill M (2016.) Toadstool footstools: are organisms manufacturing's future? [online] Phys.org. https://phys.org/news/2016-11-toadstoolfootstools-future.html

Huang J, Zhu S, Li C, Zhang C, Ji Y (2020) Cost-effective optimization of gellan gum production by *Sphingomonas paucimobilis* using corn steep liquor. Prep Biochem Biotechnol 50:1–7

Ige OO, Umoru EL, Aribo S (2012) Natural products: a minefield of biomaterials: a review article. ISRN Mater Sci 2012:1–30. https://doi.org/10.5402/2012/983062

Imran M, Revol-Junelles AM, Rene N et al (2012) Microstructure and physico-chemical evaluation of nanoemulsion-based antimicrobial peptides embedded in bioactive packaging films. Food Hydrocoll 29:407–419

Islam MA, Hadadi N, Ataman M, Hatzimanikatis V, Stephanopoulos G (2017) Exploring biochemical pathways for mono-ethylene glycol (MEG) synthesis from synthesis gas. Metab Eng 41:173–181

Jensen A, Lim LT, Barbut S et al (2015) Development and characterization of soy protein films incorporated with cellulose fibers using a hot surface casting technique. LWT-Food Sci Technol 60:162–170

Jones M, Mautner A, Luenco S, Bismarck A, John S (2020) Engineered mycelium composite construction materials from fungal biorefineries: a critical review. Mater Des 187:108397. https://doi.org/10.1016/j.matdes.2019.108397

Kofuji K, Huang Y, Tsubaki K, Kokido F, Nishikawa K, Isobe T, Murata Y (2010) Preparation and evaluation of a novel wound dressing sheet comprised of -glucan–chitosan complex. React Funct Polym 70:784–789

Li Y, Li G, Zhao X, Shao Y, Wu M, Ma T (2019) Regulation of hyaluronic acid molecular weight and titer by temperature in engineered *Bacillus subtilis*. 3 Biotech 9:225

Ling Lin F, Zi Rong X, Wei Fen L, Jiang Bing S, Ping L, Chun Xia H (2007) Protein secretion pathways in *Bacillus subtilis*: implication for optimization of heterologous protein secretion. Biotechnol Adv 25:1–12

Lopez O, Garcia MA, Villar MA, Gentili A, Rodriguez MS, Albertengo L (2014) Thermo-compression of biodegradable thermoplastic corn starch films containing chitin and chitosan. LWT-Food Sci Technol 57(1):106–115

Luengo MJ, Garcia B, Sandoval A, Nharro G, Olivera RE (2003) Bioplastics from microorganisms. Curr Opin Microbiol 6:251–260

Martinez E, Salvay AG, Ludemann V, Peltzer MA (2017) In Proceedings of the XVI congress CYTAL 2017. Argentinean congress in food science and technology, Mar del Plata, Argentina, 18–20. ISBN 978-987-22165-8-0

Marzena Z, Kulawik P, Tkaczewska J, Migdał W, Filipczak-Fiutak M, Fiutak G (2016) The effect of hyaluronic acid addition on the properties of smoked homogenized sausages. J Sci Food Agric 97:2316–2326

Necas J, Bartosikova L, Brauner P, Kolar J (2008) Hyaluronic acid (hyaluronan): a review. Vet Med 53:397–411

Pellicer E, Nikolic D, Sort J, Baro MD, Zivic F, Grujovic N, Grujic R, Pelemis S (2017) Advances in applications of industrial biomaterials. https://doi.org/10.1007/978-3-319-62767-0

Sanjaykumar MS, Kumar BAV, Yogesh B, Pradeep S (2020) Applications of biomaterials in automobile, medicine and food industries: a review. Int J Recent Sci Res 11(12): 40321–40328. https://doi.org/10.24327/ijrsr.2020.1112.5674

Santamaría ME, Diaz-Mendoza M, Diaz I, Martinez M (2014) Plant protein peptidase inhibitors: an evolutionary overview based on comparative genomics. BMC Genomics 15:812

Santhosh BS, Bhavana DR, Rakesh MG (2018) Mycelium composites: an emerging green building material. Int Res J Eng Technol (IRJET) 05(06):3066–3068

Sharma N (2019) Polyhydroxybutyrate (PHB) production by bacteria and its application as biodegradable plastic in various industries. Acad J Polym Sci 2:1–3. https://doi.org/10.19080/AJOP.2019.02.555587

Sheng J, Ling P, Wang F (2014) Constructing a recombinant hyaluronic acid biosynthesis operon and producing food-grade hyaluronic acid in *Lactococcus lactis*. J Ind Microbiol Biotechnol 42: 197–206

Shoda M, Sugano Y (2005) Recent advances in bacterial cellulose production. Biotechnol Bioproc E 10:1–10

Song Y, Nikolof JM, Zhang D (2015) Improving protein production on the level of regulation of both expression and secretion pathways in *Bacillus subtilis*. J Microbiol Biotechnol 25:963–977

Steinbüchel A (2003) Production of rubber-like polymers by microorganisms. Curr Opin Microbiol 6(3):261–270:261–270. https://doi.org/10.1016/s1369-5274(03)00061-4

Sun W, Tajvidi M, Hunt CG et al (2019) Fully bio-based hybrid composites made of wood, fungal mycelium and cellulose Nanofibrils. Sci Rep 9:3766. https://doi.org/10.1038/s41598-019-40442-8

Szałapata K, Osińska-Jaroszuk M, Kapral-Piotrowska J, Pawlikowska-Pawlęga B, Łopucki R, Mroczka R, Jarosz-Wilkołazka A (2020) Serine protease inhibitors—new molecules for modification of polymeric biomaterials. Biomol Ther 10(1):82. https://doi.org/10.3390/biom10010082

Tawakkal ISMA, Cran MJ, Miltz J, Bigger SW (2014) A review of poly (lactic acid)-based materials for antimicrobial packaging. J Food Sci 279:R1477–R1490

Thomas PE (2021) Designing enzyme-responsive biomaterials. In: Peptide-based biomaterials. The Royal Society of Chemistry, pp 76–125. https://doi.org/10.1039/9781839161148-00076

Trakunjae C, Boondaeng A, Apiwatanapiwat W et al (2021) Enhanced polyhydroxybutyrate (PHB) production by newly isolated rare actinomycetes *Rhodococcus* sp. strain BSRT1-1 using response surface methodology. Sci Rep 11:1896. https://doi.org/10.1038/s41598-021-81386-2

Wang Q, Padua GW (2005) Properties of zein films coated with drying oils. J Agric Food Chem 53 (9):3444–3448

Wang J, Wang Y, Wang X, Zhang D, Wu S, Zhang G (2016) Enhanced thermal stability of lichenase from *Bacillus subtilis* 168 by SpyTag/SpyCatcher mediated spontaneous cyclization. Biotechnol Biofuels 9(79):1–7

Yuan S, Liu C, Fang H, Zhang D (2020) *Bacillus subtilis*: a universal cell factory for industry, agriculture, biomaterials and medicine. Microb Cell Factories 19(173):1–12

Zeller P, Zocher D (2012) Ecovative's breakthrough biomaterials. Fungi Mag 5:51–56

Zhang K, Su L, Wu J (2020) Recent advances in recombinant protein production by *Bacillus subtilis*. Annu Rev Food Sci Technol 11:295–318

Chapter 16
Advanced Recombinant DNA Technology (RDT) for Improved Microbial Product Formation

Puja Dokania, Samadrita Roy, Sohom Roy Chawdhury, and Angana Sarkar

Abstract Recombinant DNA technology as a novel innovation has been around for quite some time now. Since its inception, it has revolutionized the molecular biology field, enabling us to modify and edit genomes according to our needs. We have used genome editing tools to procure novel microbial products, in medical applications, as well as to produce novel recombinant vaccines. For the improvement of microbial strains, using recombinant techniques, several methods have been undertaken, such as genetic transformation, insertion, and production of single-cell proteins. Techniques like CRISPR (clustered regularly interspaced short palindromic repeat), ZFN (zinc-finger nuclease), TALEN (transcription activator-like effector nuclease), and gene therapy have all been used for the advancement of RDT. Despite the major advancements and various applications of recombinant techniques over the years, there still, however, are various cons associated, which have to be overcome. Handling risky pathogens and creating unwanted genetic defects are some of the major concerns, along with the associated ethical dilemma. For this mainly, competent bodies have been created, and clearance from the respective authorities is essential for performing recombinant experiments.

Keywords Recombinant DNA technology (RDT) · CRISPR · ZFN · TALEN · Advancement of RDT · Advantages and disadvantages · Strain development

16.1 Introduction

Production of desirable antigenic protein, glycoprotein or a peptide, genome manipulation of the host bacteria, yeast, plant, and viruses has played a very important role, and this process is termed recombinant DNA technology (RDT). Advancement in

P. Dokania · S. Roy · S. R. Chawdhury · A. Sarkar (✉)
Department of Biotechnology and Medical Engineering, National Institute of Technology, Rourkela, India
e-mail: sarkara@nitrkl.ac.in

A. Sarkar, I. A. Ahmed (eds.), *Microbial products for future industrialization*, Interdisciplinary Biotechnological Advances,
https://doi.org/10.1007/978-981-99-1737-2_16

recombinant DNA technology is useful in the field of downstream applications such as vaccine production, therapeutics, and diagnostic which is a result of improved microbial product formation (Mustafa 2001). The base of recombinant DNA technology can be traced to Thomson's discovery in 1887 of the electron, which gave rise to the foundations of biochemistry and contributions in physical and organic chemistry and to the development of genetics and microbiology, Miescher's isolation of DNA in 1869, Avery's demonstration of DNA as genetic material in 1942, Watson and Crick's proposal "The double helix structure of DNA" in 1953, and plasmidology by Bayer and Cohen in 1974. Genetic engineering now plays an integral role in our lives. DNA within the cells contains complex twisted molecules that cannot be manipulated in their native form. It must be first purified, extracted, and precipitated from cellular constituents and cut into desired lengths (Gessler and Patocchi 2007). DNA is made up of functional units called genes. RDT depends upon a series of respective enzymes such as ligase, polymerase, nucleases, reverse transcriptase, and restriction endonuclease, and some of these enzymes are used in molecular cloning and likewise help in the improvement of microbial product formation. In genetic engineering, plasmids have been one of the most important discoveries; they are extrachromosomal genetic elements found in various bacteria. The process of genetic engineering includes a vector (plasmid) that has been taken from a bacterial cell (*E. coli*) and has been digested using restriction endonuclease. On the other hand, the human genome undergoes endonuclease cleavage and generates a DNA insert that contains desired gene of interest. After that, the DNA insert got ligated with the digested (cloning) vector with the help of the DNA ligase enzyme. Now the recombinant DNA plasmid containing the desired gene of interest is introduced into the bacterial host cell by the process called transformation. Recombinant DNA technology has proved its potential enormously, and also dispersal of the products by this technology is limitless and useful to society. The first phenomenal product of this technology was human insulin and other peptides used in human therapy. Genetic engineering has given rise to transgenic animals—an appropriate route for the production of erythropoietin, human albumin, clotting factors, growth factors, regulating factors, and tissue plasminogen activators (Glick and Patten 2017). This technology has been also useful in the field of agriculture, pollution control, oil recovery, chemical industry, human and animal medicine, diagnostic coal and plant diseases, technology, gene complementation therapy, food and animal industry, and desulfurization of oil. This chapter includes basic as well as advancement of recombinant DNA technology, development of strains using RDT, techniques involved in RDT, application of recombinant DNA technology in the development of improved microbial products, and some other applications for the production of recombinant products including vaccines, antibodies, etc., and this chapter also presents advantages and disadvantages of RDT and ethical clearance as well as some future prospects of recombinant DNA technology. The main focus is on improved microbial products and application and advanced RDT techniques for the improvement of microbial product formation.

16.2 Basic Steps and Advancement of Recombinant DNA Technology

Recombinant DNA technology, known as genetic engineering, is a useful technique that changes the phenotype of the host organism by splicing, insertion, and integrating the desired gene into the cloning vector, further carried by the vector into the host, and produces multiple copies by amplification. Depending on the requirement, various expression systems can be employed for the production of recombinant proteins. These may be bacterial host systems, the yeast host system, mammalian cell expression system, plant-based expression system, etc. The RDT is also known as a cornerstone of the biotechnology industry. *Bacillus thuringiensis* (Bt) toxin is one example of a genetically engineered plant that produces insect toxin. This Bt gene produces the toxin which can disrupt gut functions in the larvae of certain insects (caterpillars) which are crop pests, and this gene has been derived from a bacterium, namely, *Bacillus thuringiensis* (Prokop and Bajpai 1991). Using RDT, this development had a great economic impact by reducing the expenses of pesticides used annually and plays a role in increasing the longevity as well as the success of several crops. Besides this, RDT also has applications in vaccine production, prevention, and cure of such diseases as cystic fibrosis, production of clotting factors, and production of recombinant insulin. This process follows as per the propagation (cloning) produces many copies of recombinant bacterial host system containing DNA of interest. Restriction endonucleases do DNA manipulation which occurs in a wide range of bacteria. In the digestion of double-stranded (ds) DNA at high specific rates, each enzyme identifies a precise nucleotide sequence. Genes are interspaced with spacer regions containing signals that control gene expression regulation. Recombinant DNA techniques are carried out by recognizing specific sequences within double-stranded DNA by endonucleases. The most commonly used plasmid pBR322 are ampicillin and tetracycline antibiotic-resistant. They have many unique sites that can be conveniently used to clone suitable DNA fragments.

Molecular cloning and transformation, techniques used in genetic engineering, are less time-consuming and increase the yield of desirable products. Various factors threatened human life greatly, for example, lethal diseases, malnutrition caused by food limitations, environmental problems due to dramatic industrialization, and many others. To overcome such challenges, conventional strategies have been replaced by genetic engineering. Advancement in RDT has played a tremendous role in terms of improving product formation as well as enhancing desired characteristics in living organisms by altering genetic material in vitro.

RDT involves a series of various steps including identification of the gene of interest, amplification of the same by using polymerase chain reaction (PCR), and insertion of the gene in the desired vector; after that the expression takes place in a suitable expression host system. The selection of the expression system includes some critical factors such as immunologically protective epitope production, affordable as well as efficient downstream extraction and clean-up procedures, minimal

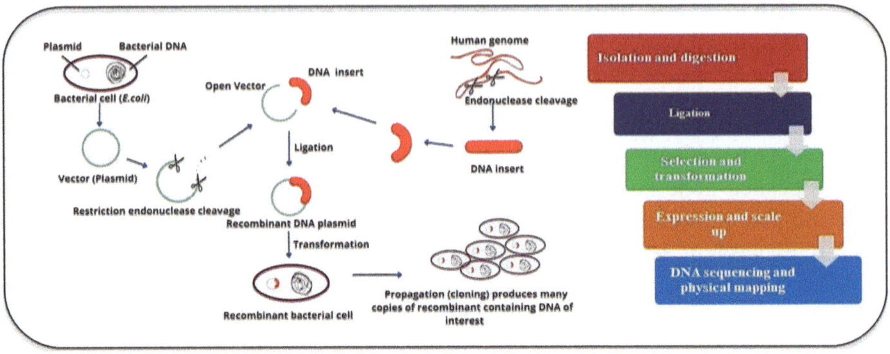

Fig. 16.1 Process of genetic engineering (cloning) in bacteria (*E. coli*) and basic steps of recombinant DNA technology (Shinde et al. 2018)

endotoxin contamination, and minimal immunological interference from host proteins (Rogan and Babiuk 2005) (Fig. 16.1).

Genetic engineering technology is maintained consecutively, to generate the desired product. The basic steps for recombinant DNA technology are the following:

Step1: Isolating and splicing at the specific site of DNA: The foreign DNA molecule from various viruses or cells need to be isolated, purified, and cleaved by specific restriction endonuclease (REN) to produce DNA fragments of various sizes. The plasmid DNA also undergoes cleavage by REN.

Step 2: Ligation of insert DNA fragments into a cloning vector: The cleaved foreign DNA fragment gets covalently attached to the cloning vector such as a plasmid molecule with the help of DNA ligase.

Step 3: Selection and transformation into host organism: Transformation is a process in which the complex ligated DNA, circular (produced in Step 2), is presented to a competent bacterial cell that takes up the DNA. The selection and reorganization of the desired clone will be tested by antibiotic resistance method followed by alternative methods like colony/plaque hybridization, determination of the size of cloned DNA, mutational complementation of the gene, immunologic screening, and DNA-RNA hybridization.

Step 4: DNA sequencing and physical mapping to confirm the cloned gene: Physical mapping including several restriction enzymes as well as DNA sequence analysis has been used for the identity and structural confirmation of the desired clone which was selected using abovementioned methods (Step 3).

Step 5: Expression and scale-up of the cloned gene to produce the desired product: The final target is large-scale protein production using cloned gene. To increase the efficiency of transcription and translation, a suitable promoter sequence and ribosome binding site are required which is joined to a cloned gene resulting in as many as two lakh molecules of the desired protein from a single cell (Johnson 1983).

16.3 Recombinant Microbial Products and Their Applications

The advancement of recombinant DNA technology has played a very important role in the genetic modification of microbes and as a result of which much recombinant microbial product has been developed. These microbial products have a tremendous role in industries. For example, microorganism such as *E. coli* has been known to produce many products like phenylpropanoids, alkaloids, ethanol, D-mannitol, glucosamine, and maltose-binding protein. Lactic acid using the advanced modification in its genome such as TAL (tyrosine ammonia lyase) from *Rhodotorula glutinis* has been heterological expressed. Gene knockout, insertion, the fdh gene encoding formate dehydrogenase from *Mycobacterium vaccae*, and the mdh gene encoding mannitol dehydrogenase from *Leuconostoc pseudomesenteroides* were co-expressed in *E. coli* together with the glf gene encoding the *Z. mobilis* glucose facilitator and overexpression of heterologous β-glucosamine acetyltransferase, fusion tags, deletion, and mutation, respectively. Apart from this, other microorganisms such as *S. aureus*, *Salmonella infantis*, *Synechocystis* sp. *PCC6803*, *Pseudomonas cellulose*, *Schistosoma japonicum*, *Clostridium butyricum*, *Corynebacterium ammoniagenes*, *Erwinia herbicola*, and *Erwinia uredovora* are also utilized to produce the microbial products like lysozyme, isoprene, cellulase, glutathione S-transferase (GST), 1,3-propanediol, vitamin C (ascorbic acid), and carotenoids, respectively.

These microorganisms undergo certain other genetic modifications. They are deletion, transformation, fusion tags, overexpression of the *E. coli* deoxyxylulose phosphate synthase gene in case of *Erwinia uredovora*, mutation, cloning, and overexpression of organism's gene like gene57 in case of microorganism *Corynebacterium ammoniagenes* as well as the expression of the gene encoding 2,5-diketo-D-gluconate reductase from *Corynebacterium* sp. into *Erwinia citreus* and GST tags (Table 16.1).

16.4 Strain Development Using RDT for Improved Microbial Product Formation

There have been several techniques developed which microbial product formation has been improved. For example, genetic transformation: The gene-based strain improvement technique showed an increased production of cephalosporin C. The genes involved in this production process in *A. chrysogenum* were cefG and cefEF encoding acetyltransferase and expandase-hydroxylase, respectively. Gene dosage in higher amount raises the production of cephalosporin and decreases the production of the intermediates. *A. chrysogenum* produce cephalosporin C along with the intermediate at 1–2% of cephalosporin level undesirably and recombinant overproducing strains of *Corynebacterium glutamicum* for L-threonine and L-phenylalanine production. Cloning the fragment containing the gene for streptomycin

Table 16.1 List of genetically modified microbial products and their applications

Sl no.	Products	Microorganism	Genetic modification	Application	References
1.	Lysozyme	*S. aureus* and *Salmonella infantis*	Deletion	Food and agriculture	Khan et al. (2016)
2.	Isoprene	*Synechocystis* sp. PCC6803	*Pueraria montana* (kudzu) isoprene synthase gene has been heterologically expressed.	Production of rubber, pesticides, synthetic oil additives, and biofuels	Marienhagen and Bott (2013)
3.	Phenylpropanoids	*E. coli* and *S. cerevisiae*	TAL (tyrosine ammonia lyase) from *Rhodotorula glutinis* has been heterologically expressed.	Antioxidants, antibacterial, anticancer agents, and antivirals	
4.	Alkaloids	*E. coli* strain	Gene knockout	Medicinal uses	
5.	Cellulase	*Pseudomonas cellulosa*	Transformation	Animal feed processing, textile processing food and brewery production, detergent production, and laundry and paper pulp manufacture	Juturu and Wu (2014)
6.	Glutathione s-transferase (GST)	*Schistosoma japonicum*	GST tags	Protein-DNA interactions, purification, antigens for vaccine studies, and protein-protein interactions	Young et al. (2012)
7.	Ethanol	*E. coli*	Insertion	Manufactures drugs and polishes, cosmetics	Adrio and Demain (2010)
8.	1,3-Propanediol	*Clostridium butyricum*	Insertion	Used as a solvent and polyglycol-like lubricant synthesizes polymers	
9.	D-Mannitol	*E. coli*	Cloning and co-expression of fdh gene and the mdh gene in *E. coli* along with the glf gene	Food, chemical, and pharmaceutical industries	
10.	Riboflavin	*Corynebacterium ammoniagenes*	Cloning and overexpressing the organism's promoter sequence as well as its riboflavin biosynthesis genes57	Antioxidants	

11.	Vitamin C (ascorbic acid)	*Erwinia herbicola* strain contains a gene from *Corynebacterium* sp.	Cloning and expression of 2,5-diketo-D-gluconate reductase encoding gene from *Corynebacterium* sp. into *Erwinia citreus*	Developing agents and preservatives in photo production, and helps in water purification	
12.	Lactic acid	Recombinant *E. coli* strain	Deletion and mutation	Food, chemical, cosmetic, and pharmaceutical industries	
13.	Glucosamine	*E. coli*	Overexpression of a heterologous glucosamine6-P-N acetyltransferase	Treatment of osteoarthritis	
14	Carotenoids	*Erwinia uredovora*	Overexpression of the *E. coli* deoxyxylulose phosphate synthase gene	Additives in food, feed, and pharmaceutical products	
15	Maltose-binding protein	*Escherichia coli K12*	Fusion tags	Improve purification purity	Young et al. (2012)

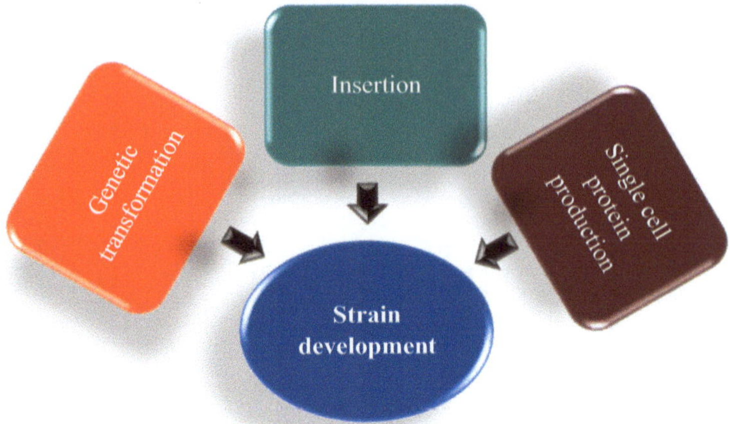

Fig. 16.2 Strain development using RDT

production on a multicopy plasmid resulted in an increase of sevenfold the titer. Cloning also occurs in fungal antibiotic producers between the genes of *Cephalosporium acremonium* and *Penicillium chrysogenum*.Inserting two extra copies of expandase-hydroxylase gene in the desired strain, modification has been done. Now without affecting cephalosporin C, the strain has half as much of the intermediate. Insertion: On multicopy plasmids, benzylpenicillin acylase gene cloning of *E. coli* and addition of penicillin V amidase gene into *Fusarium oxysporum* give rise to a 45-fold increase and 130-fold increase, respectively. These two enzymes are required for converting (Fig. 16.2) penicillin G into a valuable intermediate for synthesizing semisynthetic penicillin (6-aminopenicillanic acid).

Single-cell protein production: This is another as well as one of the first examples of a genetically engineered organism. Assimilating ammonia in *Methylotrophus methylophilus*, the energy-consuming pathway got replaced by the *E. coli* glutamate dehydrogenase pathway, and as a result of which, the yield has improved. In the case of industrial enzymes, the recombinant method has been shown to increase productivity as compared with the traditional method. For example, twice the amount of alpha-amylase is reported compared to traditional methods where the *Bacillus subtilis* strain contains a recombinant plasmid with the *Bacillus amyloliquefaciens* alpha-amylase gene. The industrial strain *Streptomyces violaceonige* carries the xylose isomerase gene in multiple copies and an integrated state. It is used to produce xylose isomerase in large tanks, increasing by 3–5-fold. Genes encoding biosynthetic pathways of several amino acids cloned from different microbial strains that are overproduced either by dosing individual genes or the entire set of genes that are responsible for their respective biochemical pathway.

16.5 Techniques Involved in the Advancement of Recombinant DNA Technology (RDT)

The advancement of recombinant DNA technology involves several techniques such as clustered regularly interspaced short palindromic repeats (CRISPR/CAS-9), zinc-finger nucleases (ZFN), and transcription activator-like effector nucleases (TALENs) which can improve the formation of recombinant microbial products.

The steps of the techniques are as follows:

CRISPR/Cas-9: The genome editing by CRISPR/Cas-9 mechanism is categorized broadly into a few steps. The first step is to select an organism that is going to be edited (Fig. 16.3),

and this organism should be related to humans. Once the model organism got selected, the gene or DNA sequences are selected on which the knockout or deletion or the study is being happening. This step will be helpful for the knowledge of the sequence length, SNPs (single-nucleotide polymorphisms), any mutations, and gene expression, and this is required for the designing of guide RNA (gRNA), the short

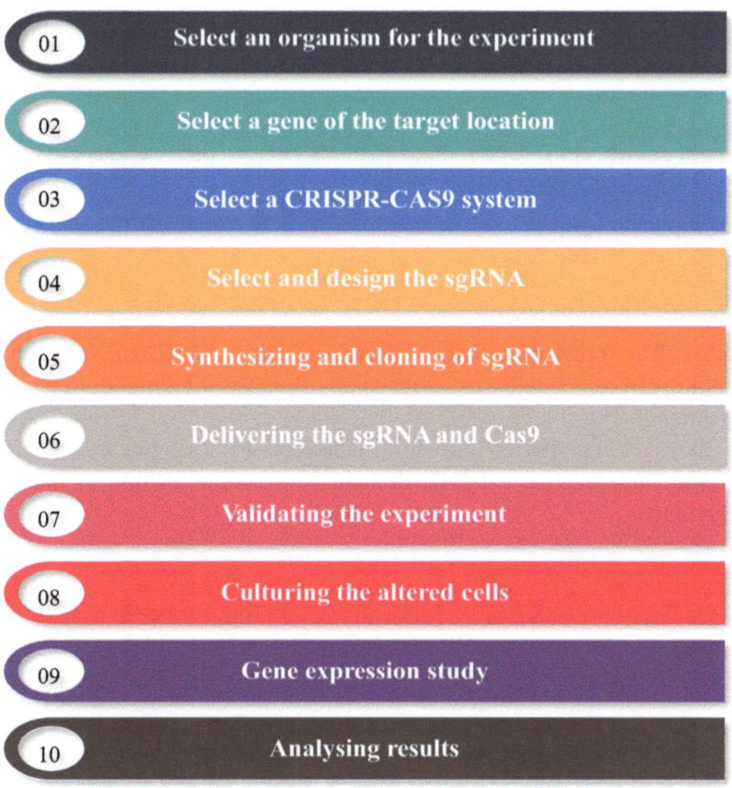

Fig. 16.3 Basic steps of CRISPR/Cas-9 system

RNA sequence which helps in gene editing through targeting a location which is specific and also in terms of editing a gene through locating the PAM (protospacer adjacent motif) sequence (Asmamaw and Zawdie 2021). After that, the CRISPR/Cas-9 system is selected based on the CRISPR and Cas-9 sequences. Based on the data collected from the sequences gRNA is being selected and designed. The next step is the synthesis and making of the clones of gRNA by selecting a specific plasmid. After that, delivery of the sgRNA as well as Cas-9 inside the target cell using the electroporation method has been widely used in the development of recombinant DNA technology. Other CRISPR/Cas-9 delivery systems are gene guns, microinjection, chemical modifications, and sonication. To confirm the accuracy of the knockout of the experiment, validation has been done with the use of polymerase chain reaction and DNA sequencing. The next step is to culture the altered cell, and this CRISPR/Cas-9 system has been performed using the following last two steps which include the study of gene expression studies with the help of quantitative PCR or RT-PCR. Computation and physical examination have been used to analyze the result which is the very last step of this advanced recombinant DNA technology (Nemudryi et al. 2014).

Zing-finger nucleases (ZFNs): The genome editing takes place by creating double-stranded breaks at specified locations. ZFNs contain two domains—DNA binding domain and the DNA cleavage domain. These two domains altogether form "genomic scissors" which are specific. The process occurs by delivery of ZFN into the target cell by electroporation or transfection which is followed by recognition and heterodimerization of ZFN pair around the target site. Then double-stranded breaks are made by ZFNs which then dissociate from DNA. After the dissociation, two types of the template are formed as follows:

(a) No repair template co-transfection with ZFN pair leads to gene deletion by homologous end joining.

(b) Repair template co-transfection with ZFN repair leads to gene integration by homologous recombination (Fig. 16.4).

TALENs: The precise genome editing is done by TALENs (transcription activator-like effector nucleases) consisting of DNA binding domain and Fokl catalytic domain (Fig. 16.5).

Once they get transported into the nucleases, artificial nucleases bind to the desired target site, and the catalytic domain causes dimerization to make double-stranded breaks.

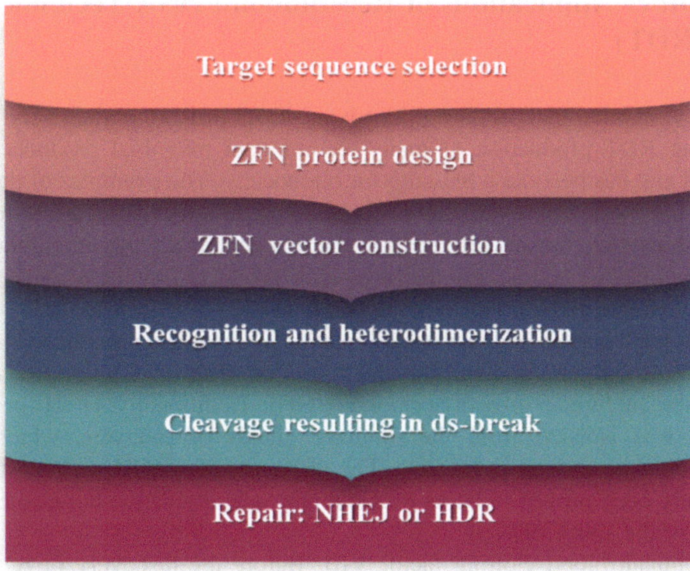

Fig. 16.4 Basic steps of ZFN

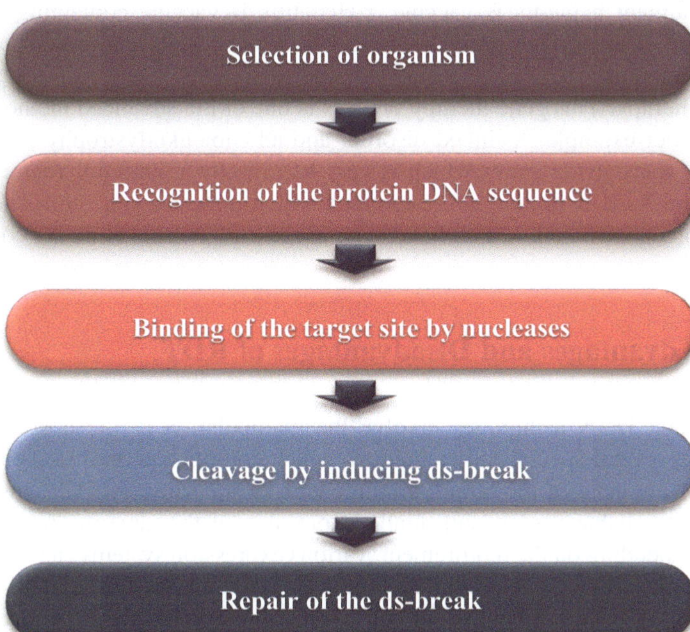

Fig. 16.5 Basic steps of TALEN

16.6 Other Applications of Recombinant DNA Technology (RDT)

For improving microbial products, the advancement of RDT plays a very important role. Using RDT production of many recombinant microbial products has been developed and has become a blessing for our society. The examples of the applications of RDT in the improvement of microbial product formation are as follows:

Recombinant chymosin: Chymosin is the first genetically modified food additive used to manufacture cheese. For the large-scale production of the enzyme, scientists genetically engineered a nonpathogenic strain (K-12) of *E. coli* bacteria.

Recombinant insulin: Human recombinant insulin has been produced by inserting the human insulin gene into species like *E. coli* or yeast.

Recombinant human growth hormone (HGH): Traditionally, HGH for therapeutic usage was isolated from the pituitary glands of cadavers, which lead to Creutzfeldt-Jakob disease in humans. The discovery of r-HGH such as somatropin has not only been proved to be an efficient and safe alternative but also eliminated the cause of the above disease.

Recombinant blood clotting factor VIII: Before the discovery of recombinant blood clotting factor VIII, the protein administered with the forms of bleeding disorder hemophilia carried a tremendous risk of transmission of blood-borne infectious diseases.

Recombinant hepatitis B vaccine: r-Hepatitis B vaccine containing a form of hepatitis B virus surface antigen produced in yeast cells controlled hepatitis B vaccine infection.

Gene therapy: This gene therapy process is one of the advanced methods used in the RDT. For example, X-linked disorder and adrenoleukodystrophy (X-ALD) are treated by transferring genes which is very specific by a lentiviral vector based on HIV-1. The X-ALD protein gets expressed which indicates gene correction of true hematopoietic stem cells which have been successful (Verma et al. 2000).

16.7 Advantages and Disadvantages of RDT

Recombinant DNA techniques involve the manipulation of the host genome by employing several specialized techniques which comprise several steps like identification of the gene of the target, amplification of the same by PCR, inserting the gene into a suitable vector, and finally expression of the gene in the desired host system. Depending on the requirement, various expression systems can be employed for the production of the recombinant products. These may be bacterial host systems, the yeast host system, mammalian cell expression system, plant-based expression system, etc. Recombinant techniques have had a positive impact on the agricultural sector as well, like the introduction of the BT gene isolated from the bacterium *Bacillus thuringiensis* which improved the shelf life and quality of several food crops by inducing insect resistance. Besides this, RDT techniques also have applications in vaccine production, prevention, and cure of cystic fibrosis, production of clotting factors, and production of recombinant insulin.

Recombinant DNA technology techniques have a variety of advantages, which even though are not as apparent as the cons provide ample opportunities to advance this particular field of study. These include the following:

- *Detecting and curing diseases*: The diseases originating from genetic mutations cannot only be tackled using active intervention in the genetic makeup of organisms, like cystic fibrosis, which is a disease with no apparent cure, which can be efficiently cured by using recombinant DNA approaches.
- *Prediction and elimination of any possible diseases:* By studying the genetic makeup, any possibility of disease in the unborn child can be predicted and successfully cured using gene therapy. Genetic engineering can also be used to help people who have the risk of passing on degenerative diseases like Huntington's disease to their offspring. Recombinant techniques can thus be used to virtually ensure a disease-free and healthy life for these individuals and also effectively increase their life expectancy.
- *Effectively increasing the life span:* By using recombinant techniques, the life span of humans can be effectively increased, by inducing modifications or mutations at the genetic level to undo or reverse the specific causes for causing age-related complications.
- *Production of recombinant vaccines:* Vaccines are administered for the prevention of any possible disease. The production strategies for vaccines have found a major boost with the introduction of recombinant techniques. Recombinant vaccines could offer a breakthrough in the medical science approach.
- *Making of tailor-made organisms to suit required demand*: Using genetic engineering and recombinant techniques, it is possible to make specific mutations and alterations in the organisms to show desirable characteristics. For example, it is possible to induce genetic mutations in trees to accelerate their carbon dioxide absorption rate, thus having a positive impact on the environment.
- *Enhanced production rate and durability in plants and animals:* By using these recombinant techniques, it is also possible to effectively alter the maturation rate in both plants and animals. For example, increased durability can be induced in plants to help them survive in unfavorable conditions such as high heat or low light or to increase crop yield. Animals can also be genetically modified, by inducing mutations to ensure improved production.

  Perhaps far more obvious than the advantages are the multifold cons or disadvantages associated with recombinant DNA techniques.
- *The ethical dilemma*: With the breakthrough boom of genetic engineering and recombinant techniques, and its associated promises, also came to the ethical dilemma that it was, in some way interfering with the divine plan and some religions arguing whether or not human beings have the right to alter what has already been planned for them by God.

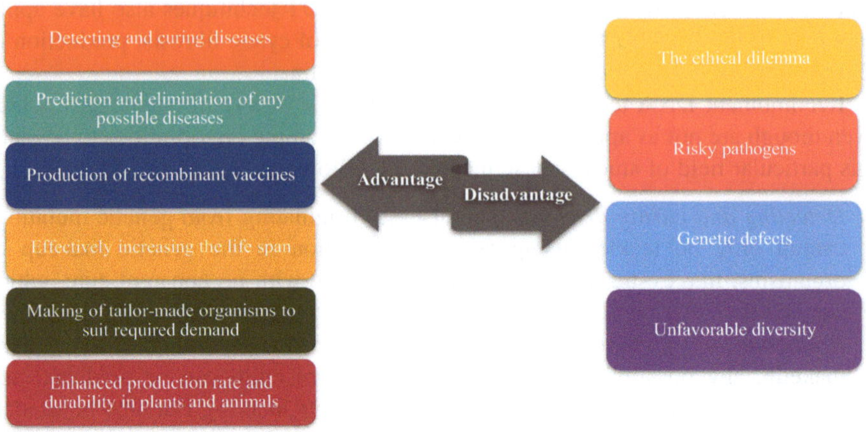

Fig. 16.6 Advantages and disadvantages of RDT

- *Risk of creating genetic defects:* Recombinant DNA techniques pose a serious question mark regarding the safety of making cellular level modifications. Applying genetic engineering techniques to babies, while still in the womb, could present dangers and complications, including miscarriages or even leading to a still baby being born in severe cases. Further adding to this apprehension is that the success rate of genetic engineering and recombinant DNA techniques still leaves a lot of room for improvement.
- *Involvement of risky pathogens:* Recombinant DNA experiments are bound to involve the use of risky pathogens. Failure to infect the genetically engineered recombinant organisms may get even stronger and infect the nongenetically engineered organisms.
- *Can lead to unfavorable diversity:* Genetically engineered recombinant species, when left in the wild, can lead to unfavorable diversity. This is because the recombinant species will have greater resistance against the pathogens or the harsh conditions in the wild, leading to greater susceptibility for the wild variants (Fig. 16.6).

16.8 Ethical Clearance

As a result of all the ethical issues associated with recombinant DNA experiments, various laws and acts have been formulated; failing the clearance of which, genetic engineering or recombinant experiments cannot be performed. Back in 1989, a series of laws and rules were formulated by the Environment Protection Act (EPA) that mandated the best practices needed to ensure a safe working environment with genetically modified organisms and recombinant species. Over the years, a series of other organizations have reformulated and updated the guidelines on the current

Table 16.2 List of competent authorities and their roles

Competent authorities	Role
Genetic Engineering Appraisal Committee (GEAC)	Approval/regulation
District Level Committee (DLC)	Monitoring
Review Committee on Genetic Manipulation (RCGM)	Approval/regulation
Institutional Biosafety Committee (IBSC)	Approval/regulation
Recombinant DNA Advisory Committee (RDAC)	Advisory

ways to work with these products and organisms, like the Review Committee for Genetic Manipulation (RCGM), administered by the DBT (Department of Biotechnology) that administered the recombinant DNA safety guidelines in 1990, or the regulations and guidelines on RDT research and biocontainment in 2017.

The rules of 1989 put forward several points to ensure the safe handling of genetically modified organisms and the smooth running of recombinant experiments. These were the following:

1. Control and regulation of large-scale genetically engineered organisms for any kind of production activities
2. Effective regulation and containment of research activities with genetically engineered and recombinant organisms
3. The release of genetically modified organisms into the environment and monitoring of their effect and environmental applications
4. Monitoring of the import, export, and transfer of genetically modified organisms, hazardous microorganisms, and other recombinant products (Table 16.2)

16.9 Future Prospects of RDT

Recombinant DNA technology has made life for humans smoother and easier in multifold ways as it has proved to be a cornerstone development in science. In the past few years, it has transformed the prevalent treatment strategies for diseases like cancer, diabetes, genetic diseases, and several viruses and fungus-induced disorders in plants, by inducing resistance. RDT techniques also have significant applications in altering the genomes of plants and microorganisms to directly or indirectly facilitate the human cause and improve the quality of life for humans. For the improvement of the RDT products in the future, there are several impending challenges at the genetic level which are required to be addressed. As in the field of pharmaceuticals, there are several grave issues in the way of producing quality products as the mutation produced in the gene could be rejected by the immune system of the body. Also, to increase the formation of products to suit the scale-up requirements, positive results are not always obtained. In the health sector as well, RDT helps in the treatment of several diseases which are virtually untreatable by traditional approaches, though the natural immune responses of the human body hinder achieving good results in this case as well, which is again an impediment that we need to overcome.

Acknowledgment This would not have been possible without the support of the National Institute of Technology (NIT) Rourkela and the guidance from Dr. Angana Sarkar, the corresponding author, who have been supportive throughout writing this chapter and taught the authors more than they could ever give her credit for here. The authors would again like to acknowledge the help and the platform provided by the National Institute of Technology (NIT) Rourkela. Authors would also like to thank their friends and family who supported and offered deep insight into the study.

References

Adrio JL, Demain AL (2010) Recombinant organisms for production of industrial products. Bioeng Bugs 1(2):116–131

Asmamaw M, Zawdie B (2021) Mechanism and applications of CRISPR/Cas-9-mediated genome editing. Biol Targets Ther 15:353–361

Gessler C, Patocchi A (2007) Recombinant DNA technology in apple. Green Gene Technol 107: 113–132

Glick BR, Patten CL (2017) Molecular biotechnology: principles and applications of recombinant DNA, vol 34. John Wiley & Sons

Johnson IS (1983) Human insulin from recombinant DNA technology. Science 219(4585):632–637

Juturu V, Wu JC (2014) Microbial cellulases: engineering, production and applications. Renew Sust Energ Rev 33:188–203

Khan S, Ullah MW, Siddique R, Nabi G, Manan S, Yousaf M, Hou H (2016) Role of recombinant DNA technology to improve life. Int J Genomics 2016:2405954

Marienhagen J, Bott M (2013) Metabolic engineering of microorganisms for the synthesis of plant natural products. J Biotechnol 163(2):166–178

Mustafa AS (2001) Biotechnology in the development of new vaccines and diagnostic reagents against tuberculosis. Curr Pharm Biotechnol 2(2):157–173

Nemudryi AA, Valetdinova KR, Medvedev SP, Zakian SM (2014) TALEN and CRISPR/Cas genome editing systems: tools of discovery. Acta Naturae (англоязычная версия) 6(3 (22)):19–40

Prokop A, Bajpai RK (eds) (1991) Recombinant DNA technology, vol 1. Allied Publishers

Rogan D, Babiuk LA (2005) Novel vaccines from biotechnology. Rev-Off Int Epizoot 24(1): 159–174

Shinde SA, Chavhan SA, Sapkal SB, Shrikhande VN (2018) Recombinant DNA technology and its applications: a review. Int J Medipharm Res 4:79–88

Verma IM, Naldini L, Kafri T, Miyoshi H, Takahashi M, Blömer U, Somia N, Wang L, Gage FH (2000) Gene therapy: promises, problems and prospects. In: Genes and resistance to disease. Springer, Berlin, Heidelberg, pp 147–157

Young CL, Britton ZT, Robinson AS (2012) Recombinant protein expression and purification: a comprehensive review of affinity tags and microbial applications. Biotechnol J 7(5):620–634

Chapter 17
Green Synthesis of Microbial Nanoparticles

Ahmad A. L. Ahmad, Javad B. M. Parambath, and Ahmed A. Mohamed

Abstract In this chapter, we address the green routes for the synthesis of group 11 nanoparticles which include copper, silver, and gold. In our plan, we will discuss microbial routes such as bacteria and fungi, among others. The traditional precursors which are easily reducible in microbial media will be discussed. In this regard, we will also briefly mention the common techniques for the characterization of metal nanoparticles.

Keywords Group 11 nanoparticles · Microbial synthesis · Copper · Silver · Gold · Precursors

17.1 Introduction

Metal nanoparticles (MNPs) have received enormous attention in nanotechnology due to their unique electrical, magnetic, catalytical, antimicrobial, and optical properties compared to their conventional bulk materials (Ahmad et al. 2019). MNPs are used in various applications including drug delivery, nanomedicine, nanosensors, energy production, optoelectronics, environmental treatments, food control, and many others. MNPs can be produced by a wide variety of methods such as pyrolysis, laser ablation, UV irradiation, photo-induced reduction, chemical vapor deposition, microemulsion synthesis, aerosol technologies, and metal salt reduction which is utilized for the production of many MNPs like gold, silver, iron, platinum, and palladium. These preparation techniques showed several limitations because of the temperature, solution pH, high energy, pressure, and production requirements (Jamkhande et al. 2019). In addition, the release of toxic side products from using

---

A. A. L. Ahmad
Department of Chemistry, University of Maine, Orono, ME, USA

J. B. M. Parambath · A. A. Mohamed (✉)
Department of Chemistry, University of Sharjah, Sharjah, United Arab Emirates
e-mail: ah.mohamed@sharjah.ac.ae

A. Sarkar, I. A. Ahmed (eds.), *Microbial products for future industrialization*, Interdisciplinary Biotechnological Advances,
https://doi.org/10.1007/978-981-99-1737-2_17

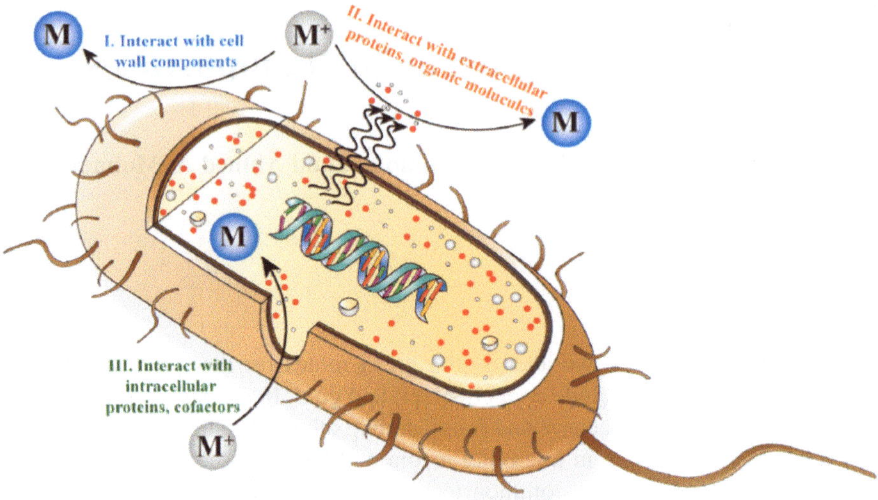

Fig. 17.1 Schematic representation of bacteria synthesis of NPs; both intracellular and extracellular processes are included (Fang et al. 2019). Reproduced with permission from the MDPI (Multidisciplinary Digital Publishing Institute) (2019)

hazardous substances such as reducing agents, stabilizers, surfactants, and organic solvents enhanced the toxicity problems. This led to preventing these nanoparticles to be used in many biological and environmental applications.

The green synthesis of MNPs was applied to address traditional preparation limitations. Natural biodegradable resources were used due to their environmental sustainability, distinctive properties, and wide applicability in biological systems. These green MNPs are prepared using available biological resources such as plants, algae, fungi, actinomycetes, yeast, bacteria, and viruses. This approach of using biodegradable resources in MNP synthesis gains researchers' attention due to reducing the toxic compounds involved in the preparation process without the need for high temperature, pressure, and energy compared with the physical and chemical methods. In addition, nanoparticles' shape and size can be controlled by dominating the cellular growth and activities of the used organism (Tan et al. 2021).

Microorganisms can convert toxic heavy metals that are present in their environment by redox reactions. Internal and external enzymatic actions can reduce the concentration of heavy metals present in the organism's environment and produce nanoparticles. The nanoparticles generated by this method showed higher catalytic activity than other techniques; also it can retain their properties in various conditions like pH, pressure, and temperature (Fang et al. 2019). Thus, MNPs generated could be highly efficient in several applications. Figure 17.1 shows the proposed reductive mechanisms in MNPs' preparation in the bacteria synthesis method. The MNPs can be fabricated both extracellularly by proteins, enzymes, and organic molecules and intracellularly by attracting metal ions inside the cell and reacting with intracellular proteins and cofactors.

The shift from chemical and physical preparation methods to the green synthesis of MNPs is preferred due to their safety and eco-friendliness. This chapter is focused on the green synthesis of MNPs of group 11 which include copper, silver, and gold. These nanoparticles showed great potential in several technologies as they provide unique properties in catalysis, biomedical, biosensors, and optical applications. The improvement and development of MNPs' preparation technology will open exciting possibilities for microbial nanoparticles to be used in every domain of life.

17.2 Green Synthesis of Group 11 Nanoparticles

A decisive, pure, biocompatible, water-mediated, and environmental-friendly method of synthesizing nanoparticles is known as green synthesis. The basic requirements of green synthesis are metal ions and a reducing agent. In general, reducing agents are biological components or biodegradable compounds and often act as stabilizing agents, so the external capping agents are not required (Tan et al. 2021; Kathiraven et al. 2015; Jain et al. 2011; Guo and Wang 2011). Herein, the focus of this chapter is to summarize the current progress in the green synthesis of group 11 nanoparticles such as copper, silver, and gold including their possible oxides. Group 11 nanoparticles are unique in their optical, electronic, and chemical properties, which leads to several applications (Guo and Wang 2011; Iravani 2011). To minimize the use of hazardous chemicals as reducing agents, green synthesis of group 11 nanoparticles was established using microorganisms, plants, and algae (Fig. 17.2a) (Senapati et al. 2005; Lee et al. 2013; Lee et al. 2020).

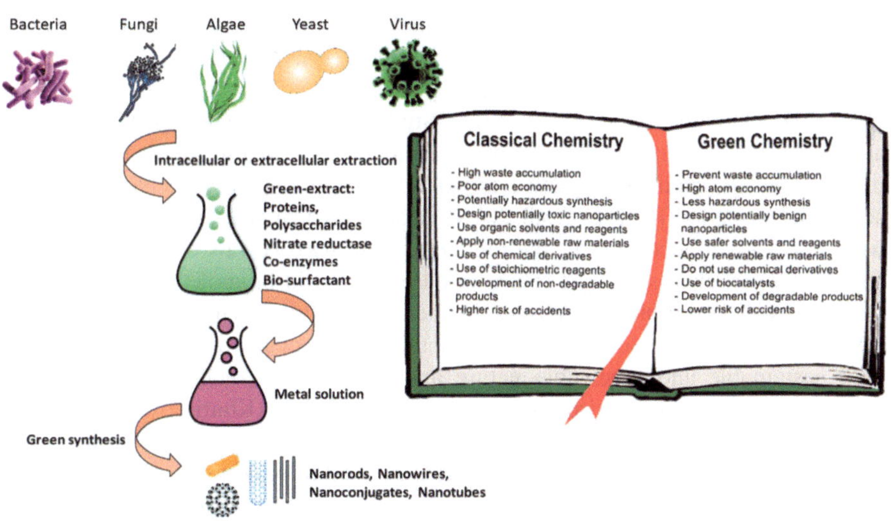

Fig. 17.2 (**a**) Schematic representation of metal nanoparticles' synthesis from microbes (Gahlawat and Choudhury 2019) (Reproduced with permission from the Royal Scientific Society) and (**b**) conceptual comparison of traditional chemistry and green chemistry (El Shafey 2020) (Reproduced with permission from De Gruyter (2020))

Biotechnology emphasizes the synthesis of these nanoparticles by a variety of plant extracts, microbes, and biomolecules (Boisselier and Astruc 2009; Saxena et al. 2012), and their advantage confides to its lesser byproducts, less toxicity, and cheaper raw materials with the minimum use of energy (Lam et al. 2018). The biogenic synthesis of nanoparticles can be classified into two categories as follows:

1. *Bioreduction:* This is a dissimilatory reduction process in which microbes are utilized to conserve energy by oxidation and transferring electrons to metal ions, thereby the metal ions are reduced and the proteins are oxidized (Deplanche et al. 2010). Thus, produced nanoparticles can be recovered from the medium.
2. *Biosorption:* The metal ions bind to the sorbent which belongs to biological origin or organism. Metal ions are bonded to cell organelles or proteins synthesized by living organisms and aggregate into stable nanoparticles (Chakdar et al. 2016). Generally, the resistance mechanism of microbes leads to the formation of nanoparticles.

As far as the mechanism of the extracellular and intracellular biogenic synthesis is concerned, it is proposed that metabolites such as enzymes, proteins, polysaccharides, flavonoids, alkaloids, and phenolic and organic acids are secreted as an outcome of the microbial resistance to metal ions when exposed to different environmental stresses, leads to the reduction of metal ions to nanoparticles (Sharma et al. 2001). Choosing biological methods for nanoparticles' synthesis depends on size, morphology, and application. Biogenic synthesis limits the choice of the organism which builds resistance against metal ions, even though several reports have shown numerous microbial resources such as algae, fungi, bacteria, viruses, and yeast for synthesizing copper, silver, and gold nanoparticles (Iravani 2011; Boisselier and Astruc 2009; Gahlawat and Choudhury 2019).

Several reports described the green synthesis using plant extracts and agricultural biomass, which offers biological media for a stable, faster reduction, and easy scale-up process in nanoparticle synthesis (Singh et al. 2016; Hussain et al. 2016; Ovais et al. 2018; Gholami-Shabani et al. 2017). Studies have shown that plant metabolites are responsible for transferring electrons to metal ions. These metabolites include phytochemicals such as terpenes, terpenoids, alkaloids, polysaccharides, amino acids, and proteins (Rastogi et al. 2018). Morphological size and shape variations were also shown according to the reducing agent used (Ovais et al. 2018). Plant-based nanoparticles' synthesis relies on the variation in biomolecules and concentration between each plant species to the metal ion solution (Gholami-Shabani et al. 2017). The advantages of plants over microbes were claimed by many authors as their sustainable and renewable nature and biomass utilization which generates further added value to environmental remediation status (Kharissova et al. 2013). Thus, the ability of the plants to synthesize nanoparticles has shown a new horizon and spectacular approach toward the natural method of scale-up.

A more convenient method of the synthesis of group 11 nanoparticles would be through traditional chemical approaches, with tunable size and morphology. However, this approach requires harmful solvents, toxic chemicals, elevated temperature, and high pressure (Fig. 17.2b). The aftermath of these synthetic strategies is environmental contamination. There we have a green synthesis of copper, silver, and gold nanoparticles as a promising method for an eco-friendly approach. This chapter contributes a comprehensive outline of a potential green synthetic approach for Cu, Ag, and Au nanoparticles.

17.3 Green Synthesis of Copper Nanoparticles

Copper nanoparticles (CuNPs) are widely used in sensor devices such as glucose sensors, amino acid detectors, etc. (Li et al. 2020; Chakrapani et al. 2014; Heli et al. 2009). Because of their high catalytic activity over a wide pH and temperature range, they are useful for various catalytic applications (Gawande et al. 2016). In addition, it is used in electronic devices due to its thermostability and high conductivity (Venkata Abhinav et al. 2015). When these devices are disposed of as electronic scrap, the nanoparticles accumulate in the soil and water. With time, the amount of such waste in the environment gradually increases. CuNPs also can be bioaccumulated in the environment.

Many methods such as electrochemical synthesis, thermal evaporation, and chemical vapor deposition, as well as atomic layer deposition, have been used in the synthesis of CuNPs (Gawande et al. 2016; John et al. 2021). However, physical processes have several disadvantages such as expensive vacuum techniques. Chemical synthesis constrains the clinical use of nanoparticles because of the use of toxic chemicals that can lead to biological and environmental hazards. In contrast, the green synthesis of CuNPs is environmentally friendly and inexpensive and does not use toxic chemicals, making CuNPs more useful for biomedical applications.

17.3.1 Plant-Mediated Synthesis of Copper Nanoparticles

Plant extracts are perfect sources for the production of MNPs which are easy, simple, and safe, consume low energy, and have better stability of the nanoparticles (Nasrollahzadeh et al. 2016). In this approach, the metal salts are mixed for several hours at room temperature with plant extracts to complete the preparation (see Fig. 17.3). Plant extracts contain many active compounds such as phenols, proteins, tannins, and flavonoids which act as reducing and stabilizing agents for the copper ions (Singh et al. 2017).

Copper and copper oxide (CuO, Cu_2O, Cu_4O_3) nanoparticles' synthesis strategy is the same in terms of preparation parameters such as temperature and pH. However, these parameters can affect the type of CuNPs' types, shapes, and properties (Singh et al. 2018). For example, using *Ixora coccinea* leaf extract in the preparation of CuNPs at 27 °C produced 80–110 nm particles (Vishveshvar et al. 2018), while using the same precursor in the preparation with *Pterolobium hexapetalum* leaf extract produced smaller nanoparticles of 10–50 nm size at

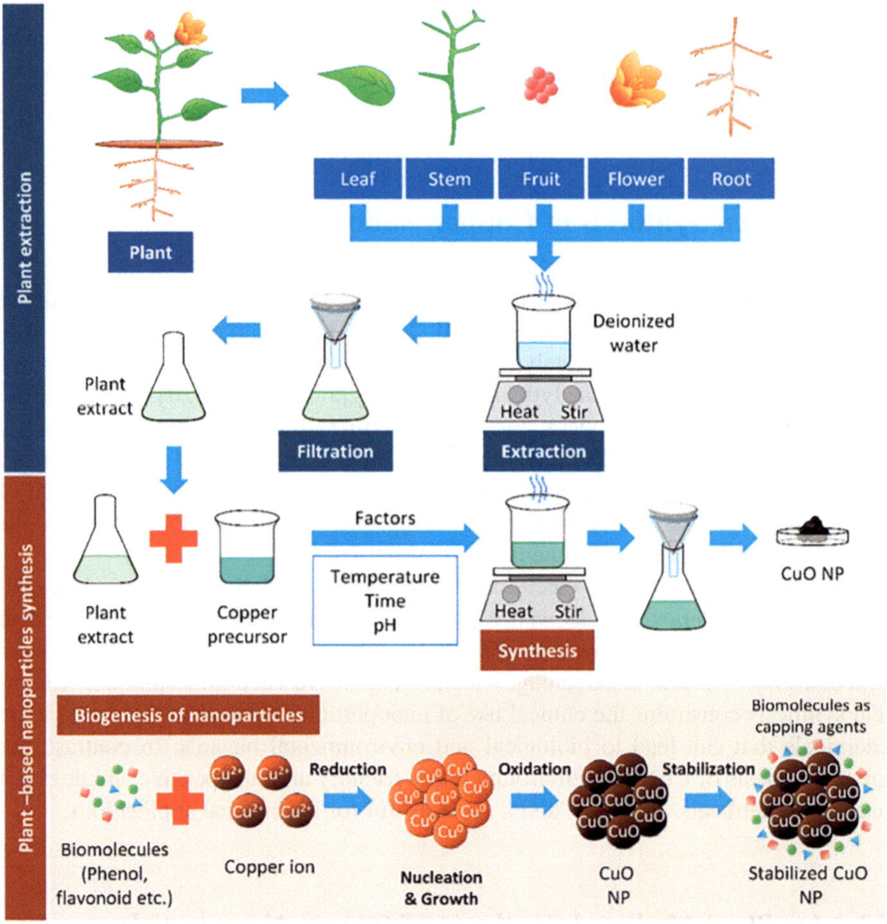

Fig 17.3 Synthesis of copper and copper oxide nanoparticles (Letchumanan et al. 2021). Reproduced with permission from MDPI (2020)

60 °C (Nagaraj et al. 2019). In addition, the texture of some biosynthesized nanoparticles was influenced by the variation in pH. In most cases, the optimal pH value for the biogenic synthesis of nanoparticles from plant extracts is 7–9 (Waris et al. 2021).

17.3.2 Fungi-Mediated Synthesis of Copper Nanoparticles

Fungi have been used for the preparation of CuNPs and other MNPs (Narayanan and Sakthivel 2010). The preparation pathways include intracellular and extracellular methods. Intracellular pathways produced smaller and good dispersibility of nanoparticles (Cuevas et al. 2015) than the extracellular methods which produced more clean nanoparticles free of cell components (Shankar et al. 2003). Copper oxide nanoparticles with a size of 110 nm were prepared using fungal strain *Trichoderma* by the reduction of $Cu(NO_3)_2$ salt overnight in the dark (Saravanakumar et al. 2019). The presence of aromatic metabolites and amide groups on the surface of the nanoparticles indicates the compounds responsible for the reduction and stabilization. White rot fungus *Stereum hirsutum* were used to prepare relatively small nanoparticles of 5–20 nm size from several copper precursors, namely, $CuCl_2$, $Cu(NO_3)_2$, and $CuSO_4$ (Cuevas et al. 2015). This synthesis method produced spherical-shaped copper oxide nanoparticles.

17.3.3 Bacteria-Mediated Synthesis of Copper Nanoparticles

Using bacteria in the production of MNPs has the advantage of their shorter synthesis times versus other microorganisms and mild experimental condition; it is easy to culture and can be modified at the genetic level (Narayanan and Sakthivel 2010). The production of MNPs occurs by reducing metal ions by bacteria metabolites such as proteins and polysaccharides. Gram-negative bacterium *Serratia* sp. was used to prepare polydisperse CuO NPs with an average size of 10–30 nm (Hasan et al. 2007). The characterization analysis showed different functional groups on the surface of the nanoparticles like primary and secondary protein amides which indicates that these compounds act as stabilizing agents. The IR (infrared spectroscopy) analysis also detected the presence of these proteins on the CuNPs' structure using *Escherichia coli* bacteria (Alasvand Zarasvand and Rai 2016). The investigation also showed the ability of these nanoparticles to inhibit the growth of *D. marinisedimins* bacteria that can cause metal corrosion.

Some parameters can affect the CuNPs produced by bacteria. For example, the initial concentration of copper ions has a high effect on the production of nanoparticles. *Penicillium aurantiogriseum*, *Penicillium waksmanii*, and *Penicillium citrinum* were used in the preparation of CuO NPs from different concentrations of $CuSO_4$ solutions at different pH values (Honary et al. 2012). The results showed a direct correlation between $CuSO_4$ concentration and diameter of the nanoparticles, whereas the lowest pH produced the smallest nanoparticles. This observation was present in the three investigated species.

17.4 Green Synthesis of Silver Nanoparticles

Silver nanoparticles (AgNPs) have shown a significant role in nanotechnology due to their repressive effect on several bacterial strains including drug-resistant bacteria (Rai et al. 2012; Durán et al. 2016). Moreover, AgNPs appeared to be catalysts, antifungal agents, and sensors (Priyaragini et al. 2014; Rizwan et al. 2014). Nano-scale biomedical properties of AgNPs have been utilized by several commodities, including clothing, medicines, and cosmetics.

Green synthesis of AgNPs has been reported using various plant extracts, microbes, cellular and agricultural biomass, and biomolecules. The plants used for AgNPs' synthesis range from phytoplankton to angiosperms; often plant extracts from the leaf, root, fruit, and stem (Rai et al. 2012; Kumar et al. 2014; Edison and Sethuraman 2012; Patil et al. 2012; Mondal et al. 2014; Ajitha et al. 2015). Mostly, these extracts contain biomolecules like proteins and phytochemicals which act as both reducing and stabilizing agents with some exceptions; in a few cases, certain nontoxic capping agents were used (Rao et al. 2014). Recent reports showed that flowering plant *Origanum vulgare* extract reduced silver ions to the dark-brown silver nanoparticles (Shaik et al. 2018). Another study reported the reduction of silver nitrate using aqueous leaf extracts of *Azadirachta indica* to yellowish-brown AgNPs (Ahmed et al. 2016).

The antimicrobial activity of AgNPs synthesized using *Tamarix gallica* plant extract was reported to be in a size range of 5–40 nm (López-Miranda et al. 2016). It was found that the color change from yellow to yellowish-brown during AgNPs' synthesis is dependent on the particle size. AgNPs synthesized using *Cleome viscosa* plant extract showed a size range of 20–50 nm in TEM (transmission electron microscopy) analysis and showed significant antibacterial activity against gram-positive bacteria (Lakshmanan et al. 2018). Further, they utilized these AgNPs to study anticancer activity on the lung A549 and ovarian PA1 cancer cell lines. Further common reports showed silver nanoparticles in cancer research (Yezhelyev et al. 2006; Ramkumar et al. 2017; Khalifa et al. 2016; Bhattacharya and Gupta 2005). A seaweed *Enteromorpha compressa* extract which reduced AgNPs was used in the cytotoxicity against Ehrlich ascites carcinoma (EAC) cells as an anticancer agent (Ramkumar et al. 2017). Algal and microalgal extract-mediated AgNPs' synthesis has also shown significant cytotoxicity in tumor cell lines. *Turbinaria turbinate*, *Anabaena oryzae*, *Nostoc muscorum*, and *Calothrix marchica* are a few algal species utilized for the synthesis of AgNPs with antibacterial and cytotoxicity against tumor cells (Khalifa et al. 2016; Bhattacharya and Gupta 2005; El Bialy et al. 2017). Balanet et al. reported AgNPs' synthesis using aqueous leaf extract of *L. japonica* and studied their antidiabetic activity (Balan et al. 2016).

Studies on AgNPs' inhibition to α-amylase, α-glucosidase, and free radical were examined as shown in Fig. 17.4a. The oligodynamic germicidal property of AgNPs is also well studied and utilized in the biomedical industry (Prasher et al. 2018). Reports showed that AgNPs from various plant extracts with variable particle sizes have potential germicidal properties (Kathiraven et al. 2015; Rahimi-Nasrabadi et al. 2014; Kulkarni et al. 2012; Nalwade et al. 2013). Although there are few reports showing silver oxide nanoparticle synthesis including plant extracts (Rashmi et al. 2020; Manikandan et al. 2017; Ravichandran et al. 2016), a few studies are reported relative to silver nanoparticles. Ravichandran et al. synthesized Ag_2O in nanoscale using *Callistemon lanceolatus* plant extract (Ravichandran et al. 2016). According to the report, the aqueous plant extract was treated with silver nitrate to produce silver

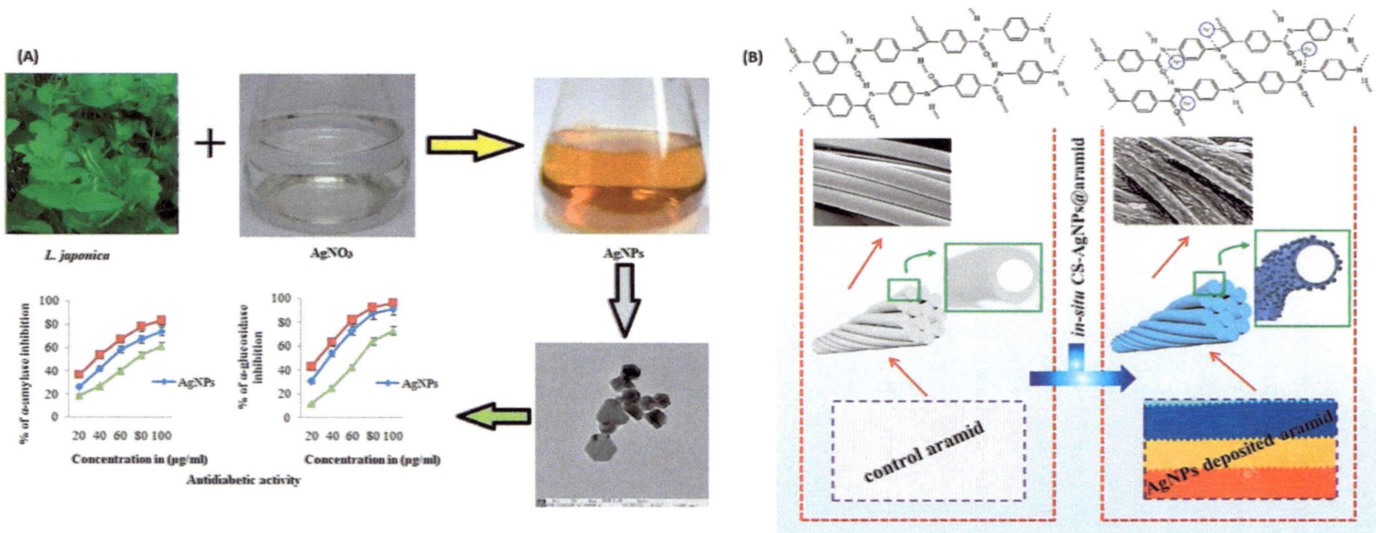

Fig. 17.4 (**a**) Schematic representation of the green synthesis of AgNPs using *L. japonica* leaf extract and antidiabetic activity showing inhibition of a-amylase and a-glucosidase (Balan et al. 2016) (Reproduced with permission from the Royal Scientific Society (2016)) and (**b**) schematic representation of in situ synthesized colorful AgNPs on aramid fabric using chitosan (Hasan et al. 2020b) (Reproduced with permission from Elsevier (2020))

hydroxides and further incubated at 37 °C to produce Ag_2O. Recently, Alghoraibia et al. studied the relation between phenolic content in the plant extracts and reducing ability (Alghoraibi et al. 2020). The authors confirmed that the increased concentration of phenolic compounds enhanced efficient metal ion reduction. Recently, Renuka et al. synthesized AgNPs with uniform size and morphology using an Indian gooseberry (*Phyllanthus emblica*) fruit extract (Renuka et al. 2020). These hexagonal AgNPs showed significant antibacterial effects against *Klebsiella pneumoniae* and *Staphylococcus aureus*.

Chitin is an abundant natural polymer similar to cellulose used for AgNPs' synthesis (Hasan et al. 2020a). Chitosan is generally used for the reduction and capping of nanoparticles as an easily available derivative of chitin. This homopolymer of β-(1–4)-linked N-acetyl-D-glucosamine was reported as a green synthetic reagent for AgNPs' preparation (Kalaivani et al. 2018; Hasan et al. 2020b). Recently, Hasan et al. synthesized chitosan-mediated AgNPs on the aramid fiber surface to give different colors, antibacterial properties, and thermal stability (Fig. 17.4b) (Hasan et al. 2020b). Several other reports showed chitosan-reduced AgNPs used for functionalizing textile fabrics (Butola and Kumar 2020; Hasan et al. 2020c; Hasan et al. 2019).

Silver nanoparticles' synthesis using microbial culture is arduous compared to plant extracts or biomass-mediated. A general method of AgNPs' synthesis using microbes involves the culturing of isolated or genetically modified microbes in media such as Luria–Bertani broth (LB), glucose, malt, and yeast extracts. Biomolecules and enzymes produced in the culture media are responsible for the resilient reduction and stabilization (Liu et al. 2014). The major difficulty in maintaining microbial growth, culture media, and inoculation is still an existing challenge in microbial synthesis. However, several reports have shown fungal and bacterial assisted AgNPs' synthesis (Gan and Li 2012; Arun et al. 2014). This may be through either utilizing extracellular secretions or the whole microbial biomass. The extracellular synthetic route is often observed, whereas few reports have shown intracellular AgNPs' formation (Mukherjee et al. 2001).

Different bacterial supernatants were used in the synthesis of AgNPs by aqueous mediated silver ion reduction. This method demonstrated a quick reduction of metal ions in cell filtrate and found that piperitone partially inhibited the AgNPs' formation (Shahverdi et al. 2007). It is proposed that biogenic synthesis might get inhibited by different gram-negative bacterial strains. However, reports using individual strains have shown successful AgNPs' synthesis using various bacterial strains. Korbekandi et al. utilized *Lactobacillus casei* for AgNPs' synthesis (Korbekandi et al. 2012). Liu et al. used dried cells of *Bacillus megaterium* (Liu et al. 2000), whereas Das et al. used *Bacillus* strain CS 11 culture media for extracellular AgNPs' formation (Das et al. 2014). Lee et al. synthesized AgNPs using bacterial cellulose and dopamine to develop magnetic nanocomposites (Lee et al. 2013). Incorporating Fe_3O_4 into bacterial cellulose and reducing silver ions using dopamine-coated Fe_3O_4 bacterial cellulose solution in basic medium produced AgNPs-Fe_3O_4 composite.

Fungal species were also used for AgNPs' synthesis. *Fusarium*, *Aspergillus*, *Trichoderma verticillium*, *Rhizopus*, and *Penicillium species* are the most explored fungi in AgNPs' synthesis (Rai et al. 2021; Philip 2009). *Trichoderma* species such as *T. reesei* have gained considerable interest due to their harmless nature, whereas several fungal species are harmful to plants and animals (Guilger-Casagrande et al. 2019). Vigneshwaran et al. used the fungus *Aspergillus flavus* to reduce silver nitrate on the cell wall (Vigneshwaran et al. 2007). Xue et al. synthesized AgNPs with antifungal properties using *Arthroderma fulvum* (Tile and Bholay 2012). Reports have shown that most of the fungal species such as *Fusarium solani*, *Humicola* sp., *Pleurotus cornucopiae*, *Aspergillus fumigatus*, and *Fusarium oxysporum* produce extracellular silver nanoparticles (Ingle et al. 2008; Csavina et al. 2014; Owaid et al. 2015; Bhainsa and D'Souza 2006).

17.5 Green Synthesis of Gold Nanoparticles

Gold nanoparticles (AuNPs) have been extensively utilized in a wide range of applications from biomedical to energy materials for the past few years. The green synthetic approach used for the synthesis of such nanoparticles involves the reduction of gold salts to gold(0) and capping reduced metallic gold in the nanoscale range with a variety of nontoxic biodegradable agents. Gold nanoparticles have been described as the most robust (Daniel and Astruc 2004). As innate nanoparticles, they are biologically inactive and stable. Functionalization of AuNPs rather enhances their reactivity, photoactivity, and other chemical activities. For example, AuNPs have photothermal properties upon near-infrared (NIR) irradiation, namely, tuning particle size and plasmonic oscillation. Specific properties such as localized surface plasmon resonance (LSPR), and the quantum size effects of AuNPs with unique stability fascinate their potential applications.

Synthesis of gold nanoparticles by microorganisms, plant extracts, biomolecules, etc. belongs to the emerging green chemistry synthetic strategies and requires further developments in understanding the mechanistic perspectives to scale up the industrial production of well-dispersed gold nanoparticles. Among the few earlier reports, AuNPs prepared by *Zingiber officinale* extract have a surface plasmon resonance (SPR) peak around 523 nm with a particle size of 4–13 nm (Kumar et al. 2011a). *Cassia auriculata* leaf extract generated different morphologies such as spherical, hexagonal, and triangular AuNPs with a size range of 15–25 nm (Kumar et al. 2011b). *Sorbus aucuparia* was used to reduce $HAuCl_4$ to generate *fcc* AuNPs with an average crystallite size of 18 nm, calculated using the Debye–Scherrer equation (Dubey et al. 2010). Mostly, AuNPs synthesized using plant extracts were monodispersed with spherical morphology and have an average particle size of 45 nm (Elbagory et al. 2016; Elia et al. 2014). As discussed earlier, plant-mediated green synthesis of AuNPs also depends on the concentration of phytochemicals as they are directly proportional to the reduction rate due to their variations in response to biotic and abiotic stresses (Lee et al. 2016).

AuNPs have been synthesized from microorganisms as an effective green synthetic strategy. *Bacillus subtilis* were used to synthesize octahedral AuNPs with a size range of 5–25 nm by reducing $HAuCl_4$ (Fortin and Beveridge 2000). Under optimum conditions, gold precursors were added to the bacterial culture medium for reduction. *Geobacillus* sp. produced purple color of AuNPs with a quasi-hexagonal morphology and 5–50 nm size (Correa-Llantén et al. 2013). Bacterial cells exposed to Au(III) ions turned the solution intense purple color indicating the accumulation of intracellular AuNPs. *Delftia acidovorans* showed resistance against toxic gold ions by generating Delftibactin, a protein that reduces Au(III) (Johnston et al. 2013). Alternatively, synthesis by the bacterium *Rhodopseudomonas capsulata* produced extracellular AuNPs with 10–20 nm size via NADH reductase (He et al. 2007).

Similarly, fungi culture media containing mycelia-free extracts were used for the green synthesis of AuNPs. Reports showed that *Penicillium aculeatum* generated spherical AuNPs with a particle size of 60 nm (Barabadi et al. 2017). *Penicillium citrinum* produced irregular-shaped AuNPs with a size range of 60–80 nm, whereas (Manjunath et al. 2017) *Aspergillus sydowii* generated AuNPs of relatively smaller sizes of 8.7–15.6 nm (Binupriya et al. 2010). There are only a handful of fungi with the potential to produce AuNPs, even though the reduced particle size makes them more reactive and convenient in electronic and catalyst applications (Mukherjee et al. 2001; Eustis and El-Sayed 2006).

Biomolecules like proteins, reducing sugars, enzymes, etc. were also used for the reduction of Au(III) to Au(0). Polysaccharides were reported to synthesize AuNPs, among which fructose was identified as the best-suited sugar for the synthesis of smaller particles with considerable stability (Panigrahi et al. 2004). Gold nanoclusters (AuNCs) were synthesized using a protein-templated method to detect Hg(II) ions (Xie et al. 2010). This was extended to a paper strip-based sensing of Hg (II) ions by dispersing AuNCs on a nitrocellulose strip. The encapsulated BSA (bovine serum albumin) scaffolds entrapped in the nitrocellulose membrane helped the rapid visual estimation of the Hg(II) ions. Recent studies showed that green synthesized AuNPs using different proteins such as BSA, collagen, zein, and lysozyme were uptaken by cancer cells (Hameed et al. 2020). Gold bioconjugates were prepared by capping AuNPs with tyrosine, tryptophan, and cysteine amino acids and successfully utilized in the detection of ranitidine drug detection in real samples. The study showed the photoluminescence quenching on different amino acid conjugated gold nanoparticles upon ranitidine interaction (Hameed et al. 2021). Compared to other approaches, utilizing low reduction potential diazonium gold salt as a precursor for this green synthetic route is novel and an alternative for conventional $HAuCl_4$ precursor (Fig. 17.5).

Other reports have also shown that proteins mediated AuNPs' synthesis and considered that an electrostatic surfactant interaction between carboxylate groups found on proteins enables capping AuNPs and prevents aggregation. Reports have shown that the amide III groups in proteins took part in the reduction of AuNPs and were confirmed by SDS (sodium dodecyl sulfate) gel electrophoresis (Mukherjee et al. 2012). Further, commercially available enzymes were also used to prepare AuNPs. Nisin peptide, a class-Ia bacteriocin obtained from *Lactococcus lactis* subsp, was used to synthesize spherical AuNPs with a size of 25 nm (Otari et al. 2017).

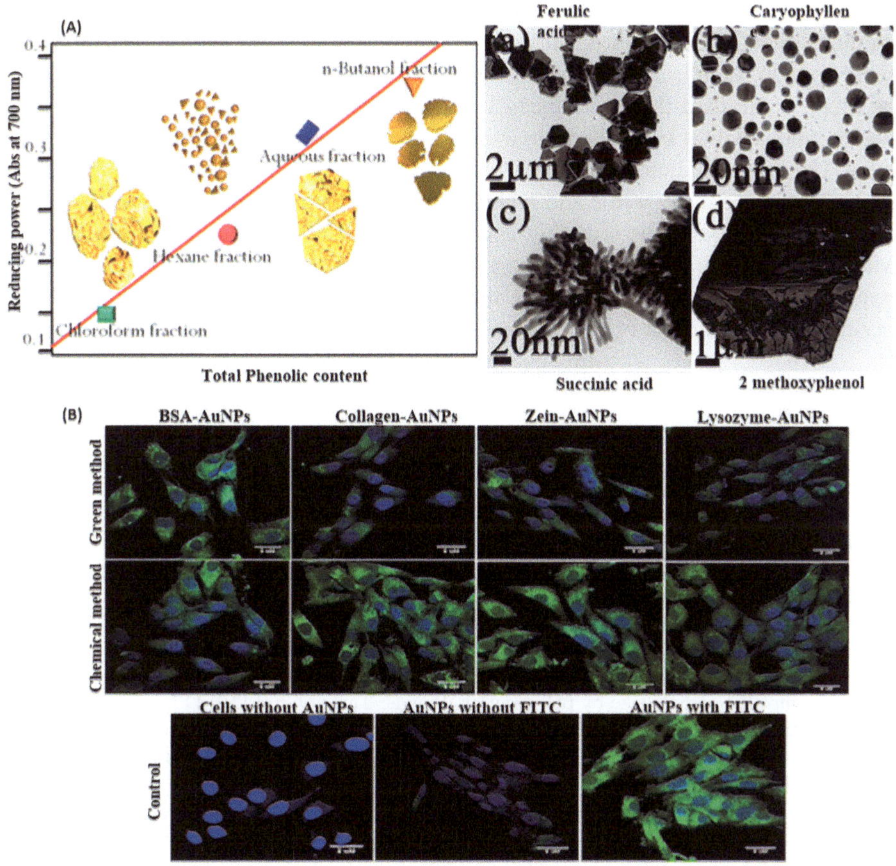

Fig. 17.5 (**a**) Schematic representation showing the relationship of total phenolic content, antioxidant activity, and reducing the power of different solvent extracts and TEM images of AuNPs synthesized by *O. sanctum* extracts by using different solvents (Lee et al. 2016). (Reproduced with permission from the American Chemical Society (2016)) and (**b**) confocal microscopy images of the AuNP-protein bioconjugates by green and chemical routes showing the higher uptake of the green synthesized AuNPs by MG-63 cells (Hameed et al. 2020). (Reproduced with permission from the American Chemical Society (2020))

17.6 Conclusions

During the last few years, increasing demand for green chemistry and nanotechnology encouraged the adoption of green synthetic routes for nanomaterial synthesis via bacteria, actinomycetes, fungi, plants, and biopolymers as competent "green" nanofactories. Understanding the susceptibility of group 11 nanoparticles CuNPs, AgNPs, and AuNPs due to their unique optical, electronic, and chemical properties, in a wide range of applications, comprehensive eco-friendly green synthetic routes are outlined in this chapter. The decisive parameters such as size, morphology, and embodiment of biogenic nanoparticles were also discussed. However, existing methods need to be modified further for making these methods cost-effective and scale up the production. Nanoparticles produced from plant extracts are observed to be more stable than those produced by microorganisms and are relatively easy to handle; moreover, a controlled phytochemical concentration can tune the reduction rates. The fungi-mediated biogenic synthesis also has a potential future in large-scale production since it can withstand high metal salt concentrations and secretes enough enzymes and proteins as reducing agents for efficient fabrication of nanoparticles. Managing the gap between different studies and identifying specific bimolecular requirements will help to overcome the laborious process of optimization and will benefit the construction of nanomaterials on an industrial scale.

References

Ahmad F, Ashraf N, Ashraf T, Zhou R-B, Yin D-C (2019) Biological synthesis of metallic nanoparticles (MNPs) by plants and microbes: their cellular uptake, biocompatibility, and biomedical applications. Appl Microbiol Biotechnol 103:2913–2935. https://doi.org/10.1007/s00253-019-09675-5

Ahmed S, Saifullah M, Ahmad BL, Swami S (2016) Ikram, green synthesis of silver nanoparticles using Azadirachta indica aqueous leaf extract. J Radiat Res Appl Sci 9:1–7

Ajitha B, Reddy YAK, Reddy PS (2015) Enhanced antimicrobial activity of silver nanoparticles with controlled particle size by pH variation. Powder Technol 269:110–117

Alasvand Zarasvand K, Rai VR (2016) Inhibition of a sulfate reducing bacterium, Desulfovibrio marinisediminis GSR3, by biosynthesized copper oxide nanoparticles, 3. Biotech 6:84. https://doi.org/10.1007/s13205-016-0403-0

Alghoraibi I, Soukkarieh C, Zein R, Alahmad A, Walter J-G, Daghestani M (2020) Aqueous extract of Eucalyptus camaldulensis leaves as reducing and capping agent in biosynthesis of silver nanoparticles. Inorg Nano-Metal Chem 50:895–902

Arun G, Eyini M, Gunasekaran P (2014) Green synthesis of silver nanoparticles using the mushroom fungus Schizophyllum commune and its biomedical applications. Biotechnol Bioprocess Eng 19:1083–1090

Balan K, Qing W, Wang Y, Liu X, Palvannan T, Wang Y, Ma F, Zhang Y (2016) Antidiabetic activity of silver nanoparticles from green synthesis using Lonicera japonica leaf extract. RSC Adv 6:40162–40168

Barabadi H, Honary S, Mohammadi MA, Ahmadpour E, Rahimi MT, Alizadeh A, Naghibi F, Saravanan M (2017) Green chemical synthesis of gold nanoparticles by using Penicillium aculeatum and their scolicidal activity against hydatid cyst protoscolices of Echinococcus granulosus. Environ Sci Pollut Res 24:5800–5810

Bhainsa KC, D'Souza SF (2006) Extracellular biosynthesis of silver nanoparticles using the fungus Aspergillus fumigates. Colloids Surf B Biointerfaces 47:160–164

Bhattacharya D, Gupta RK (2005) Nanotechnology and potential of microorganisms. Crit Rev Biotechnol 25:199–204

Binupriya AR, Sathishkumar M, Vijayaraghavan K, Yun S-I (2010) Bioreduction of trivalent aurum to nano-crystalline gold particles by active and inactive cells and cell-free extract of Aspergillus oryzae var. viridis. J Hazard Mater 177:539–545

Boisselier E, Astruc D (2009) Gold nanoparticles in nanomedicine: preparations, imaging, diagnostics, therapies and toxicity. Chem Soc Rev 38:1759–1782

Butola BS, Kumar A (2020) Green chemistry based in-situ synthesis of silver nanoparticles for multifunctional finishing of chitosan polysaccharide modified cellulosic textile substrate. Int J Biol Macromol 152:1135–1145

Chakdar H, Kumar M, Pandiyan K, Singh A, Nanjappan K, Kashyap PL, Srivastava AK (2016) Bacterial xylanases: biology to biotechnology, 3. Biotech 6:1–15

Chakrapani V, Ayaz Ahmed KB, Kumar VV, Ganapathy V, Anthony SP, Anbazhagan V (2014) A facile route to synthesize casein capped copper nanoparticles: an effective antibacterial agent and selective colorimetric sensor for mercury and tryptophan. RSC Adv 4:33215–33221. https://doi.org/10.1039/C4RA03086A

Correa-Llantén DN, Muñoz-Ibacache SA, Castro ME, Muñoz PA, Blamey JM (2013) Gold nanoparticles synthesized by Geobacillus sp. strain ID17 a thermophilic bacterium isolated from Deception Island, Antarctica. Microb Cell Factories 12:1–6

Csavina J, Field J, Félix O, Corral-Avitia AY, Sáez AE, Betterton EA (2014) Effect of wind speed and relative humidity on atmospheric dust concentrations in semi-arid climates. Sci Total Environ 487:82–90

Cuevas R, Durán N, Diez MC, Tortella GR, Rubilar O (2015) Extracellular biosynthesis of copper and copper oxide nanoparticles by *Stereum hirsutum*, a native white-rot fungus from Chilean forests. J Nanomater 2015:789089. https://doi.org/10.1155/2015/789089

Daniel M-C, Astruc D (2004) Gold nanoparticles: assembly, supramolecular chemistry, quantum-size-related properties, and applications toward biology, catalysis, and nanotechnology. Chem Rev 104:293–346. https://doi.org/10.1021/cr030698

Das VL, Thomas R, Varghese RT, Soniya EV, Mathew J, Radhakrishnan EK (2014) Extracellular synthesis of silver nanoparticles by the, Bacillus Strain CS 11 isolated from industrialized area. 3 Biotech 4:121–126

Deplanche K, Caldelari I, Mikheenko IP, Sargent F, Macaskie LE (2010) Involvement of hydrogenases in the formation of highly catalytic Pd (0) nanoparticles by bioreduction of Pd (II) using Escherichia coli mutant strains. Microbiology 156:2630–2640

Dubey SP, Lahtinen M, Sillanpää M (2010) Green synthesis and characterizations of silver and gold nanoparticles using leaf extract of Rosa rugosa. Colloids Surf A Physicochem Eng Asp 364:34–41

Durán N, Durán M, De Jesus MB, Seabra AB, Fávaro WJ, Nakazato G (2016) Silver nanoparticles: a new view on mechanistic aspects on antimicrobial activity, nanomedicine nanotechnology. Biol Med 12:789–799

Edison TJI, Sethuraman MG (2012) Instant green synthesis of silver nanoparticles using Terminalia chebula fruit extract and evaluation of their catalytic activity on reduction of methylene blue. Process Biochem 47:1351–1357

El Bialy BE, Hamouda RA, Khalifa KS, Hamza HA (2017) Cytotoxic effect of biosynthesized silver nanoparticles on Ehrlich ascites tumor cells in mice. Int J Pharmacol 13:134–144

El Shafey AM (2020) Green synthesis of metal and metal oxide nanoparticles from plant leaf extracts and their applications: a review. Green Process Synth 9:304–339

Elbagory AM, Cupido CN, Meyer M, Hussein AA (2016) Large scale screening of southern African plant extracts for the green synthesis of gold nanoparticles using microtitre-plate method. Molecules 21:1498

Elia P, Zach R, Hazan S, Kolusheva S, Porat Z, Zeiri Y (2014) Green synthesis of gold nanoparticles using plant extracts as reducing agents. Int J Nanomedicine 9:4007

Eustis S, El-Sayed MA (2006) Why gold nanoparticles are more precious than pretty gold: noble metal surface plasmon resonance and its enhancement of the radiative and nonradiative properties of nanocrystals of different shapes. Chem Soc Rev 35:209–217

Fang X, Wang Y, Wang Z, Jiang Z, Dong M (2019) Microorganism assisted synthesized nanoparticles for catalytic applications. Energies 12:190. https://doi.org/10.3390/en12010190

Fortin D, Beveridge TJ (2000) From biology to biotechnology and medical applications. Biomineralisation

Gahlawat G, Choudhury AR (2019) A review on the biosynthesis of metal and metal salt nanoparticles by microbes. RSC Adv 9:12944–12967

Gan PP, Li SFY (2012) Potential of plant as a biological factory to synthesize gold and silver nanoparticles and their applications. Rev Environ Sci Biotechnol 11:169–206

Gawande MB, Goswami A, Felpin F-X, Asefa T, Huang X, Silva R, Zou X, Zboril R, Varma RS (2016) Cu and cu-based nanoparticles: synthesis and applications in catalysis. Chem Rev 116: 3722–3811. https://doi.org/10.1021/acs.chemrev.5b00482

Gholami-Shabani M, Gholami-Shabani Z, Shams-Ghahfarokhi M, Jamzivar F, Razzaghi-Abyaneh M (2017) Green nanotechnology: biomimetic synthesis of metal nanoparticles using plants and their application in agriculture and forestry. In: Nanotechnology. Springer, pp 133–175

Guilger-Casagrande M, Germano-Costa T, Pasquoto-Stigliani T, Fraceto LF, de Lima R (2019) Biosynthesis of silver nanoparticles employing Trichoderma harzianum with enzymatic stimulation for the control of Sclerotinia sclerotiorum. Sci Rep 9:1–9

Guo S, Wang E (2011) Noble metal nanomaterials: controllable synthesis and application in fuel cells and analytical sensors. Nano Today 6:240–264

Hameed M, Panicker S, Abdallah SH, Khan AA, Han C, Chehimi MM, Mohamed AA (2020) Protein-coated aryl modified gold nanoparticles for cellular uptake study by osteosarcoma cancer cells. Langmuir 36:11765–11775

Hameed MK, Parambath JBM, Kanan SM, Mohamed AA (2021) FRET-based fluorescent probe for drug assay from amino acid@ gold-carbon nanoparticles. Anal Bioanal Chem 413:1117–1125

Hasan SS, Singh S, Parikh RY, Dharne MS, Patole MS, Prasad BLV, Shouche YS (2007) Bacterial synthesis of copper/copper oxide nanoparticles. J Nanosci Nanotechnol 8:3191–3196. https://doi.org/10.1166/jnn.2008.095

Hasan KM, Pervez M, Talukder M, Sultana M, Mahmud S, Meraz M, Bansal V, Genyang C (2019) A novel coloration of polyester fabric through green silver nanoparticles (G-AgNPs@ PET). Nano 9:569

Hasan KM, Horváth PG, Alpár T (2020a) Potential natural fiber polymeric nanobiocomposites: a review. Polymers (Basel) 12:1072

Hasan KMF, Wang H, Mahmud S, Genyang C (2020b) Coloration of aramid fabric via in-situ biosynthesis of silver nanoparticles with enhanced antibacterial effect. Inorg Chem Commun 119:108115

Hasan KMF, Wang H, Mahmud S, Taher MA, Genyang C (2020c) Wool functionalization through AgNPs: coloration, antibacterial and wastewater treatment. Surf Innov 9:25–36

He S, Guo Z, Zhang Y, Zhang S, Wang J, Gu N (2007) Biosynthesis of gold nanoparticles using the bacteria Rhodopseudomonas capsulata. Mater Lett 61:3984–3987

Heli H, Hajjizadeh M, Jabbari A, Moosavi-Movahedi AA (2009) Fine steps of electrocatalytic oxidation and sensitive detection of some amino acids on copper nanoparticles. Anal Biochem 388:81–90. https://doi.org/10.1016/j.ab.2009.02.021

Honary S, Barabadi H, Gharaei E, Naghibi F (2012) Green synthesis of copper oxide nanoparticles using Penicillium Aurantiogriseum, Penicillium Citrinum and Penicillium Waksmanii. Dig J Nanomater Bios 7:999–1005

Hussain I, Singh NB, Singh A, Singh H, Singh SC (2016) Green synthesis of nanoparticles and its potential application. Biotechnol Lett 38:545–560

Ingle A, Rai M, Gade A, Bawaskar M (2008) Fusarium solani: a novel biological agent for the extracellular synthesis of silver nanoparticles. J Nanopart Res 11:2079–2085. https://doi.org/10.1007/s11051-008-9573-y

Iravani S (2011) Green synthesis of metal nanoparticles using plants. Green Chem 13:2638–2650

Jain N, Bhargava A, Majumdar S, Tarafdar JC, Panwar J (2011) Extracellular biosynthesis and characterization of silver nanoparticles using Aspergillus flavus NJP08: a mechanism perspective. Nanoscale 3:635–641

Jamkhande PG, Ghule NW, Bamer AH, Kalaskar MG (2019) Metal nanoparticles synthesis: an overview on methods of preparation, advantages and disadvantages, and applications. J Drug Deliv Sci Technol 53:101174. https://doi.org/10.1016/j.jddst.2019.101174

John MS, Nagoth JA, Zannotti M, Giovannetti R, Mancini A, Ramasamy KP, Miceli C, Pucciarelli S (2021) Biogenic synthesis of copper nanoparticles using bacterial strains isolated from an Antarctic consortium associated to a psychrophilic marine ciliate: characterization and potential application as antimicrobial agents. Mar Drugs 19:263. https://doi.org/10.3390/md19050263

Johnston CW, Wyatt MA, Li X, Ibrahim A, Shuster J, Southam G, Magarvey NA (2013) Gold biomineralization by a metallophore from a gold-associated microbe. Nat Chem Biol 9:241–243

Kalaivani R, Maruthupandy M, Muneeswaran T, Beevi AH, Anand M, Ramakritinan CM, Kumaraguru AK (2018) Synthesis of chitosan mediated silver nanoparticles (Ag NPs) for potential antimicrobial applications. Front Lab Med 2:30–35

Kathiraven T, Sundaramanickam A, Shanmugam N, Balasubramanian T (2015) Green synthesis of silver nanoparticles using marine algae Caulerpa racemosa and their antibacterial activity against some human pathogens. Appl Nanosci 5:499–504

Khalifa KS, Hamouda RA, Hanafy D, Hamza A (2016) In vitro antitumor activity of silver nanoparticles biosynthesized by marine algae. Dig J Nanomater Biostructures 11:213–221

Kharissova OV, Dias HVR, Kharisov BI, Pérez BO, Pérez VMJ (2013) The greener synthesis of nanoparticles. Trends Biotechnol 31:240–248

Korbekandi H, Iravani S, Abbasi S (2012) Optimization of biological synthesis of silver nanoparticles using Lactobacillus casei subsp. casei. J Chem Technol Biotechnol 87:932–937

Kulkarni AP, Srivastava AA, Nagalgaon RK, Zunjarrao RS (2012) Phytofabrication of silver nanoparticles from a novel plant source and its application. Int J Biol Pharm Res 3:417–421

Kumar KP, Paul W, Sharma CP (2011a) Green synthesis of gold nanoparticles with Zingiber officinale extract: characterization and blood compatibility. Process Biochem 46:2007–2013

Kumar VG, Gokavarapu SD, Rajeswari A, Dhas TS, Karthick V, Kapadia Z, Shrestha T, Barathy IA, Roy A, Sinha S (2011b) Facile green synthesis of gold nanoparticles using leaf extract of antidiabetic potent Cassia auriculata. Colloids Surf B Biointerfaces 87:159–163

Kumar PPNV, Pammi SVN, Kollu P, Satyanarayana KVV, Shameem U (2014) Green synthesis and characterization of silver nanoparticles using Boerhaavia diffusa plant extract and their anti bacterial activity. Ind Crop Prod 52:562–566

Lakshmanan G, Sathiyaseelan A, Kalaichelvan PT, Murugesan K (2018) Plant-mediated synthesis of silver nanoparticles using fruit extract of Cleome viscosa L.: assessment of their antibacterial and anticancer activity. Karbala Int J Mod Sci 4:61–68

Lam SJ, Wong EHH, Boyer C, Qiao GG (2018) Antimicrobial polymeric nanoparticles. Prog Polym Sci 76:40–64

Lee H, Song JY, Kim BS (2013) Biological synthesis of copper nanoparticles using Magnolia kobus leaf extract and their antibacterial activity. J Chem Technol Biotechnol 88:1971–1977

Lee SY, Krishnamurthy S, Cho C-W, Yun Y-S (2016) Biosynthesis of gold nanoparticles using Ocimum sanctum extracts by solvents with different polarity. ACS Sustain Chem Eng 4:2651–2659

Lee KX, Shameli K, Yew YP, Teow S-Y, Jahangirian H, Rafiee-Moghaddam R, Webster TJ (2020) Recent developments in the facile bio-synthesis of gold nanoparticles (AuNPs) and their biomedical applications. Int J Nanomedicine 15:275–300

Letchumanan D, Sok SPM, Ibrahim S, Nagoor NH, Arshad NM (2021) Plant-based biosynthesis of copper/copper oxide nanoparticles: an update on their applications in biomedicine, mechanisms, and toxicity. Biomol Ther 11:564. https://doi.org/10.3390/biom11040564

Li J, Tang J, Wei L, He S, Ma L, Shen W, Kang F, Huang Z (2020) Preparation and performance of electrochemical glucose sensors based on copper nanoparticles loaded on flexible graphite sheet. New Carbon Mater 35:410–419. https://doi.org/10.1016/S1872-5805(20)60498-X

Liu Y, Fu J, Chen P, Yu X, Yang P (2000) Studies on biosorption of Au3+ by Bacillus megaterium. Acta Microbiol Sin 40:425–429

Liu L, Liu T, Tade M, Wang S, Li X, Liu S (2014) Less is more, greener microbial synthesis of silver nanoparticles. Enzym Microb Technol 67:53–58

López-Miranda JL, Vázquez M, Fletes N, Esparza R, Rosas G (2016) Biosynthesis of silver nanoparticles using a Tamarix gallica leaf extract and their antibacterial activity. Mater Lett 176:285–289

Manikandan V, Velmurugan P, Park J-H, Chang W-S, Park Y-J, Jayanthi P, Cho M, Oh BT (2017) Green synthesis of silver oxide nanoparticles and its antibacterial activity against dental pathogens. 3 Biotech 7:72

Manjunath HM, Joshi CG, Raju NG (2017) Biofabrication of gold nanoparticles using marine endophytic fungus–Penicillium citrinum. IET Nanobiotechnol 11:40–44

Mondal NK, Chowdhury A, Dey U, Mukhopadhya P, Chatterjee S, Das K, Datta JK (2014) Green synthesis of silver nanoparticles and its application for mosquito control. Asian Pacific J Trop Dis 4:44204–4S210

Mukherjee P, Ahmad A, Mandal D, Senapati S, Sainkar SR, Khan MI, Parishcha R, Ajaykumar PV, Alam M, Kumar R (2001) Fungus-mediated synthesis of silver nanoparticles and their immobilization in the mycelial matrix: a novel biological approach to nanoparticle synthesis. Nano Lett 1:515–519

Mukherjee S, Sushma V, Patra S, Barui AK, Bhadra MP, Sreedhar B, Patra CR (2012) Green chemistry approach for the synthesis and stabilization of biocompatible gold nanoparticles and their potential applications in cancer therapy. Nanotechnology 23:455103

Nagaraj E, Karuppannan K, Shanmugam P, Venugopal S (2019) Exploration of bio-synthesized copper oxide nanoparticles using Pterolobium hexapetalum leaf extract by photocatalytic activity and biological evaluations. J Clust Sci 30:1157–1168. https://doi.org/10.1007/s10876-019-01579-8

Nalwade AR, Badhe MN, Pawale CB, Hinge SB (2013) Rapid biosynthesis of silver nanoparticles using fern leaflet extract and evaluation of their antibacterial activity. Int J Biol Technol 4:12–18

Narayanan KB, Sakthivel N (2010) Biological synthesis of metal nanoparticles by microbes. Adv Colloid Interf Sci 156:1–13. https://doi.org/10.1016/j.cis.2010.02.001

Nasrollahzadeh M, Sajadi SM, Rostami-Vartooni A, Hussin SM (2016) Green synthesis of CuO nanoparticles using aqueous extract of Thymus vulgaris L. leaves and their catalytic performance for N-arylation of indoles and amines. J Colloid Interface Sci 466:113–119. https://doi.org/10.1016/j.jcis.2015.12.018

Otari SV, Kumar M, Kim I-W, Lee JH, Lee J-K (2017) Rapid, thermostable antimicrobial peptide-mediated synthesis gold nanoparticles as highly efficient charge trapping medium for sol-gel-derived thin film. Mater Lett 188:375–378

Ovais M, Khalil AT, Islam NU, Ahmad I, Ayaz M, Saravanan M, Shinwari ZK, Mukherjee S (2018) Role of plant phytochemicals and microbial enzymes in biosynthesis of metallic nanoparticles. Appl Microbiol Biotechnol 102:6799–6814

Owaid MN, Raman J, Lakshmanan H, Al-Saeedi SSS, Sabaratnam V, Abed IA (2015) Mycosynthesis of silver nanoparticles by Pleurotus cornucopiae var. citrinopileatus and its inhibitory effects against Candida sp. Mater Lett 153:186–190

Panigrahi S, Kundu S, Ghosh S, Nath S, Pal T (2004) General method of synthesis for metal nanoparticles. J Nanopart Res 6:411–414

Patil RS, Kokate MR, Kolekar SS (2012) Bioinspired synthesis of highly stabilized silver nanoparticles using Ocimum tenuiflorum leaf extract and their antibacterial activity, Spectrochim. Acta Part A Mol Biomol Spectrosc 91:234–238

Philip D (2009) Biosynthesis of au, ag and au–ag nanoparticles using edible mushroom extract, Spectrochim. Acta Part A Mol Biomol Spectrosc 73:374–381

Prasher P, Singh M, Mudila H (2018) Green synthesis of silver nanoparticles and their antifungal properties. Bionanosci 8:254–263

Priyaragini S, Veena S, Swetha D, Karthik L, Kumar G, Rao KVB (2014) Evaluating the effectiveness of marine actinobacterial extract and its mediated titanium dioxide nanoparticles in the degradation of azo dyes. J Environ Sci 26:775–782

Rahimi-Nasrabadi M, Pourmortazavi SM, Shandiz SAS, Ahmadi F, Batooli H (2014) Green synthesis of silver nanoparticles using Eucalyptus leucoxylon leaves extract and evaluating the antioxidant activities of extract. Nat Prod Res 28:1964–1969

Rai MK, Deshmukh SD, Ingle AP, Gade AK (2012) Silver nanoparticles: the powerful nanoweapon against multidrug-resistant bacteria. J Appl Microbiol 112:841–852

Rai M, Bonde S, Golinska P, Trzcińska-Wencel J, Gade A, Abd-Elsalam K, Shende S, Gaikwad S, Ingle A (2021) Fusarium as a novel fungus for the synthesis of nanoparticles: mechanism and applications. J Fungi 7:139

Ramkumar VS, Pugazhendhi A, Gopalakrishnan K, Sivagurunathan P, Saratale GD, Dung TNB, Kannapiran E (2017) Biofabrication and characterization of silver nanoparticles using aqueous extract of seaweed Enteromorpha compressa and its biomedical properties. Biotechnol Reports 14:1–7

Rao P, Chandraprasad MS, Lakshmi YN, Rao J, Aishwarya P, Shetty S (2014) Biosynthesis of silver nanoparticles using lemon extract and its antibacterial activity. Int J Multidiscip Curr Res 2:165–169

Rashmi BN, Harlapur SF, Avinash B, Ravikumar CR, Nagaswarupa HP, Kumar MRA, Gurushantha K, Santosh MS (2020) Facile green synthesis of silver oxide nanoparticles and their electrochemical, photocatalytic and biological studies. Inorg Chem Commun 111:107580

Rastogi A, Singh P, Haraz FA, Barhoum A (2018) Biological synthesis of nanoparticles: an environmentally benign approach. In: Fundamentals of nanoparticles. Elsevier, pp 571–604

Ravichandran S, Paluri V, Kumar G, Loganathan K, Kokati Venkata BR (2016) A novel approach for the biosynthesis of silver oxide nanoparticles using aqueous leaf extract of Callistemon lanceolatus (Myrtaceae) and their therapeutic potential. J Exp Nanosci 11:445–458

Renuka R, Devi KR, Sivakami M, Thilagavathi T, Uthrakumar R, Kaviyarasu K (2020) Biosynthesis of silver nanoparticles using Phyllanthus emblica fruit extract for antimicrobial application. Biocatal Agric Biotechnol 24:101567

Rizwan M, Singh M, Mitra CK, Morve RK (2014) Ecofriendly application of nanomaterials: nanobioremediation. J Nanoparticles 2014:1–7

Saravanakumar K, Shanmugam S, Varukattu NB, MubarakAli D, Kathiresan K, Wang M-H (2019) Biosynthesis and characterization of copper oxide nanoparticles from indigenous fungi and its effect of photothermolysis on human lung carcinoma. J Photochem Photobiol B Biol 190:103–109. https://doi.org/10.1016/j.jphotobiol.2018.11.017

Saxena A, Tripathi RM, Zafar F, Singh P (2012) Green synthesis of silver nanoparticles using aqueous solution of Ficus benghalensis leaf extract and characterization of their antibacterial activity. Mater Lett 67:91–94

Senapati S, Ahmad A, Khan MI, Sastry M, Kumar R (2005) Extracellular biosynthesis of bimetallic Au–Ag alloy nanoparticles. Small 1:517–520

Shahverdi AR, Minaeian S, Shahverdi HR, Jamalifar H, Nohi A-A (2007) Rapid synthesis of silver nanoparticles using culture supernatants of Enterobacteria: a novel biological approach. Process Biochem 42:919–923

Shaik MR, Khan M, Kuniyil M, Al-Warthan A, Alkhathlan HZ, Siddiqui MRH, Shaik JP, Ahamed A, Mahmood A, Khan M (2018) Plant-extract-assisted green synthesis of silver nanoparticles using Origanum vulgare L. extract and their microbicidal activities. Sustain 10: 913

Shankar SS, Ahmad A, Pasricha R, Sastry M (2003) Bioreduction of chloroaurate ions by geranium leaves and its endophytic fungus yields gold nanoparticles of different shapes. J Mater Chem 13: 1822–1826. https://doi.org/10.1039/B303808B

Sharma SK, Singh DP, Shukla HD, Ahmad A, Bisen PS (2001) Influence of sodium ion on heavy metal-induced inhibition of light-regulated proton efflux and active carbon uptake in the cyanobacterium Anabaena flos-aquae. World J Microbiol Biotechnol 17:707–711

Singh P, Kim Y-J, Zhang D, Yang D-C (2016) Biological synthesis of nanoparticles from plants and microorganisms. Trends Biotechnol 34:588–599

Singh S, Kumar N, Kumar M, Jyoti A, Agarwal BM (2017) Electrochemical sensing and remediation of 4-nitrophenol using bio-synthesized copper oxide nanoparticles. Chem Eng J 313:283–292. https://doi.org/10.1016/j.cej.2016.12.049

Singh J, Dutta T, Kim K-H, Rawat M, Samddar P, Kumar P (2018) 'Green' synthesis of metals and their oxide nanoparticles: applications for environmental remediation. J Nanobiotechnology 16: 84. https://doi.org/10.1186/s12951-018-0408-4

Tan KB, Sun D, Huang J, Odoom-Wubah T, Li Q (2021) State of arts on the bio-synthesis of noble metal nanoparticles and their biological application, Chinese. J Chem Eng 30:272–290. https://doi.org/10.1016/j.cjche.2020.11.010

Tile VA, Bholay AD (2012) Biosynthesis of silver nanoparticles and its antifungal activities. J Env Res Dev 7:338–345

Venkata Abhinav K, Venkata Krishna Rao R, Karthik PS, Singh SP (2015) Copper conductive inks: synthesis and utilization in flexible electronics. RSC Adv 5:63985–64030. https://doi.org/10.1039/C5RA08205F

Vigneshwaran N, Ashtaputre NM, Varadarajan PV, Nachane RP, Paralikar KM, Balasubramanya RH (2007) Biological synthesis of silver nanoparticles using the fungus Aspergillus flavus. Mater Lett 61:1413–1418

Vishveshvar K, Aravind Krishnan MV, Haribabu K, Vishnuprasad S (2018) Green synthesis of copper oxide nanoparticles using Ixora coccinea plant leaves and its characterization. Bionanoscience 8:554–558. https://doi.org/10.1007/s12668-018-0508-5

Waris A, Din M, Ali A, Ali M, Afridi S, Baset A, Ullah Khan A (2021) A comprehensive review of green synthesis of copper oxide nanoparticles and their diverse biomedical applications. Inorg Chem Commun 123:108369. https://doi.org/10.1016/j.inoche.2020.108369

Xie J, Zheng Y, Ying JY (2010) Highly selective and ultrasensitive detection of Hg^{2+} based on fluorescence quenching of Au nanoclusters by Hg^{2+}–Au^{+} interactions. Chem Commun 46:961–963

Yezhelyev MV, Gao X, Xing Y, Al-Hajj A, Nie S, O'Regan RM (2006) Emerging use of nanoparticles in diagnosis and treatment of breast cancer. Lancet Oncol 7:657–667

Chapter 18
Electroactive Microorganisms Involved in Power Generation in a Microbial Fuel Cell

Barun Kumar, Harshika Varshney, Kalpana Sharma, Ankit Kumar, and Soumya Pandit

Abstract Bio-electrocatalysts (microbes) are the most important components of microbial fuel cells (MFCs). Microbes act as a biocatalyst to generate reducing/ oxidising power. Primarily extremophiles are used as biocatalyst to obtain the desired redox reactions. They show electrocatalytic action by oxidation-reduction process via metabolic reaction to transport the electrons. The strategy involves formation of electrochemical cell with electrode chambers which may or may not be separated by the semipermeable permeable membrane. Electroactive biofilm formation takes place over these electrodes to harness the reducing power (reduction involves generation of NADH and $FADH_2$). Concomitantly, metabolic reactions are necessary for production of high-energy electron (e^{-1}) for generation of electricity in MFCs. Microbes for the above purpose are isolated from highly polluted areas such as wastewater, lake sediment and soil. These microbes range from algae, bacteria, cyanobacteria, fungi, eubacteria, etc. Microbial selection is based on their ability to consume a wide range of substrates, such as fatty acid, alcohols, gases, cellulose, etc. in MFC. Microbes contain specialised protein carrier which help in redox on respective electrode. This chapter discusses the variety of biocatalysts (microbes) that are applied in MFCs.

Keywords Microbial fuel cells · Biocatalyst · Electrode · Redox · Biofilm

18.1 Introduction

MFCs bear similarities with chemical fuel cells which generate energy through fuels via chemical reaction without involving combustion. Like them, MFCs have anode-cathode, electrolyte and catalyst. While other components of MFCs are likely their chemical counterparts, the catalysts are of microbial origin. Electrocatalytic

B. Kumar · H. Varshney · K. Sharma · A. Kumar · S. Pandit (✉)
Department of Life Science, School of Sciences and Basic Research, Sharda University, Greater Noida, India

A. Sarkar, I. A. Ahmed (eds.), *Microbial products for future industrialization*, Interdisciplinary Biotechnological Advances,
https://doi.org/10.1007/978-981-99-1737-2_18

microorganisms generate H^+ and e^{-1} from their metabolism and transport them to suitable redox agent (Logan and Rabaey 2012). Electrode compartments may or may not be partitioned by semipermeable membrane in MFCs (Butti et al. 2016). The large range of electroactive microorganisms comes from eubacteria (Liu et al. 2002), archaea (Dopson and Holmes 2014), fungi (Fernández de Dios et al. 2013), algae (Chandra et al. 2012), photobacteria (Venkidusamy and Megharaj 2016) and cyanobacteria (Lea-Smith et al. 2016). They are isolated from diverse range of environments and have capability to oxidise a wide variety of organic molecules to carbon dioxide and water (Venkata Mohan et al. 2014a). Yet understanding the relation between electroactive microbes and electrodes is a major challenge in developing efficient MFCs. This could be accomplished by understanding the basic biochemistry of biocatalyst fixing, electrode interactivity and electrode charge transport. Apart from power generation, MFCs can be applied to synthesise variety of useable product while addressing the problem of sewage treatment, bioremediation, eutrophication, biomass generation, etc. The application and efficiency of MFCs greatly depend on nature of microorganism (biocatalyst) used, and this chapter focuses on them. It is important to note that these biocatalysts are majorly of bacterial origin.

18.2 Electroactive Microorganisms Involved in MFCs

Electroactive microorganisms oxidise variety of organic substrates and generate power via extracellular electron transfer (EET) when grown in electrochemical cell with electrodes. The real challenge in establishing MFCs is to discover microbes that metabolise organic matter while electron transfer to the anode (Debabov 2008). The approach for isolating microbes from different sources that can contribute to power generation is by collecting environmental samples and selecting them based on their ability to colonise electrode surfaces, that is, they should be able to reduce/oxidise minerals, while the enrichment steps involve their colonisation of the electrode surface in the form of biofilm (Jung and Regan 2007).

18.3 Biocatalyst

18.3.1 Biocathode

In contrast to bacteria that grow on the anode, microorganisms may use the cathode, which acts as an electron donor in a MET, as a source of electrons. It is also known as electrode oxidising bacteria. The principle cathode oxidising bacterium studied was *Geobacter*. *Clostridium ljungdahlii*, *Clostridium metallireducens*, *Clostridium sulfurreducens* and other microorganisms have also shown electrode oxidation. It is mostly used to produce biofuels, H_2, CH_4, formate and other liquid organic fuels. By employing carbon substrates such as acetate or CO_2 and using H_2 or formate created by the cathode as an electron donor, electron oxidising bacteria may make CH_4 and higher organic fuels such as 2–3 butanediol, butyrate, butanol and so on.

Following that, various studies looked at how acetate changed into medium-chain fatty acids like caproate and caprylate. The coordination of diverse species with their distinct roles is extremely necessary for the continuous process of electron consumption and fuel generation by microorganisms. With the aid of electron donor bacteria, uranium, selenium, tellurium, chlorinated compounds and other elements are reduced as well. Biogeochemical cycles and bioremediation by leaching minerals are critical applications. Though many bacteria engage in cathodic reactions, the cathode mechanism in biocatalysts is significantly less well understood.

18.3.2 Bioanode

Electroactive bacteria grown with an anode may oxidise a large amount of organic substrate, and EET can be used to reduce the electrode. The finding that nonfermentable substrates including such acetate may be oxidised into carbon dioxide by directly moving an electron to an electrode that acts as the only electron acceptor was considered when electron transfers to electrodes. Because acetate, as an electron donor, has no chance of phosphorylation at the substratum phase under anaerobic circumstances, electron transfer to an electrode has been proved to be a respiratory process.

Bacteria may absorb and effectively transfer electrons from an organic source to electricity, according to the findings of this and other related investigations. One hundred years ago, every electron emitted during microbial metabolism could be recovered using electrodes, but discovering electrical bacteria that consume organic matter and deliver electrons to the anode is a significant step forward. Discovering bacteria that preferentially colonise anode surfaces and bacteria that may reduce mineral ores is one method for identifying bacteria from various sources that contribute to power generation. The enhanced circumstances promote the growth of certain bacteria on the anode surface.

Because the lowest substrate redox potential in the anode chamber is −0.4 V and the oxide with the highest potential for use is oxygen (+0.8 V (versus SHE), the maximum possible difference in potential for bacteria is +1.2 V, which provides the maximum theoretical benefit for the colonisation of the specific bacteria on the anode surface. Another disadvantage of anode biocatalyst is the rate of electron transfer to the anode, which is restricted by the biocatalyst's metabolic rate and distance from the anode. Genetically engineered findings, such as cloning genes, can increase electron flow through EET routes, overcoming these restrictions.

18.4 Pure Cultures

18.4.1 Pure Cultured Microorganisms as Electricigens in the Anode

Electricigens are essential in MFCs as the biocatalyst. Large number of electricigens have been isolated and used in MFCs up to this point. Proteobacteria and firmicutes account for the majority of these electricigens. Recent research (Cao et al. 2019) has showed that the electricigens in microbial fuel cell have a wide range of behaviours. Microorganisms with the ability to generate electricity have yet to be discovered. It is important to summarise the emerging electricity-producing microorganisms in terms of understanding the diversity and similarity of electricigens.

18.5 Mixed Microbiome and Communities

MFCs must have at least one electricigen active in order to generate power. A variety of electricigens, on the other hand, can help to produce current and, in most situations, may be more efficient. In the recent decade, the concept of mixed communities has been presented. Pure cultures can help explain the electron transfer process at the microbe's level and minimise the number of different microbial strains in mixed cultures. Pure cultured electricigens, on the other hand, require very rigid working conditions and can only use certain substrates, whereas diverse consortiums are more suited to the use of complex substrates. In MFCs with the best performance, mixed communities, such as sewage or activated sludge, are often employed as the anodic biocatalyst (Cao et al. 2019).

The electrical activity of the biocatalyst is responsible for the development of MET as bioenergy or any other biological-based product. The biocatalyst for MET might be a single strain or a mixed inoculum. In some circumstances, a single strain acting as a biocatalyst can reduce the expense of maintaining culture purity. When wastewater is utilised as an anolyte, mixed biocatalyst syntrophy enables to break down a range of substrates, efficiently removing organic waste. Despite this, there are a large number of bacteria in the mixed crop that utilise organic substrates without the need of electricity and include non-electrogenes. This constraint can be overcome by doing pretreatment prior to reactor initiation.

18.6 Extracellular Electron Transfer

As worldwide worries about energy security, change in climate and issue related to sustainability had increased. Thus, the requirement for an alternative energy-efficient wastewater treatment has raised. Anaerobic wastewater treatment systems that able to recover the energy or value-added products from organic wastes and wastewaters are more common as a result. Among these technologies are anaerobic digestion, dark biohydrogen fermentation and microbial electrochemical cells. MxCs (microbial electrochemical technologies) are a new type of anaerobic biotechnology that involves the metabolism of anode-respiring bacteria (ARB) with electrochemistry to harvest directly from decomposing organic compounds present in wastewater. This capability might be used to turn the recovered electrons into energy and other materials.

Microorganisms in MxCs absorb energy by electron transfer to the anode from a donor substrate. Microorganisms associated with the energy transfer process often develop a biofilm on the surface of the anode, oxidising a donor substrate and releasing electrons, protons and CO_2. Diffusible internal electron transporters (i.e. NADP/NADH) and membrane-bound electron transport systems transmit the electron released to outer membrane proteins (OMPs). The transport of electrons from the e^-donor to the OMPs is known as intracellular electron transfer (IET). The technique by which electrons are transmitted from the OMPs to the negative electrode is known as extracellular electron transfer (EET).

According to previous study, EET can occur through direct contact of OMPs with the electrode via mediators. Several researches have highlighted the relevance of conducting EET in biofilm anodes that provide maximum current density, yet conduction-based EET is remain unclear. Microbial nanowires (or pili, in particular for *Geobacter* species), redox conductivity (or electron hopping) and conductive biofilm matrix have all been proposed as potential paths for this conductive EET. Although extensive research is being conducted to investigate these conductive EET routes, nothing is known about the biofilm electrical conductivity that could influence EET kinetics in biofilm anodes.

18.6.1 Conduction-Based EET in Biofilm Anode

Biofilms present in negative electrode have previously been used to describe conduction-based EET as a conductive biofilm matrix. At least two mechanisms for conductive EET have been proposed or proven so far: (1) metal-like conduction via microbial nanowires and (2) redox conduction through biofilm-bound extracellular redox cofactors.

ARB manufacture and employ conductive nanowires to transmit EET to conductive solid or insoluble electron acceptors in metallic-like conduction which had been explained. Although all EET methods depend on electrical conduction, the potential gradient and related properties (e.g. concentration of extracellular co-factors and conductivity of nanowires or the co-factors) differ, and some are unknown.

18.6.2 Metal-Like Ohmic Conduction

ARB's production of nanowires is a novel approach to EET conduction. In both *Geobacter* and *Shewanella* species, nanowires were formerly thought to be mostly pilin-based current in MxCs. For several *Geobacter* species, Malvankar et al. found a relationship that is linear between pili A and Kbio abundance. Pili filaments isolated from biofilm anodes also exhibited 'metal-like' properties, such as a reduced in conductivity as the temperature increased above 25 °C. Geobacter species' nanowires have been recognised as type IV pili; however, *Shewanella* species' nanowires have been recognised as an outgrowth of cytochrome-rich outer membranes and periplasm, instead of pili. Pili filaments produced by *Geobacter* species were primarily too accountable for high conductivity biofilms, while pilin-deficient mutants were still capable of producing a minimal amount of conductive biofilm.

18.6.3 Redox Conduction

Redox chains generated through multi-heme c-type cytochromes arranged across nanowires are discovered as contributing to nanowire conductivity and high current density in *Shewanella* species. This finding implies that c-type cytochromes are capable of redox conduction. Spatial structuring of cytochromes within a certain distance may allow for redox-driven EET via multi-step electron hopping (cytochrome-to-cytochrome). Similar to Shewanella species, a redox conduction model for Geobacter species has been proposed. However, it has been discovered that the distance between c-type cytochromes in *Geobacter* pili is too great to allow electron hopping. Furthermore, experimental measurements revealed that cytochromes had a negligible contribution to Kbio in *Geobacter* biofilms, although cytochrome-dependent redox conduction was demonstrated in *Geobacter* sulfurreducens biofilm anode.

Because of biophysical changes in nanowires or cytochromes, the EET approach for redox conduction may differ across *Geobacter* and *Shewanella* species. Because the technique frequently used for Kbio measurement (two- or four-probe tests with electrodes divided by a non-conductive gap) may be particular to *Shewanella* species due to poor biofilm growth, no Kbio of *Shewanella oneidensis* MR-1 has been published to date. As a result, further research on redox conduction in the context of biofilm conductivity has been undertaken using the *Geobacter* genus. A redox conduction concept was expanded from pili to full biofilms by including extracellular redox co-factors: electron hopping between membrane-bound and extracellular co-factors may occur throughout anode biofilms. In the *Geobacter* genus, extracellular co-factors help overcome the long-distance obstacle of electron hopping. For this improved redox conduction, EET kinetics is dependent on the potential gradient formed in redox co-factors across biofilm anodes (concentration of oxidised and reduced forms of redox co-factors in the biofilms). The redox gradient of extracellular c-type cytochromes was detected throughout a complete biofilm anode using confocal Raman imaging. Temperature and water content have an effect on the redox conduction characteristics of *Geobacter* biofilms. However, further research on extracellular cofactors and their concentrations (e.g. reduced and oxidised forms) in relation to EET kinetics is required.

18.7 Extremophilic Bacteria

Extremophiles are unique microbes that live and develop in extreme environments that are detrimental to others. The situation might be life-threatening. Extremophilic organisms have since been discovered to exist in all three biological domains, providing a feasible procedure for a variety of environmental applications, including waste treatment. On the basis of the stress circumstances in which they develop, extremophiles are classified as acidophiles, thermophiles, alkaliphiles, halophiles, psychrophilic and hyperthermophiles. Extremes can occur naturally, such as cold arctic areas, desert zones, hydrothermal waves and hyperalkaline lakes, or they could be artificial, such as acidic mine wetlands, nuclear waste sites and so on.

18.7.1 Thermophilic

The thermophiles are creatures that thrive in hot environments (45 °C 122 °C). Cellular and molecular systems evolved to allow them to grow at such high temperatures, which may also be found in geologically active areas. Improved salt bridges and a hydrophobic core protein are among the approaches utilised to improve thermostability and boost membrane saturation to make them compact and stiff. The high amount of guanine-cytosine, which improves the DNA's thermal stability, and the positive DNA superstructure, which stabilises high-temperature DNA, are two examples of adaptations.

The capacity to operate the MET in thermophile circumstances, which contributes to a rise in catalytic activity, is one of the benefits of using the MET thermophile. High temperatures also increase the substrate's transmission mass and solubility. With mixed culture (55 °C), MET was 100 days stable with high columbic productivity. According to the findings of the 16S rRNA clone library, 80% of Firmicutes microorganisms are endospore-forming and electricity-producing. With concomitant sulphate reduction, increased output power and columbic efficiency were obtained in MFC operated under thermophilic conditions.

18.7.2 Psychrophiles

Psychrophiles can live in temperatures ranging from 15 to 0 °C. For growth, a temperature of 15 °C is ideal. Temperatures below 15 °C are ideal for reproduction. Guanine-cytosine content, developed protein flexibility, decreased cold shock thermostability and physiological membrane changes such as cryoprotectants, anti-freezing proteins or extracellular polysaccharides have all been shown to promote low-temperature growth and survival. However, it is unclear why life is hampered at such a low temperature. Although all aspects of MFC operation will decline, such as energy output, substrate conversion, start-up time and so on, the use of low-temperature advantage is high columbic efficiency MET.

18.7.3 Halophilic

Halophiles flourish in either naturally occurring hypersaline settings or in man-made operations. These bacteria are abundant in all three life domains. Halophiles are found in various phyla of Eubacteria, including Proteobacteria, Firmicutes, Actinobacteria, Bacteria residues and spirochaetes, whereas high salt species in Archaea are observed in a class of halo bacteria that need 100–150 g/l of salt for growth and structural integrity.

In metabolic form, halophiles include nitrogen and oxygen phototrophic compounds, aerobic heterotrophic fermenters, denitrifies, sulphate-reducing agents and methanogens. Due to high saline circumstances and limited levels of contact to non-point toxicity sources, biologic treatment of typical microbial excessive saline wastewaters may result in poor degradation efficiency. MFCs with varying salt concentrations and pH conditions demonstrated excellent COD (chemical oxygen demand), high power density and columbic performance. Using high levels of salt in MFCs has been found to be useful since it improves proton transfer and conductivity, which helps to raise power output.

18.7.4 Alkaliphiles

Extremophiles with excessive alkalinity are known as alkaliphiles (pH 8.5–11). A few alkaliphilic species are also recognised as haloalkaliphilic at high salt concentrations, in addition to severe pH conditions. The qualifiers rely on the sodium motive power of the Mrp Na/H antiporter, a tensioned sodium channel, NaVBP and others to sustain an internal pH of minimum two units, more than the surrounding environment in order to sustain their biological components at such severe pH circumstances. Enhanced cell voltages were attained in both alkaline and acidic environments in the anode and cathode. The various alkalophilic strains geoalkalibacter sp., alkalophilic bacillus and *S. oneidensis* have showed good current density generation in MFCs as a biocatalyst. The *Corynebacterium* and *Bacillus pseudofirmus* isolated from humic degradation MFC have been found to increase electron transport to the insoluble mode of transportation by using humic agents as redox shuttles. In alkaline circumstances, *Bacillus pseudofirmus* may also reduce extracellular electron acceptors [AQDS (anthraquinone-2, 6-disulfonate, humus analogue), humic acids (HA) and Fe (III) oxides] as well as a decolourising azo dye (Orange I). *Pseudomonas* alkaliphilic, on the other hand, has been shown to generate a lot of electricity by excreting phenazine-1-carboxylic acid, which acts as an electron shuttle during the oxidation of the substrate.

18.7.5 Acidophilic

Acidophiles are bacteria that can thrive in extremely acidic environments. *Acidophilus cryptum* was the first acidophile to show a pH less than four redox reactions, which was generated by simultaneous oxidation of glucose and ferric iron reduction. Acidophilic species oxidise metal compounds and produce an e^- donor in the absence of an electron donor. For example, ferrous iron sulphide and sulphur compounds, as well as CO_2 in some instances, can be oxidised. To survive in these acidic environments, microorganisms developed a variety of strategies, including membrane modifications to prevent proton intake and the usage of primary and secondary carriers. Acidophiles of varying acidities develop at both the anode (*Acidiphilium* sp.) and cathode (*Acidithiobacillus ferrooxidans*) of METs.

Low pH provides an advantage, in addition to reducing the cathode's diffusion for catalysed oxygen reduction. Nonetheless, the organism's proton outflow energy will be utilised by the low pH circumstances, resulting in reduced coulombic efficiency. As a result, the virtually neutral cytoplasm is preserved. Acidophiles employ DET, or electron transfer elements, for their e^- transfer mechanism. The mediator of ferrooxidans, the passage of an e^- from the electrode to the microbe, was considered to be cytochrome 2 present on the outer membrane. The electrons were immediately transported to the anode in *Acidophilus* species at pH 2.5 and in the presence of oxygen.

18.8 Metal-Reducing Bacteria

The capacity to utilise ferric iron [Fe (III)] as an electron acceptor has been used to isolate and describe a variety of bacteria. Fe (III)-reducing bacteria are known to play a crucial role in the iron and organic matter cycling in sediments and other anaerobic environments. To better understand the processes of Fe (III) reduction, bacteria such as *Shewanella putrefaciens* and *Geobacter metallireducens* have been researched. Anaerobic respiration necessitates physical and direct contact between the bacterial cell and Fe (III) minerals since Fe (III) is virtually insoluble in water at neutral pH. It has been reported that the outer membrane of anaerobically grown *S. putrefaciens* contains approximately 80% of membrane-bound cytochromes.

Transposon insertion metagenesis was used to identify a Fe (III) reduction mutant of *S. putrefaciens* SR-21. The mutant lacked MtrB, which is demanded for ferric reductase to reduce Fe (III) in vivo. Electrochemical approaches, in addition to biochemical research, may be the best way to investigate electron transfer processes in the Fe (III)-reducing bacteria. Researchers discovered that anaerobically grown cells of the iron-reducing bacterium *S. putrefaciens* IR-1 are electrochemically active and that the bacterium can be grown in an electrochemical cell without the addition of any terminal electron acceptors such as oxygen or Fe (III) or electrochemical mediators using cyclic voltammetry.

18.9 Biocatalysts of Prokaryotic Origin

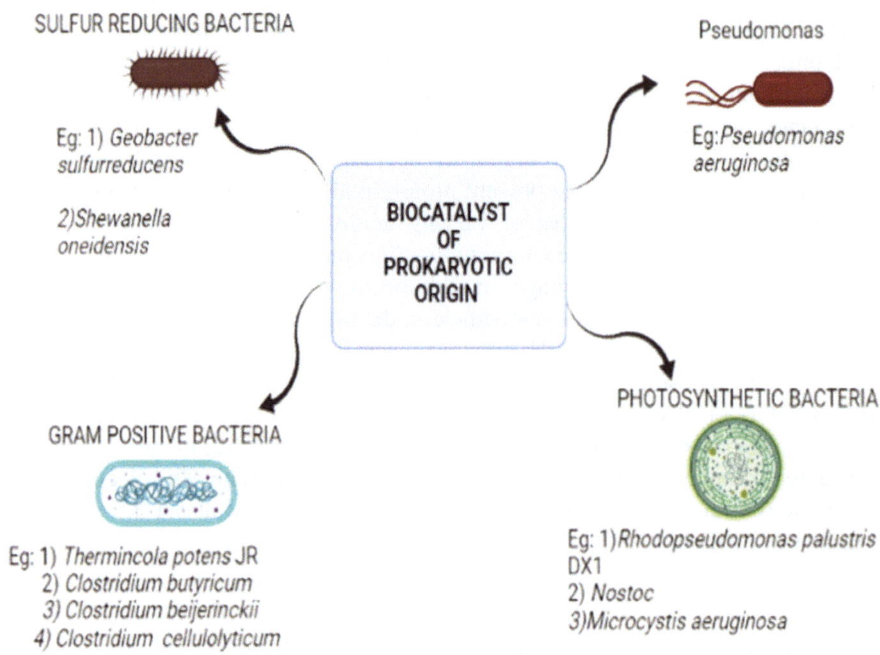

18.9.1 Sulphur-Reducing Bacteria

Bacteria used in MFC can survive on oxygen-rich or oxygen-insufficient condition by using different e^{-1} acceptors. Oxygen acts as terminal e^{-1} acceptor in aerobic bacteria during energy generation, while anaerobic bacteria can use NO^{3-}, SO^{4-}, metals, etc. as a terminal e^{-1} acceptor (Venkata Mohan et al. 2014b). Few bacterial species are also able to modify oxidation state of metals. These electroactive bacteria can modify oxidation state of metals and hence are used for immobilisation of radionucleotides to prevent their spread and degrade pollutants present in a wide range of environments such as sea, aquifers, lakes, etc. (Kumar et al. 2017). *Geobacter sulfurreducens* and *Shewanella oneidensis* represent the examples of electroactive bacteria that not only use oxygen and nitrates as terminal e^{-1} acceptors but can also use iron, manganese, sulphur and nickel. From the large community of electroactive bacteria, only few have been isolated. Once isolated, they are analysed for growth, role in environment and their gene markers. These microbes can be applied in biochemical cycle and pollutant removal process (Ilbert and Bonnefoy 2013). *Geobacter sulfurreducens* and *Shewanella oneidensis* are the most investigated electroactive bacteria. Though many bacteria engage in redox reactions, the mechanism is far less well known on the biocatalyst side.

The transport of electrons in *Geobacter sulfurreducens* and *Shewanella oneidensis* is facilitated by specialised electrical canals that carry the electrons through cell membranes. But still these bacteria face several challenges while acting as extracellular e^{-1} acceptors or donors during metabolism. To exemplify, neutrophilic bacteria need to avoid accumulation of insoluble ferric and ferrous form on their surface (Bond et al. 2012). Bacteria in MFC, which are placed near electrode, have the greatest accessibility to terminal e^{-1} acceptor but have lesser reach to nutrient. On the contrary, those placed on the surface of biofilm are in lesser contact with electrode surface but have higher accessibility to the nutrients. Surface-based biofilm bacteria have several pathways for transferring electrons from the biofilm surface to the electrode surface, including direct and induced (indirect) processes (Fig. 18.1) (Cao et al. 2019). Direct electron transfer, as the name suggests, is mediated by e^{-1} transporting cytochromes (Bond et al. 2012) and by pili which possess nanowires to transfer e^{-1} from the surface to electrode. The main advantage of it is that it can conduct electricity to longer distance (Kracke et al. 2015). Direct electron transport is mediated by MtrAB porin cytochrome complex which exist in e^{-1} active bacteria such as *Geobacter sulfurreducens* and *Shewanella oneidensis* (White et al. 2016) (Fig. 18.2).

Comparatively, less challenging is the e^{-1} transfer between electrode and bacteria through soluble redox mediators. The cellular membrane is crossed by the mediators facilitating e^{-1} transport between electrodes and electroactive bacteria. The examples of mediators are flavins, phenazines and siderophores. These three mechanisms work in accordance with environmental pressures (Richardson et al. 2012; White et al. 2016).

Genetics of the external e^{-1} transfer is well studied in *Shewanella oneidensis* MR 1. Metal-reducing (Mtr) pathway present in it uses iron and manganese as a terminal e^{-1} acceptor (Shi et al. 2012). C-type cytochrome complex oxidises quinol present in cytoplasmic membrane, which is shuttled to insoluble e^{-1} acceptors present outside via periplasmic membrane through channel proteins (Pawar et al. 2022).

Nanowires in *Geobacter sulfurreducens* are physically attached to the electrode by pili-like organelle. The e^{-1} transfer mechanism through nanowires is under examination, and two hypotheses are being put forward to explain it. According to first hypothesis, aromatic amino acid allows delocalisation of e^{-1} by π stability

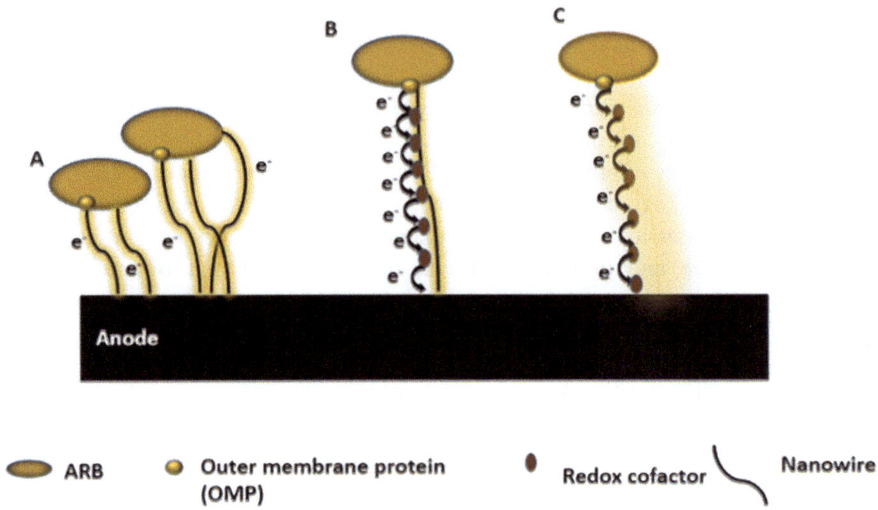

Fig. 18.1 EET via (**a**) nanowire or conductive pili, (**b**) hopping of electrons through redox cofactor aligned through microbial nanowire and (**c**) electron hopping via biofilm-bound redox cofactors

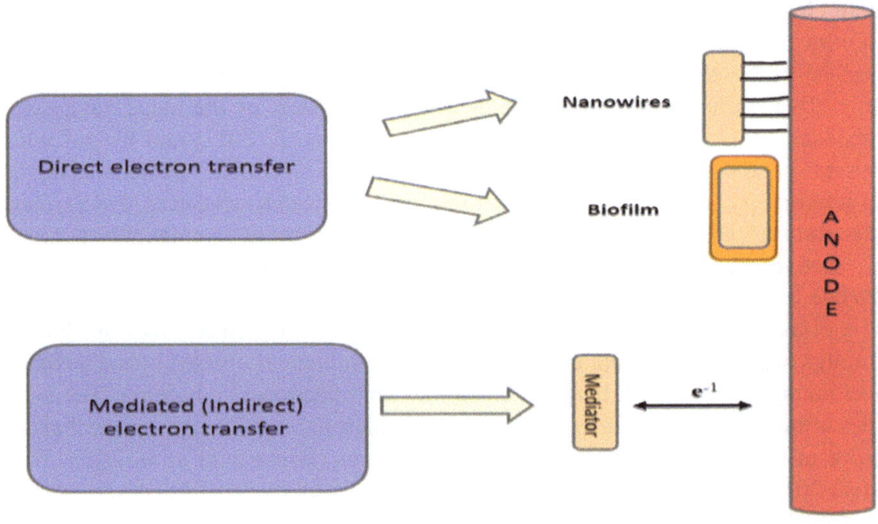

Fig. 18.2 Direct and mediated electron transfer in MFCs. (Adapted from Cao et al. (2019))

(Malvankar et al. 2015). While second hypothesis proposes 'super exchange model', according to which e^{-1} transfer happens by direct transfer to redox proteins (Shi et al. 2016). This system also possesses multiheme C-type cytochrome along with other channel proteins which assist in donating an e^{-1} to external electrode (Pawar

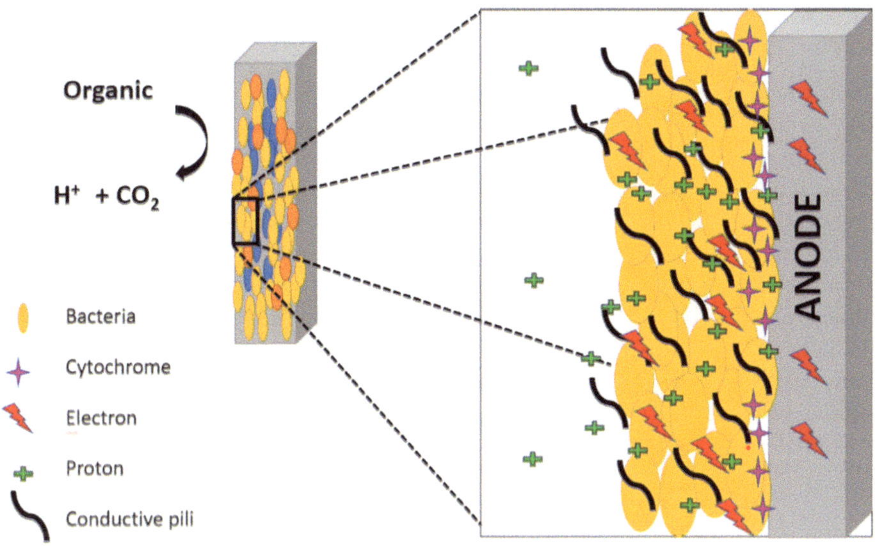

Fig. 18.3 'Super exchange model' of *Geobacter sulfurreducens*

et al. 2022). *Geobacter sulfurreducens* form thick biofilm and superior extracellular e^{-1} transport which make it one of the best electroactive bacteria (Fig. 18.3).

18.9.2 Pseudomonas

For cell-to-cell communication, one of the vital signalling mechanisms in bacteria is quorum sensing. This mechanism is a universal regulator in accordance with cell density, and advantage of this mechanism is that individual cells show coordination as a single unit with surroundings. In simpler words, quorum sensing aids in collective decision-making like pathogenesis and biofilm generation. In previous decade, quorum sensing proteobacteria, *Pseudomonas* was tested in MFC for bio-electricity generation (Yong et al. 2011). Electricity generation by these bacteria is mediated by the synthesis of phenazines, which transport e^{-1} between bacteria and negative electrode. Synthesis of phenazines is coordinated by quorum sensing pathway while PqsE can upregulate phenazine production in absence of quorum sensing system (Wang et al. 2013). Extracellular polymeric substance (EPS) reduces the conductivity of biofilm resulting in lesser generation of energy and therefore suppression of quorum sensing to downregulate EPS production in biofilm while enhancing power generation (Wang et al. 2013).

18.9.3 Gram-Positive Bacteria

Gram-positive bacteria rarely display extracellular e^{-1} transport as indicated by very few literature reports perhaps due to the presence of a cell wall, consequently preventing porin exposure to milieu (Carlson et al. 2012). The mechanism of extracellular e^{-1} transport in gram-positive bacteria is yet to be understood. Nevertheless, *Thermincola potens* JR is a gram-positive bacteria, forms anode biofilm and is efficient in power generation (Wrighton et al. 2008). Numerous C-type cytochrome coding genes are investigated from this gram-positive bacterium. Thermophilic gram-positive bacteria mostly help in reduction of insoluble iron, and this is mediated by cytochrome and conductive pilli (Gavrilov et al. 2012). Other gram-positive bacteria biocatalysts tested for power generation via MFC are *Clostridium butyricum*, *Clostridium beijerinckii* and *Clostridium cellulolyticum* (Cao et al. 2019).

18.9.4 Photosynthetic Bacteria

Researchers are looking forward for using photosynthetic bacteria as biocatalyst in MFC owing to two reasons. First, they are more beneficial compared to other microorganisms as they help in deducting carbon dioxide from environment (He et al. 2009). Additionally, mutualistic relation between photosynthetic and heterotrophic bacteria can enhance power generation without the involvement of supplementary e^{-1} acceptor or oxygen supply (Xiao and He 2014).

The photosynthetic biocatalysts applied in MFC include purple non-sulphur-reducing bacteria and cyanobacteria. Photosynthetic purple non-sulphur-reducing bacteria like *Rhodopseudomonas palustris* DX-1 have shown promising result as biocatalyst in MFC (Xing et al. 2008). The other photosynthetic purple non-sulphur-reducing bacteria investigated for MFC application which have shown promising results include *Rhodospirillum*, *Rhodobacter* and *Rhodovulum* (Qi et al. 2018).

Cyanobacteria represent another group of photosynthetic bacteria which has been applied as biocatalyst in MFCs.

High bioelectricity output is demonstrated by *Synechocystis* in a microfluidic cell (Bombelli et al. 2015). *Nostoc* immobilised on anode, fabricated with carbon nanotubes, has shown light-dependent energy generation in MFC (Sekar et al. 2014). As far as mechanism of power generation is concerned, *Microcystis aeruginosa* is a cyanobacteria species that produces reactive oxygen species helping in the production of electricity (Cai et al. 2013), while in another report, it has shown dual role of power generation and wastewater treatment (Ali et al. 2020). Evidence suggests that the mechanism of cyanobacteria to generate electricity is different from the other electroactive organism (Pisciotta et al. 2010).

18.10 Biocatalysts of Eukaryotic Origin

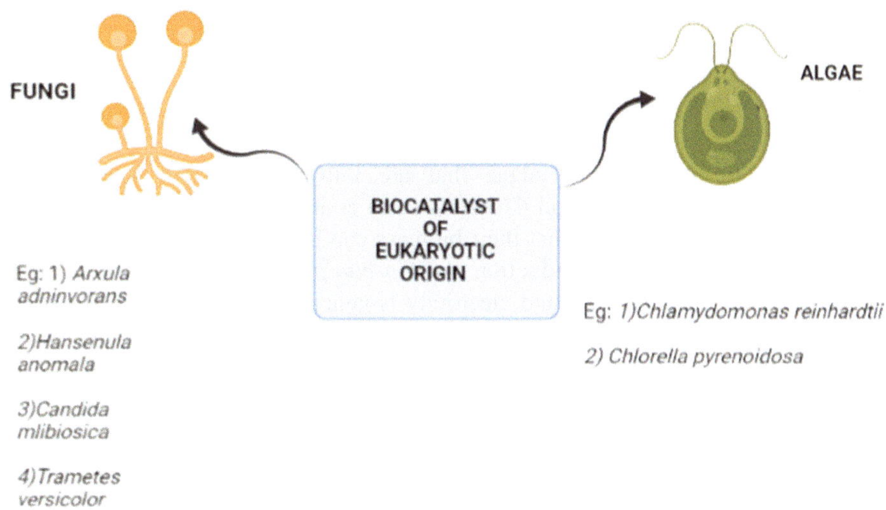

18.10.1 Fungi

Yeast species such as *Arxula adeninivorans (*Haslett et al. 2011)*, Hansenula anomala* (Prasad et al. 2007) and *Candida melibiosica (*Hubenova and Mitov 2010*)* can be used in MFC as biocatalyst. The advantages of using yeast are that they are non-pathogenic and easy to manage and have the ability to take up maximum substrates for growth. Moreover, they can manage to survive in diverse condition (aerobic or anaerobic). Initially, yeast biocatalyst showed lower output in comparison to bacteria. It could be due to lower catabolic rate owing to complication in accessing e^{-1} transfer mediators. Further, internal redox mediators are not present in *Saccharomyces cerevisiae*. Therefore, microorganism which is used in fuel cell requires exogenous or external mediator for external e^{-1} transfer. In an investigation, methylene blue/ferricyanide were used as the e^{-1} transfer mediator (Gunawardena et al. 2008) which showed improved output at same growth rates. Similarly, improvements were also observed for *Candida melibiosica* (Hubenova and Mitov 2010). Subsequent improvement in electrode fabrication also increased the output in *Candida melibiosica* yeast MFCs (Hubenova et al. 2011).

Filamentous fungi, like *Trametes versicolor*, are also investigated for their application in MFC. It possesses long thread-like structure known as hyphae (Fernández de Dios et al. 2013). The hyphae cannot help directly as biocatalyst as it does not make e^{-1} exchange with electrode but provides connectivity between them. Fungi can survive in diverse condition and produce oxidative enzymes which provides reasonable opportunity to act as co-biocatalyst in MFC applications which can be coupled with wastewater treatment contaminated with harmful chemicals (Fernández de Dios et al. 2013).

18.10.2 Algae

Few eukaryotic members of algae that are known to power MFC include *Chlamydomonas reinhardtii* and *Chlorella* (Cao et al. 2019), and major advantage of using these biocatalysts is that their biomass can be harvested for biofuels. The electricity is generated by introduction of *Chlorella pyrenoidosa* at anode of MFC, where it acts as an e^{-1} donor and electricity is generated by adjusting the oxygen content, light intensity and algal cell density with no need of externally added substrates (Xu et al. 2015).

18.11 Impacts of Electroactive Bacteria in MFC

In biofuel cells, either microorganisms or enzymes can be utilised; however, using microorganisms allows for the use of various enzymes and as a result of many substrates (or mixed substrates). Pure enzymes are used in specific and well-defined processes. Biosensor technology has been utilised to build this form of biofuel cell. As previously stated, two types of bacteria are used in MFC: those that require a mediator and those that do not. Bacteria that require mediators include *Saccharomyces cerevisiae*, *E. coli*, *Actinobacillus succinogenes*, *Proteus mirabilis*, *Proteus vulgaris* and *Pseudomonas fluorescens*. Some bacteria, such as *Shewanella putrefaciens*, a Geobacteraceae and Geobacter family member, *Rhodoferax ferrireducens*, and *D. desulfuricans*, mediators, are not required and have recently attracted attention. Kim et al. discovered that anaerobically grown *S. putrefaciens* cells had electrochemical activities, but aerobically produced *S. putrefaciens* cells have none. MFC can be injected with either pure or mixed bacteria cultures. Because of their nutritional flexibility and stress resilience, mixed cultures are more helpful than pure cultures.

The presence of various types of bacteria, including electrogenic, in mediator-less MFCs derived from a rich and diverse source of bacteria, such as wastewater, activated sludge, soil or sediments, is more beneficial in wastewater treatment because various types of bacteria, including electrogenic, are available, resulting in a high-power density. Exoelectrogenic bacteria may oxidise organic molecules and transfer e^{-} to the anode. They can be found in wastewater. To convert, MFCs must have at minimum one electrogenic microorganism. In biofuel cells, either microorganisms or enzymes can be utilised; however, using microorganisms allows for the use of different enzymes and, as a result, many different substrates (or mixed substrates). While pure enzymes can be utilised in particular and defined reactions, biofuel cells of this sort have been produced in biosensor technology.

Because of their nutritional flexibility and stress resilience, mixed cultures are more helpful than pure cultures. For MFCs to convert organic substrate to electricity, at least one electrogenic microbe must be present. Because of their high columbic efficiency and long-term sustainability, these bacteria are more useful in MFCs because they save energy for upkeep and development by transferring electron to the anode. The first metabolism of the organic fuel is carried out by fermentative or methanogenic bacteria in mixed cultures. Following that, at least one electricigen will be required to recover electrons from the metabolic products of the nonelectricigen and achieve maximal electron recoveries as electricity. Numerous studies have used mixed cultures as inoculums, including activated sludge, anaerobic sludge and domestic wastewater. As inoculums for MFC, Yates et al. investigated wastewater and bog sediment. They observed that bog-inoculated MFC produced more power density than wastewater-inoculated MFC in the first 20 cycles, but after that, the voltage and maximum power density were comparable. Anaerobic sewage sludge, which contains fermentative bacteria, methanogens and sulphate-reducing bacteria, might be a good choice for MFC inoculums. However, when anaerobic sludge was employed as an inoculum, methane generation was found on a regular basis, showing that the bacteria followed the anaerobic method.

Enrichment tactics have been recommended in this scenario to limit methanogen development while increasing the growth of electrochemically active bacteria in the anodic chamber. Mathuriya demonstrated that the type of inoculum used in MFCs has a substantial impact on power generation and waste disposal. Mathuriya examined at the self-microbial flora of normal wastewater, activated sludge and single and combined electrogenic microbial strains acquired from a self-tannery. The mixed electrogenic strain inoculum was shown to have a greater current yield while also having a higher substrate clearance efficiency. In the anode-associated community of MFCs, proteobacteria are the most common microbes. *Propionibacterium* and *Clostridium* were the most common bacteria in soil-inoculated two-chamber MFCs, according to Futamata et al. Nevertheless, they discovered that during the first week after inoculation, several prominent bacteria closely similar to *Pseudomonas* strains were found, implying that these bacteria served as oxygen scavengers. Rismani-yazdi et al. employed rumen microorganisms as inoculums for cellulose-based energy generation. According to the researchers, the major bacteria in the anode-attached biofilm were linked to *Clostridium* and *Sedimentibacter*, with sequences similar to *Desulfotomaculum* and Ruminococcus, while *Comamonas* was prominent in the suspended consortia.

According to Chen et al., *Ochrobactrum intermedium*, *Delftia acidovorans* and *Citrobacter freundii* may be novel bioelectricity-producing proteobacteria lineage species. According to a microbial culture investigation of MFC treating organic wastewater with high sulphate, *Thiobacillus* dominated the adhering biofilm on the anode, and sulphate-reducing bacteria such as Desulfobacteraceae were prevalent in suspended microorganisms. Nor et al. found that using pure *Pseudomonas aeruginosa* as an inoculum in a double-chambered membrane MFC to generate electricity resulted in maximum power density and current density than using anaerobic palm oil mill effluent sludge as an inoculum.

18.12 Challenges

Electrochemical studies in complex systems such as (electroactive bacteria) EABs or exoelectrogens need additional physical and chemical information in order to evaluate if an observed electrochemical response was caused by changes in an experimental parameter. Furthermore, molecular methods such as the development of mutants with varying amounts of gene/protein expression give a biological link to electrochemical studies. In the presence of EABs, electrochemical methods such as CV (cyclic voltammetry) and SWV (square wave voltammetry) clearly define the concept of the electron transfer event (reversibility, mass transfer limitations, redox couple properties and reaction steps); moreover, they do provide evidence on the how EABs participate in transfer of electrons or even what aspect of EABs promotes electron transfer. As a result, only because EABs have redox reactions does not mean they are active in electron transport. Furthermore, the presence or lack of electrochemical activity (current peaks in CV or SWV) in EABs does not always suggest that it is or is not a source of long-term, stable current. One example is the ability of some microbes to use soluble exogenous electron transporters for extracellular electron transfer in specific environments.

Some microbes, for example, can use soluble exogenous electron transporters in their environment for extracellular electron transfer. Despite the fact that this was demonstrated over a century ago, a lot of scientists are still interested in establishing that energy can be created using diverse substrates and bacteria. Despite the fact that numerous investigations continue focusing on MFC innovation, the existing state of knowledge shows clearly that BESs are not capable of meeting future energy demands. BESs, on the other side, can be used to discover new bacterial abilities in a variety of situations that could lead to some interesting applications.

18.13 Future Aspects

EAB research is still in its early stages, but there have already been some exciting breakthroughs in the field of electron transfer mechanisms. Many innovative methods were developed to understand better electron transport mechanisms in EABs. This study, which focused on these methods, displayed the data obtained in the researchers' laboratories. While our understanding of EAB electron transport processes has improved dramatically in the past five years, it has yet to permeate the whole BES literature. The scientists hope that their explanation of how EAB extracellular electron transport is studied would help the BES literature. In the future, they will be on the search for BES experiments that maximise both the BES design and the projected electron transfer from the EABs. To address some of the challenges surrounding EAB extracellular electron transfer, the researchers also wish to see BES studies that evaluate ideas made in nonpractical EAB extracellular electron transfer investigations.

18.14 Conclusion

The interest in MFCs is not limited to power generation and has diverse application like wastewater treatment, desalination, biomass generation, etc. which hold the key in achieving circular bioeconomy. Moreover, MFCs can aid in reducing carbon footprint globally. Hence, there have been consistent efforts to improve MFC efficiency, and they range from electrode fabrication, electrode compartmentalisation and biocatalyst development. It is important to note that these biocatalysts have been isolated from a wide range of environments (halophiles, psychrophiles, thermophiles, acidophile, etc.) and have been co-cultured or in consortia to enhance the efficiency of MFC. We have not covered these aspects in this chapter, while discussion on a variety of biocatalysts is also far from over. But the above discussion will aggrandise readers knowledge on MFCs and their biocatalyst. In spite of their promising advantages and efforts to improve them, they are marred by low power generation capacity. However, we expect that interdisciplinary research will improve efficiency of MFCs to the application level. Development in engineering and synthetic biology tools hold the key of MFC future along with knowledge of biocatalyst.

References

Ali J, Wang L, Waseem H, Song B, Djellabi R, Pan G (2020) Turning harmful algal biomass to electricity by microbial fuel cell: a sustainable approach for waste management. Environ Pollut 266(Pt 2):115373. https://doi.org/10.1016/j.envpol.2020.115373

Bombelli P, Müller T, Herling TW, Howe CJ, Knowles TP (2015) A high power-density, mediator-free, microfluidic biophotovoltaic device for cyanobacterial cells. Adv Energy Mater 5(2):1–6. https://doi.org/10.1002/aenm.201401299

Bond DR, Strycharz-Glaven SM, Tender LM, Torres CI (2012) On electron transport through Geobacter biofilms. Chem Sus Chem 5(6):1099–1105. https://doi.org/10.1002/cssc.201100748

Butti SK, Velvizhi G, Sulonen MLK, Haavisto JM, Koroglu EO, Cetinkaya AY et al (2016) Microbial electrochemical technologies with the perspective of harnessing bioenergy: maneuvering towards upscaling. Renewable Sustain Energy Rev 53:462–476

Cai PJ, Xiao X, He YR, Li WW, Zang GL, Sheng GP et al (2013) Reactive oxygen species (ROS) generated by cyanobacteria act as an electron acceptor in the biocathode of a bio-electrochemical system. Biosens Bioelectron 39(1):306–310. https://doi.org/10.1016/j.bios.2012.06.058

Cao Y, Mu H, Liu W, Zhang R, Guo J, Xian M, Liu H (2019) Electricigens in the anode of microbial fuel cells: pure cultures versus mixed communities. Microb Cell Factories 18(1):39. https://doi.org/10.1186/s12934-019-1087-z

Carlson HK, Iavarone AT, Gorur A, Yeo BS, Tran R, Melnyk RA et al (2012) Surface multiheme c-type cytochromes from Thermincola potens and implications for respiratory metal reduction by Gram-positive bacteria. Proc Natl Acad Sci U S A 109(5):1702–1707. https://doi.org/10.1073/pnas.1112905109

Chandra R, Venkata Subhash G, Venkata Mohan S (2012) Mixotrophic operation of photo-bioelectrocatalytic fuel cell under anoxygenic microenvironment enhances the light dependent

bioelectrogenic activity. Bioresour Technol 109:46–56. https://doi.org/10.1016/j.biortech.2011.12.135
Debabov VG (2008) Electricity from microorganisms. Mikrobiologiia 77(2):149–157
Dopson M, Holmes DS (2014) Metal resistance in acidophilic microorganisms and its significance for biotechnologies. Appl Microbiol Biotechnol 98(19):8133–8144. https://doi.org/10.1007/s00253-014-5982-2
Fernández de Dios M, del Campo AG, Fernández FJ, Rodrigo M, Pazos M, Sanromán M (2013) Bacterial-fungal interactions enhance power generation in microbial fuel cells and drive dye decolourisation by an ex situ and in situ electro-Fenton process. Bioresour Technol 148:39–46. https://doi.org/10.1016/j.biortech.2013.08.084
Gavrilov SN, Lloyd JR, Kostrikina NA, Slobodkin AI (2012) Fe(III) oxide reduction by a Gram-positive thermophile: physiological mechanisms for dissimilatory reduction of poorly crystalline Fe(III) oxide by a thermophilic gram-positive bacterium Carboxydothermus ferrireducens. Geomicrobiol J 29(9):804–819. https://doi.org/10.1080/01490451.2011.635755
Gunawardena A, Fernando S, To, F (2008) Performance of a yeast-mediated biological fuel cell. Int J Mol Sci 9(10):1893–1907. https://doi.org/10.3390/ijms9101893
Haslett ND, Rawson FJ, Barriëre F, Kunze G, Pasco N, Gooneratne R, Baronian KH (2011) Characterisation of yeast microbial fuel cell with the yeast Arxula adeninivorans as the biocatalyst. Biosens Bioelectron 26(9):3742–3747. https://doi.org/10.1016/j.bios.2011.02.011
He Z, Kan J, Mansfeld F, Angenent LT, Nealson KH (2009) Self-sustained phototrophic microbial fuel cells based on the synergistic cooperation between photosynthetic microorganisms and heterotrophic bacteria. Environ Sci Technol 43(5):1648–1654. https://doi.org/10.1021/es803084a
Hubenova Y, Mitov M (2010) Potential application of Candida melibiosica in biofuel cells. Bioelectrochemistry 78(1):57–61. https://doi.org/10.1016/j.bioelechem.2009.07.005
Hubenova YV, Rashkov RS, Buchvarov VD, Arnaudova MH, Babanova SM, Mitov MY (2011) Improvement of yeast–biofuel cell output by electrode modifications. Ind Eng Chem Res 50(2): 557–564. https://doi.org/10.1021/ie1000949
Ilbert M, Bonnefoy V (2013) Insight into the evolution of the iron oxidation pathways. Biochim Biophys Acta 1827(2):161–175. https://doi.org/10.1016/j.bbabio.2012.10.001
Jung S, Regan JM (2007) Comparison of anode bacterial communities and performance in microbial fuel cells with different electron donors. Appl Microbiol Biotechnol 77(2):393–402. https://doi.org/10.1007/s00253-007-1162-y
Kracke F, Vassilev I, Krömer JO (2015) Microbial electron transport and energy conservation - the foundation for optimizing bioelectrochemical systems. Front Microbiol 6:575. https://doi.org/10.3389/fmicb.2015.00575
Kumar A, Hsu LH-H, Kavanagh P, Barrière F, Lens PNL, Lapinsonnière L et al (2017) The ins and outs of microorganism–electrode electron transfer reactions. Nature Rev Chem 1(3):0024. https://doi.org/10.1038/s41570-017-0024
Lea-Smith DJ, Bombelli P, Vasudevan R, Howe CJ (2016) Photosynthetic, respiratory and extracellular electron transport pathways in cyanobacteria. Biochim Biophys Acta 1857(3):247–255. https://doi.org/10.1016/j.bbabio.2015.10.007
Liu WT, Chan OC, Fang HH (2002) Characterization of microbial community in granular sludge treating brewery wastewater. Water Res 36(7):1767–1775. https://doi.org/10.1016/s0043-1354(01)00377-3
Logan BE, Rabaey K (2012) Conversion of wastes into bioelectricity and chemicals by using microbial electrochemical technologies. Science 337(6095):686–690. https://doi.org/10.1126/science.1217412
Malvankar NS, Vargas M, Nevin K, Tremblay PL, Evans-Lutterodt K, Nykypanchuk D et al (2015) Structural basis for metallic-like conductivity in microbial nanowires. mBio 6(2):e00084. https://doi.org/10.1128/mBio.00084-15

Pawar A, Karthic A, Lee S, Pandit S, Jung S (2022) Microbial electrolysis cells for electromethanogenesis: materials, configurations and operations. Published online: December 24, 2020. Environ Eng Res 27(1):53–73

Pisciotta JM, Zou Y, Baskakov IV (2010) Light-dependent electrogenic activity of cyanobacteria. PLoS One 5(5):e10821. https://doi.org/10.1371/journal.pone.0010821

Prasad D, Arun S, Murugesan M, Padmanaban S, Satyanarayanan RS, Berchmans S, Yegnaraman V (2007) Direct electron transfer with yeast cells and construction of a mediatorless microbial fuel cell. Biosens Bioelectron 22(11):2604–2610. https://doi.org/10.1016/j.bios.2006.10.028

Qi X, Ren Y, Liang P, Wang X (2018) New insights in photosynthetic microbial fuel cell using anoxygenic phototrophic bacteria. Bioresour Technol 258:310–317. https://doi.org/10.1016/j.biortech.2018.03.058

Richardson DJ, Butt JN, Fredrickson JK, Zachara JM, Shi L, Edwards MJ et al (2012) The 'porin-cytochrome' model for microbe-to-mineral electron transfer. Mol Microbiol 85(2):201–212. https://doi.org/10.1111/j.1365-2958.2012.08088.x

Sekar N, Umasankar Y, Ramasamy RP (2014) Photocurrent generation by immobilized cyanobacteria via direct electron transport in photo-bioelectrochemical cells. Phys Chem Phys 16(17):7862–7871. https://doi.org/10.1039/c4cp00494a

Shi L, Rosso KM, Zachara JM, Fredrickson JK (2012) Mtr extracellular electron-transfer pathways in Fe(III)-reducing or Fe(II)-oxidizing bacteria: a genomic perspective. Biochem Soc Trans 40(6):1261–1267. https://doi.org/10.1042/BST20120098

Shi L, Dong H, Reguera G, Beyenal H, Lu A, Liu J et al (2016) Extracellular electron transfer mechanisms between microorganisms and minerals. Nat Rev Microbiol 14(10):651–662. https://doi.org/10.1038/nrmicro.2016.93

Venkata Mohan S, Velvizhi G, Modestra JA, S Srikanth S (2014a) Microbial fuel cell: critical factors regulating bio-catalyzed electrochemical process and recent advancements. Renew Sustain Energy Rev 40:779–797

Venkata Mohan S, Velvizhi G, Vamshi Krishna K, Lenin Babu M (2014b) Microbial catalyzed electrochemical systems: a bio-factory with multi-facet applications. Bioresour Technol 165: 355–364. https://doi.org/10.1016/j.biortech.2014.03.048

Venkidusamy K, Megharaj M (2016) A novel electrophototrophic bacterium Rhodopseudomonas palustris strain RP2, exhibits hydrocarbonoclastic potential in anaerobic environments. Front Microbiol 7:1071. https://doi.org/10.3389/fmicb.2016.01071

Wang VB, Chua SL, Cao B, Seviour T, Nesatyy VJ, Marsili E et al (2013) Engineering PQS biosynthesis pathway for enhancement of bioelectricity production in pseudomonas aeruginosa microbial fuel cells. PLoS One 8(5):e63129. https://doi.org/10.1371/journal.pone.0063129

White GF, Edwards MJ, Gomez-Perez L, Richardson DJ, Butt JN, Clarke TA (2016) Mechanisms of bacterial extracellular electron exchange. Adv Microb Physiol 68:87–138. https://doi.org/10.1016/bs.ampbs.2016.02.002

Wrighton KC, Agbo P, Warnecke F, Weber KA, Brodie EL, DeSantis TZ et al (2008) A novel ecological role of the Firmicutes identified in thermophilic microbial fuel cells. ISME J 2(11): 1146–1156. https://doi.org/10.1038/ismej.2008.48

Xiao L, He Z (2014) Applications and perspectives of phototrophic microorganisms for electricity generation from organic compounds in microbial fuel cells. Renew Sustain Energy Rev 37:550–559

Xing D, Zuo Y, Cheng S, Regan JM, Logan BE (2008) Electricity generation by Rhodopseudomonas palustris DX-1. Environ Sci Technol 42(11):4146–4151. https://doi.org/10.1021/es800312v

Xu C, Poon K, Choi MM, Wang R (2015) Using live algae at the anode of a microbial fuel cell to generate electricity. Environ Sci Pollut Res Int 22(20):15621–15635. https://doi.org/10.1007/s11356-015-4744-8

Yong YC, Yu YY, Li CM, Zhong JJ, Song H (2011) Bioelectricity enhancement via overexpression of quorum sensing system in Pseudomonas aeruginosa-inoculated microbial fuel cells. Biosens Bioelectron 30(1):87–92. https://doi.org/10.1016/j.bios.2011.08.032

Chapter 19
Techno-economic Feasibility Analysis of Microbial Product Commercialization

Aparupa Das, Anuradha A, Vinod Kumar Nigam, and Muthu Kumar Sampath

Abstract Microbial products are typically the products derived from various microscopic organisms. These products may consist of the organism themselves and/or the metabolites they produce. However, the cost and efficiency of the lab-scale microbial-based developed processes remain a significant bottleneck when it comes at the industrial level. In this chapter, we wish to address a case study mainly focusing on the techno-economic analysis of the industrial-scale production of microbial alkaline protease enzyme.

Enzymes are the proteins that function as catalyst that assist entire biological and metabolic response. Alkaline proteases are the hydrolytic enzymes that perform as biocatalyst for the separation of protein into smaller peptides and amino acid. Alkaline protease possesses wide application in leather industry, food processing, and detergent industries. With an expansion for significant demand for alkaline protease, it is essential to survey the details on techniques prior to rise commercialization manufacture originating from segregated strain.

Techno-economic analysis (TEA) is performed for one similar method generating alkaline protease from sugar through industrial scale to acquire economic viability. The change of leading machinery to commercialization product is thought to be highly cost-effective, time consuming, and hazardous attempt. Therefore, balance on advance process and study should be formulated on techno-economic analysis. Because of the absence of materials on the required process design, limitation, and relevant theories in the initial phase of process advancement, TEA is bounded to contributing a rough evaluation of the funding for a theoretical alkaline protease composition. However, the TEA emphasizes the division among initial phases of research and commercialization such that in addition to process analysis and advancement feasibly conducted in the better productive aspects. This will surely help to understand whether the developed lab-scale microbial-based process is feasible or economical at the industrial scale.

A. Das · Anuradha A · V. K. Nigam · M. K. Sampath (✉)
Department of Bioengineering and Biotechnology, Birla Institute of Technology Mesra, Ranchi, India
e-mail: muthubio@bitmesra.ac.in

A. Sarkar, I. A. Ahmed (eds.), *Microbial products for future industrialization*, Interdisciplinary Biotechnological Advances,
https://doi.org/10.1007/978-981-99-1737-2_19

Keywords Techno-economic analysis (TEA) · Process design · Alkaline protease · Economic viability

19.1 Introduction

Enzymes have extensively been employed as substitute to chemical for advancement of the ability and are cost productive in a broad range of industrial structures and activities. Enzymes are highly efficient environment-friendly protein catalysts, synthesized by living systems. They can increase the reaction rates through factor of nearly a million or more and also are distinctly specific to their substrate. They have important advantages over the chemical catalysts, of which the most important are specificity, high catalytic activity, ability to work at moderate as well as extreme temperatures, and the ability to produce in large qualities. Moreover, due to their biological nature, enzyme usually does its work and is active in an aqueous media under moderate temperature and pH condition and is biodegradable, and these characteristics assemble them commercially and environmentally interesting.

Enzyme is a crucial resource exploited over the chemicals, food, and an integrated industries to produce extended biotechnology outputs and possess a known beneficial catalyst for several organic modifications and also produce superior chemicals and pharmaceuticals (Bhunia et al. 2012). Enzymes play a wide role of application in various industrial sectors such as food and beverage industries, pulp and paper industries, textile and leather industries, and detergent and biofuel production industries. Global market for industrial enzyme was extended about $5.5 billion in 2018 and is expected to expand at a compound annual growth rate (CAGR) of about 6.5% through the period of 2020–2025. The food and beverage sectors hold the largest share of around 38%. This growth is mainly induced by increasing demand of food enzyme in developing markets and by the development of biofuel production. Enzymes are also widely utilized in agro-biotechnological methods, and their chief function is to produce feed supplement to boost feed productivity. Phytase is involved in feed enzymes to enhance plant phosphorus uptake through monogastric animal (Demain and Adrio 2008) (Fig. 19.1).

Protease exhibits proteolytic action and high cohesion that sustain high intensity of cross-linking properties of H-bond and are essential tools of all life forms. Microbial protease involved in various cellular metabolic processes performs one of the main hydrolytic enzymes having several industrial applications. Alkaline protease holds an enormous share in an enzyme market. Moreover, studies have exhibited that nutritional factors such as sources of carbon and nitrogen can affect the protease production along with the physical parameter such as temperature, pH incubation time, and inoculum concentration and can also influence protease production (Boominadhan et al. 2009).

These enzymes have been described to work under harsh physiological condition of temperature of 20–70 °C and pH range up to 11 and have been probable in detergent formulation and certain other functions in the presence of organic solvent. Globally, it pays 60% of entire worldwide enzyme. The amount of protease

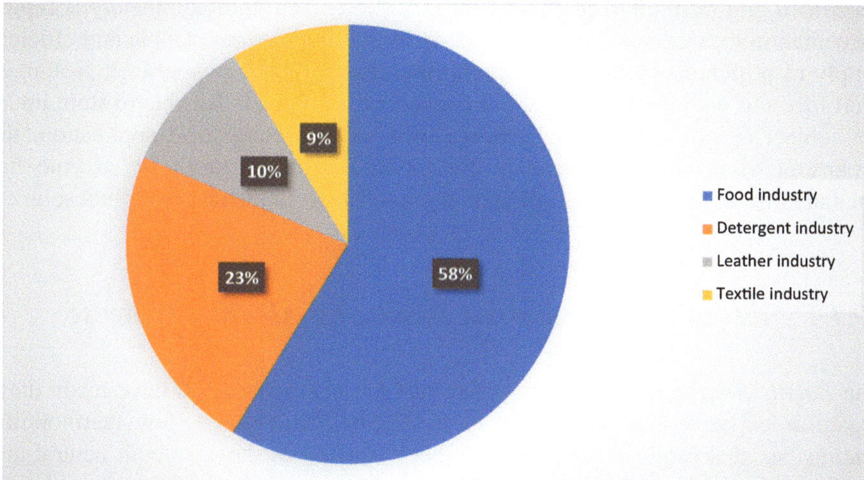

Fig. 19.1 Distribution of alkaline protease in different industrial applications

produced on a commercial range widespread is greater than any other enzymatic class of biotechnological significance. In recent years, biotechnology is achieving promptly because of the several benefits that it proposes across current chemical methods mainly concerning surrounding and capital cost association. This is directed by advance expansion in pharmaceutical enzyme desire, food and beverage industries, superior chemical manufacture, and detergent and leather production industries. Alkaline protease constitutes immense group of industrial enzymes. Nowadays, the majority of an enzyme applied in the industrial processes are prone for the degradation of several natural processes since they perform a key function as metabolic catalyst. As reported by EC (Enzyme Commission), protease refers to class 3 and subdivision 4 that hydrolyze peptide bond (Sharma et al. 2017), although divided on the basis of site of isolation (animal or microbial) and catalytic activity (endopeptidase which divide inner peptide bond and exopeptidase which divide C-N end peptide bond). Microbial protease referred to be extracellular in nature, hence facilitating downstream operation of the enzyme. The alkaline proteases have various industrial applications such as chemical industry, leather industry, food and feed industry, waste treatment industry, etc. By the progress of current protein engineering methodology, alkaline protease production is developed in multifold ways such as rare thermostability, knowledge of structure, and function relation of alkaline protease (Verma et al. 2011a).

Proteases occur naturally in all organisms and constitute 1–5% of their gene content. Additionally, protease is an enormous family of enzyme which holds 2% of human genome. And because of its configuration and functional variation, protease plays a wide extent of intracellular protein recycling and rapids of nutrition digestion along with development of immune system (Naveed et al. 2021). In contrast to plants and animals, microorganisms are studied as a main source of protease as they can be performed and purified in large amount in a short-lived with completely accepted fermentation technique (Varia et al. 2019). They can contribute ample and efficient supply of required product. In addition, microbial protease possesses a prolonged shelf life, and without any substantial damage of action, it is feasible to store under the subject optimal condition. Consequently, in commercialized application, the expansion of production boundary for the cost-efficient profit of enzyme has obtained limitation significance of alkaline protease from distinct microbial sources.

19.1.1 List of Microorganisms Producing Alkaline Protease

The *Bacilli* contribute 70% of protease; thus the distinct sources have made these organism exposures in biotechnology field. So far, although, few thermophilic *Bacillus* sp. that produce protease have been isolated as they are both neutral and alkaline. Mostly, bacterial neutral protease is active in pH scale 5–8 and holds low heat resistance. Because of their transition rate of reaction, these neutral proteases cause less acidity; hence, they are reasonable for food industry. And their low heat resistance is important for regulating the reactivity within the producing of food hydrolysates. Likewise, bacterial alkaline protease is distinct by their action at pH 10 such as *Bacillus* sp. SSR1. Their ideal temperature is about 60 °C and is applicable in detergent industries. Detergent industry is the huge single market for enzymes at 25–30% of total marketing, and alkaline protease is used as cleansing additive in detergent industry because of stability toward surfactant and oxidant and high pH. Their ideal temperature is about 60 °C and is applicable in detergent industries. This consists of various alkaline protease-producing bacteria such as *Bacillus* sp. JB-99, *Bacillus* sp. B001, and *Halobacterium* sp. The most broadly studied alkaline protease producer among fungi is the genera of *Rhizopus* and *Conidiobolus*, within actinomycetes is *Streptomyces* species and among yeast is the genus *Conidia*. In *Bacillus pumilus* MK6–5, alkalophilic isolates are discovered in the utilization of UF membrane cleaning (Kumar 2002). However, the survey of research specifies that the *Bacillus* sp. is certainly the most attractive source of commercial alkaline protease. *Bacillus* species is a familiar producer of extracellular protease and holds 99% of this extracellular protease that has the benefit of ease of improvement and is widely utilized in several industrial sectors (Kumari and Reshma 2021). The elicited protease from *Aspergillus oryzae* CH93 owned biochemical property that would be effective in several function utilizations (Salihi et al. 2017). Most alkaliphilic microorganisms produce alkaline proteases, but only those with substantial production are considered for commercial production. *Aspergillus oryzae* U1521 has the possibility to be an accomplished nominee for large-scale production of alkaline protease enzyme (Samarntarn et al. 1999) (Table 19.1).

The carbon and nitrogen are vital components which prompt growth of microorganism and enzyme production. Glucose, yeast extract, distilled water, and water are the raw materials required for the alkaline protease manufacturing. In contrast, glucose and fructose showed the prime carbon source for enhancing the efficiency

Table 19.1 Diverse sources of alkaline protease producing microorganism strain

Microorganism	Source	Incubation temperature (°C)	References
Bacillus cereus KM05	Contaminated soil	70	Jayakumar et al. (2021)
Bacillus sp. *BBXS-2*	Molasses	45	Qureshi et al. (2018)
Bacillus cereus RS3	Desert soil	45	Shine et al. (2016)
Halobacterium sp.	Soil	37	Vijayaraghavan et al. (2012)
Bacillus subtilis	Hot spring	60	Kembhavi et al. (1993)
Bacillus firmus 7728	Soil sample extract from leather industries	40	Rao and Narasu (2007)
Pseudomonas aeruginosa	Sea water	37	Marathe et al. (2018)
Micrococcus sp.	Marine sediment sample	50	Hou et al. (2017)
Bacillus clausii	Fish waste sample	25–35	Reshma (2021)
Bacillus subtilis KO	Molasses	45	Younis et al. (2010)
Bacillus licheniformis	Agroindustrial by-products	50	Aguilar et al. (2019)
Thermoactinomyces sp. *RM4*	Soil sample	80	Verma et al. (2011b)
Bacillus cereus	Wastewater sample	50	Ayantunji et al. (2020)
Aspergillus versicolor	Poultry farm	40	Choudhary (2013)
Bacillus pumilus MP 27	Ocean water	50	Mehak Baweja et al. (2016)

of protease. High concentration of glucose suppresses the enzyme production using *Streptomyces* sp. The maximum quantity of protease manufacturing by *Bacillus cereus* 146 was examined bearing beef extract being nitrogen source in growth media. Peptone developed is ideal for protease produced by *Bacillus licheniformis* BBRC 100053 which also shows higher efficiency of carbon source in culture media of protease. *Streptomyces ambofaciens* is too an optimal nitrogen source for protease production. Dextrose was noted to be a relevant carbon source for protease production by *Bacillus subtilis*.

19.1.2 Application of Alkaline Protease Enzyme Detergent Industry

Utilization of enzyme in detergent is possibly the massive innovation of the detergent industry during the past quarter era. Protease is the primary ingredient of multiple detergent varying from household laundry to industrial cleaning purpose, and it accounts for around 25% of the entire global sales of enzyme. They attribute about two-thirds share of the worldwide detergent enzyme sales.

Owing to high alkalinity of the detergent, pancreatic enzyme did not work effectively. The detergent industry utilizes various hydrolytic enzymes which work under alkaline pH scale. In 1956, the first detergent involves the bacterial enzyme found into the market under the trade name Bio-40 (Rao et al. 1998). In 1963, alcalase, an alkaline protease, was completely assimilated in detergent market and was marketed by Novo industry; Denmark is under the trade name Biotex. Nowadays, detergent industry accounts for 81% of the entire protease sales in the world from many *Bacillus* sp. Protease is discovered for a vital application as detergent additive as it hydrolyzes and separates proteinaceous stain like blood, milk, egg, and also protein from body secretion. The application of alkaline protease in detergent formulation is its significant expansion because of its huge effect on quality of detergent. In present market trends and consumer needs, they are affecting the manufacturer to develop new enzymes having effective cost and enhance performance across staining. For selection of detergent protease, isoelectric point (PI) is the chief parameter. It is studied that detergent protease function best when the pH value of the detergent solution to which it works is approximately same as the PI value for the enzyme. There are many further parameters affected in the selection of an ideal detergent protease like compatibility with detergent ingredients, for example, surfactant, perfumes, and bleached, good action at applicable washing temperature and pH, compatibility with ionic strength of the detergent solution, and strain degradation.

At 50 °C, washing performance produces superior destaining. Detergents such as Ariel, Rin, and Surf excel have most protease activities at 50 °C and exhibit poor protease function at 40 °C. In the *Bacillus cereus*, protease produce superior stability at alkaline pH scale with commercial detergent because it has high action at either 40–50 °C. Additionally, *Bacillus clausii* is widely known to produce an industrially chief alkaline serine protease which is employed as detergent additive to separate protein consisting spots from laundry and shows low activity as well as stability with respect to anionic surfactant such as SDS along with oxidant, for instance, H_2O_2 that has been the usual element in current bleach from detergent expression (Joo et al. 2003). The major disadvantage to their use is that it is essential to acquire microbe-free enzyme which needs high-cost filtration technique. Although if in fungi the mycelia are perhaps freely separated by filtration process, there is constantly demand for newer enzyme with unique properties that can promote the increase of the wash performance of presently used enzyme from detergent.

19.1.2.1 Leather Industry

Alkaline protease has ample application in leather industry. It consists of distinct steps during the leather preparation such as soaking, dehairing, bating, and tanning. The utilization of an enzyme has been conducted to better leather quality and reduce pollution. Mainly, dehairing preparation utilizes the exploit chemicals like lime and sodium sulfide which are highly expensive and pollution hazardous in nature (Sharma et al. 2019). Hence, these chemicals are not advised to eco-friendliness owing to the issue came across with sewage discarding.

Although protease enzyme from definite bacterial species *Thermoactinomyces* sp. RM4 had an optimum pH (Marathe et al. 2018), temperature of 80 °C has been confirmed to be potential in hair removing step. Traditionally, dehairing process affected the utilization of these chemicals which resulted to pollution; thus, aims were made to reduce the use of chemicals and establish natural techniques. Collagen protein is the main entity which is present in the skin, and it exhibits superior strength properties and also shows greater surface area (Arunachalam and Saritha 2009). Alkaline protease accelerates the dehairing steps since roots swell up in alkaline state and there is a consecutive attack by an enzyme on hair follicle protein and allow discharging of hair.

Mainly, dehairing is processed by bating and later tanning. In dehairing method of buffalo hide, alkaline protease enzyme was used and was precipitated with ammonium sulfate. At medium temperature between 25 and 35 °C, this ammonium-precipitated enzyme was stable. Within next 12 h of incubation, the enzyme tends to appear clear dehairing action, and after 21 h, proper dehairing of buffalo hide occurs. Then, it reveals clear hair follicle and terminating total removal of hair.

In bating steps, its purpose is to remove undesirable inter-fibrillary protein causing soft, fine, flexible leather. It is studied that alkaline protease isolated from particular microbial sources causes efficient bating. The key objective of bating processing is to cause inner segregation of collagen by lowering keratin and hence exhibit extreme reactive plane for tanning reagent to influence. Some disadvantage like enzymatic method is most costly and then the chemical processing since it needs control and underperformance in dehairing.

19.1.2.2 Silver Extraction

Industrially, silver is the major metal used in various regions like X-ray film and photographic film, silver wares, jewelry wares, and some electronic devices. Seeing its incomparable characters being light-sensitive matter, it is employed for photographic and X-ray film. Throughout photography, silver is not crashed and hence can be regained and recycled. X-ray film holds around 1.5–2% of silver in gelatin layer. Alkaline protease isolated from *Conidiobolus coronatus* detaches the gelatin layer in a period of 40 °C and pH of 9 in 6 min. This enzyme plays an important aspect during the bioprocessing of utilized silver and also extracts gelatin layer fixed with silver in X-ray film. Traditionally, silver recovering affects burning of X-ray which directed environmental pollution. However, the enzymatic method is low-cost and pollution-free and has a slow process. Around 20% of the global silver demand is entirely filled by photographic waste reclaiming. It was noticed that the gelatin surface totally decayed within 20 min at pH of 9 and 50 °C and approximately 0.135 g of silver was recovered with 0.335% output. Adopting protease treatment, it is feasible to obtain refined silver from applied X-ray along with photographic films (Gemechu et al. 2020).

19.1.2.3 Waste Management

Several fibrous proteins such as hair, feather, horns, and nails are rich in keratin particularly maintained by hydrophobic interaction and have hydrogen and disulfide bond and are abundant as waste in the environment. The breakdown of keratin waste is done by both chemical and mechanical disintegrations, yet it is not biodegradable. Additionally, bioconversion of wastewater sludge toward value supplement product is an economical and environment-friendly method. *Bacillus licheniformis* is employed for wastewater sludge for manufacture of alkaline protease and is now fruitfully employed in laboratory (Bezawada et al. 2011). Alkaline protease is employed in the conversion of waste to effective biomass. Protease dissolved the proteinaceous waste, which decreases the BOD (biochemical oxygen demand). It plays a huge important application in waste management by degradation wastes from industries and household. *Bacillus* species is considered as an extensively bacterial source of keratinases for feather deterioration. *Streptomyces* sp., *Pseudomonas* sp. MS21, and *Microbacterium* sp. are more bacterial sources against degeneration of keratin.

19.1.2.4 Medical Sector

Researchers have also innovated the wide use of protease in medical area. For medicinal purpose, various techniques, soft gel medical formula, gauze, nonwoven tissue, and ointment configuration involve alkaline protease isolated from *Bacillus subtilis*. It has been studied that fibrin degeneration has been attained by alkaline fibrinolytic protease. The utilization of alkaline fibrinolytic protease advises its future function in an anticancer drug treatment. The elastolytic characteristics of serine protease isolated from *Bacillus subtilis* have been known to develop elastase. This expression is employed in the burn surgery, abscesses, and other injuries. Alkaline protease has also been practiced in the medicinal field. *Aspergillus oryzae* aids in detection and helps to precise specific enzyme deficiency syndrome. Additionally, in normal cells, protease plays a vital function to perform biological processes inside the body and when in interference in stability of living system and exceeds to cause diseases such as cancer which causes tumor growth (Singh 2021).

19.1.2.5 Food Industry

Due to their potential to segregation of protein, alkaline protease has been conventional for producing high nutritional protein hydrolysates. There are several kinds of protein hydrolysates which are perhaps produced from food protein. Protein hydrolysates play an important function in the control of blood pressure and utilizes the synthesis of infant food, certain therapeutic dietary products, and fortification of soft drink and fruit juice. These hydrolysates generate from a diversity of substrates such as meat, casein, etc. Generally, meat hydrolysates have bitter flavor during the hydrolysis level which is more than 10%.

Throughout the cheese manufacture, protease plays a vital role in the food industry. Protease enzyme hydrolyzes a definite peptide bond in K-casein and thus permits milk coagulation, an important part in cheese manufacture, and produces yoghurt using lactic acid bacteria.

Lactic acid bacteria rely on protease enzyme which hydrolyze casein and attain nitrogen source demand for growth purpose. These bacteria have two main roles being starter bacteria in the production of cheese:

(a) Advancement of flavor during ripening
(b) Acidification of milk

Lactobacillus sp., *Streptococcus cremoris*, *Amycolatopsis* sp., and *Amycolata* sp. have been applied for cheese manufacturing.

Protease enzymes are applied in the alternation and treatment of protein which cause advanced stability and heat solubility and oppose precipitation in acidic surrounding. However, protein hydrolysates act as an emulsifying agent in several applications such as ice cream, coffee whitener, and sausages. In addition, debittering of hydrolysates contains particular dissociation by several processes like removal of alcohol, covering of bitter flavor, analysis of activated carbon, and chromatography above the silica gel. *Lactobacillus* performs as a debittering starter supplement. The usage of laccase enzyme for clarification of juices and taste improvement in beer are currently accepted applications inside the beverage industries (Kirk et al. 2002).

In general, milk protein acquires a high nutritional value, and it consists of two main proteins: casein and whey. Out of whole milk protein, casein contains around 80% and whey contains around 20%. Industrially, after precipitation of casein, rennet or acid remains soluble, and this milk protein is termed as whey protein and comprises maximum biological value in contrast to other proteins. It has an abundant sulfur consisting of amino acid that aids in antioxidant activity. Hence, in the dairy industry, it plays a vital part in the alteration of flavor as well as lactose depletion.

19.1.3 Techno-economic Analysis

Techno-economic analysis (TEA) helps to integrate operation modeling and design with economic estimation. As it is expected, there is expansion in demand of protease or other industrial enzymes in coming years due to the increase in demand of food enzyme in advance market. Because of these, TEA is carried out to develop the alkaline protease production by balancing capital expenditure and time consumption and to attain economic feasibility. A defined structure of a techno-economic model permits to reduce error in the manufacture phase and to evaluate average result. Considering technical criterion in the techno-economic structure aids to deploy cost-effective approach network construction with essential technical acts (Kantor et al. 2010). TEA evaluates predictable production and capital expenditure of a specific process at productive scale in order to produce a quantitative information of the operation economic viability (Sun et al. 2020). TEA is also applied to accomplish economic assessment of the method to provide input and output flow data required for environmental estimation (da Gama Ferreira et al. 2018).

Techno-economic analysis (TEA) is performed at various stages of the modification process to give complete recognition of the tools, labor, valuable amount, and manufacturing price depending on current productivity quantity. Throughout the development phase, the only main action of altered product or supply is the assessment of their economic viability which is accomplished with the techno-economic analysis method. TEA further gives observation into regions where R & D would be trained to attain the maximum vital increase in the manufacture, in case by the substitution of high-cost raw component. TEA is carried out for developing alkaline protease from sugar by industrial method to obtain economic viability. Such simulation recognizes us to reexamine the extent of upcoming research technique. Production value is highly affected by process design and optimization of the process (Macrelli et al. 2012).

In this chapter, the subsequent materials are executed: ideal design of batch fermentation of 100 m^3 operation depends on the research study and bench measure evaluation. SuperPro Designer has enhanced a crucial software for bioprocess simulation and permits environmental studies as proof. So it is regarded as a primary implementation for bioprocess simulation (Niño and Gelves 2021). A structure of parallel plant-size design using SuperPro Designer software is used which aids in estimation of capital expenditure and manufacturing value along with productive calculation to permit the chosen appropriate process design (Wu and Maravelias 2018). A combination of techno-economic along with environmental measure was determined to recognize finest substitute (Gómez et al. 2020). Material and energy balance and flow rate details are utilized for size of an apparatus and to evaluate capital cost (Tao et al. 2017).

19.2 Process Description

SuperPro Designer v9.5 was selected to model and simulate the protease production operation and diversity of this operation in relation to economic and technical criterion. This software is applied to predict plant structure before implementing its construction in order to assume that plant would be cost-effective or not prior to viable expansion process (Ramadhani et al. 2021). SuperPro Designer v9.5 software is often applied in biotechnology industries as well as in pharmaceutical. And it is also used to approximate information for the large-scale commercial study.

The process flow diagram of the alkaline protease enzymes was separated into three vital parts: media composition, fermentation section, and downstream segment. A segment in SuperPro Designer consists of a unit procedure, task, and many other icons for processing, and each new flow diagram holds a main section by default in which each new unit procedure automatically authorized.

19.2.1 Media Composition

In the software, materials are classified into two classes: components and mixture. Mixed materials comprising a list of elements in it are generally termed as mixture. However, pure raw elements, for simulation design, are known as components. Whole materials utilized were defined inside the software.

The operation requires three vital medium mixtures: glucose, ammonium hydroxide, and salts which constitute a mixture of mineral with other elements required for microbial growth. It would be highlighted that the media for the industrial enzyme ought to be easy and low cost, specified that the enzyme product should be sold at moderately inexpensive prize. Other elements above the ones defined as component of mixtures are yeast extract and biomass. However, glucose, NH_4OH, and salt are the three compounds mixed with water in a distinct blending tank to acquire suitable stock solution.

19.2.2 Fermentation Section

The fermentation media is prepared in a stainless-steel tank (V-101) and sterilized in a continuous sterilizer (ST-101). A compressor (G-101) and an absolute air filter (AF-101) provide sterile air to the fermenter (FR-101).

19.2.3 Primary Recovery and Purification Section (Downstream Section)

Because industrial enzyme is in moderately simple purity demand, the product recovery as well as purification method is periodically easy. The initial step is the removal of microbial cell out of the extracellular solution carrying the enzyme of interest, be usually attained by centrifugation or cloth filtration. After removal of biomass, the filtrate is assembled in a storage tank. To purify and concentrate the enzyme solution, two crossflow filters are employed: a microfilter and an ultrafilter. The microfilter acts as polishing filter discharging residual cells as well as cell debris which are not extracted by the rotary vacuum filter. This is carried out by applying a filter membrane to which pores are smaller than the single cell, yet larger protein is determined by the microorganism. An effluent of the microfilter that holds the impurities maintained by the membrane is conducted for waste treatment. The permeate of the microfilter which includes the enzyme as well as unreacted solutes is pumped into the ultrafilter. Enzyme is then absorbed by the ultrafilter membrane and yet permits water and minute molecule impurities to pass over.

The first step of the primary recovery section is cell harvesting to reduce the volume of the broth and remove extracellular impurities; it is carried out by a membrane microfilter (MF-101). Here the broth is concentrated twofold. Since the protease is an intracellular product, the next step is cell disruption, performed in a high-pressure homogenizer (HG-101). After homogenization, a disk stack centrifuge (DS-101) is used to remove most of the cell debris particles, followed by a dead-end polishing filter (DE-101) that removes the remaining cell debris particles. Finally, the resulting protein solution is concentrated twofold by an ultrafilter (UF-101) as shown in Fig. 19.2. The mass balance for the simulation results was shown in Table 19.2. The batch production of the protease enzyme was found to be 119.352 kg.

19.2.4 Annual Operating Cost

Annual operating cost was estimated and balanced the sum of the consecutive aspects as described in the software: raw materials, facility dependent, labor dependent, laboratory /QC/QA (*QC*, quality control, *QA*, quality analysis), and wastewater treatment/disposal utilities and consumables.

Eight distinct raw materials are considered in the process for this chapter and economic estimation purpose. Two distinct utilities were estimated in the production method such as electricity and heat transfer agents consisting of steam, cooling water and chilled water.

The total labor cost (TLC) is estimated as the sum of the labor demand per type (LDT) multiplied with the labor rate per type (LRT). So in this chapter, total annual labor cost is $5774,00. Additionally, laboratory/QC/QA describes the costs of off-line analysis along with quality control cost. Further, the waste developed in entire process is split in solid and liquid waste materials. The default estimation of the waste material is derived as $149,000.

Once the mentioned cost items are calculated, the total operation costs were obtained as detailed in Table 19.3. Hence, the annual operating cost was found to be $3,415,593,000. The cost breakdown of the annual operating cost for the process scheme was shown in Fig. 19.3.

19.3 Process Scheduling

The formulation of final product firmly depends on the specified enzyme properties and its considered application. The batch time can be defined as the time passed from the beginning of a given batch, that is, the preparation of the initial seed fermenter to the ending of that certain batch, that is, the formation of the final product. The total batch time in this process exceeds ten days. Despite that, the definite process of cycle time is 24 h. Since the separate procedure in this process is far briefer rather than the complete batch time, multiple equipment things are utilized in numerous steps of the process. The introduction of staggered tools decreases the productive cycle—time of

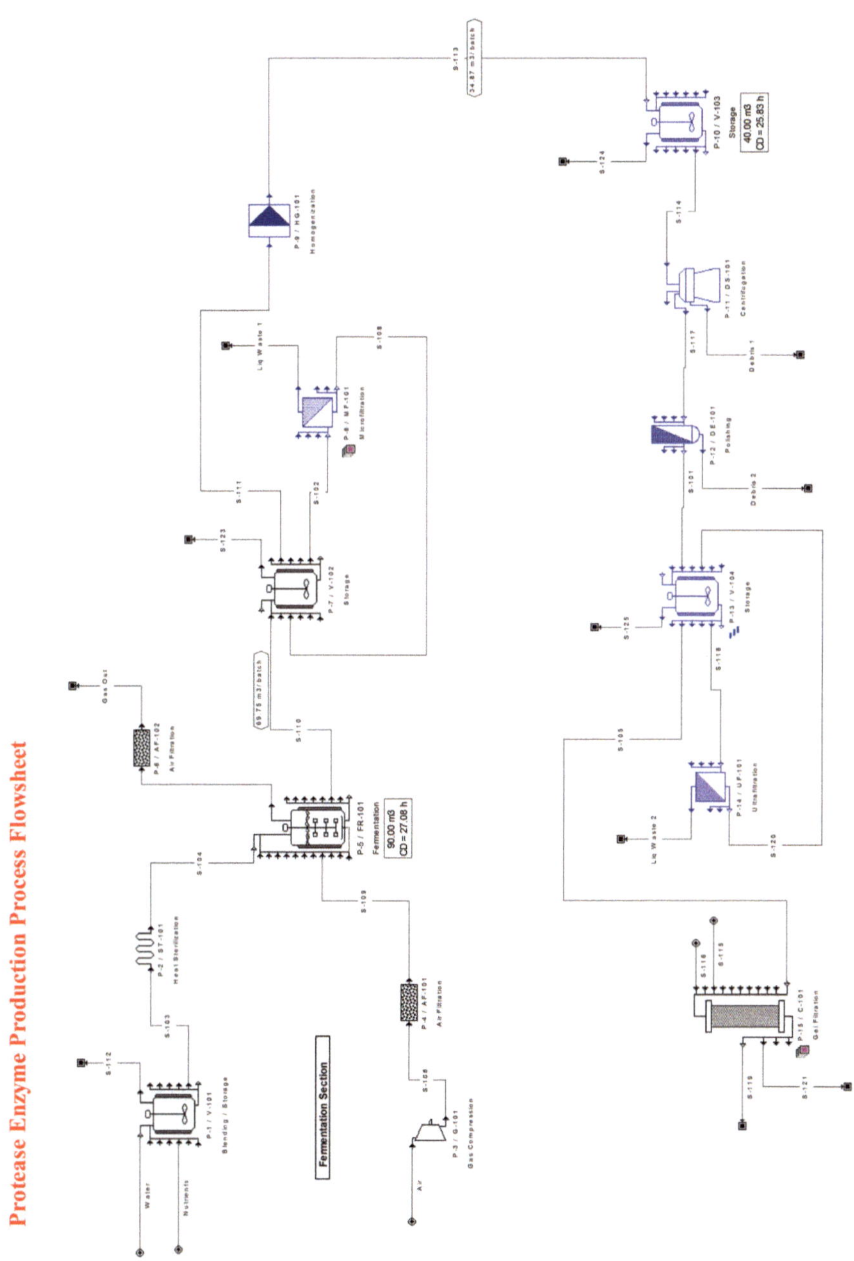

Fig. 19.2 Alkaline protease enzyme production process flow sheet

Table 19.2 Overall component balance (kg/batch)

Component	Initial	Input	Output	Final	In-out
Biomass	0.000	0.000	12.759	0.000	−12.759
Carbon dioxide	0.000	0.000	4238.806	2.010	−4240.816
Debris	0.000	0.000	761.870	0.000	−761.870
Glucose	0.000	6143.000	122.930	0.000	6020.070
Nitrogen	298.519	36,989.523	37,176.948	237.183	−126.090
Nucleic acids	0.000	0.000	483.538	0.000	−483.538
Oxygen	90.624	11,229.302	8337.219	70.606	2912.102
Phosphoric acid	0.000	422.915	422.915	0.000	0.000
Process water	0.000	73,398.119	76,540.398	0.000	−3142.279
Protease	0.000	0.000	119.352	0.000	−119.352
Proteins	0.000	0.000	1088.282	0.000	−1088.282
Salts	0.000	1026.000	229.768	0.000	796.232
Sodium chloride	0.000	2246.941	2246.941	0.000	0.000
Sodium hydroxide	0.000	282.107	282.107	0.000	0.000
Water	0.000	14,111.112	14,111.112	0.000	0.000
WFI (water for injection)	0.000	792,235.481	792,235.481	0.000	0.000
Total	389.143	938,084.500	938,410.427	309.799	246.583
				Overall error	0.026%

Table 19.3 Annual operating cost-process summary

Cost item	$	%
Raw materials	22,067,000	0.65
Labor-dependent	5774,000	0.17
Facility-dependent	504,342,000	14.77
Laboratory/QC/QA	1,504,000	0.04
Consumables	2,881,339,000	84.36
Waste treatment/disposal	149,000	0.00
Utilities	417,000	0.01
Transportation	0	0.00
Miscellaneous	0	0.00
Advertising/selling	0	0.00
Running royalties	0	0.00
Failed product disposal	0	0.00
Total	3,415,593,000	100.00

the entire process—and expand the production of the plant. Since biochemical processes, the production fermentation stage is generally the strategy with the prolonged duration. Figure 19.4 shows a section of the Gantt chart representing the scheduling of working in the Media Prep segment. In addition, the brown bar at the upmost shows the duration of the whole recipe, although the dark blue along with cyan bars exhibit the duration of method as well as working, respectively. This chart facilitates user to show the implementation of a batch process thoroughly and also promote editing of batch method.

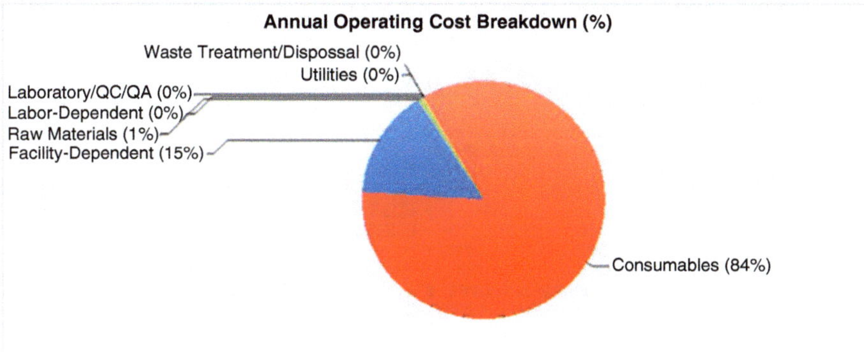

Fig. 19.3 Cost breakdown for the annual operating cost

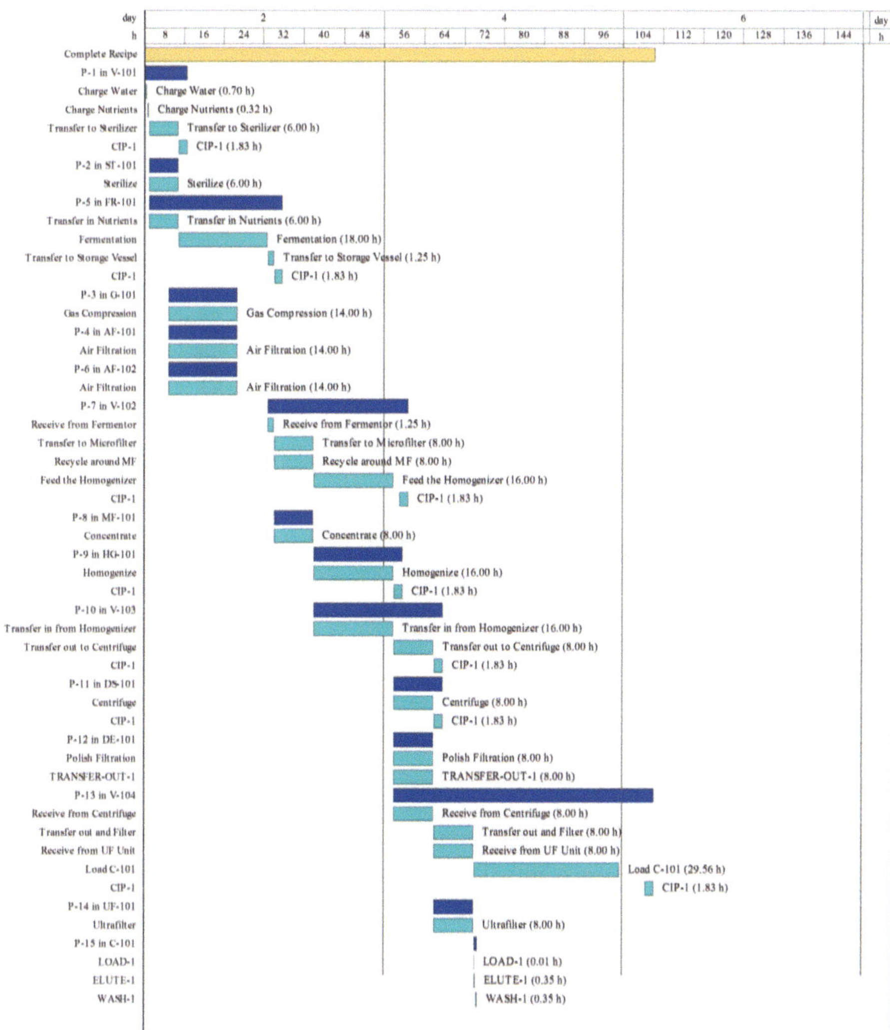

Fig. 19.4 Section of the operation Gantt chart

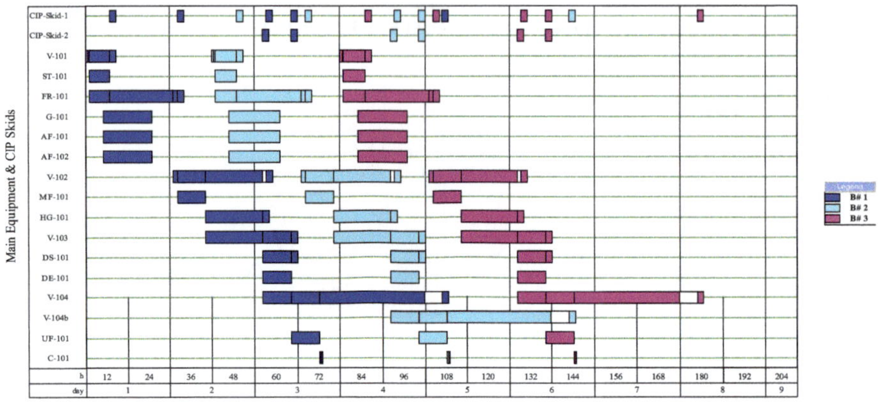

Fig. 19.5 Section of the equipment occupancy chart

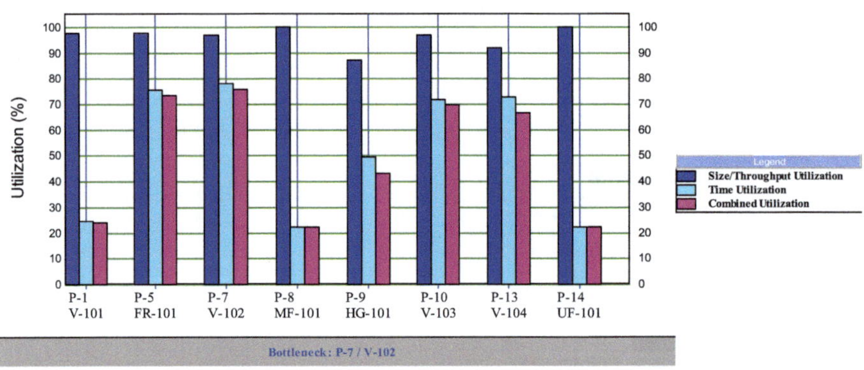

Fig. 19.6 Utilization of the equipment for the production process

The equipment occupancy chart was shown in Fig. 19.5 for the three batches of the production process. It will helps to identify the occupancy of the equipment for the batch to decide whether the equipment is occupied for the batch for the particular time. The (%) utilization of the equipment and the time of utilization are also shown in Fig. 19.6.

19.4 Conclusion

The present study focused on the production of alkaline protease enzyme with simulations using the software SuperPro Designer. Altogether, this work provides a comprehensive analysis of the factors affecting the cost associated with the industrial-scale production of alkaline protease enzyme. The simulation of whole process which includes downstream segregation, balancing of mass and energy estimation, and economic assessment whole can be executed in SuperPro Designer along with integral techno-economic criterion that was evaluated.

References

Aguilar JGDS, de Castro RJ, Sato HH (2019) Alkaline protease production by Bacillus licheniformis LBA 46 in a bench reactor: effect of temperature and agitation. Braz J Chem Eng 36:615–625

Arunachalam C, Saritha K (2009) Protease enzyme: an eco-friendly alternative for leather industry. Indian J Sci Technol 2(12):29–32

Ayantunji YJ, Omole RK, Olojo FO, Awojobi KO (2020) Optimization of alkaline protease production in submerged fermentation using Bacillus cereus isolated from an abattoir wastewater in Ile-Ife, Nigeria. J Adv Biol Biotechnol 23(3):1–15

Baweja M, Tiwari R, Singh PK, Nain L, Shukla P (2016) An alkaline protease from Bacillus pumilus MP 27: functional analysis of its binding model toward its applications as detergent additive. Front Microbiol 7:1195

Bezawada J, Yan S, John RP, Tyagi RD, Surampalli RY (2011) Recovery of Bacillus licheniformis alkaline protease from supernatant of fermented wastewater sludge using ultrafiltration and its characterization. Biotechnol Res Int 2011:238549

Bhunia B, Basak B, Dey A (2012) A review on production of serine alkaline protease by Bacillus spp. J Biochem Technol 3(4):448–457

Boominadhan U, Rajakumar R, Sivakumaar PKV, Joe MM (2009) Optimization of protease enzyme production using Bacillus sp. isolated from different wastes. Bot Res Int 2(2):83–87

Choudhary V (2013) Recovery of silver from used X-ray films by Aspergillus versicolor protease. J Acad Ind Res 2:39–41

da Gama Ferreira R, Azzoni AR, Freitas S (2018) Techno-economic analysis of the industrial production of a low-cost enzyme using E. coli: the case of recombinant β-glucosidase. Biotechnol Biofuels 11(1):1–13

Demain AL, Adrio JL (2008) Contributions of microorganisms to industrial biology. Mol Biotechnol 38(1):41–55

Gemechu G, Masi C, Tafesse M, Kebede G (2020) A review on the bacterial alkaline proteases. J Xidian Univ 14(11):632–634

Gómez JA, Sánchez ÓJ, Correa LF (2020) Techno-economic and environmental evaluation of cheesemaking waste valorization through process simulation using superpro designer. Waste Biomass Valorization 11(11):6025–6045

Hou E, Xia T, Zhang Z, Mao X (2017) Purification and characterization of an alkaline protease from Micrococcus sp. isolated from the South China Sea. J Ocean Univ China 16(2):319–325

Jayakumar D, Sachith SK, Nathan VK, Rishad KSM (2021) Statistical optimization of thermostable alkaline protease from Bacillus cereus KM 05 using response surface methodology. Biotechnol Lett 43(10):2053–2065

Joo HS, Kumar CG, Park GC, Paik SR, Chang CS (2003) Oxidant and SDS-stable alkaline protease from Bacillus clausii I-52: production and some properties. J Appl Microbiol 95(2):267–272

Kantor M, Wajda K, Lannoo B, Casier K, Verbrugge S, Pickavet M et al (2010) General framework for techno-economic analysis of next generation access networks. In: 2010 12th international conference on transparent optical networks. IEEE, pp 1–4

Kembhavi AA, Kulkarni A, Pant A (1993) Salt-tolerant and thermostable alkaline protease from Bacillus subtilis NCIM no. 64. Appl Biochem Biotechnol 38(1):83–92

Kirk O, Borchert TV, Fuglsang CC (2002) Industrial enzyme applications. Curr Opin Biotechnol 13(4):345–351

Kumar CG (2002) Purification and characterization of a thermostable alkaline protease from alkalophilic Bacillus pumilus. Lett Appl Microbiol 34(1):13–17

Kumari PS, Reshma R (2021) Effect of alkaline protease produced from fish waste as substrate by bacillus clausii on destaining of blood stained fabric. J Trop Life Sci 11(1)

Macrelli S, Mogensen J, Zacchi G (2012) Techno-economic evaluation of 2nd generation bioethanol production from sugar cane bagasse and leaves integrated with the sugar-based ethanol process. Biotechnol Biofuels 5(1):1–18

Marathe SK, Vashistht MA, Prashanth A, Parveen N, Chakraborty S, Nair SS (2018) Isolation, partial purification, biochemical characterization and detergent compatibility of alkaline protease produced by Bacillus subtilis, Alcaligenes faecalis and Pseudomonas aeruginosa obtained from sea water samples. J Genet Eng Biotechnol 16(1):39–46

Naveed M, Nadeem F, Mehmood T, Bilal M, Anwar Z, Amjad F (2021) Protease—a versatile and ecofriendly biocatalyst with multi-industrial applications: an updated review. Catal Lett 151(2): 307–323

Niño L, Gelves G (2021) Industrial-scale bioprocess simulation of polyphenol production using superpro designer. J Eng Sci Technol 16(3 (2021)):2100–2113

Qureshi AS, Simair AA, Ali CH, Khushk I, Khokhar JA, Ahmad A et al (2018) Production, purification and partial characterization of organo-solvent tolerant protease from newly isolated Bacillus sp. BBXS-2 Ferment Technol 7(151):2

Ramadhani, N. H., Darmawan, M. A., Harahap, A. F. P., Ramadhan, M. Y. A., Silalahi, U. C., & Gozan, M. (2021). Simulation of Illipe butter purification originated from West Kalimantan by SuperPro designer. In Journal of Physics: Conference Series (1726, 1, 012013). IOP Publishing

Rao K, Narasu ML (2007) Alkaline protease from Bacillus firmus 7728. Afr J Biotechnol 6(21): 2493–2496

Rao MB, Tanksale AM, Ghatge MS, Deshpande VV (1998) Molecular and biotechnological aspects of microbial proteases. Microbiol Mol Biol Rev 62(3):597–635

Reshma R (2021) Effect of alkaline protease produced from fish waste as substrate by Bacillus clausii on destaining of blood stained fabric. J Trop Life Sci 11(1)

Salihi A, Asoodeh A, Aliabadian M (2017) Production and biochemical characterization of an alkaline protease from Aspergillus oryzae CH93. Int J Biol Macromol 94:827–835

Samarntarn W, Cheevadhanarak S, Tanticharoen M (1999) Production of alkaline protease by a genetically engineered Aspergillus oryzae U1521. J Gen Appl Microbiol 45(3):99–103

Sharma KM, Kumar R, Panwar S, Kumar A (2017) Microbial alkaline proteases: optimization of production parameters and their properties. J Genet Eng Biotechnol 15(1):115–126

Sharma M, Gat Y, Arya S, Kumar V, Panghal A, Kumar A (2019) A review on microbial alkaline protease: an essential tool for various industrial approaches. Ind Biotechnol 15(2):69–78

Shine K, Kanimozhi K, Panneerselvam A, Muthukumar C, Thajuddin N (2016) Production and optimization of alkaline protease by Bacillus cereus RS3 isolated from desert soil. Int J Adv Res Biol Sci 3(7):193–202

Singh D (2021) Protease enzyme: a safe and effective industrial and commercial alternative. Int J Adv Multidiscip Topic 2(7):302–306

Sun C, Theodoropoulos C, Scrutton NS (2020) Techno-economic assessment of microbial limonene production. Bioresour Technol 300:122666

Tao L, Markham JN, Haq Z, Biddy MJ (2017) Techno-economic analysis for upgrading the biomass-derived ethanol-to-jet blendstocks. Green Chem 19(4):1082–1101

Varia, A. D., Shukla, V. Y., & Tipre, D. R. (2019). Alkaline protease-a versatile enzyme

Verma J, Modi DR, Sharma R, Saxena S (2011a) Vital role of alkaline protease in bio-industries: a. Plant Archives 11(2):1083–1092

Verma A, Pal HS, Singh R, Agarwal S (2011b) Potential of alkaline protease isolated from Thermoactinomyces sp. RM4 as an alternative to conventional chemicals in leather industry dehairing process. Int J Agric Environ Biotechnol 4(2):173–178

Vijayaraghavan P, Jebamalar TRJ, Vincent SGP (2012) Biosynthesis optimization and purification of a solvent stable alkaline serine protease from Halobacterium sp. Ann Microbiol 62(1): 403–410

Wu W, Maravelias CT (2018) Synthesis and techno-economic assessment of microbial-based processes for terpenes production. Biotechnol Biofuels 11(1):1–14

Younis MA, Hezayen FF, Nour-Eldein MA, Shabeb MS (2010) Optimization of cultivation medium and growth conditions for Bacillus subtilis KO strain isolated from sugar cane molasses. Am Eurasian J Agric Environ Sci 7(1):31–37

Chapter 20
Ethical Issues of Microbial Products for Industrialization

Idris Adewale Ahmed

Abstract The applications and roles of microbial biotechnology in industrialization are enormous with engineered microbial systems among the emerging technologies being predicted to revolutionize various industries such as cosmetics, environmental bioremediation, food and agriculture, health care, paper, pharmaceutical, nutraceuticals, and textile, to mention but a few. Nevertheless, the assessment of long-term safety, efficacy, and rationale behind pursuing microbial products for industrialization in the presence of other alternatives are essential especially in post-COVID-19 pandemic. There are other salient ethical, environmental, economic, social, scientific, and religious concerns not only to the scientists and regulators but also to the users. Such issues include, but are not limited to, intellectual property rights, accidental release and environmental implication, biosafety and bioterrorism concern, economic concern, public trust (societal concern), and consequences of exposure to pathogens by humans and animals both in the developing and developed nations. This work is aimed at acknowledging the role of ethics and proffers some suggestions to ensure the sustainable use and development of microbial products for industrialization. In addition, a framework for microbial products' effectiveness has been suggested which comprises five distinct phases, namely, access, exploration, knowledge acquisition, adoption, and innovation and transformation. Cooperation, equity, and intersubjectivity are the fundamental tripods and relational dimensions for effective communication of ethical issues. The proper regulatory mechanisms, the formation of a professional and ethical society, and the development of relevant checks and balances are essential to circumventing all malicious attempts to misuse these technologies. Regulatory bodies, however, must also consider a wide range of other perspectives such as the economic impact, scientific progress, and risk and moral reasoning for the industrialization and monetization of microbial products.

I. A. Ahmed (✉)
Department of Biotechnology, Faculty of Applied Science, Lincoln University College, Petaling Jaya, Malaysia
e-mail: idrisahmed@lincoln.edu.my

A. Sarkar, I. A. Ahmed (eds.), *Microbial products for future industrialization*, Interdisciplinary Biotechnological Advances,
https://doi.org/10.1007/978-981-99-1737-2_20

Keywords Ethical and legal issues · Ethics communication · Intellectual property rights · Microbial biotechnology · Regulatory mechanisms

20.1 Background

Natural products, with their chemically and structurally diverse molecular nature as well as remarkable therapeutic properties, are mostly secondary metabolites from a myriad of sources such as plants, animals, microorganisms, and marine organisms. The earliest application of natural products to enhance human health dates back to Mesopotamia's period from 2900 to 2600 BCE. However, the shift from plants to microorganisms as a natural product's source became imminent in 1928 with the discovery of penicillin from *Penicillium notatum* by Alexander Fleming. Several compounds derived from microorganisms have, then, been utilized in various fields including agriculture, food industry, medicine, and other scientific endeavors. Other antibiotics such as streptomycin, chloramphenicol, chlortetracycline, cephalosporin C, erythromycin, and vancomycin were also discovered from *Streptomyces griseus*, *Streptomyces venezuelae*, *Streptomyces aureofaciens*, *Cephalosporium acremonium*, *Saccharopolyspora erythraea*, and *Amycolatopsis orientalis*, respectively (Pham et al. 2019). Microorganisms from different agro-industrial residues have been also used for the production of various categories of enzymes for various environmental and industrial applications (Lekshmi et al. 2021).

Despite the widespread use of myriad of industrial microbes in biotechnology, *Escherichia coli* and *Saccharomyces cerevisiae*, in particular, have been well studied by the various research communities such as microbial model organisms with model organism databases and extensive functional annotation (Dikicioglu et al. 2014). Recently, *Bacillus subtilis* and *Pichia pastoris* have also become promising heterologous expression systems for both prokaryotic and eukaryotic candidate genes (Kumar et al. 2018). Microorganisms are very useful for the production of oxygen, decomposition of organic material, provision of nutrients to plants, and maintenance of human health, to mention but a few. However, they can also be pathogenic causing a worldwide socio-economical loss due to various diseases in plants and animals (Das et al. 2022).

There is no doubt that the application of biotechnology in different human endeavors offers a wide-ranging diversity of scientific methods and solutions to both productivity and sustainability (Lokko et al. 2018). The advent of advanced biotechnologies involving gene manipulation such as gene editing, synthetic biology, and transgene in microbial research and development has led to the domestication of several wild microbial spp., thus culminating in increasing concerns about the associated risks regarding the safety of the environment, food, and the management of genetically modified agricultural organisms (He et al. 2020).

Genetic engineering, in contrast to conventional breeding, allows the transfer of genes across taxonomic boundaries as well as between closely related species and completely different species (Kumar et al. 2018). The domestication process is an artificial selection approach and breeding of wild but industrially important microbial spp. to obtain several cultivated variants that can thrive in man-made niches to meet both industrial and human requirements. Microbial diversity has, thus, been somewhat shaped by the emergence of highly specific and novel man-made environments that allow microbes to gain the increasing capacity to cope with a host of industry-specific stress factors by efficiently consuming particular nutrients for the production of desirable compounds. Though domestication may result in phenotypically and genetically distinct strains, it often comes at the cost of a reduction in fitness in the microbial original and natural environments as different lineages of the same spp. are mostly adapted to the highly diverse artificial niches (Steensels et al. 2019). The safety regulations and monitoring of industrial microbial products should not be only be provided by the government, but such regulations should evolve continuously and remain more science-based with greater flexibility and transparency (Neill 2008). In the light of this, this paper is aimed at critically discussing the impact of microbes in society, the triumphing microbial industry, the framework for microbial products effectiveness, ethical issues, and communication.

20.2 Impact of Microbes in the Society

Microorganisms have been explored from antiquity for the transformation of raw foods through fermentation as well as various upstream and downstream processes (Kumar et al. 2018). It is also used in recent times in the pharmacological industry for the synthesis of several secondary metabolites such as anticancer molecules, enzymes, antibiotic drugs, growth promoters, hormones, immuno-modulating inhibitors of specific enzymatic activities, and primary metabolites such as CO_2 or ethanol. Other uses of bacteria include pollutant biodegradation, land bioremediation, and wastewater treatments (La Maestra et al. 2021).

Undoubtedly, microbes offer certain beneficial effects to the hosts such as the acquisition of beneficial microbes, resilience to community perturbations, maintenance of homeostasis in the digestive tract, energy provision, and regulation of the immune system. Nevertheless, horizontal microbial transmission particularly between the abiotic and biotic environments provides a crucial route and a risk for pathogen exposure (Kuthyar et al. 2019). The use of microorganisms and renewable feedstock in the biological approach is also used for several industrial chemicals (Table 20.1). For instance, microbial fermentation is being used for the production of succinic acid used in several industries such as agriculture, food, and pharmaceuticals. It is a critical element used in chemical, food, and pharma industries. It is also a precursor of several industrial bio-based chemicals such as 1,4-butanediol, detergents, flavors, fungicides, herbicides, perfumes, plasticizers, polyester polyols, polybutylene succinate (biodegradable plastic), and surfactants (Thakur et al. 2022). The microbial production of lactic acid, depending on the enzyme lactate dehydrogenase used, produces only pure L-lactic acid or D-lactic acid. Lactic acid is extensively used as an essential raw material for different commercial value-added

Table 20.1 Common commercial products being produced from microbes

Products	Microbe sources	References
Succinic acid	*Recombinant Escherichia coli, Actinobacillus succinogenes, Mannheimia succiniciproducens, Basfia fragilis, Basfia succiniciproducens, Corynebacterium* spp., *Aspergillus* spp., *Penicillium simplicissimum*, and *Fusarium* spp.	Ahn et al. (2020) and Thakur et al. (2022)
Lactic acid	Lactic acid bacteria and *Corynebacterium* spp.	Camesasca et al. (2021), Sudhakar and Dharani (2021), and Wuyts et al. (2020)
Citric acid	*Aspergillus niger* or *A. wentii*	Żywicka et al. (2020)
Glutamic acid	*Bacillus* spp.	Azarhava et al. (2020)
Gluconic acid	*Aspergillus niger*	Fernandes et al. (2021)
Cheeses, yogurts, and fermented vegetables	Lactic acid bacteria	Wuyts et al. (2020)
Amylases	Thermophilic *Bacillus* spp.	Ullah et al. (2021)
Cellulases	*Bacillus pacificus* and *Pseudomonas mucidolens*	Krishnaswamy et al. (2022)
Pectinase and protease	*Aspergillus niger* and *A. aureus*	da Câmara Rocha et al. (2020)
Proteases (thermolysin, sub-tilisin, aqualysin)	*Thermus aquaticus* and *Bacillus* spp.	Caballero et al. (2020)
Glucose isomerase	*Streptomyces rubiginosus, E. coli BL21, B. subtilis, Thermobifida fusca, Streptomyces lividans RSU26, Caldicoprobacter algeriensis, Caldicellulosiruptor bescii, Saccharomyces cerevisiae*, and *Streptomyces murinus*	Jin et al. (2021)
Beta-galactosidase	*Aspergillus* spp. and *Kluyveromyces* spp.	Oliveira et al. (2011)
Cobalamin (cyanocobalamin)	*Propionibacterium peterssoni* and *Propionibacterium pituitosum*	Bodur et al. (2021)
Vinegar	Acetic acid bacteria: *Acetobacter* spp. and *Gluconacetobacter* spp.	Ai et al. (2019) and Zhang et al. (2020)
Monosodium glutamate	*Corynebacterium glutamicum*	Nakayama et al. (2018)
Ethanol (nonbeverage)	*Saccharomyces cerevisiae*	Walker and Stewart (2016)
Alcohols	*Clostridium acetobutylicum*	Dai et al. (2016) and Sreekumar et al. (2015)
Acetone	*Clostridium acetobutylicum*	Sreekumar et al. (2015)
2-Ketogluconic acid	*Pseudomonas* spp.	Sun et al. (2020)
Dextran	*Leuconostoc mesenteroides*	Esmaeilnejad-Moghadam et al. (2019)
Sorbose	*Gluconobacter suboxydans*	Macauley-Patrick et al. (2005)

chemical compounds in the industry. It is also used in the pharmaceutical and textile industries. It is equally used in the food industry as an acidifier, flavoring, and preservative (Camesasca et al. 2021).

On the other hand, the uncontrolled and overuse of antibacterial drugs against bacteria has also led to the emergence and development of various resistance strategies by the microbes as a means of survival. Antibacterial resistance is reported to cause over 700,000 annual deaths and is estimated to rise to 20 million and cost over $2.9 trillion by 2050, according to the World Health Organization (Uddin et al. 2021). Furthermore, while microbial-based disinfectants or cleaning products might offer solutions or attractive alternatives to specific problems, the potential risks of being opportunistic pathogens to the environment and human health after exposure are a cause of concern and thus require specific regulatory systems to monitor their use, production, and sale through adequate risk assessment (La Maestra et al. 2021). Risk assessment is strictly a scientific process that requires risk management and communication. Though risk assessments could be influenced by ethical, economic, scientific, social, and political information, a holistic risk assessment strategy should, however, include at least the identification of the hazard or risk as well as its characterization and estimation (Myhr and Dalmo 2005).

20.3 Triumphing Microbial Industry

Globally, there are increasing concerns among economists, industrialists, and scientists on the depleting level of clean water and limited energy resources. This has culminated in a paradigm shift from conventional to renewables. One of such strategic scientific approaches is the evolution and emergence of microbial fuel cell technology to deal with the depleting level of clean water and limited energy resources (Mehrotra et al. 2021).

Microbial fermentation with the use of diverse renewable feedstock bio-resources is, thus, gaining more traction due to the concerns about the depletion of biomass assets, environment, eco-friendly, and renewable biomass-based chemical extraction. It also reduces the ecological consequences from improper disposal and, thus, repurposes them into more valuable products. Bio-resources are not only biodegradable and useful for various biomedical applications but also biocompatible and non-toxic (Thakur et al. 2022).

The application of whole-genome shuffling or DNA shuffling and genome recombination has also been greatly explored to successfully improve the titers of different varieties of microorganisms. For instance, two rounds of genome shuffling have been used to generate two strains of *Streptomyces fradiae* to produce up to a ninefold increase in the production of antibacterial tylosin as compared to the initial strain (Pham et al. 2019).

There are also different microbe varieties with an overabundance of abilities to thrive and grow in different environments and produce a variety of enzymes (Shukla et al. 2021). Microbial pigments, in addition to algae, plants, fungi, and yeast, are

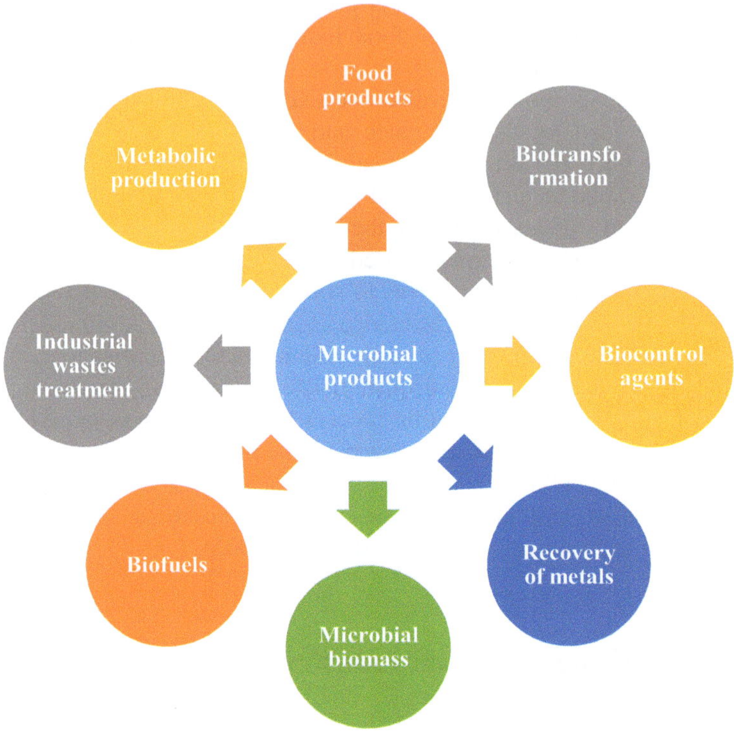

Fig. 20.1 Categories of microbial product

one of the key source materials used in the synthesis of pigment-mediated nanoparticles for prospective varieties of applications such as antibacterial agents, biodegradation, biofuels, biosensors, cancer treatment, drug delivery, and wastewater treatment (Venil et al. 2021). Various microbial cells have also been employed in the bioconversion of sugars for the production of industrial enzymes (such as lipase and protease) and value-added bioproducts such as bioethanol, butyric acid, docosahexaenoic acid, fructooligosaccharides, and pullulan, all of which have vital applications in bioenergy, cosmetics, food, and pharmaceutical industries (Zhang et al. 2021). Microbes are also very useful in other big industries such as textile (Samuchiwal et al. 2021), wastewater treatment (Ismail and Habeeb 2017; Mandeep et al. 2020), and enzyme technology (Handa et al. 2020), to mention but a few. In general, about eight broad categories of microbial products have been identified (Fig. 20.1), namely, biotransformation, fermentation of food products, metabolic production, production of biofuels, production of microbial biomass for food and feed, production of biocontrol agents, recovery of metals, and treatment of organic and industrial wastes (Abdel-Aziz et al. 2017). Common industrial applications of microbes are, however, briefly discussed as follows.

20.3.1 Bioenergy Industry

Microbial fuel cells (MFC) are an efficient technology for electricity generation from the biodegradation of several pharmaceutical drugs and other pollutants in a wide range of wastewater. The technology also produces other valuable chemicals like hydrogen and methane making it more sustainable and environmentally friendly than most other conventional technologies employed for the treatment of wastewater. MFC produces bioenergy using active microorganism biocatalysts in an anaerobic environment to oxidize the organic substrates or carbon sources (pollutants) for the generation of protons and electrons (Aguilera Flores et al. 2021; Ismail and Habeeb 2017). The electrons are usually passed through the external circuit, while the protons, through the proton exchange membrane, are carried to the cathode chamber. The reaction of the electrons and protons in the cathode chamber also results in a parallel reduction of oxygen to water. Microalgae are increasingly being used as a biocathode while other natural sources such as bamboo charcoal, corn straw, coconut shell, coffee waste, and onion peels are transformed into biochar to build anode electrodes in MFC (Aguilera Flores et al. 2021). Certain microbial phyla such as Bacteroidetes, Firmicutes, and Proteobacteria have been used in osmotic membrane bioreactors for the production of hydrogen and methane in simultaneous wastewater treatment and bioenergy production (Hosseinzadeh et al. 2021).

20.3.2 Microbes in Cosmetics

The unique properties of bacterial cellulose (such as its biocompatibility, chemical controllability, high purity, high water uptake, mechanical strength, morphology, and nontoxicity) make it a promising alternative material for different cosmetic and biomedical applications. It is synthesized in different forms such as aerogel, fiber, film, hydrogel, membrane, orderlies, random, and tube for various applications (Mbituyimana et al. 2021). Several metabolic engineering strategies have also been combined to transform microbes into cell factories for de novo biosynthesis of high-value natural products using cheap recyclable starting materials like waste cooking oils, thus offering a more economical and sustainable approach than traditional production methods. For instance, geographical and seasonal variability mostly affects the production of monoterpenoids that exhibit a broad variety of biological activities and are primarily obtained through the extraction of plant biomass. The alternative chemical or synthetic workflows are not only energy-intensive but also generate a huge lot of organic waste (Zhu et al. 2021). Microbes have also been reportedly engineered into the production host for the biosynthetic pathways of mushroom-originated natural compounds (Yang et al. 2021).

20.3.3 Food Industry

Human beings have explored the fermentation process for the production of various fermentations all around the world. Plant-based fermentations, especially their microbes, are rich sources of novel biotechnological applications for the production of various novel prebiotic, probiotic, synbiotic and other industrial applications. The lactic acid bacteria from these plant-based fermentations has also been explored for the isolation of a cellulase enzyme in specific *Lactobacillus plantarum* group members (Wuyts et al. 2020). Biosensors with nano-additives are commonly applied in nano-delivery systems for food applications. Microbes are also widely used in the applications of nanotechnology in agriculture through the design of nano-pesticides and the development of nano-fertilizers to enhance crop production with no harm to the soil (Badawy et al. 2022). Microbes are also beneficial in the area of food waste management. Rumen contents with complex microbial interactions of about 19 existing phyla and 180 genera are used in the degradation of particulate matter such as celluloses (Yan et al. 2014). Microbial transglutaminase is equally gaining acceptance and being involved in protein cross-linking and protein modification technologies which have proven to help in diversifying consumers' food needs. The common advantages of enzymatic techniques for protein modification include high reaction specificity; lack of a need for chemical solvents, high-temperature, and high-pressure conditions; and low frequency of side reactions (Miwa 2020). On the other hand, fungi and certain bacteria produce spores, which are dormant, tough, and nonreproductive structures responsible for most of the common hazards in foodstuffs such as food-borne illnesses or food spoilage. Bacterial spores are found in different types of food such as dairy products, grains, meat, vegetables, etc. while fungal spores mostly produce aflatoxins (mycotoxins) (Farag et al. 2021).

20.3.4 Pharmaceutical Industry

Microbial cell factories have been reportedly explored for the production and supply of polyunsaturated fatty acids (PUFAs) especially docosahexaenoic acid and eicosapentaenoic acid to meet the demand for PUFA-rich aquafeed, medical formulations, and superfoods (Jovanovic et al. 2021). For instance, L-malate, a C4 dicarboxylic acid and a crucial intermediate of the tricarboxylic acid cycle naturally occurring in various microorganisms, is widely used in agricultural, beverage and food, textile, polymer, and pharmaceutical industries (Jiang et al. 2021). On the other hand, there are increasing concerns about emerging pollutants from an unregulated discharge of pharmaceuticals into the water streams. Pharmaceutical industry wastewater is considered as one the most challenging industrial and recalcitrant wastewater types with complex composition, high chemical oxygen demand (COD), and high toxicity, containing a variety of organic and inorganic constituents (Ismail and Habeeb 2017). Pharmaceutical fragments and their transformed metabolites may be potentially toxic on nontarget organisms even at very low concentrations due to various biological, chemical, and physical processes in treatment facilities. Adequate biodegradation of pharmaceutical industry wastewater, however, has been reported for some microbes such as *Micrococcus yunnanensis* (Sharma et al. 2020).

20.3.5 Textile Industry

The textile industry deals with multiple processes, thus generating significant pollution from the complex chemical processes of textile pretreatment, coloring, and decolorization. With about 20–50% of the dyes staying in the water phase (Yurtsever et al. 2017), the coloring process uses a wide range of synthetic dyes such as direct dyes, reactive dyes, and vat dyes, which in addition to other unit processes (such as bleaching, de-sizing, and scouring) generate a large amount of discharged water containing auxiliaries, chemicals, and synthetic dyes. These inadvertently result in effluents with alkaline pH, high temperature, high COD (chemical oxygen demand), biochemical oxygen demand (BOD), total suspended solids (TSS), total dissolved solids (TDS), total nitrogen (TN), and relatively high salinity (Samuchiwal et al. 2021).

Microorganisms, however, are being used at several stages in the textile industry as complimentary green chemistry to curtail contaminant production. Different varieties of enzymes such as amylase, cellulose, ligninase, lipase, protease, pectinase, and xylanase are produced by several varieties of microbes with unique abilities to thrive well and grow in different environments. These enzymes are exploited as alternative and sustainable substitutes for the conventional hazardous chemicals for the pretreatment of textile in the textile industry. Microbes also produce a diversity of pigments that could be used as alternatives to persistent azo dyes. Microbes are also very useful in the treatment process of the effluent, remediation of heavy metals, and degradation of dyes to bring down their COD and BOD levels (Shukla et al. 2021).

Complete dye color removal has been reported using an anaerobic membrane bioreactor (AnMBR) while a slight color increase was detected in the anaerobic membrane bioreactor (AeMBR). The dominant sulfide-oxidizing and sulfate-reducing bacteria in both AeMBR and AnMBR were *Desulfuromonas thiophila* and *Thioalkalivibrio sulfidiphilus* (Yurtsever et al. 2017). Nazari and Jookar Kashi also reported 100% efficiency of immobilized silver nanoparticles in the decolorization of Disperse Blue 183 (Nazari and Jookar Kashi 2021).

20.3.6 Enzyme Technology

A plethora of microbes generates potent biocatalysts as enzymes, thus offering variations in their chemical properties, microbial sources, and mechanisms.

Microbial enzymes are usually involved in the catalysis of various oxidation, reduction, or hydrolysis reactions. Microbial enzymes are produced using solid-state fermentation and submerged fermentation. They target different substrates owing to differences in their active site motifs (Liu and Kokare 2017). The advancement in biotechnology is mostly responsible for the increasing commercial synthesis of various enzymes with wide applications in food preservation, food processing, and other commercial products such as detergents, leather, paper, pharmaceutical, and textile industries. Microbial phytase, for instance, is very useful in food and feed industries as well as for the bioremediation and removal of marine eutrophication (Handa et al. 2020). The metabolic regulatory processes and synthesis pathways of riboflavin in microorganisms have also been reported. Riboflavin and its derivatives are highly needed in the food, cosmetic, pharmaceutical, and nutraceutical industries (You et al. 2021). Bacteria, fungi, and yeast are the major potential producers of lipases, with microbial lipases gaining superior industrial attention due to their broad substrate specificity, selectivity, and stability (Bharathi and Rajalakshmi 2019).

20.4 Framework for Microbial Products' Effectiveness

A complex regulatory pathway involving several genes is usually involved in the synthesis of bioactive compounds in all living systems. Thus, microbial systems would inadvertently be subjected to some manipulations and optimizations of their metabolic networks and sometimes an addition of new genes or novel pathways to reprogram and maximize the production and isolation of relevant and industrially important compounds in more economically efficient ways (Meena and Mohanty 2020). The breakthrough in next-generation sequencing and genetic engineering technologies, in the last few decades, also enables the scientists with the required genome information and methodology for the engineering of both prokaryotic and eukaryotic cells (He et al. 2020; Lokko et al. 2018). There is, thus, an enormous potential in microbial engineering to offer a remarkable change in human lives and facilitate sustainable living through the reduction in health-care cost, the higher yield from various agricultural practices, the provision of cheaper and greener renewable fuel sources to meet the energy demand of the increasing human population, and the resource recovery from various waste materials (Belcher et al. 2020; Wang et al. 2021). The five stages (Fig. 20.2) that engineered microbes should pass through to guarantee their safety and facilitate regulatory decision-making: laboratory research, pilot testing, environmental release field testing, preproduction testing, and biosafety certificates' application (Kumar et al. 2018).

There are, however, certain serious ethical, legal, political, scientific, religious, and social issues regarding the application of genetic engineering principles to living systems either to reconstruct the existing microorganisms or to pursue a tailor-made or complete design of novel biological entities (Kumar et al. 2018). It is, therefore, necessary and as a matter of urgency to revisit the various aspects of ethical,

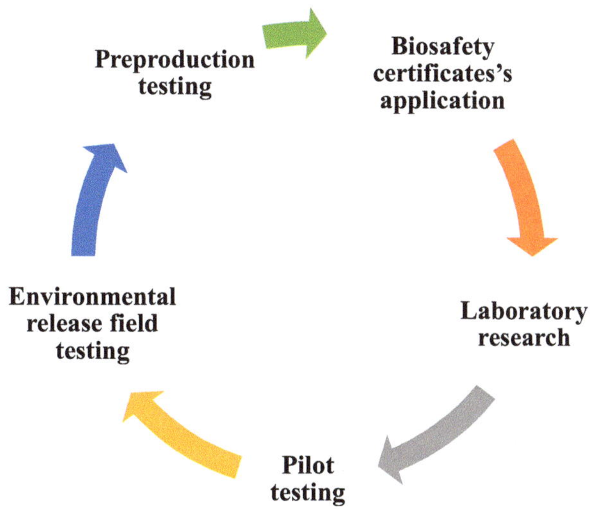

Fig. 20.2 Stages for safety and regulation of biosafety

biosafety, intellectual property rights, political, and regulatory issues concerning microbial engineering technology.

20.5 Ethical Concerns

Microbial cells are engineered and designed during microbial engineering using several manipulations to enhance the quality and yields of disease-fighting agents and other industrial chemicals such as antibiotics, alcohol, amino acids, biopolymers, enzymes, probiotics, vaccines, and vitamins (Meena and Mohanty 2020). Such engineered microbial systems are not only designed to solve specific problems and commercialized for revenue earning but also bring prosperity to the economy and impetus to R&D activities as several industries such as agroindustry, biomedical, cosmetics, food processing, energy sector, and pharmaceuticals industries are benefiting from the plethora of value-added products obtained from microbial systems. Microbial engineering, however, raises many issues of ethical, legal, political, and societal concerns owing to its ability to alter life forms or even create or develop novel life forms and turn them into profitable business ventures (Mehta and Gair 2001). Another highly contentious issue is "owning a life" which is also related to other questions concerning transfer, use, and dissemination (Meena and Mohanty 2020). In general, the ethical issues about the prospects and concerns of microbial engineering should be addressed in the light of three main categories of ethics vis-à-vis consequential ethics, virtue ethics, and deontological ethics (Vallero 2020). The preservation of intellectual property can also inadvertently breed a confidentiality culture within the scientific community, thus limiting data sharing, which, in turn, may lead to unnecessary duplication of genetically engineered lines or strains (Kumar et al. 2018).

20.5.1 Societal Concern and Public Trust

Owing to the potential of genetic engineering to create novel life forms, there are ultimately heightened ethical safety and security concerns as these powerful technologies might distort the line between what is natural and what is not otherwise (Meena and Mohanty 2020). Genetic manipulation has been either supported or opposed, due to mostly ethical and religious reasons. Thus, it is often difficult to balance public perception. Nevertheless, the concerted interests of academia, environments, governments, and industries have a great influence on public opinions and policymakers on biotechnology (Kumar et al. 2018).

Transparency greatly helps to ensure the integrity of the process involved such that a sound decision is made by the stakeholders after adequate data review. The nonavailability of assessment information and lack of transparency and openness in the approval process for some GM products, in part, due to the need to protect the proprietary interests of the party involved, however, may affect public confidence and credibility in future decisions (Myhr and Dalmo 2005).

20.5.2 Biosafety and Bioterrorism Concern

With the creation of cDNA of poliovirus and influenza virus inside the laboratories by scientists in the early twenty-first century, there are palpable concerns about the adoption of engineered pathogenic microorganisms for both legitimate and illegitimate (bioterrorism) purposes, if such technologies are made accessible to rogues and biohackers who could create more virulent strains of the pathogen and use them for nefarious purposes or to unleash terror attacks (Kambouris et al. 2020; Myhr and Dalmo 2005). There are also other health concerns about the use of the engineered microbial system in therapeutic clinical trials (Meena and Mohanty 2020).

20.5.3 Accidental Release and Environmental Implication

The effect of the release of engineered microbes into the environment lingers, unlike air or water pollution which dissipates over time. Most microbes multiply; thus, engineered microbes could pose a grave threat to naturally occurring microbes in different parts of the ecosystem. Severe health and environmental hazards as well as many ecological disasters may result from the potential release of these engineered microorganisms because of their emergent and unpredictable properties. It is,

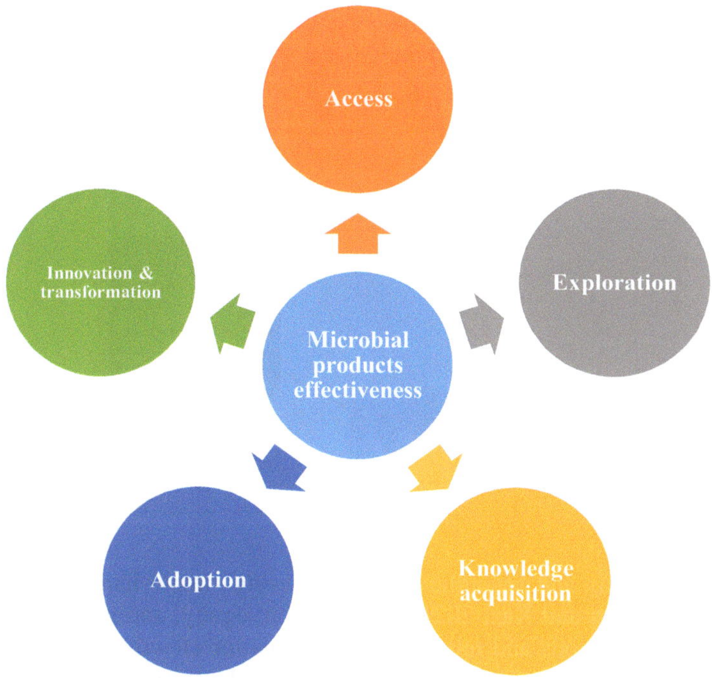

Fig. 20.3 Framework for microbial products' effectiveness

therefore, necessary to carry out a thorough assessment of every genetically modified microorganism and its effect on the environment before its release. Furthermore, adequate long-term trials must be conducted, and regulations must be put in place for a thorough evaluation of the biosafety trials to mitigate general public fear (Meena and Mohanty 2020). Genetic engineering may cause some unintended changes such as immunological activity, allergy, or pathological effects; thus, the adequate characterization of the engineered microbes and products should be properly investigated following ingestion, processing, and storage (Myhr and Dalmo 2005).

In addition, a framework for microbial products effectiveness has been suggested which comprises five phases, namely, access, exploration, knowledge acquisition, adoption, and innovation and transformation (Fig. 20.3).

20.6 Ethical Issues of Communication

Science communication requires better strategies and approaches especially during this age of increasing mistrust and misinformation. Building synergy between audience connection, content, and medium as well as a positive relationship between the general public and science is highly important in addition to avoiding

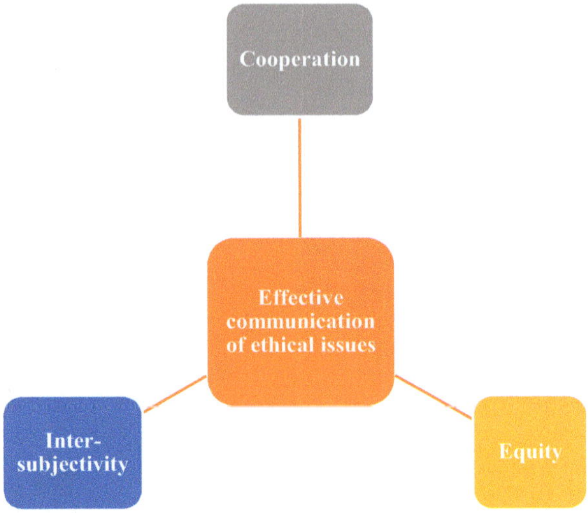

Fig. 20.4 Relational dimensions for effective communication of ethical issues

unnecessary details and scientific jargon but using simple language and diverse mediums (Mushtaq and Kazi 2021). Cooperation, equity, and intersubjectivity are the fundamental tripods and relational dimensions for effective communication of ethical issues (Fig. 20.4). There is somewhat a high level of interdependency in the public culture. Nevertheless, the tempo for information exchange and intense collaboration among scientists should be sustained for strains and culture collections (Dijkshoorn et al. 2010). The intellectual property should also be properly managed so as not to lead to data sharing limits and unnecessary strain duplication (Kumar et al. 2018).

20.7 Conclusion

The potential of microbial engineering to revolutionize agriculture, health-care cost, greener fuels, and bioremediation is indisputable. However, adequate regulatory mechanisms and checks and balances must be put in place to prevent any malicious attempt to misuse or abuse these technologies through effective laboratory screening and through biosafety documentation before patents are granted. Relevant professional and regulatory bodies should be involved in the decision-making process with civic society members as stakeholders before patents are granted for the engineered microbial system. The "one-size-fits-all" approach would not be appropriate with diverse technologies and systems. Thus, the regulatory bodies must give adequate consideration to a wide range of ethical, legal, and social issues as well as the scientific progress and moral reasoning for the industrialization and monetization of microbial products.

References

Abdel-Aziz SM, Abo Elsoud MM, Anise AAH (2017) Chapter 2 - microbial biosynthesis: a repertory of vital natural products. In: Grumezescu AM, Holban AM (eds) Food biosynthesis. Academic Press, pp 25–54

Aguilera Flores MM, Ávila Vázquez V, Medellín Castillo NA, Cardona Benavides A, Carranza Álvarez C, Ocampo Pérez R (2021) Biodegradation of carbamazepine and production of bioenergy using a microbial fuel cell with bioelectrodes fabricated from devil fish bone chars. J Environ Chem Eng 9(6):106692. https://doi.org/10.1016/j.jece.2021.106692

Ahn JH, Seo H, Park W, Seok J, Lee JA, Kim WJ et al (2020) Enhanced succinic acid production by Mannheimia employing optimal malate dehydrogenase. Nat Commun 11(1):1970. https://doi.org/10.1038/s41467-020-15839-z

Ai M, Qiu X, Huang J, Wu C, Jin Y, Zhou R (2019) Characterizing the microbial diversity and major metabolites of Sichuan bran vinegar augmented by Monascus purpureus. Int J Food Microbiol 292:83–90. https://doi.org/10.1016/j.ijfoodmicro.2018.12.008

Azarhava H, Bajestani MI, Jafari A, Vakilchap F, Mousavi SM (2020) Production and physicochemical characterization of bacterial poly gamma- (glutamic acid) to investigate its performance on enhanced oil recovery. Int J Biol Macromol 147:1204–1212. https://doi.org/10.1016/j.ijbiomac.2019.10.090

Badawy MT, Mostafa M, Khalil MS, Abd-Elsalam KA (2022) Chapter 18 - Agri-food and environmental applications of bionanomaterials produced from Agri-waste and microbes. In: Abd-Elsalam KA, Periakaruppan R, Rajeshkumar S (eds) Agri-waste and microbes for production of sustainable nanomaterials. Elsevier, pp 441–463

Belcher MS, Mahinthakumar J, Keasling JD (2020) New frontiers: harnessing pivotal advances in microbial engineering for the biosynthesis of plant-derived terpenoids. Curr Opin Biotechnol 65:88–93. https://doi.org/10.1016/j.copbio.2020.02.001

Bharathi D, Rajalakshmi G (2019) Microbial lipases: an overview of screening, production and purification. Biocatal Agric Biotechnol 22:101368. https://doi.org/10.1016/j.bcab.2019.101368

Bodur S, Erarpat S, Balçık U, Bakırdere S (2021) A rapid, sensitive and accurate determination of cobalamin with double monitoring system: HPLC-UV and HPLC-ICP-OES. Food Chem 340:127945. https://doi.org/10.1016/j.foodchem.2020.127945

Caballero P, Macías-Benítez S, Revilla E, Tejada M, Parrado J, Castaño A (2020) Effect of subtilisin, a protease from bacillus sp., on soil biochemical parameters and microbial biodiversity. Eur J Soil Biol 101:103244. https://doi.org/10.1016/j.ejsobi.2020.103244

Camesasca L, de Mattos JA, Vila E, Cebreiros F, Lareo C (2021) Lactic acid production by Carnobacterium sp. isolated from a maritime Antarctic lake using eucalyptus enzymatic hydrolysate. Biotechnol Rep 31:e00643. https://doi.org/10.1016/j.btre.2021.e00643

da Câmara Rocha J, da Silva Araújo J, de Paiva WKV, Ribeiro ESS, de Araújo Padilha CE, de Assis CF et al (2020) Yellow mombin pulp residue valorization for pectinases production by Aspergillus Niger IOC 4003 and its application in juice clarification. Biocatal Agric Biotechnol 30:101876. https://doi.org/10.1016/j.bcab.2020.101876

Dai Z, Dong H, Zhang Y, Li Y (2016) Elucidating the contributions of multiple aldehyde/alcohol dehydrogenases to butanol and ethanol production in clostridium acetobutylicum. Sci Rep 6(1):28189. https://doi.org/10.1038/srep28189

Das K, Chetri S, Khadgawat P, Minocha S, Sengupta A, Sharma BK, Kumar A (2022) Chapter 18 - an ecofriendly approach toward waste management and environmental safety through microorganisms. In: Shah MP, Rodriguez-Couto S, Kapoor RT (eds) Development in wastewater treatment research and processes. Elsevier, pp 401–417

Dijkshoorn L, De Vos P, Dedeurwaerdere T (2010) Understanding patterns of use and scientific opportunities in the emerging global microbial commons. Res Microbiol 161(6):407–413. https://doi.org/10.1016/j.resmic.2010.06.001

Dikicioglu D, Wood V, Rutherford KM, McDowall MD, Oliver SG (2014) Improving functional annotation for industrial microbes: a case study with Pichia pastoris. Trends Biotechnol 32(8): 396–399. https://doi.org/10.1016/j.tibtech.2014.05.003
Esmaeilnejad-Moghadam B, Mokarram RR, Hejazi MA, Khiabani MS, Keivaninahr F (2019) Low molecular weight dextran production by Leuconostoc mesenteroides strains: optimization of a new culture medium and the rheological assessments. Bioact Carbohydr Diet Fibre 18:100181. https://doi.org/10.1016/j.bcdf.2019.100181
Farag MA, Mesak MA, Saied DB, Ezzelarab NM (2021) Uncovering the dormant food hazards, a review of foodborne microbial spores' detection and inactivation methods with emphasis on their application in the food industry. Trends Food Sci Technol 107:252–267. https://doi.org/10.1016/j.tifs.2020.10.037
Fernandes S, Belo I, Lopes M (2021) Highly aerated cultures boost gluconic acid production by the yeast-like fungus Aureobasidium pullulans. Biochem Eng J 175:108133. https://doi.org/10.1016/j.bej.2021.108133
Handa V, Sharma D, Kaur A, Arya SK (2020) Biotechnological applications of microbial phytase and phytic acid in food and feed industries. Biocatal Agric Biotechnol 25:101600. https://doi.org/10.1016/j.bcab.2020.101600
He R, Cao Q, Chen J, Tian J (2020) Perspectives on the management of synthetic biological and gene edited foods. Biosafety Health 2(4):193–198. https://doi.org/10.1016/j.bsheal.2020.07.003
Hosseinzadeh A, Zhou JL, Navidpour AH, Altaee A (2021) Progress in osmotic membrane bioreactors research: contaminant removal, microbial community and bioenergy production in wastewater. Bioresour Technol 330:124998. https://doi.org/10.1016/j.biortech.2021.124998
Ismail ZZ, Habeeb AA (2017) Experimental and modeling study of simultaneous power generation and pharmaceutical wastewater treatment in microbial fuel cell based on mobilized biofilm bearers. Renew Energy 101:1256–1265. https://doi.org/10.1016/j.renene.2016.10.008
Jiang Y, Ye X, Zheng T, Dong W, Xin F, Ma J, Jiang M (2021) Microbial production of L-malate from renewable non-food feedstocks. Chin J Chem Eng 30:105–111. https://doi.org/10.1016/j.cjche.2020.10.017
Jin L-Q, Jin Y-T, Zhang J-W, Liu Z-Q, Zheng Y-G (2021) Enhanced catalytic efficiency and thermostability of glucose isomerase from Thermoanaerobacter ethanolicus via site-directed mutagenesis. Enzym Microb Technol 152:109931. https://doi.org/10.1016/j.enzmictec.2021.109931
Jovanovic S, Dietrich D, Becker J, Kohlstedt M, Wittmann C (2021) Microbial production of polyunsaturated fatty acids — high-value ingredients for aquafeed, superfoods, and pharmaceuticals. Curr Opin Biotechnol 69:199–211. https://doi.org/10.1016/j.copbio.2021.01.009
Kambouris ME, Manoussopoulos Y, Patrinos GP, Velegraki A (2020) Chapter 7 - microbial genomics in public health: a translational risk-response aspect. In: Patrinos GP (ed) Applied genomics and public health. Academic, pp 131–148
Krishnaswamy VG, Sridharan R, Kumar PS, Fathima MJ (2022) Cellulase enzyme catalyst producing bacterial strains from vermicompost and its application in low-density polyethylene degradation. Chemosphere 288:132552. https://doi.org/10.1016/j.chemosphere.2021.132552
Kumar A, D'Souza R, Asthana M (2018) Chapter 12 - genetic engineering and public perception. In: Singh RL, Mondal S (eds) Biotechnology for sustainable agriculture. Woodhead Publishing, pp 335–373
Kuthyar S, Manus MB, Amato KR (2019) Leveraging non-human primates for exploring the social transmission of microbes. Curr Opin Microbiol 50:8–14. https://doi.org/10.1016/j.mib.2019.09.001
La Maestra S, D'Agostini F, Geretto M, Micale RT (2021) Microbial-based cleaning products as a potential risk to human health: a review. Toxicol Lett 353:60–70. https://doi.org/10.1016/j.toxlet.2021.09.013
Lekshmi R, Arif Nisha S, Thirumalai Vasan P, Kaleeswaran B (2021) A comprehensive review on tannase: microbes associated production of tannase exploiting tannin rich agro-industrial wastes

with special reference to its potential environmental and industrial applications. Environ Res 201:111625. https://doi.org/10.1016/j.envres.2021.111625

Liu X, Kokare C (2017) Chapter 11 - microbial enzymes of use in industry. In: Brahmachari G (ed) Biotechnology of microbial enzymes. Academic, pp 267–298

Lokko Y, Heijde M, Schebesta K, Scholtès P, Van Montagu M, Giacca M (2018) Biotechnology and the bioeconomy—towards inclusive and sustainable industrial development. New Biotechnol 40:5–10. https://doi.org/10.1016/j.nbt.2017.06.005

Macauley-Patrick S, McNeil B, Harvey LM (2005) By-product formation in the d-sorbitol to l-sorbose biotransformation by Gluconobacter suboxydans ATCC 621 in batch and continuous cultures. Process Biochem 40(6):2113–2122. https://doi.org/10.1016/j.procbio.2004.07.014

Mandeep L, H., Luo, J., & Shukla, P. (2020) Effluents detoxification from pulp and paper industry using microbial engineering and advanced oxidation techniques. J Hazard Mater 398:122998. https://doi.org/10.1016/j.jhazmat.2020.122998

Mbituyimana B, Liu L, Ye W, Ode Boni BO, Zhang K, Chen J et al (2021) Bacterial cellulose-based composites for biomedical and cosmetic applications: research progress and existing products. Carbohydr Polym 273:118565. https://doi.org/10.1016/j.carbpol.2021.118565

Meena SS, Mohanty A (2020) Ethical, patent, and regulatory issues in microbial engineering. In: Engineering of microbial biosynthetic pathways. Springer, pp 133–142

Mehrotra S, Kumar VK, Mohan KM, Gajalakshmi S, Pathak B (2021) Bioelectrogenesis from ceramic membrane-based algal-microbial fuel cells treating dairy industry wastewater. Sustain Energy Technol Asses 48:101653. https://doi.org/10.1016/j.seta.2021.101653

Mehta MD, Gair JJ (2001) Social, political, legal and ethical areas of inquiry in biotechnology and genetic engineering. Technol Soc 23(2):241–264. https://doi.org/10.1016/S0160-791X(01)00012-4

Miwa N (2020) Innovation in the food industry using microbial transglutaminase: keys to success and future prospects. Anal Biochem 597:113638. https://doi.org/10.1016/j.ab.2020.113638

Mushtaq A, Kazi F (2021) World microbe forum 2021. Lancet Infect Dis 21(8):1075. https://doi.org/10.1016/S1473-3099(21)00409-6

Myhr AI, Dalmo RA (2005) Introduction of genetic engineering in aquaculture: ecological and ethical implications for science and governance. Aquaculture 250(3):542–554. https://doi.org/10.1016/j.aquaculture.2004.12.032

Nakayama Y, Hashimoto K-I, Sawada Y, Sokabe M, Kawasaki H, Martinac B (2018) Corynebacterium glutamicum mechanosensitive channels: towards unpuzzling "glutamate efflux" for amino acid production. Biophys Rev 10(5):1359–1369. https://doi.org/10.1007/s12551-018-0452-1

Nazari N, Jookar Kashi F (2021) A novel microbial synthesis of silver nanoparticles: its bioactivity, ag/ca-Alg beads as an effective catalyst for decolorization disperse blue 183 from textile industry effluent. Sep Purif Technol 259:118117. https://doi.org/10.1016/j.seppur.2020.118117

Neill MA (2008) Chapter 8 - food safety in the industrialized world. In: Mayer KH, Pizer HF (eds) The social ecology of infectious diseases. Academic, San Diego, pp 215–240

Oliveira C, Guimarães PMR, Domingues L (2011) Recombinant microbial systems for improved β-galactosidase production and biotechnological applications. Biotechnol Adv 29(6):600–609. https://doi.org/10.1016/j.biotechadv.2011.03.008

Pham JV, Yilma MA, Feliz A, Majid MT, Maffetone N, Walker JR et al (2019) A review of the microbial production of bioactive natural products and biologics. *10* (1404). 10. https://doi.org/10.3389/fmicb.2019.01404

Samuchiwal S, Gola D, Malik A (2021) Decolourization of textile effluent using native microbial consortium enriched from textile industry effluent. J Hazard Mater 402:123835. https://doi.org/10.1016/j.jhazmat.2020.123835

Sharma K, Kaushik G, Thotakura N, Raza K, Sharma N, Nimesh S (2020) Enhancement effects of process optimization technique while elucidating the degradation pathways of drugs present in pharmaceutical industry wastewater using Micrococcus yunnanensis. Chemosphere 238:124689. https://doi.org/10.1016/j.chemosphere.2019.124689

Shukla A, Parmar P, Goswami D, Gehlot Y, Vala J, Parmar N, Saraf M (2021) Chapter 13 - microbial technologies in textile industries: an elixir for the greener environment. In: Ibrahim N, Hussain CM (eds) Green chemistry for sustainable textiles. Woodhead Publishing, pp 173–189

Sreekumar S, Baer ZC, Pazhamalai A, Gunbas G, Grippo A, Blanch HW et al (2015) Production of an acetone-butanol-ethanol mixture from clostridium acetobutylicum and its conversion to high-value biofuels. Nat Protoc 10(3):528–537. https://doi.org/10.1038/nprot.2015.029

Steensels J, Gallone B, Voordeckers K, Verstrepen KJ (2019) Domestication of industrial microbes. Curr Biol 29(10):R381–R393. https://doi.org/10.1016/j.cub.2019.04.025

Sudhakar MP, Dharani G (2021) Evaluation of seaweed for the production of lactic acid by fermentation using lactobacillus plantarum. Bioresour Technol 17:100890. https://doi.org/10.1016/j.biteb.2021.100890

Sun L, Wang D-M, Sun W-J, Cui F-J, Gong J-S, Zhang X-M et al (2020) Two-stage semi-continuous 2-keto-Gluconic acid (2KGA) production by pseudomonas plecoglossicida JUIM01 from rice starch hydrolyzate. 8:120. https://doi.org/10.3389/fbioe.2020.00120

Thakur S, Chaudhary J, Singh P, Alsanie WF, Grammatikos SA, Thakur VK (2022) Synthesis of bio-based monomers and polymers using microbes for a sustainable bioeconomy. Bioresour Technol 344:126156. https://doi.org/10.1016/j.biortech.2021.126156

Uddin TM, Chakraborty AJ, Khusro A, Zidan BMRM, Mitra S, Emran TB, Koirala N et al (2021) Antibiotic resistance in microbes: history, mechanisms, therapeutic strategies and future prospects. J Infect Public Health 14:1750–1766. https://doi.org/10.1016/j.jiph.2021.10.020

Ullah I, Khan MS, Khan SS, Ahmad W, Zheng L, Shah SUA et al (2021) Identification and characterization of thermophilic amylase producing bacterial isolates from the brick kiln soil. Saudi J Biol Sci 28(1):970–979. https://doi.org/10.1016/j.sjbs.2020.11.017

Vallero DA (2020) Prologue: ethical challenges posed by using emerging technologies and genetically modified organisms to remove gaseous pollutants. In: Soreanu G, Dumont É (eds) From biofiltration to promising options in gaseous fluxes biotreatment. Elsevier, pp xxvii–liv

Venil CK, Usha R, Devi PR (2021) Chapter 12 - green synthesis of nanoparticles from microbes and their prospective applications. In: Kumar RP, Bharathiraja B (eds) Nanomaterials. Academic Press, pp 283–298

Walker GM, Stewart GGJB (2016) Saccharomyces cerevisiae in the production of fermented beverages. Beverages 2(4):30

Wang Y, Liu Y, Zheng P, Sun J, Wang M (2021) Microbial Base editing: a powerful emerging Technology for Microbial Genome Engineering. Trends Biotechnol 39(2):165–180. https://doi.org/10.1016/j.tibtech.2020.06.010

Wuyts S, Van Beeck W, Allonsius CN, van den Broek MFL, Lebeer S (2020) Applications of plant-based fermented foods and their microbes. Curr Opin Biotechnol 61:45–52. https://doi.org/10.1016/j.copbio.2019.09.023

Yan BH, Selvam A, Wong JWC (2014) Application of rumen microbes to enhance food waste hydrolysis in acidogenic leach-bed reactors. Bioresour Technol 168:64–71. https://doi.org/10.1016/j.biortech.2014.03.085

Yang L-Y, Gong Q, Guo J-Q, Li G-L (2021) Microbes as a production host to produce natural active compounds from mushrooms: biosynthetic pathway elucidation and metabolic engineering. Chin J Nat Med 19(8):580–590. https://doi.org/10.1016/S1875-5364(21)60058-4

You J, Pan X, Yang C, Du Y, Osire T, Yang T et al (2021) Microbial production of riboflavin: biotechnological advances and perspectives. Metab Eng 68:46–58. https://doi.org/10.1016/j.ymben.2021.08.009

Yurtsever A, Calimlioglu B, Sahinkaya E (2017) Impact of SRT on the efficiency and microbial community of sequential anaerobic and aerobic membrane bioreactors for the treatment of textile industry wastewater. Chem Eng J 314:378–387. https://doi.org/10.1016/j.cej.2016.11.156

Zhang Q, Fu C, Zhao C, Yang S, Zheng Y, Xia M et al (2020) Monitoring microbial succession and metabolic activity during manual and mechanical solid-state fermentation of Chinese cereal vinegar. LWT 133:109868. https://doi.org/10.1016/j.lwt.2020.109868

Zhang S, Wang J, Jiang H (2021) Microbial production of value-added bioproducts and enzymes from molasses, a by-product of sugar industry. Food Chem 346:128860. https://doi.org/10.1016/j.foodchem.2020.128860

Zhu K, Kong J, Zhao B, Rong L, Liu S, Lu Z et al (2021) Metabolic engineering of microbes for monoterpenoid production. Biotechnol Adv 53:107837. https://doi.org/10.1016/j.biotechadv.2021.107837

Żywicka A, Junka A, Ciecholewska-Juśko D, Migdał P, Czajkowska J, Fijałkowski K (2020) Significant enhancement of citric acid production by Yarrowia lipolytica immobilized in bacterial cellulose-based carrier. J Biotechnol 321:13–22. https://doi.org/10.1016/j.jbiotec.2020.06.014

MIX
Papier aus verantwortungsvollen Quellen
Paper from responsible sources
FSC® C105338

If you have any concerns about our products,
you can contact us on
ProductSafety@springernature.com

In case Publisher is established outside the EU,
the EU authorized representative is:
Springer Nature Customer Service Center GmbH
Europaplatz 3, 69115 Heidelberg, Germany

Printed by Libri Plureos GmbH
in Hamburg, Germany